Climate Change and Global Food Security

BOOKS IN SOILS, PLANTS, AND THE ENVIRONMENT

Soil Biochemistry, Volume 1, edited by A. D. McLaren and G. H. Peterson

Soil Biochemistry, Volume 2, edited by A. D. McLaren and J. Skujins

Soil Biochemistry, Volume 3, edited by E. A. Paul and A. D. McLaren

Soil Biochemistry, Volume 4, edited by E. A. Paul and A. D. McLaren

Soil Biochemistry, Volume 5, edited by E. A. Paul and J. N. Ladd

Soil Biochemistry, Volume 6, edited by Jean-Marc Bollag and G. Stotzky

Soil Biochemistry, Volume 7, edited by G. Stotzky and Jean-Marc Bollag

Soil Biochemistry, Volume 8, edited by Jean-Marc Bollag and G. Stotzky

Soil Biochemistry, Volume 9, edited by G. Stotzky and Jean-Marc Bollag

Climate Change and Global Food Security

Edited by

Rattan Lal
The Ohio State University
Columbus, Ohio, U.S.A.

B. A. Stewart
West Texas A&M University
West Canyon, Texas, U.S.A.

Norman Uphoff
Cornell University
Ithaca, New York, U.S.A.

David O. Hansen
The Ohio State University
Columbus, Ohio, U.S.A.

CRC Press
Taylor & Francis Group
Boca Raton London New York

CRC Press is an imprint of the
Taylor & Francis Group, an **informa** business
A TAYLOR & FRANCIS BOOK

CRC Press
Taylor & Francis Group
6000 Broken Sound Parkway NW, Suite 300
Boca Raton, FL 33487-2742

First issued in paperback 2019

ISBN-13: 978-0-8247-2536-5 (hbk)
ISBN-13: 978-0-367-39275-8 (pbk)

Library of Congress Cataloging-in-Publication Data

Climate change and global food security / edited by R. Lal ... [et al.].
        p. cm. -- (Books in soils, plants, and the environment ; v. 96)
    Includes bibliographical references and index.
    ISBN 0-8247-2536-0
    1. Climate changes. 2. Crops and climate. 3. Crop yields. 4. Food supply. I. Lal, R. II. Series.

S600.7.C54C6524 2005
338.1'4--dc22   2004063439

Visit the Taylor & Francis Web site at
http://www.taylorandfrancis.com

and the CRC Press Web site at
http://www.crcpress.com

Preface

Anthropogenic perturbation of the global carbon cycle has increased the atmospheric concentration of carbon dioxide, and decreased the carbon pool in the world's agricultural soils. Since the industrial revolution, the atmospheric concentration of carbon dioxide has increased by about 30% from 280 parts per million by volume (ppmv) to 370 ppmv. This increase is attributed to emissions of carbon from fossil fuel combustion estimated at 270 Pg (gigatons), and from land use change and soil cultivation estimated at 136 Pg. Conversion of natural to agricultural ecosystems, with attendant soil erosion and rapid mineralization of soil organic matter, has depleted the carbon pool by 66 to 90 Pg for global soils, and 3 to 5 Pg for soils in the United States. Depletion of the soil organic carbon pool has adverse impacts on soil quality leading to increase in risks of soil erosion, decline in aggregation and soil structure, reduction in plant available water capacity, decline in activity and species diversity of soil fauna, and overall decline in agronomic/biomass productivity. The decline in soil quality is more severe in soils of the tropics than temperate regions, and in soils managed for low-input subsistence farming than those under intensive commercial agriculture. Soils of Sub-Saharan Africa, Central and South Asia, and tropical

America are severely depleted of their organic matter pool, prone to degradation by erosion and other processes, do not respond to inputs, and have low productivity.

The world population of 6.06 billion in 2000 will increase to 7.2 billion in 2012, 8.3 billion in 2030, and 9.3 billion in 2050. Practically all the increase in the world population will occur in the developing countries, where soils are severely depleted of their organic carbon pool and have low productivity. The population of developing countries will increase by 35% from 4.9 billion in 2000 to 6.6 billion in 2025. The required increase in cereal production by 2025 will be 778 million MT, an average of 31 million MT per year. The required increase in 2050 will be 1519 million MT, an average of 30 million MT per year. The required cereal production in developing countries will be more than double by 2050, mainly because of the projected rapid increase in population. The increase in food production will have to come from increasing production per unit area from existing land, because there is little if any potential for bringing new land under cultivation. Therefore, restoring the quality of degraded soils is essential, for which enhancing soil organic carbon pool is a principal prerequisite.

Restoring the depleted organic carbon pool in soils of developing countries of the tropics and subtropics is a challenging task for several reasons. First, the resource poor farmers may not be able to afford the inputs needed to attain the required increase in crop yield even if the inputs were made available. Second, crop residues and other bio-solids that must be returned to the soil are invariably used for other purposes, such as household fuel, fodder, fencing and construction material, and so on. Third, the decomposition rate of organic matter may be four to five times higher in the tropics than in temperate climates. Thus, there is a need to develop appropriate farming systems to cater to the multifaceted demands of the resource-poor small landholders of the tropics.

Encouraging adoption of recommended management practices for enhancing the organic carbon pool is not a simple task for the soils of temperate climates of the developed economies either. There is a strong need to provide incentives and commodify soil carbon, which can then be traded like any other farm commodity. While the Clean Development Mechanism (CDM) under the Kyoto Protocol and the BioCarbon Fund of the World Bank may be policy tools for providing incentives to farmers of developing countries, international emissions trading joint implementation, among others, may be useful tools for those in developed countries.

This book addresses six complex and interactive themes as follows:

1. The impact of projected climate change on soil quality, water resources, temperature regime, and growing season duration on net primary productivity of different biomes
2. Soil carbon dynamics under changing climate
3. The impact of changes in carbon dioxide and ecological environments on agronomic yields and food production in various world regions
4. World food demands and supply during the 21st century
5. Policy and economic issues related to carbon trading and enhancing agricultural production
6. Research and development priorities for enhancing soil carbon pool and food security.

Contents

SECTION I Global Food Security

SECTION II   Climate Change and Net Primary Productivity

SECTION III *Climate Change and Agronomic Production*

SECTION IV   Soil Carbon Dynamics and Farming/Cropping Systems

SECTION V   Policy and Economic Issues

SECTION VI   Toward Research and Development Priorities

Contributors

Miguel M. Acosta
Instituto de Recursos Naturales
Colegio de Postgraduados
Montecillo, México

Arthur L. Allen
Department of Agriculture
University of Maryland
  Eastern Shore
Princess Anne, Maryland

John M. Antle
Department of Agricultural
  Economics and Economics
Montana State University
Bozeman, Montana

Jacques Antoine
Land and Soil Fertility
  Management Service
Food and Agriculture
  Organization
Rome, Italy

Martial Bernoux
Institut de Recherche pour le
  Dévelopment
Paris, France
and
Centro de Energia Nuclear na
  Agricultura
Universadade de São Paulo
São Paulo, Brazil

Catherine S. Bolek
Office of Sponsored Research
  and Programs
University of Maryland
  Eastern Shore
Princess Anne, Maryland

Norman E. Borlaug
Sasakawa Africa Association
International Maize and Wheat
  Improvement Center
México City, México

Tanveer Butt
Department of Agricultural
  Economics
Texas A&M University
College Station, Texas

Carlos C. Cerri
Centro de Energia Nuclear na
  Agricultura
Universidade de São Paulo
São Paulo, Brazil

Carlos E.P. Cerri
Centro de Energia Nuclear na
  Agricultura
Universidade de São Paulo
São Paulo, Brazil

Albert Chalabesa
Mount Makulu Research
  Station
Chilanga, Zambia

Jose I. Cortés
Instituto de Recursos Naturales
Colegio de Postgraduados
Montecillo, México

Robert B. Dadson
Department of Agriculture
University of Maryland
  Eastern Shore
Princess Anne, Maryland

Roy Darwin
U.S. Department of
  Agriculture–Economic
  Research Service
Washington, D.C.

Evan H. DeLucia
Department of Plant Biology
  and
Program in Ecology and
  Evolutionary Biology
University of Illinois
Urbana, Illinois

Prócoro Diaz
Instituto de Recursos Naturales
Colegio de Postgraduados
Montecillo, México

Mamadou Doumbia
Laboratoire de Sol, Eau et
  Plante
Institut d'Economie Rurale
Bamako, Mali

Christopher R. Dowswell
Sasakawa Africa Association
International Maize and Wheat
  Improvement Center
México City, México

John M. Duxbury
Department of Crop and Soil
  Science
Cornell University
Ithaca, New York

Hallie Eakin
Centro de Ciencias de la
 Atmósfera
Universidad Nacional
 Autónoma de México
México City, México

William Easterling
Penn State Institutes of the
 Environment
The Pennsylvania State
 University
University Park, Pennsylvania

Jorge D. Etchevers
Instituto de Recursos Naturales
Colegio de Postgraduados
Montecillo, México

Arjan J. Gijsman
University of Hawaii
Honolulu, Hawaii
and
Institute of Food and
 Agricultural Sciences
University of Florida
Gainesville, Florida
and
Central Internacional de
 Agricultura Tropical
Cali, Colombia

Dennis J. Greenland
Department of Soil Science
University of Reading
Reading, United Kingdom

Jason G. Hamilton
Department of Biology
Ithaca College
Ithaca, New York

David O. Hansen
International Programs in
 Agriculture
The Ohio State University
Columbus, Ohio

Fawzy M. Hashem
Department of Agriculture
University of Maryland
 Eastern Shore
Princess Anne, Maryland

Robert D. Havener
Winrock International (ret.)
Sacramento, California

Ernesto Hernández
Instituto de Socioeconomía
 Estadística e Informática
Colegio de Postgraduados
Campus Puebla, México

Daniel Hillel
Center for Climate Systems
 Research
Columbia University
New York, New York

R. César Izaurralde
Joint Global Change Research
 Institute
College Park, Maryland

Leobardo Jiménez
Instituto de Socioeconomía
 Estadística e Informática
Colegio de Postgraduados
Montecillo, México

Anthony Joern
School of Biological Sciences
University of Nebraska-Lincoln
Lincoln, Nebraska

James W. Jones
Institute of Food and
  Agricultural Sciences
University of Florida
Gainesville, Florida

Jagmohan Joshi
Department of Agriculture
University of Maryland
  Eastern Shore
Princess Anne, Maryland

Anthony W. King
Environmental Science Division
Oak Ridge National
  Laboratory
Oak Ridge, Tennessee

Parviz Koohafkan
Sustainable Development
  Department
Rural Development Division
Food and Agriculture
  Organization
Rome, Italy

Rattan Lal
Carbon Management and
  Sequestration Center
School of Natural Resources
The Ohio State University
Columbus, Ohio

J. David Logan
Department of Mathematics
University of Nebraska–Lincoln
Lincoln, Nebraska

Matthias K. Magunda
Kawanda Agricultural Research
  Institute
Kampala, Uganda

Bruce McCarl
Department of Agricultural
  Economics
Texas A&M University
College Station, Texas

Ricardo Mendoza
Instituto de Socíoeconomía
  Estadística e Informática
Colegio de Postgraduados
Campus Pueblo, México

Carlos M. Monreal
Agriculture and Agri-Food
Ottawa, Ontario, Canada

David J. Moore
Department of Plant Biology
Program in Ecology and
  Evolutionary Biology
University of Illinois
Urbana, Illinois

Steven W. Muliokela
Golden Valley Research Trust
Lusaka, Zambia

Ranjan S. Muttiah
Department of Geology
GIS & Remote Sensing Center
Texas Christian University
Fort Worth, Texas

Majaliwa Mwanjalolo
Department of Soil Science
Makerere University
Kampala, Uganda

Richard J. Norby
Environmental Science Division
Oak Ridge National
  Laboratory
Oak Ridge, Tennessee

Lloyd B. Owens
U.S. Department of
  Agriculture–Agricultural
  Research Service
North Appalachian
  Experimental Watershed
Coshocton, Ohio

Zaitao Pan
Department of Earth and
  Atmospheric Sciences
St. Louis University
St. Louis, Missouri

Keith Paustian
Department of Soil and Crop
  Science and Natural Resource
  Ecology Laboratory
Colorado State University
Fort Collins, Colorado

Wilfred M. Post
Environmental Science Division
Oak Ridge National
  Laboratory
Oak Ridge, Tennessee

Ana Rey
Food and Agriculture
  Organization
Rome, Italy
and
The University of Edinburgh
Edinburgh, Scotland

Stacey Rosen
U.S. Department of
  Agriculture–Economic
  Research Service
Washington, D.C.

Cynthia Rosenzweig
NASA Goddard Institute for
  Space Studies
New York, New York

Pedro A. Sanchez
Tropical Agriculture
The Earth Institute at
  Columbia University
Palisades, New York

G. Edward Schuh
Hubert H. Humphrey Institute
  of Public Affairs
Minneapolis, Minnesota

Shahla Shapouri
U.S. Department of
  Agriculture–Economic
  Research Service
Washington, D.C.

Martin J. Shipitalo
U.S. Department of
  Agriculture–Agricultural
  Research Service
North Appalacian
  Experimental Watershed
Coshocton, Ohio

Clint J. Springer
Department of Biology
West Virginia University
Morgantown, West Virginia

B.A. Stewart
Dryland Agricultural Institute
West Texas A&M University
Canyon, Texas

Eugene S. Takle
Department of Agronomy
Department of Geological and
  Atmospheric Sciences
Iowa State University
Ames, Iowa

Godfrey Taulya
Department of Soil Science
Makerere University
Kampala, Uganda

Moses M. Tenywa
Department of Soil Science
Makerere University
Kampala, Uganda

Richard B. Thomas
Department of Biology
West Virginia University
Morgantown, West Virginia

Petra Tschakert
Department of Biology
McGill University
Montreal, Canada

Antonio Turrent
Campo Experimental Valle de
  México
Instituto Nacional de
  Investigaciones Forestales,
  Agrícolas y Pecuarias
México

Luther Tweeten
The Ohio State University
Columbus, Ohio

Norman Uphoff
Cornell International Institute
  for Food, Agriculture and
  Development
Cornell University
Ithaca, New York

Miguel A. Vergara
Instituto de Recursos
  Naturales
Colegio de Postgraduados
Montecillo, México

Valerie Walen
Institute of Food and
  Agricultural Sciences
University of Florida
Gainesville, Florida

William Wolesensky
Program in Mathematics
College of St. Mary's
Omaha, Nebraska

Ralph A. Wurbs
Department of Civil
  Engineering
Texas A&M University
College Station, Texas

Linda M. Young
Department of Agricultural
  Economics and Economics
Montana State University
Bozeman, Montana

Section I

Global Food Security

1

Reducing Hunger in Tropical Africa while Coping with Climate Change

PEDRO A. SANCHEZ

CONTENTS

Most of the world has witnessed dramatic increases in per capita food production over the last 30 years. However, the opposite occurred in Sub-Saharan Africa. Per capita food production in this region continues to decline, and hunger, largely due to insufficient food production, affects about 200 million people, 34% of the region's population (Table 1.1). Projections to 2015 suggest that hunger in Asia and Latin America is likely to decline with continued economic growth, while in Africa it is likely to remain constant (Dixon et al., 2001). The difference is that enough food is produced in countries like India and China. Hunger in these nations is primarily caused by unemployment and a corresponding lack of income-generating capacity. Africa simply does not produce enough food. The lack of a major impact of the Green Revolution in this region is one key reason for this difference.

The Green Revolution is one of the major accomplishments of the past 30 years. During this period, the number of rural poor decreased by half, the proportion of malnourished people in the world dropped from 30% to 18%, and the real prices of main cereal crops decreased by 76%. It was initiated by a small group of determined scientists and policymakers who identified a need for high-yielding varieties of rice and wheat. Then enabling government policies, fertilizers and irrigation, better marketing, infrastructure, national research institutions, strong agricultural universities, the

Table 1.1 Basic Hunger Statistics in Developing Regions of the World

Region	Per Capita Food Production Index 1999/1961	Caloric Intake 1999 cal/person/day	Undernourished 1999 %
Sub-Saharan Africa	87	2195	34
South Asia	120	2403	12
East Asia	149	2921	6
Latin America	133	2824	6

Source: Food and Agriculture Organization. 2003. FAOSTAT: FAO Statistical Database. Available at: http://apps.fao.org/

international agricultural research system, and other necessary factors were put in place. However, the contribution of improved varieties to crop yield increases has been 70% to 90% in Asia, Latin America, and the Middle East, but only 28% in Africa (Evenson and Gollin, 2003).

A major biophysical reason and a major economic reason help explain this discrepancy. The major biophysical reason is that unlike other developing regions, soil nutrient depletion is extreme in Africa. Therefore, the key need is not to improve varieties, but rather to replenish soil fertility at the lowest possible cost (Sanchez, 2002). Closely related to improving soil fertility is the need to improve small-scale water management, provide small-scale rain-fed farms with critical life-saving irrigation, and grow high-value crops. Soil fertility goes hand in hand with water in many regions. Even with excellent genetic improvements, crops cannot grow well without sufficient nitrogen, phosphorus, or water. These are biological imperatives that transcend socioeconomic and political ones. The major economic constraint is poor rural infrastructure in Africa. Road density for rural dwellers in Africa is only one-sixth the average of Asia (Paarlberg, 2002). Hence, access to markets is difficult; fertilizer prices are two to six times higher at the farm gate in Africa than they are in the rest of the world; health, education, and sanitation are often appalling; access to information is poor; and prices drop precipitously when crop surpluses occur.

Research scientists have also learned that community participation in research and development can work. A new paradigm, based on natural resource management, has emerged. It addresses soil and water issues as well as pest management constraints in ways that minimize tradeoffs with environmental services (Izac and Sanchez, 2001). Furthermore, an enormous biotechnology potential exists to address these issues through crop genetic improvement (Wambugu, 1999). We know more about the crucial need for functional markets for the poor, farm diversification, trade imbalances, environmental services, and a reawakening of the importance of agriculture as the engine of economic growth. It is very positive to see agricultural scientists interacting with counterparts who focus on environmental, macroeconomic policy, health, education, gender, water and sanitation, energy, and other development sectors.

1.1 REDUCING HUNGER IN AFRICA

The time is right to drastically increase the productivity of African agriculture and to improve human nutrition, with a new and highly focused action plan, called the Doubly Green Revolution in Africa. "Doubly green" means increasing productivity in environmentally sustainable ways (Conway, 1997). In response to a request from the UN Secretary General in February 2003, the U.N. Millennium Project's Task Force on Hunger is developing a plan to attain the Millennium Development Goal of cutting hunger in half by 2015 (Millennium Project, 2005). The emerging plan is based on (1) moving from political commitments to concrete actions, (2) policy reforms that give high priority to investments in agriculture, nutrition, rural infrastructure, marketing, and rural women, and (3) three key interventions at the community level. The latter interventions include (a) improving agricultural productivity on smallholder farms through investments in soil fertility restoration and small-scale water management; (b) making markets work for the rural poor through storage facilities, feeder roads, market information systems, and other interventions; and (c) providing school lunches with locally produced food in order to increase school attendance, especially by girls, to enable learning, improve nutrition, and increase local demand for food production.

These three synergistic community-based actions and overarching policy reforms can break the log jam of inaction in the short term, and open the way for other necessary actions to take place if there is political commitment. However, the specter of climate change will make this task even more daunting. The remainder of this paper addresses some additional priority interventions that will facilitate coping with climate change in Africa as well as in other tropical regions.

1.2 COPING WITH CLIMATE CHANGE

The Third Assessment Report of the Intergovernmental Panel on Climate Change (IPCC) stated for the first time that scientific evidence of human-induced global warming is unequivocal, and that the latest predictions are much worse than previous estimates (Houghton et al., 2001). The last 100 years have been the warmest on record. Furthermore, warming during the last 50 years has a clear human signature. Global temperatures will increase by 1.4°C to 5.8°C by 2100; sea levels are rising and are expected to increase by 14 to 88 cm by 2100, flooding low-lying areas and displacing hundreds of millions people. Rainfall patterns are changing, El Niño events are increasing in frequency and intensity, Arctic ice is thinning, and tropical mountain glaciers are retreating.

The consequences of these changes are also dire according to this report. Agricultural productivity in Africa and Latin America could decrease by as much as 30% during this century. Severe droughts will occur in Southern Africa and Southeast Asia. Wetter climates and more floods are predicted for parts of East Africa and Latin America. And more smoke and haze problems are predicted for Southeast Asia and Central America. Higher worldwide food prices are likely to result, negatively affecting the urban poor.

Major changes are also predicted in critical ecosystems, particularly coral reefs and tropical forests. The geographic spread of malaria and increased crop pest and disease pressure in wetter climates are also predicted. The IPCC reported global economic losses of around $40 billion due to existing global warming in 1999, of which 25% occurred in the tropics (Houghton et al., 2001). The capacity of people to adapt to these global changes is correlated with poverty level. Countries with the least diversified agriculture, forestry, and fisheries will suffer the most. Africa is considered to be the region most vulnerable to global warming (Houghton et al., 2001). A major discrepancy exists between developed and developing countries in terms of human-induced global warming and who pays for the consequences. About 75% of anthropogenic CO_2 emissions are due to fossil fuel burning, mainly from the

North, while the remaining 25% is due to changes in tropical land use, especially deforestation in the South. While contributing the least to global warming, the tropical countries will suffer the most from it.

The following section includes a discussion of some key research issues identified by the Consultative Group on International Agricultural Research (CGIAR) Inter-Center Working Group on Climate Change (2001). The tropics will face a special challenge in coping with climate change. Issues discussed are arranged in terms of impact, adaptation, and mitigation of climate change.

1.2.1 Impact

Research about the projected impacts of climate change provides a predictive understanding of the processes involved and their consequences. Many models used to predict impacts of climate change are based on obsolete primary tropical data sets. These data often keep being recycled in climate change studies, creating self-evident truths by continued quoting. Some studies acknowledge that such data sets are admittedly inadequate, but researchers continue to use them because they are unable to find better ones. Many models also express results in spatial scales that are of little use to national scientists and decision makers. The following three examples illustrate how some of these limitations can be overcome.

1.3 ESTIMATING BIOMASS OF YOUNG TROPICAL VEGETATION

Allometric equations for estimating tropical forest biomass (Brown et al., 1989) were developed for mature forests by the IPCC. The equations provide the basis for estimating the impact of tropical deforestation on the global carbon cycle. But such equations significantly underestimate biomass carbon in young tree vegetation that occupies about 72% of the original tropical forest area. New allometric equations developed by Ketterings et al. (2001) for young secondary forests and fallows in Indonesia result in biomass estimates that only approximate those obtained using the equation by Brown et al. (1989). The use of these equations plus new hard data have changed the image of Indonesia, which has to be regarded a net carbon sink instead of a net carbon source (Van Noordwijk et al., 1995).

1.4 HOW TO MEASURE SOIL CARBON

The IPCC special report titled *Land Use, Land Use Change and Forestry* (Watson et al., 2000) notes that the inability to accurately measure changes in soil carbon at low cost is a major impediment for carbon sequestration projects. It is equally applicable to projects related to the Clean Development Mechanism (CDM) in Article 12 of the Kyoto Protocol. Soil carbon has enormous spatial variability and accurate measurement requires the collection of multiple soil samples at different times, as well as laboratory analyses to obtain estimate of changes in soil carbon. Based on extensive sampling in Africa, Shepherd and Walsh (2002) have developed a promising approach that estimates several soil properties simultaneously using diffuse reflectance spectroscopy in rapid nondestructive ways. The measurement of soil carbon, as well as other soil properties, can be predicted using soil reflectance spectra with accuracy similar to that of duplicate laboratory determinations.

Because this spectral technique allows large numbers of samples to be quickly analyzed, it can be used to thoroughly characterize the soil and its spatial variability within a CDM project. The problem of large spatial variability in soil carbon determinations is addressed by making many measurements, each of which only takes nanoseconds. By returning to the same site at a later date, it is possible to quantify the amounts of soil carbon sequestered or released consistent with CDM verification requirements. This technique can use spectral bands from satellite imagery, thus permitting remote sensing analysis (Shepherd and Walsh, 2002).

1.5 SHARPENING PREDICTIVE TOOLS FOR KEY AGROECOSYSTEMS

Most modeling to forecast climate change impacts on crop yields uses average weather data adjusted for forecasted national variations. Using average data ignores both the inherent variability of weather and its effect on crop yield, and hence food security, as the climate changes. A methodology has been developed and tested that allows outputs from global circulation models (GCMs) to be downscaled and applied to point simulation models (Jones and Thornton, 2001). It is possible to model and map the impacts of climate change on poor farmers at the subnational scale by using different GCM scenarios of climate change; land-use change scenarios; crop, livestock, and tree production models; and maps of the distribution of the world's poor and their crop and livestock resources.

Figure 1.1 shows an example of this approach. It predicts the maize caloric deficit in Southern Africa caused by both climate change and population growth. The subnational level of resolution, as well as the easily interpretable nature of such maps, makes this alternative more useful to policymakers than country-level resolution, single-factor assessments.

1.6 ADAPTING TO THERMAL DAMAGE

The mean maximum temperature for much of the tropics where crops are grown is 34°C. The IPCC Third Assessment Report indicates that temperatures are going to increase throughout the tropics, regardless of changed rainfall regimes. J. Sheehy of the International Rice Research Institute (2003) has observed that the fertility of rice flowers falls from 100% at 34°C to near zero at 40°C, regardless of CO_2 levels in the atmosphere (Figure 1.2). Any increase in temperature due to global climate change is potentially damaging to rice. Yields decrease by about 10% for each 1°C increase in temperature. Similar trends have been found in wheat, maize, beans, soybeans, and peanuts.

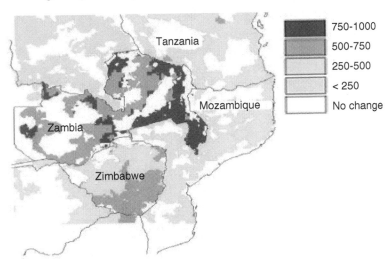

Figure 1.1 Maize calorie deficit (kcal/person/day) in Southern Africa caused by climate change and human population growth in 2050 (From Jones and Thornton, 2001).

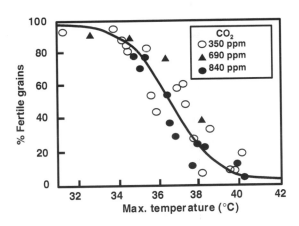

Figure 1.2 Relationship between grain sterility and maximum temperatures at rice flowering time at different CO_2 concentrations in the atmosphere. (From J. Sheehy, personal communication, 2003).

Large increases in the sterility of cereal and legume crops are related to temperature increases. They represent an alarming food security issue, which increases challenges that the world faces to feed itself in the coming decades. The extent of this threat to root and tuber crops, pasture, and tree species in unknown. If the rates of rice yield decline due to thermal stress are broadly validated, and assuming that temperature increases consistent with the latest IPCC data (0.14°C to 0.58°C per decade) tropical grain crop yields may decrease by 2% to12% by the year 2020, and by 7% to 29% by the year 2050. The IPCC Third Assessment Report does not consider thermal damage to grain crops in its predictions, but according to J. Jones (2003) some models are now incorporating thermal damage into their predictions.

A full assessment of this threat needs to be done. Genetic manipulation offers several approaches, including breeding for resistance to higher temperatures during flowering time; shifting the time of day at which crops flower to avoid the hottest hours; and gene transfers from crops that tolerate higher temperatures, such as sorghum and millet, to rice and maize. It is also possible to manipulate the microclimate. An example is the marked reduction in air temperature when sorghum and millet are grown under *Faidherbia albida* trees in the Sahel (Vandenbeldt, 1992).

1.7 MITIGATION

Since most carbon is emitted through fossil fuel combustion, mitigating global warming will logically depend on what happens in the Northern Hemisphere. One key exception is the important role of carbon sequestration in the tropics.

1.7.1 High Carbon Sequestration Potential of Tropical Agroecosystems

A recent IPCC study on land use, land-use change, and forestry (Watson et al., 2000) documented the large potential for tropical agroecosystems to sequester carbon. The tropics have two major advantages over the temperate regions. Trees grow faster under high year-round temperatures and high solar radiation. In addition, many tropical soils are depleted of carbon because of unsustainable land use practices. Table 1.2 illustrates the potential for carbon sequestration in the tropics.

Land use intensification practices usually start from a high carbon stock base, resulting in annual sequestration

Table 1.2 Carbon Sequestration Rates and Annual Potential of Agricultural Practices by 2010

Practice	Carbon Sequestration Rate (tons C/ha/year)	Annual Potential by 2010 (million tons C/year)
Land Use Intensification (Global)		
Croplands (reduced tillage, rotations, cover crops, fertilization, and irrigation)	0.36	50
Forest lands (forest regeneration, better species, silviculture)	0.31	69
Grasslands (better herds, woody plants, and fire management)	0.80	168
Lowland rice production	0.10	<1
Land Use Change (Tropics)		
Agroforestry (conversion from unproductive croplands and grasslands at humid tropical forest margins, and by replenishing soil fertility in subhumid tropical Africa)	3.10	390
Improved pastures in subhumid tropical South America (conversion from native pasture to deep-rooted improved grasses and legumes)	2.8 without a legume, 7.0 with a legume	75
Tropical deforestation	—	−1644

Source: From Watson, R.T., I.R. Noble, B. Bolin, N.H. Ravindranath, D.J. Verardo, and D.J. Dokken. 2000. *Land Use, Land-Use Change and Forestry.* Cambridge University Press, London; New York. With permission. Data also provided by M. Fisher, personal communication, 2003.

rates of tenths of a metric ton per hectare, in both tropical and temperate regions. Because of the large areas, the total carbon sequestration potential by 2010 ranges from 50 to 168 Mt (million metric tons) per year, except for paddy rice production because of methane emissions associated with it.

Transforming unproductive tropical croplands or grasslands into highly productive agroforestry and improved pasture systems results in annual carbon sequestration rates of a higher order of magnitude. Trees are periodically harvested in agroforestry systems. Thus, these calculations refer to time-averaged carbon, which takes into account carbon removals associated with harvesting (Palm et al., 1999). The high sequestration rates in these land-use change categories are due to a drastic increase in biomass production. Either originally fallow lands have lost much of their system carbon stock in the agroforestry systems (Sanchez and Jama, 2002) or a new sink of carbon has been developed in the subsoil (Fisher et al., 1994). Given the large areas to which these conditions apply, the overall potential for additional carbon sequestration is huge. Conversion to tropical agroforestry has the potential to soak up 390 Mt of carbon per year, equal to about one-fifth of annual carbon emissions of the United States from all sources.

The importance of avoiding further deforestation is evident from data shown in Table 1.2. The magnitude of emitted carbon is enormous, and avoidance of such emissions by preventing deforestation will play a major role in the global carbon cycle.

The magnitude of carbon sequestration in developing countries through systems described in Table 1.2 depends to a major extent on rainfall regimes. The carbon sequestration potential per hectare of such systems is lowest in the semiarid tropics and highest in the humid tropics, with the subhumid tropics in between (Schroeder, 1994). Hotspots could be identified at a similar scale of resolution as the well-established biodiversity hotspots. However, there are tradeoffs between carbon sequestration per hectare, and the number of hectares that can be put into such systems. The carbon sequestration potential in the Sahel is in the range of 0.25 to 0.05 tons of carbon per hectare per year (Sloger, 2003), or one-tenth of what land use change with legume-based pastures or agroforestry can yield in the subhumid and humid tropics. However, there are large areas of degraded lands in the dry areas stretching from Morocco to Mongolia for which land use change could make a major difference, even if the sequestration rates are low on a per hectare basis.

1.7.2 Carbon Sequestration by Smallholder Farming Communities

Most carbon offset projects involve a large carbon emitter in an industrialized nation contracting with a tropical government to protect a large area of primary forest from deforestation, thereby avoiding the emission of more carbon. This process excludes many farmers from the process. Given the large emissions due to tropical deforestation, shown in Table 1.2, such protection is very important. But this does not result in substantial carbon reductions in the atmosphere, since primary forests are mature, and most of the carbon sequestered by photosynthesis in them is lost by respiration. To sequester large quantities of carbon, one must work with young secondary forests or agroforests.

Poor farmers in the tropics could benefit financially by sequestering carbon. It is a product they provide to the global community when using the other practices described in Table 1.2. This idea was proposed by the CGIAR Inter-Center Working Group on Climate Change at a meeting of the subsidiary bodies of the UN Framework Convention on Climate Change in Lyon, France in September 2000 (*www.iisd.ca/climate/sb13/side/enbots11mon.htm*). The idea was well received by developing country representatives and donor agencies. It represents a potential integrated approach to food security/poverty alleviation issues because it would also involve carbon sequestration as a "no-regrets" option. Research needs to be done to determine how the sequestered carbon can be accounted for in a heterogeneous landscape that includes hundreds of small farms, and about how benefits could accrue to farmers. Payments for sequestering carbon can help alleviate poverty. Assuming a price of $10/ton of carbon, a 1-ha farm could generate an additional income of $10 to $30 per year, which is significant to farmers who earn less than $1 per day.

1.8 CONCLUSION

Cutting hunger and malnutrition in half in Africa is a perfectly feasible goal in the next 10 years, if the world community decides to do it, and if it can address major climate change threats such as thermal damage. The international community has successfully tackled hunger before, when widespread famine was averted in Asia. It can do it again in Africa. Hunger has been defeated twice in the last 50 years — the Marshall Plan after World War II and the Green Revolution during the 1960s through the 1980s. It can be done again in Africa. We have the breakthrough technology, the know-how, and an emerging political will. This effort can be facilitated by linking it to carbon sequestration payments, thus helping to mitigate climate change as well.

ACKNOWLEDGMENT

This chapter is based partly on presentations made at the mid-term meeting of the CGIAR in Durban, South Africa, May 2001, as chair of the Inter-Center Working Group on Climate Change, and to the UN Economic and Social Council in New York, March 2003.

REFERENCES

Brown, S., A.J.R. Gillespie, and A.E. Lugo. 1989. Biomass estimation methods for tropical forests with applications to forest inventory data. *For. Sci.*, 35:881–902.

Conway, G. 1997. *The Doubly Green Revolution: Food for All in the 21st Century.* Penguin Books, London.

Dixon, J., A. Gulliver, and D. Gibbon. 2001. *Farming Systems and Poverty. Improving Farmers' Livelihoods in a Changing World.* Food and Agriculture Organization, Rome.

Evenson, R.E. and D. Gollin. 2003. *The Green Revolution at the End of the Twentieth Century.* CAB International, Wallingford, United Kingdom.

Fisher, M. 2003. Centro Internacional de Agricultura Tropical, personal communication.

Fisher, M.J., I.M. Rao, M.A. Ayarza, C.E. Lascano, J.I. Sanz, R.J. Thomas, and R.R. Vera. 1994. Carbon storage by introduced deep-rooted grasses in the South American savannas. *Nature,* 371:236–238.

Food and Agriculture Organization. 2002–2003. FAOSTAT: FAO Statistical Database. Available at: http://www.apps.fao.org/

Houghton, J.T., Y. Ding, D.J. Griggs, et al., Eds. 2001. *Climate Change 2001: The Scientific Basis* Cambridge University Press, London; New York.

Inter-Center Working Group on Climate Change. 2001. Beating the Heat: Climate Change and Rural Prosperity. Report to Consultative Group on International Agricultural Research Mid-Term Meeting, Durban, South Africa. World Agroforestry Centre, Nairobi, Kenya.

Izac, A.-M.N. and Sanchez, P.A. 2001. Towards a natural resource management paradigm for international agriculture: the example of agroforestry research. *Agric. Syst.,* 69:5–25.

Jones, J. Distinguished professor, Agricultural and Biological Engineering Department, University of Florida. Personal communication, 2003.

Jones, P.G. and P.K. Thornton. 2001. Spatial modelling of risk in natural resource management: applying plot-level, plant-growth modelling to regional analysis. *Conserv. Ecol.* 5(2), 27. URL: http://www.consecol.org/vol5/iss2/art27, January 2002.

Ketterings, Q.M., R. Coe, M. van Noordwijk, Y. Ambagau, and C.A. Palm. 2001. Reducing uncertainty in the use of allometric biomass equations for predicting aboveground tree biomass in mixed secondary forests. *For. Ecol. Manage.,* 146:201–211.

Millennium Project, 2003. Hunger Task Force. 2003. Interim Report, Millennium Project. U.N. Development Programme, New York.

Paarlberg, R. 2002. Governance and Food Security in an Age of Globalization. Vision Discussion Paper 36. International Food Policy Research Institute, Washington, D.C.

Palm, C.A., P.L. Woomer, J.C. Alegre, et al. 1999. Carbon Sequestration and Trace Gas Emissions in Slash-and-Burn and Alternative Land Uses in the Humid Tropics. Alternatives to Slash-and-Burn Climate Change Working Group, Final Report, PHASE II. ASB Coordination Office, ICRAF, Nairobi, Kenya. http://www.asb.cgiar.org/ViewEntry.asp?pubid=203

Sanchez, P.A. 2002. Soil fertility and hunger in Africa. *Science,* 295:2019–2020.

Sanchez, P.A. and B.A. Jama. 2002. Soil fertility replenishment takes off in East and Southern Africa. In J.D.B. Vanlauwe, N. Sanginga, and R. Merckx, Eds. *Integrated Nutrient Management in Sub-Saharan Africa.* CAB International, Wallingford, United Kingdom, pp. 23–45.

Schroeder, P. 1994. Carbon storage benefits of agroforestry systems. *Agrofor. Syst.,* 27:89–97.

Sheehy, J. 2003. International Rice Research Institute, personal communication.

Shepherd, K.D. and M.G. Walsh. 2002. Development of reflectance spectral libraries for characterization of soil properties. *Soil Sci. Soc. Am. J.,* 66:988–998.

Sloger, C. 2003. U.S. Agency for International Development, personal communication.

U.N. Millennium Project 2005, Task Force on Hunger. *Halving Hunger: It Can Be Done.* London, Sterling, Va: Earthscan.

Van Noordwijk, M., T.P. Tomich, R. Winahyu, D. Murdiyarso, S. Suyanto, S. Partoharjono, and A.M. Fagi. 1995. *Alternatives to Slash and Burn in Indonesia.* Summary Report of Phase 1. ASB–Indonesia and ICRAF SE Asia, Bogor, Indonesia.

Vandenbeldt, R.J., Ed. 1992. Faidherbia albida *in the West African Semi-Arid Tropics.* Crops Research Institute for Semi-Arid Tropics/International Center for Research in Agroforestry, Nairobi, Kenya.

Wambugu, F. 1999. Why Africa needs agricultural biotech. *Nature (London),* 400:15–16.

Watson, R.T., I.R. Noble, B. Bolin, N.H. Ravindranath, D.J. Verardo, and D.J. Dokken. 2000. *Land Use, Land-Use Change and Forestry.* Cambridge University Press, London; New York.

2

World Food Security: Perspectives Past, Present, and Future

DENNIS J. GREENLAND

CONTENTS

2.1 INTRODUCTION

Food security has been a problem of concern to humanity from the beginning of time. Cohen (1977) summarized much of the earlier literature, and Dyson (1996), Hillel (1991), Evans (1998), and Wild (2003) have discussed it in the light of more recent findings. Dyson, a demographer, provides a remarkably well-balanced account of the information currently available, while stressing the inadequacies of the environmental data;

Hillel, a soil scientist, emphasizes the long-term damaging effects of poor soil management on sustainable production; and Evans, a plant physiologist, stresses the resilience of plants to changing conditions. But all four conclude on an optimistic note.

While noting that "hunger has probably been the lot of most people, in most places, at most times," Dyson concludes that by 2020 there will be rises in "food consumption per head in most regions" with Sub-Saharan Africa being the exception. Of Sub-Saharan Africa, his conclusion is that "demographic, socio-economic and political conditions ... may be so difficult that there will be little change." He might well have concluded that in many countries in Sub-Saharan Africa, it will be difficult to maintain the improving but still inadequate levels that prevailed in the early 1990s. Demographic, socioeconomic, and political conditions have continued to deteriorate in most Sub-Saharan countries.

Hillel says that "as an agricultural and environmental scientist, I am convinced that we have the essential knowledge and capability to manage soil and water efficiently enough to feed all of humanity even allowing for the unavoidable measure of expectable population growth."

Evans is a little more cautious. His book starts with the assumption that the world population will follow the high rate of increase of UN projections, and reach 10 billion before 2050 (Figure 2.1). In fact, since Evans completed his book, the rate of increase has been closer to the medium than the high variant. Evans adds that not only does global climate change pose a major threat, but also shortages of water and land resources, and the fact that "the genetic yield potentials of our staple crops may be approaching their limits, unless their capacity for photosynthesis and growth can be substantially improved." His final conclusion is that "[f]eeding the ten billion can be done, but to do so sustainably, in the face of climatic change, equitably in the face of social and regional inequalities, and in time when few seem concerned, remains one of humanity's greatest challenges."

Following a meeting at the Royal Society in England (Greenland et al., 1998) entitled "Land Resources: on the Edge

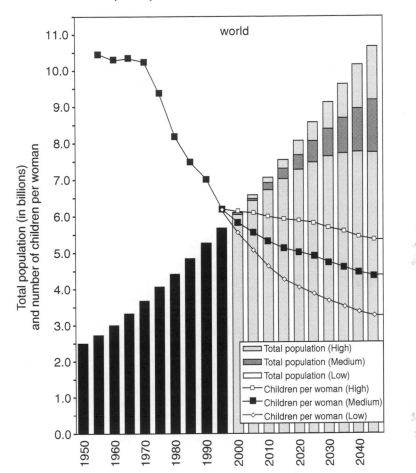

Figure 2.1 Total world population 1950–2050 (in billions), and average number of children per woman (total fertility rate). High, medium, and low variant, UN projections 1996. (From UN Population Division.)

of the Malthusian Precipice?", the organizers of the meeting concluded, *inter alia*, "If all resources are harnessed and adequate measures taken to minimise soil degradation, sufficient food to feed the population in 2020 can be produced, and probably sufficient for a few billion more." They added, "Production increases will only be achieved in resource-poor countries as a result of increased understanding of the basic principles of crop production and sustainable land management. The necessary knowledge will only be gained through improved education at all levels of society, and by close and continued collaboration between scientists, extension workers, and farmers (male and female) in research and development activities."

Wild (2003) wrote that most developing countries have the potential to feed their growing populations if "water is available, the land is properly managed, inputs are used efficiently and crop varieties with higher yield potential are developed." Young (1999) drew attention to the inadequacy of current knowledge of land resources, as illustrated by the conclusion "and probably a few billion more" drawn from the Royal Society meeting. In *Land Resources, Now and for the Future* (Young, 1998), which is based on his own experience and knowledge of Malawi and other parts of Africa, Young stressed the need to observe and map where further productive land is to be found. He is very critical of the Food and Agriculture Organization (FAO) data (Alexandratos, 1995), which are claimed to show that there is much "unused" land at present. Most of this is thought to be in Africa. Young may well be right, as Africa is the continent where food shortages are being most acutely felt at present. Much of the unused land is unused because it is difficult to manage sustainably without easy and economic access to the necessary inputs. And without security of tenure, all farmers are reluctant to invest in land.

Thus, the riders added to the conclusions of these books are sufficiently serious to make it doubtful whether there should be any complacency about future world food security, in spite of current surpluses in many developed countries.

2.2 PAST PERSPECTIVES

At least in the English-speaking world, most credit for drawing attention to the problems posed by potential food insecurity is given to the Reverend Thomas Robert Malthus, who lived from 1776 to 1834. He was preceded by many others, including Greeks, Romans, and Chinese (Evans, 1998); and Saether (1993) notes that there was a Dane, Otto Diederich Lutken, who wrote on the same theme 40 years earlier than Malthus. The Greeks, like Malthus, were mostly philosophical in their approach, whereas the Romans and Chinese were mostly practical, and tended to stress methods of sustainable and productive land use. In China as well as Egypt and other Middle Eastern countries where irrigation was essential to produce crops, the emphasis was also on the practicalities of sustainable crop production.

But it was Malthus in his series of essays published from 1798 who wrote most eloquently about the dangers posed by failure to grow sufficient food on a sustainable basis. He concentrated on "checks to population," and says little about food production. The checks he mostly discusses are warfare, pestilence and disease, storms and floods, as well as famines, and he also includes several mentions of countries and regions where infanticide was practiced.

Malthus put considerable emphasis on China and its needs, although the reliability of the sources of his "evidence," mostly reports from Jesuit missions, have often been questioned. The Jesuit priests in China probably obtained much of their information from the essays of Hung Liang-Chi, published in 1793 in China shortly before the first and anonymous version of Malthus' essay was published in London in 1798. Ho Ping-ti, in his book on the population of China published by Harvard University Press in 1959, refers to Hung Liang-Chi as "the Chinese Malthus." It might be more appropriate in terms of precedence to refer to Malthus as "the English Hung." They apparently arrived at similar gloomy prognostications, namely that population would exceed the means of production, and therefore that some disastrous end to growth of the population must occur.

In fact, during the eighteenth century Ho notes that there were several major disasters in China with population falling by 30 to 40 million on each occasion. In Hubei Province in the central Yangzi Valley, floods and other catastrophes have been a major problem for many years, and continue to be so. But there were also serious problems further north in

China, where there are huge areas of easily, and now badly, eroded loessal soils that form the headwaters of what used to be called the "Yellow River," because of discoloration due to the massive amounts of eroded sediments it carried. It has now been renamed the Huanghe, but is also known as "The Sorrow of China" (Hillel, 1991). The floods are of course also the source of the water and the sediments on which the sustainability of much rice farming depends (Greenland, 1997). The floods feed the lakes and ponds where the fish that supplement the rice diets are raised. Comparison of the Huanghe with the Ganges in India is interesting. The Ganges is venerated as "The Mother of India," while the Huanghe is "The Sorrow of China."

It is not surprising that the Chinese government is pressing ahead with the "three Gorges dams" in an effort to reduce the flood problems of central China, in spite of the publicity given to problems that will arise from the need to rehouse those whose villages and farms will be flooded by the new water storages. The Chinese government is of course balancing the (easier) problem of managing the welfare of the people under controlled conditions against those created by unpredictable flooding that are more difficult to manage.

The Taiping Wars were some of the most disastrous civil wars ever, with deaths certainly exceeding 30 million. The area of the lower Yangzi in which they occurred is among the most fertile, and heavily populated, in China (Thorpe, 1936; Ho, 1959). Ho Ping-ti attributes the large loss of life to the Taiping and other wars, the destruction of farmland, famines, epidemics, and the evils of opium.

Malthus collected his data from many parts of the world (Table 2.1) and many different sources. Patricia James (1989) — like Malthus, a Cambridge don — includes 104 pages devoted to "authorities quoted or cited by Malthus." These refer only to those that she felt required comment or amplification, beyond the reference details given in the text. Malthus argued that because population increases by geometrical progression, but food supply increases arithmetically, sooner or later population numbers must exceed food supply.

Table 2.1 Summary of Malthus' Studies on Population

Country	Reference
In James (1989), vol. 1, book 1	
American Indians	Chapter 4
Islands of the South Seas	Chapter 5
Ancient inhabitants of the north of Europe	Chapter 6
Modern Pastoral Nations	Chapter 7
Different parts of Africa	Chapter 8
Siberia, northern and southern	Chapter 9
Turkish Dominions and Persia	Chapter 10
Indoostan and Tibet	Chapter 11
China and Japan	Chapter 12
Greeks	Chapter 13
Romans	Chapter 14
In James (1989), vol. 1, book 2	
Norway	Chapter 1
Sweden	Chapter 2
Russia	Chapter 3
Middle parts of Europe	Chapter 4
Switzerland	Chapter 5
France	Chapters 8 and 9
England	Chapters 10 and 11
Scotland and Ireland	Chapter 12

China appeared to provide strong arguments in favor of his hypothesis.

It now appears unfortunate that Malthus did not consult the scientists of his time as carefully as he read the literature before publication of the anonymous version of his essay in 1798. As James noted, he made many important changes in the later versions of his essay. These showed that he was becoming aware of the effects that population pressure has on technical and scientific innovation. There were certainly some scientists working in 1798 who could have provided a scientifically based opinion about the potential for new lands to increase food production, and balanced this against population growth, although it is very doubtful if anyone could have foreseen the extent to which food production has increased during the subsequent 200 years. Fortunately, as

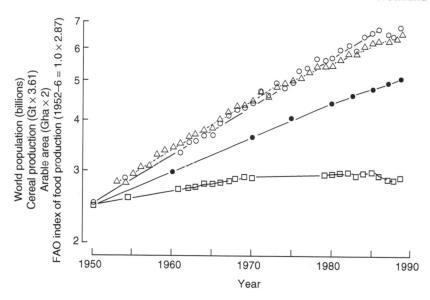

Figure 2.2 Increases since 1950 (on a logarithmic vertical scale) in world population (●), total cereal production (○), arable area (□) and the Food and Agriculture Organization index of total food production (△), all scaled to equality in 1948–1952. (Data from FAO Production Yearbooks.)

Evans (1998) has shown (Figure 2.2), production of cereals per caput has closely paralelled population growth.

Ester Boserup (1965, 1981), formerly of the University of California, took a very different view to those expressed in the original version of Malthus' essay. She believed that population growth drives technical and scientific progress of agriculture, rather than that population growth was checked by failures in agricultural production. In accordance with her hypothesis, she argued that China's rapid growth of population had driven the extensive development of water distribution systems, the establishment of various systems of multiple cropping, the breeding of dwarf and early maturing rice varieties, and she could now add the more recent successes in the breeding and use of hybrid rice (Virmani, 1988 and 1994). All of these advances originated in China, and have been important in sustaining the continuing population increase. Comments in the later editions of Malthus' essay show that Malthus was aware of this possibility. As Evans put it, "[W]e face both a Malthusian time bomb and a Boserupian treadmill."

The first chapter in the 11th edition of Russell's *Soil Conditions and Plant Growth* (Wild, 1988) is an excellent account of the studies of plant nutrition that were being conducted at that time. Liebig (1840), Boussingault (for whom McCosh [1984] provides a useful bibliography), and others were already producing the evidence on which the fertilizer industry was based, and it was only a few years before the establishment of Sir John Lawes' superphosphate factory in London (Dyke, 1993). Malthus must also have been familiar with the journals of James Cook (Williams, 1997) concerning the exploration of the South Pacific and the coasts of Australia and New Zealand. Cook returned from his first travels in the South Pacific in 1771. Sir Joseph Banks, eminent botanist who was to be elected president of the Royal Society in 1778, was the leading scientist who accompanied Cook. He would certainly have been very willing to inform Malthus on the extent of new lands and the productivity of their soils. Malthus must also have had some knowledge of the vigorous discussions about plant nutrition that were being conducted at that time.

Malthus occasionally mentions soils, and the advantages of virgin land in growing crops, and his account of the suitability for crop production of the chernozems of the Ukraine (chapter 9 of James' vol. 1) makes it clear that he was well aware of the role that soil fertility could play in supporting larger populations.

2.3 PRESENT PERSPECTIVES

To a far too great an extent, current debates on world food security have been focused on environmental problems. Many were misled by the publications of the "Club of Rome." Possibly the greatest and most misleading mistake contained in that publication was the assertion that chemical fertilizers were "poisoning the land," so that food production must be set on a downward trend. Fortunately, some have repudiated this thinking. Borlaug (1971), in his lecture on receipt of the Nobel Peace Prize, demonstrated the fallacies in the arguments of the Club of Rome. There were also experiments in the United States and at Rothamsted where plots had received heavy dressings of fertilizer for very many years and continued to produce wheat and other crops. Nevertheless, the arguments of the Club of Rome have persisted. For instance, they are found in the current discussions in the United Kingdom and other European countries, centering on the dangers of food-stuffs that contain genetically modified organisms (GMOs). Dangers to human health have been a serious part of the arguments. Wrecking of field experiments by environmentalists dressed in protective clothing have been televised, and helped to give credence to the arguments about the dangers of GMOs to health. In spite of strenuous denials by the Royal Society and other esteemed scientific bodies, strong support from the British government, and the lack of any evidence of risks from countries such as the United States and Brazil where large areas of GMOs have been grown for several years, the propaganda has attracted much public attention. The misleading arguments have been widely publicized by organizations such as the Soil Association in the United Kingdom, and Greenpeace and other international environmentalist groups. The success of the publicity is nowhere better demonstrated than in recent meetings of the Consultative Group for International Agricultural Research (CGIAR), where greater attention has been given to those who attack the achievements of the International Agricultural Research Centers than to the spokespersons of the Centers.

After a careful analysis of the arguments of Brown (1988), Ehrlich and Ehrlich (1990), Myers (1991), and other environmentalists, Dyson (1996) concluded that "their assertions during the recent past regarding the relationship between population and food can be firmly rejected." Unfortunately, his arguments have received too little public attention.

2.4 FUTURE PERSPECTIVES

There is another part of the equation relating food production and population growth that remains a puzzle: how much food, and how varied a diet, is required for each member of the population? This is not a topic on which I can claim any authority. Nevertheless, I feel that it merits attention in relation to future perspectives. There is now an extensive literature, provoked initially in regard to protein, following the large reductions in the FAO figures for essential quantities required, spurred by John Waterlow's work at the Rowett Institute in Aberdeen. The debate on protein quality, and the need for high levels of animal rather than vegetable protein, has distorted the discussions of food security for many years and is still not finally resolved (Waterlow et al., 1998). I will not rehearse the arguments here, other than to say that livestock produce about 30% of current food consumption, and that it is important that the need for this is recognized. The success of the early semidwarf, high yielding varieties of rice and wheat was such that many nutritionists feared that a serious "protein gap" would emerge, because increases in carbohydrates would not be balanced by increases of protein. Both the amounts of protein and its quality in terms of essential amino acids present were queried. More careful analysis of the problem has subsequently shown that many aspects required further attention (McLaren, 1974; Waterlow and Payne, 1975; Young, 1997). Livestock products account for more than half of current protein and essential amino acid needs. Thus, for a balanced diet it appears at present that it is important that there is an increase in animal foodstuffs as well as cereals and root crops. Changing to a vegetarian diet, as has been advocated by some, appears not to be an answer to the food security problem, and falling production of pulses and other high-protein crops suggests that a different answer must be sought.

Much more land is required to produce equivalent quantities of calories from meat and milk than from cereals, so that there is a significant disadvantage to animal-based diets in terms of quantities of land required, although there are advantages when food quality is considered. Fitzhugh (1998), director general of the International Livestock Research Institute (ILRI), also believes that the land required for animal production has tended to be overestimated, because much land used for animal production is rough grazing, unsuited to arable use.

Fortunately, much can be done to improve the quality of higher-yielding cereal varieties. In the course of the next two decades, adjustments in diets to bring them more in line with real human nutritional needs may become more important than the simple increase in food production. IRRI's original mandate was often simply expressed by Nyle Brady as "more rice." In the last 20 years, that has changed to "more rice with fewer adverse environmental effects." In the course of the next 20 years that may well change to "more rice of more appropriate nutritional value."

The first steps toward that end may well have been taken in the breeding of "golden rice" (Potrykus, 2001) in which genetic engineering has been used to enhance vitamin A content. Release by the International Centre for the Improvement of Maize and Wheat (CIMMYT) of high-protein maize varieties is another example of the route to better diets that will change the need for additional land resources. However, a major advance in education regarding human nutrition will be needed before such dietary changes will be accepted by the public. But as a greater understanding of the relationship between food and body size and shape is gained, people should be better able to understand what they should eat, and how much is necessary. At the present time, dietary requirements are based on old estimates that around 2,500 calories per person per day are needed. This is still well below current U.S. and European consumption.

The importance of global food security seems largely to be forgotten at the present time, with most developed countries more concerned with profitable disposal of surpluses than with the reduction of hunger in the Third World. African calamities continue to occur, as in Rwanda, Ethiopia, and much of Southern Africa where there has been below-average rainfall for several years. But while the majority of those living outside Africa see increases in population numbers as a major hindrance to African economic development, that view is not shared by all African scientists.

At the conference on population organized in India by the Royal Society in association with 60 other national academies in 1994, the final recommendation was that demographic increases should be minimized, and that a path to greater food production must be found that respects the environment and natural resources as well as economic, social, and cultural conditions (Graham-Smith, 1994). But there were two dissenting voices. One that could have been anticipated came from the Vatican, concerned by the prominence given to the need for birth control. Sadly the other came from the African academies of science. They added their own amendment, which stated that "care must be taken to acknowledge that while current rates of population growth and even absolute rates of (increase in) population sizes may be and are a problem for particular countries, for Africa population remains an important resource for development without which the continent's natural resources will remain latent and unexplored." And "[w]hether or not the earth is finite will depend on the extent to which science and technology (are) able to transform the resources available to humanity" (Graham-Smith, 1994:391).

We should pause here to consider for a moment the time scales of the perspectives of different writers. As noted earlier, the main theme of Malthus' essay was that population increases by a geometrical progression, but food supply increases arithmetically. Therefore, at some point population numbers must exceed food supply. The examples Malthus used were mostly of events within living memory, and so the timescale he was considering was that of a human lifetime. Politicians' timescales are largely limited to the days or months to the next election, whereas scientists' timescales tend to be much longer. Hence, in drawing attention to the problems of food security, scientists are mostly struggling to convince politicians of the immediacy of the need to provide further financial support.

As noted earlier, had Malthus consulted more widely among the scientists of his time he might have been less adamant in his arguments. Better and more scientifically based opinions about the potential to increase food production were becoming available. And the success of John Lawes' superphosphate factory opened in London in 1840 provided excellent evidence of the potential to increase food production. Nevertheless, Malthus recognized the importance of improvements in crop production methods. For example, he praised the industry of the Chinese peasants in manuring, watering, and cultivating their land. But in spite of their efforts, the pressure of population in the Yangzi Valley was the major contributor to the causes of the Taiping Wars. Between 1850 and 1867, more than 30 million died, and much of the infrastructure created to support rice production was destroyed (Ho Ping-ti, 1959). Remedies proposed and used included infanticide, at least of female children, compulsory sterilization, and heavy taxation of families having more than one child. In spite of the continued use of some of these drastic measures by recent communist governments, the Chinese population has continued to grow, and is now believed to exceed 1.3 billion.

Tim Dyson concludes his recent book, *Population and Food: Global Trends and Future Prospects*, with a section headed "Tempered Hope." He suggests that "there is fair reason to expect that in the year 2020 world agriculture will be feeding the larger global population no worse — and probably a little better — than it manages to do today." Just over a decade ago there was serious concern that the "population momentum" created by the dominant numbers of women of childbearing age would lead to a surge in the rate of population growth. In fact, there has been a significant decrease, to which AIDS and warfare have made only small contributions. Of the problems recognized by Dyson (1996), Evans (1993, 1998), and Hillel (1991), it is perhaps most difficult to predict how fast changes in climate are likely to proceed, and what can be done about them. We do not know how far human activity is to be blamed as the causal agent of the current phase of climate warming. Reduction of emissions of carbon dioxide and methane and nitrous oxide may help to slow the rate of change, and there is still hope for much to be done to increase carbon sequestration, as Lal et al. (1995a, 1995b, 1998, 1999) have shown, but if the effects of the tilt of the earth are the main cause, changes in human activities may have little influence. Immediate and long-term effects on food production are difficult to predict, particularly the balance between warming of the colder regions to allow cereal production to be extended, and increasingly frequent drought in many warmer regions, causing major losses of productivity.

But at the same time, the population growth rate has slowed significantly. Personally, I believe that the major contribution to this recent decrease has come from better education. Various threats to food security remain, and not least are those posed by political factors related to access to water and oil, and the long-term problems associated with continuing soil degradation. To these I would add the threat that comes from present-day disbelievers in sound science, who, at least in the United Kingdom, receive too much favorable publicity in the press.

REFERENCES

Alexandratos, N., Ed. 1995. *World Agriculture: Towards 2010.* Food and Agriculture Organization, Rome; John Wiley & Sons, Chichester, United Kingdom; New York.

Borlaug, N.E. 1971.The Green Revolution, Peace and Humanity. Nobel Peace Prize Speech. The Nobel Foundation, Stockholm.

Boserup, E. 1965. *The Conditions of Agricultural Growth.* Allen and Unwin, London.

Boserup, E. 1981. *Population and Technology.* Basil Blackwell, Oxford.

Brown, L.R. 1988. *The Changing World Food Prospect: The Nineties and Beyond.* Worldwatch Paper 85. Worldwatch Institute, Washington, DC.

Cohen, M.N. 1977. Food Crises in Prehistory. Overpopulation and the Origins of Agriculture. Yale University Press, New Haven, CT.

Dyke, G.V. 1993. *John Lawes of Rothamsted: Pioneer of Science, Farming and Industry.* Hoos Press, Harpenden, United Kingdom.

Dyson, T. 1996. *Population and Food: Global Trends and Future Prospects.* Routledge, London and New York.

Ehrlich, P.R. and Ehrlich, A.H. 1990. *The Population Bomb.* Simon and Schuster, New York.

Evans, L.T. 1993. *Crop Evolution, Adaptation and Yield.* Cambridge University Press, London; New York.

Evans, L.T., 1998. *Feeding the Ten Billion: Plants and Population Growth.* Cambridge University Press, London; New York.

Fischer, G. and Heilig, G.K. 1998. Population momentum and the demand on land and water resources. In *Land Resources: On the Edge of the Malthusian Precipice?* CAB International, Wallingford, United Kingdom; The Royal Society, London, pp. 9–28.

Fitzhugh, H.A. 1998. Competition between livestock and mankind for nutrients: let ruminants eat grass. In *Feeding a World Population of More Than Eight Billion People: A Challenge to Science.* In Waterlow, J.C., Armstrong, D.G., Fowden, L., and Riley, R., Eds. Oxford University Press, New York, pp. 223–231.

Graham-Smith, F., Ed. 1994. *Population, the Complex Reality,* Royal Society, London; North American Press, Golden, CO.

Greenland, D.J. 1997. *The Sustainability of Rice Farming.* CAB International, Wallingford, United Kingdom.

Greenland, D.J., Gregory, P.J., and Nye, P.H., Eds. 1998. Land Resources: on the Edge of the Malthusian Precipice? CAB International, Wallingford, United Kingdom; The Royal Society, London.

Hillel, D. 1991. *Out of the Earth: Civilisation and the Life of the Soil.* University of California Press, Berkeley.

Ho, Ping-ti. 1959. *Studies on the Population of China, 1368–1953.* Harvard University Press, Cambridge, MA.

Hung, Liang-chi. 1793. Reign of Peace, and Livelihood. Essays published by Chuan-shih-ko wen-chi (SPTK Ed.) Series A, pp. 8–106. (Reference from Ho, Ping-ti, 1959.)

James, P., Ed. 1989. *Malthus, T.R. An Essay on the Principle of Population, or, A View on its Past and Present Effects on Human Happiness*, 1803 edition with varioria of 1806, 1807, 1817, and 1826. 2 vols. Cambridge University Press, London; New York.

Lal, R., Kimble, J.M., Follett, R., and Stewart, B.A., Eds. 1995a. *Soil Processes and the Carbon Cycle*. CRC Press, Boca Raton, FL.

Lal, R., Kimble, J.M., Levine, E., and Stewart, B.A., Eds. 1995b. *Soil Management and Greenhouse Effect*. CRC Press, Boca Raton, FL.

Lal, R., Kimble, J.M., Follett, R., and Stewart, B.A., Eds. 1998. *Management of Carbon Sequestration in Soil*. CRC Press, Boca Baton, FL.

Lal, R., Kimble, J.M., and Stewart, B.A., Eds. 1999. *Global Climate Change and Tropical Ecosystems*. CRC Press, Boca Raton, FL.

Liebig, J. von. 1840. *Organic Chemistry in Its Application to Agriculture and Physiology*. Taylor and Walton, London.

McCosh, F.W.J. 1984. *Boussingault: Chemist and Agriculturist*. Reidel, Dordrecht, Netherlands.

McLaren, D.S.. 1974. The great protein fiasco. *Lancet*, 2:93–96.

Malthus, T.R. 1798. *An Essay on the Principle of Population as it Affects the Future Improvement of Society with Remarks on the Speculations of Mr. Godwin, M. Condorcet and Other Writers*. Printed for J. Johnson in St. Paul's Churchyard, London.

Myers, N. 1991. *Population, Resources and the Environment: The Critical Challenges*. UN Population Fund, New York.

Potrykus, I. 2001. Golden rice and beyond. *Plant Physiol.*, 125, 1157–1161.

Saether, A. 1993. Otto Diederich Lutken: 40 years before Malthus? *Population Stud.*, 47, 511–513.

Thorpe, J. 1936. *Geography of the Soils of China*. National Geological Survey of China, Nanjing, China.

Virmani, S.S., Ed. 1988. *Hybrid Rice*. International Rice Research Institute, Manila.

Virmani, S.S. 1994. *Heterosis and Hybrid Rice Breeding*. International Rice Research Institute, Manila; Springer Verlag, Berlin.

Waterlow, J.C., Armstrong, D.G., Fowden, L., and Riley, R., Eds. 1998. *Feeding a World Population of More Than Eight Billion People: A Challenge to Science.* Oxford University Press, New York.

Waterlow, J.C. and Payne, R. 1975. The protein gap. *Nature (London),* 258:113–117.

Wild, A. 2003. *Soils, Land and Food: Managing the Land during the Twenty-first Century.* Cambridge University Press, London; New York.

Wild, A., Ed. 1988. *Russell's Soil Conditions and Plant Growth.* 11th ed. Longman Scientific and Technical, Harlow, United Kingdom.

Williams, G. 1997. *Captain Cook's Voyages, 1768–1779.* Folio Society, London.

Young, A. 1998. *Land Resources, Now and for the Future.* Cambridge University Press, London; New York.

Young, A. 1999. Is there really spare land? A critique of estimates of available, cultivable land in developing countries. *Environ. Dev. Sustainability,* 1:3–18.

Young, V.R. 1997. Human amino acid requirements: a revaluation. *Food Nutr. Bull.,* 17:191–203.

3

Changing Times and Directions

ROBERT D. HAVENER, CHRISTOPHER R. DOWSWELL, AND
NORMAN E. BORLAUG

CONTENTS

3.1 INTRODUCTION

For centuries the Chinese have used an appropriate adage
that can be given either as a curse or a blessing: "May you
live in interesting times." It is clear to us that someone has
put just such a curse and blessing on humankind today, and
we are reaping the harvests of both.

Never in the history of the human race have so many
people been so well off. Our capacity for rapid and inexpensive
transportation and communications is expanding dramatically and for much of the world's population, food has never
been more abundant and more affordable. Modern medicine
can at times perform miracles. On average, we are living
longer and healthier lives. Any unbiased study will indicate
that on balance, human drudgery has been greatly reduced,
and that we have more time and other resources for the
pursuit of leisure, cultural, and civic activities. In OECD
(Organization for Economic Cooperation and Development)
countries, advanced education is available to more young people than ever before. Clothes, watches, cameras, and even cars
have become expendable items of consumption. Not long ago,
they were passed from one generation to the next! But the

great material advances of the 20th century have not been equitably shared. We have the dilemma of great wealth and excessive consumption in the OECD countries and extreme poverty, underproduction, and overexploitation of the environment in the less developed countries. These diverging welfare streams are leading to abundant living standards coexisting with lives of crushing poverty and desperation.

We all will pay a price for such disparities in income and well-being. For one thing, hungry nations are unstable nations, which become fertile breeding grounds for violence and terrorism, on both a local and international scale. Lest we forget, in the midst of global plenty, more than 800 million people still go to bed hungry every night (Food and Agriculture Organization [FAO], 2003). Extreme poverty is also no friend of the environment. Indeed, the most serious environmental damage occurring in the world at present is both rural and poverty based.

Over the past 50 years, world population has grown from 2.5 billion to 6.2 billion people. At present, some 79 million people join us on spaceship Earth each year, and largely in countries that are too poor to provide the newborn with adequate food, health care, education, and employment, let alone a vision of a better life tomorrow. Feeding the growing world population of the 20th century has been a challenging proposition. However, farmers and ranchers — and the research and production infrastructure that supports them — have proven to be up to the task. Since 1950, world cereal production has increased threefold to 2 billion metric tons annually, with only a 10% increase in land area (Figure 3.1). This has resulted in a 20% increase in per capita food production and more than a 50% decline in real food prices (FAO, 2003).

3.2. THE GREEN REVOLUTION — WHAT HAS BEEN ACHIEVED?

High-yield agriculture is a 20th-century invention. The first major impact was the hybrid maize revolution that began in the United States in the late 1930s, and was driven by the need to help feed allies during World War II. Fueled by the

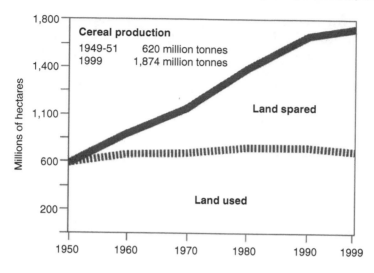

Figure 3.1 Land that farmers throughout the world spared in cereal production by raising yields. (From authors' calculations based on selected Food and Agriculture Organization Production Yearbooks, 1950–1960 and FAOSTAT.)

combination of high-yielding hybrid varieties, markedly increased fertilizer use, chemical pest control, and improved mechanization, maize production in the United States and later Europe and Canada, expanded at unprecedented rates. Attempts after WWII to transfer the high-yield temperate-zone maize technology to other nations met with mixed results, especially in tropical and subtropical environments where the U.S. technology performed poorly. However, pioneering agricultural research and development programs of the Rockefeller Foundation, which began in Mexico in 1943 and later spread to other developing countries during the 1950s, began supplying new classes of varieties that were adapted to subtropical and tropical production conditions.

By the mid-1960s, International Center for the Improvement of Maize and Wheat and International Rice Research Institute (IRRI) scientists — successor organizations to former bilateral programs supported by the Rockefeller and Ford Foundations — had developed rust-resistant, semidwarf wheat and rice varieties with radically improved yields. The new short wheat varieties drew on dwarfing genes found in the Japanese Norin wheat germ plasm. When crossed to the high-yielding, disease-resistant tall wheat varieties developed in Mexico, the new semidwarf progenies were much more efficient than their tall predecessor varieties in converting sunlight and nutrients into grain production. Furthermore, their superior plant architecture provided resistance against lodging (falling over) in heavy winds and under improved conditions of soil fertility and moisture. In the case of rice, IRRI scientists used dwarf strains first selected by Japanese breeders in the 1920s in Taiwan (then a Japanese colony) to shorten plant height. By crossing tropical (indica) and temperate (japonica) rice germ plasm, they created new types of high-yielding varieties that combined the best traits of both gene pools.

Under most conditions, even when farmers used traditional methods of cultivation without fertilizer, the new wheat and rice varieties yielded more grain than the traditional local cultivars. However, when these new varieties were grown with adequate moisture and under higher soil fertility, they yielded up to four times as much. Nevertheless, many agricultural scientists and extension workers were skeptical about the willingness of farmers to accept the new varieties. Not only did they look different from traditional types, they required substantially different care, especially in depth of planting, fertilizer application, water management, and weed control.

Despite the misgivings of many national researchers, government leaders in India and Pakistan (and somewhat later in Turkey and China), who faced desperate and deteriorating food situations and growing prospects of widespread famine, decided to introduce the new varieties and crop management techniques as quickly as possible. Extensive farm demonstrations were established, and once farmers saw the results themselves, they became the major spokesmen for the new methods.

In the beginning, Pakistani and Indian leaders authorized the purchase of large quantities of the new seeds and massive amounts of fertilizer. But they also radically changed national investment policies to build up domestic seed and fertilizer production facilities and to ensure farmers a harvest price for their grain similar to that prevailing on the international market. All these policy changes were critical for sustained growth in food production. The gamble taken by courageous leaders in India and Pakistan — and later in other Asian countries — paid off handsomely. Within 10 years, wheat and rice production had increased by 50% (FAOSTAT, 2003).

Over the past four decades, FAO reports that in the developing countries of Asia, irrigated area more than doubled to 175 million hectares, fertilizer consumption increased more than 30-fold and now stands at about 70 million metric tons of nutrients, and tractors in use increased from 200,000 to 4.8 million (Table 3.1).

Many Green Revolution observers have tended to focus too much on the high-yielding wheat and rice varieties, as if they alone can produce miraculous results. Certainly, modern varieties can shift yield curves higher due to more efficient plant architecture and the incorporation of genetic sources of disease and insect resistance. However, modern varieties can only achieve their genetic yield potential if systematic changes are also made in crop management, such as in dates and rates of planting, fertilization, water management, and weed and pest control. Moreover, many of these changes must be applied simultaneously if the genetic yield potential of modern varieties is to be fully realized.

In describing the rapid spread of the new wheat and rice technology across Asia, William Gaud, administrator for the U.S. Agency for International Development, in a talk given in March 1968 to the Society for International Development in Washington, D.C., said: "These and other developments in the field of agriculture contain the makings of a new revolution. It is not a violent Red Revolution like that of the Soviets or the White Revolution in Iran. But rather, I call it a Green Revolution."

Thus, the term "Green Revolution" was coined. To us, it symbolizes the process of applying agricultural science to develop modern techniques for Third World food production conditions.

Table 3.1 Changes in Factors of Production in Developing Asia

	Adoption of Modern Varieties		Irrigation (million ha)	Fertilizer Nutrient Consumption (million metric tons)	Tractors (millions)
	Wheat (million ha/% area)	Rice (million ha/% area)			
1961	0/0%	0/0%	87	2	0.2
1970	14/20%	15/20%	106	10	0.5
1980	39/49%	55/43%	129	29	2.0
1990	60/70%	85/65%	158	54	3.4
2000	70/84%	100/74%	175	70	4.8

Source: From FAOSTAT, April 2000; adoption of modern varieties based on authors' calculations derived from International Center for the Improvement of Maize and Wheat and International Rice Research Institute data.

3.2.1　Criticisms of the Green Revolution

The Green Revolution has been a much-debated subject. The initial euphoria during the late 1960s of the high-yielding wheat and rice varieties — and more intensive crop production practices — was followed by a wave of criticism. Some criticism reflected a sincere concern about social and economic problems in rural areas that were not — and cannot — be solved by technology alone. Some criticism was based on premature analyses of what was actually happening in areas where the Green Revolution technologies were being adopted. Some criticism focused on issues of environmental damage and sustainability. Many of these criticisms have some element of truth to them. Obviously, wealth has increased in irrigated areas, relative to less-favored rainfed regions, thus increasing income disparities. Cereals, with their higher yield potential, have displaced pulses and other lower-yielding crops. Farm mechanization has displaced low-paid laborers. High-yielding cereal varieties have replaced lower-yielding land races.

For those whose main concern is protecting the "environment," what would the world have been like without the technological advances that have occurred? Had the global cereal yields of 1950 still prevailed in 1999, we would have needed nearly 1.2 billion ha of additional land of the same quality — instead of the 660 million that was used.

Obviously, such a surplus of land was not available, and certainly not in Asia, where the population increased from 1.2 to 3.8 billion over the time period. Moreover, if more environmentally fragile land had been brought into agricultural production, the impact on soil erosion, loss of forests, grasslands, and biodiversity, and extinction of wildlife species would have ensued.

The debate on benefits and shortcomings of the Green Revolution must be framed within the larger context of population growth. The continuing decline in the real price of cereals also needs to be considered. Lower food costs benefit everybody in society, but especially the poor consumer. Finally, the very strong growth linkages between Green Revolution technology and industrial development are also apparent. Indeed, much of Asia's spectacular economic development in industry and services over the past 20 years has followed in the wake of the agricultural revolutions that preceded them.

3.2.2 Hunger Still Stalks Asia

Despite the ability of smallholder Asian farmers to triple cereal production since 1961, the battle to ensure food security for millions of miserably poor people is far from won. Huge stocks of grain have accumulated in India, while tens of millions need more food but do not have the purchasing power to buy it. China has been more successful in achieving broad-based economic growth and poverty reduction than India. Nobel Economics Laureate, Amartya Sen (2000), attributes China's achievement to the greater priority its government has given to investments in rural education and healthcare services, compared to India. Hopefully, the Indian government will make the necessary investments in the rural sector to bring food security to the millions of hungry poor that still populate this great country.

Unfortunately, smallholder farmers in higher-risk tropical and subtropical production environments — those characterized by nutrient-depleted soils, unreliable water supply, steeper slopes, and remote locations — have benefited little from Green Revolution technologies to date. In many cases, improved varieties have been available, but other factors have constrained their effectiveness. These include climatic stresses, low and declining soil fertility, and low levels of adoption of fertilizers or other soil nutrient-restoring technologies, ecosystem degradation associated with intensified crop production, poor access to markets and inputs, and lack of political will, to name a few. Sub-Saharan Africa, in particular, has such higher-risk production environments, where a new paradigm is developing, focused more on the interplay between agricultural technologies and the need to improve natural resource management.

3.3 FOOD PRODUCTION CHALLENGES AHEAD

3.3.1 Future Demand and Supply Projections

The United Nations' medium projection is for world population to reach about 7.9 billion by 2025 before *hopefully* stabilizing at about 10 to 11 billion toward the end of the 21st century. At least in the foreseeable future, plants — and especially cereals — will continue to constitute much of food supply, both for direct human consumption and as livestock feed to satisfy the rapidly growing demand for meat and milk products in the newly industrializing countries. It is likely that an additional 1 billion metric tons of grain will be needed annually by 2030 — a 50% increase over world cereal production in 2000 (FAO, 2003). Developing countries of Asia — because of rapid economic growth, urbanization, and large populations — will account for half of the increase in global demand for cereals. Most of this increase must be supplied through yield improvements on lands already in production, although the agricultural area is expected to expand in tropical lands in South America (*cerrados*) and Sub-Saharan Africa, and in temperate zones, mainly in North America.

There is concern in some circles that crop yields in the world's more intensively cultivated areas may be approaching their physical limitations. Many advances — such as introduction of fertilizer-responsive varieties, higher planting density, multiple cropping, and better management practices — may have represented one-time gains. Environmental constraints add additional uncertainty, as some of the most intensively cultivated areas now suffer repercussions from intensive use of irrigation, fertilizers, and crop protection chemicals.

3.3.2 Livestock Demand — The Global Wealth Effect

Higher incomes and urbanization are leading to major changes in dietary patterns. A massive increase is foreseen in per capita consumption of fish, meat, and milk products, especially in populous and increasingly prosperous Asia (Delgado et al, 1999). Expanding poultry and livestock demand, in turn, will result in major increases in the share of cereal production consumed by livestock, a trend which runs the risk of reducing cereal availability for the very poor and food insecure in coming decades. Rural to urban migrations will also affect farm production in several ways. First, with an out-migration of labor, more farm activities will have to be mechanized to replace labor-intensive practices. Second, large urban populations, generally close to seaports, are likely to buy more food from the lowest-price producer, which for certain crops may very well mean importing from abroad. Domestic producers, therefore, will have to compete in price and quality with imported foodstuffs.

3.3.2.1 Marginalized People and Lands

Of the 800 million hungry and malnourished people in the developing world, 232 million are in India, 200 million in Sub-Saharan Africa, 112 million in China, 152 million elsewhere in Asia and the Pacific, 56 million in Latin America, and 40 million in the Near East and North Africa (UN Development Programme, 2003b). Of the total number of hungry, about 214 million or 26%, have caloric intakes so low that they are unable to work or care for themselves. The 800 million hungry people can be grouped into four broad types, based on their household means of obtaining food. Roughly 50% of the hungry are farm households in higher-risk environments for crop production, such as low, highly unreliable, or excessive rainfall; inherently poor or degraded soils; steep topography; and remoteness from markets and public services. Another 22% live in non-farm rural households, and another 20% in poor urban households. The remaining 8% are herders, fishermen, and forest-dependent households.

These statistics on hunger point to the need to improve drastically the food security of farmers in higher-risk environments and remote regions — to bring the Doubly Green Revolution that Gordon Conway, Rockefeller Foundation president, talks about (Conway, 1999). The statistics also point to the need to develop poverty-reduction strategies that will provide employment options for marginal farmers, especially in marginal lands, in sectors other than agriculture. Clearly, too many people in the developing world are trying to gain their livelihoods through agriculture, with too few resources. Reducing agricultural populations — and increasing the land and water resources available to those that remain — will be one of our greatest challenges in the 21st century.

3.3.3   Africa Is the Major Challenge

More than any other region of the world, food production south of the Sahara is in crisis. High rates of population growth and little application of improved production technology resulted in declining per capita food production, escalating food deficits, and deteriorating nutritional levels, especially among the rural poor during the past two decades. While there was some indication during the 1990s that smallholder food production was beginning to turn around, this recovery is still very fragile.

Sub-Saharan Africa's extreme poverty, poor soils, uncertain rainfall, increasing population pressures, changing ownership patterns for land and cattle, political and social turmoil, shortages of trained agriculturalists, and weaknesses in research and technology delivery systems all make the task of agricultural development more difficult. But we should also realize that to a considerable extent, the present food crisis is the result of the neglect of agriculture by political leaders, even though agriculture provides the livelihood for 60% to 75% of the people in most African countries. Investments in agricultural research and education and in input distribution and food marketing systems have been woefully inadequate. Furthermore, many governments have pursued a policy of providing cheap food — often imported from abroad — for politically volatile urban dwellers at the expense of production incentives for farmers.

Many of the lowland tropical environments — especially the forest and transition areas — are fragile ecological systems with deeply weathered, acidic soils that lose fertility rapidly under repeated cultivation. Traditionally, slash-and-burn shifting cultivation and complex cropping patterns permitted low-yielding, but relatively stable, food production systems. Expanding populations and food requirements have pushed farmers onto more marginal lands, and also have led to a shortening in the bush/fallow periods previously used to restore soil fertility. With more continuous cropping on the rise, organic material and nitrogen are being rapidly depleted, while phosphorus and other nutrient reserves are slowly but steadily becoming depleted. This is having disastrous environmental consequences, such as serious erosion and weed invasions leading to impoverished fire-climax vegetations.

What is needed in African agriculture is a broader range of technological options. For the high-potential production areas with better moisture availability, soils, and access to markets, and especially for higher-value crops, the priority must be to get modern factors of production such as fertilizers, improved seeds, crop protection chemicals, and machinery into the hands of farmers, so that they can make the transformation from subsistence to commercial agriculture. However, for those poor farmers with extremely marginal production circumstances, the first priority should be to attain household food security. Since many of these areas have severely degraded environments, resource-improving social capital investments will be needed in soil fertility restoration, microscale water management, tree establishment, and livestock and forest/range rehabilitation. These practices lead to increased carbon sequestration, which can ameliorate the impact of global warming.

In many cases, because of their extreme lack of purchasing power and frequent remoteness from markets, low external input crop production technologies are likely to be the most appropriate. Once farmers have achieved household food security, they can then consider using additional external inputs to boost production and generate small surpluses to sell into the market.

The second step in many cases for such very poor farmers in low-potential areas is to develop off-farm employment opportunities for them, either within the area itself or somewhere else in the country. In many cases, the carrying capacity of the land simply will not support such large numbers of smallholder farmers. Thus, the solution to their poverty must be found outside of agriculture, with the land returning to permanent pastures and forests. This should have a favorable impact on the environment.

3.4 SCIENTIFIC CHALLENGES AND OPPORTUNITIES

3.4.1 Raising Maximum Genetic Potential

The slowing of gains in maximum genetic yield potential is a matter of considerable concern. Continued genetic improvement of food crops — using both conventional as well as biotechnology research tools — is needed to shift the yield frontier higher and to increase stability of yield. In rice, wheat, and maize research, three distinct, but interrelated strategies are being pursued to increase genetic maximum yield potential: changes in plant architecture, hybridization, and wider genetic resource utilization. Significant progress has been made in all three areas. IRRI remains optimistic that it will be successful in developing the new "super rice," with fewer — but highly productive — tillers. IRRI claims that this new plant type, in association with direct seeding, could increase rice yield potential by 20% to 25%, although it is still probably 10 to 12 years away from widespread impact on farmers' fields (Khush, 1995). New wheat plants with an architecture similar to the "super rices," including larger heads, more grains, and fewer tillers, could lead to an increase in yield potential of 10% to 15% above the best current germ plasm (Rajaram and Borlaug, 1997). Introducing genes from related wild species into cultivated wheat can introduce important sources of resistance to several biotic and abiotic stresses, and perhaps permit higher yield potential as well, especially if the synthetic wheat is used as parent material in hybrid wheat production.

The success of hybrid rice in China, which now covers more than 50% of irrigated area, has led to a renewed interest in hybrid wheat. Recent improvements in chemical hybridization agents, advances in biotechnology, and the emergence of the new wheat plant type have made a reassessment of hybrids worthwhile. With better heterosis and increased grain filling, the yield frontier of the new plant material could be 25% to 30% above the current germ plasm base.

In maize, yield increases have been achieved by breeding plants that can withstand higher planting densities, as well as the shift to single cross hybrids.

Maize production has really begun to take off in many Asian countries, especially China. It now has the highest average yield of all cereals in Asia, with much of the genetic yield potential yet to be exploited. Recent developments with high-yielding quality protein maize (QPM) varieties and hybrids will also improve the nutritional quality of the grain without sacrificing yields. This achievement offers important nutritional benefits for livestock and humans.

3.4.2 Sustaining Yields in Irrigated Areas

Water covers about 70% of the Earth's surface. Of this total, only about 2.5% is fresh water, and most of this is frozen in the ice caps of Antarctica and Greenland, in soil moisture, or in deep aquifers not readily accessible for human use. Indeed, less than 1% of the world's freshwater — found in lakes, rivers, reservoirs, and underground aquifers shallow enough to be tapped economically — is readily available for direct human use. Irrigated agriculture, which accounts for 70% of global water withdrawals, covers some 17% of cultivated land (about 275 million ha) yet accounts for nearly 40% of world food production.

The rapid expansion in world irrigation and in urban and industrial water uses has led to growing shortages. The UN's Comprehensive Assessment of the Freshwater Resources of the World estimates that "about one third of the world's population lives in countries that are experiencing moderate-to-high water stress, resulting from increasing demands from a growing population and human activity (World Meteorological

Organization, 1997). By the year 2025, as much as two thirds of the world's population could be under stress conditions.

In order to expand food production for a growing world population within the parameters of likely water availability, the inevitable conclusion is that humankind in the 21st century will need to bring about a "Blue Revolution" to complement the "Green Revolution" of the 20th century. In the new Blue Revolution, water-use productivity must be wedded to land-use productivity. New science and technology must lead the way. Clearly, we need to rethink our attitudes about water, and move away from thinking of it as nearly a free good, and a God-given right. Pricing water delivery closer to its real costs is a necessary step to improving use efficiency. Farmers, irrigation officials, and urban consumers will need incentives to save water. Moreover, management of water distribution networks, except for the primary canals, should be decentralized and turned over to the users.

Many technologies exist to improve the efficiency of water use. Wastewater can be treated and used for irrigation. This could be an especially important source of water for periurban agriculture, which is growing rapidly around many of the world's megacities. By using modern technologies such as drip irrigation systems, water can be delivered much more efficiently to the plants and largely in ways that avoid soil waterlogging and salinity. Changing to crops that require less water (and/or new improved varieties), together with more efficient crop sequencing and timely planting, can achieve significant savings in water use.

Proven technologies such as drip irrigation, which saves water and reduces soil salinity, are suitable for much larger areas than currently used. Various new precision irrigation systems are also on the horizon, which will supply water to plants only when they need it. There is also a range of improved low-cost, small-scale, and supplemental irrigation systems to increase the productivity of rainfed areas, which offer much promise for smallholder farmers.

An outstanding example of new Green/Blue Revolution technology in irrigated wheat production is the "bed planting system," which has multiple advantages over conventional planting systems. Plant height and lodging are reduced, leading to 5% to 10% increase in yields and better grain quality. Water use is reduced 20% to 25%, a spectacular savings! Input efficiency (fertilizers and crop protection chemicals) is also greatly improved, which permits total input reduction by 25%. After growing acceptance in Mexico and other countries, India, Pakistan, and Shandong Province and other parts of China are now preparing to rapidly extend this technology.

Conservation tillage (no tillage, minimum tillage) is another technology that has important "water harvesting" and soil conservation characteristics. About 100 million ha are planted in such systems throughout the world. By reducing and/or eliminating the tillage operations, conservation tillage reduces turnaround time on lands that are double and triple cropped annually, which adds significantly to total yield potential, especially rotations such as rice/wheat and cotton/wheat. This leads to higher production and lower production costs. Conservation tillage controls weed populations and greatly reduces the time that small-scale farm families must devote to this backbreaking work. The mulch left on the ground reduces soil erosion, lowers soil temperatures, builds up organic matter, increases moisture retention in the soil profile, and reduces moisture loss through evaporation. These "water harvesting" aspects of conservation tillage may prove increasingly important in the future as water stress becomes more prevalent in many parts of the world.

3.4.2.1 Addressing Marginal Lands and Environments

As noted earlier, many of the world's poorest and socially and nutritionally disadvantaged people live in marginal environments and seek to make their living from marginal lands. Historical geologic events can substantially affect soil quality as can inappropriate agricultural practices characteristic of more recent times. Also, because of low levels of precipitation or cold temperatures, it is possible, for instance, to have a poor agricultural environment associated with fertile soils. For purposes of this discussion, however, we shall deal with three general topics, namely drought, problem soils, and low soil fertility. Once again, these are frequently but not always associated.

As noted elsewhere in this document, biotechnology offers many new and exciting opportunities to make more rapid progress in developing higher-yielding varieties better suited to harsh production environments. It appears that, with the aid of new scientific techniques and knowledge, considerable progress in being made in the areas of drought tolerance and avoidance, heat tolerance, tolerance to acid and/or saline soils, and mineral toxicities. There is also evidence that varieties can be developed that are more efficient at extracting otherwise unavailable soil nutrients or able to produce well with lower levels of nutrients or at higher levels of soil toxicities. This research needs to be substantially expanded and accelerated.

Particularly in disadvantaged areas, increasing farmer incomes depends on lowering fertilizer prices and enhancing the availability of cropping patterns and crop production systems that provide higher and more consistent returns to cultivators in high-risk environments. Restoring soil organic matter through various means, such as off-season green manure/cover crops, crop rotations with legumes, and livestock production in association with cultivated crops, need to be further investigated and where appropriate, promoted.

Better and more productive and economical systems of water harvesting and distribution are needed. Microirrigation systems, which use limited water supplies more advantageously and that are used along with the production of high-value crops must be developed.

Only through the public sector or public–private sector partnerships will the required investments in such research activities be made. Without these investments, many more poor people will be crowding the city slums in Third World countries and, from there, migrating to the city streets of the industrialized world.

3.4.2.2 Improving Nutritional Quality

Plant breeding can improve nutritional quality of staple foods. The development of quality protein maize (QPM) at CIMMYT during 1970 to 1990 is one such example. QPM carries the opaque-2 gene that doubles the levels of lysine and tryptophan — two essential amino acids needed to build proteins — over normal maize. It offers considerable nutritional advantages to humans and monogastric animals. QPM is also an excellent maize-based weaning food for poor people and reduces feed costs (by reducing the amount of protein components) for swine and poultry production.

Plant breeding offers the potential to reduce micronutrient deficiencies by increasing the concentration of micronutrients within staple food crops, either to reduce inhibitors of micronutrient absorption or to raise the levels of amino acids that promote micronutrient absorptions. Natural genetic variation in many crops, including rice, wheat, maize, and beans, shows a wide range of concentration of iron, zinc, and other micronutrients. In addition, through biotechnology, pro-vitamin A can be introduced into rice, white maize, and other food crops.

This could have profound positive results for millions of people too poor to have access to balanced diets and food supplements.

3.4.2.3 What to Expect from Biotechnology

What began as a biotechnology bandwagon nearly 20 years ago has developed invaluable new scientific methodologies and products that need active financial and organizational support to bring them to fruition in food and fiber production systems. Initially, biotechnology had the greatest impact in medicine and public health. However, now several fascinating developments are apparent in agriculture. In animal biotechnology, bovine somatotropin (BST) is now widely used to increase milk production. In plants, transgenic varieties and hybrids of cotton, maize, and potatoes containing genes from

Bacillus thuringiensis, which effectively control a number of serious insect pests, are now being successfully introduced commercially in the United States and elsewhere. The use of such varieties will greatly reduce the need for insecticide sprays and dusts. Considerable progress also has been made in the development of cotton, maize, oilseed rape, soybeans, sugar beets, and wheat, with tolerance to several herbicides, which can lead to a reduction in overall herbicide use through much more specific interventions and dosages. Thus far, it is estimated that in the United States alone, pesticide use has been reduced by 21,000 metric tons per year in cotton, maize, and soybeans (Gianessi, 2002).

Substantial progress has been made in developing cereal varieties with greater tolerance for soil alkalinity, free aluminum, and iron toxicities. These varieties will help to ameliorate the soil degradation problems that have developed in many existing irrigation systems. They will also allow agriculture to succeed into acidic soil areas, such as the *cerrados* in Brazil and in central and southern Africa, thus adding more arable land to the global production base. Greater tolerance of abiotic extremes, such as drought, heat, and cold, will benefit irrigated areas in several ways. First, we will be able to achieve "more crop from every drop" through designing plants with reduced water requirements and adoption of moisture-conserving crop/water management systems. Second, enhanced drought tolerance can reduce the risks of food production in more marginal rainfed environments, thus significantly contributing to food security of some of the world's poorest people. Recombinant DNA techniques can speed up the development process.

There is growing evidence that genetic variation exists within most cereal crop species for genotypes that are more efficient in the use of nitrogen, phosphorus, and other plant nutrients than are currently available in the best varieties and hybrids. Scientists from the University of Florida and Monsanto have been working on the genetic engineering of wheat and other crops that have high levels of glutamate dehydrogenase (GDH). Transgenic wheat varieties with high

GDH are reported to yield up to 29% more with the same amount of fertilizer than do the normal crop (Smil, 1999).

Transgenic plants that can control viral and fungal diseases are not nearly so developed. Nevertheless, there are some promising examples of specific virus coat genes in transgenic varieties of potatoes and rice that confer considerable protection. Other promising genes for disease resistance are being incorporated into other crop species through transgenic manipulations.

The biofortification work mentioned above will also be greatly enhanced and accelerated through the tools of genetic engineering, which allow us to reach beyond a particular species to other taxonomic groups, orders, and kingdoms. More nutritionally balanced foods can be expected in the future. They will have more balanced amino acids and higher levels of essential vitamins and micronutrients. Significant advances will also occur in what are often labeled "nutriceuticals," plants that can help control and cure diseases, and also serve to deliver vaccines to humans and livestock.

Since much biotechnology research is being done by the private sector, which patents its inventions, agricultural policymakers must face up to a potentially serious problem of access, especially for resource-poor farmers. How long, and under what terms, should patents be granted for bioengineered products? Further, the high cost of biotechnology research is leading to rapid consolidation in the ownership of agricultural life science companies. Is this desirable? These issues are matters for serious consideration by national, regional, and global governmental organizations.

National governments need to be prepared to work with — and benefit from — the new breakthroughs in biotechnology. First and foremost, governments must establish a regulatory framework to guide the testing and use of genetically modified crops. These rules and regulations should be reasonable in terms of risk aversion, and cost-effective to implement to avoid tying the hands of scientists through excessively restrictive regulations. Since much of the biotechnology research is underway in the private sector, the issue of intellectual property rights must be addressed, and accorded adequate safeguards by national governments.

3.4.2.4   Coping with Climate Change

Although considerable differences of opinion continue to exist as to the timing, severity, and differential effect of the actual climate change associated with global warming, a consensus seems to have emerged about three important aspects.

The first is that catastrophic weather events are likely to increase, taking the form of more severe storms, more flooding, and, of most concern for agriculture production, more frequent and severe droughts. Second, it appears possible that favored lands will experience even more favorable growing conditions but that areas that are currently subject to periodic flooding and, more particularly, drought, are likely to experience increased devastation. Third, virtually all agricultural research directed at overcoming the effects of heat, drought, and associated biotic and abiotic stresses will be of high potential benefit to ameliorating the likely negative effects of global warming.

It is fortuitous indeed that these same research priorities coincide with those most valuable and urgent in a "pro-poor" agricultural research agenda.

3.5   THE FUTURE OF PUBLIC GOODS RESEARCH

3.5.1   Intellectual Property Rights, the Regulatory Process, and Scientific Innovation

Much has been written and discussed about the way in which intellectual property rights legislation and patents has distorted research objectives, raised product prices, slowed the exchange of scientific findings and genetic materials, and otherwise negatively impacted research results reaching the consumer in a timely and economical manner. Each of these concerns has some validity. It is also true, however, that insuring exclusive control over — or exploitation of — for innovators and developers of their innovation has led to much larger investments in research and development directed at marketable products. Without these investments, many profit-enhancing, risk-avoidance products that are now in farmers' fields or in the research pipeline would not have been developed.

However, a much more dangerous barrier to the delivery of new science is now coming to the forefront. During the early to mid-1990s, regulation of GMOs in the United States was relatively streamlined and not strongly focused on the process of transgenic transformation *per se*. Rather, it focused on the effect that a particular trait, such as insect resistance or herbicide tolerance, might have on consumers or the environment. But starting in the late 1990s, the regulatory process moved ever more toward review of the transformation process itself. Originally, large multinational companies welcomed the increasing complexity of regulation, both to protect themselves and to serve as a type of screening mechanism to control smaller, perhaps less rigorous, research initiatives. The cost of regulation would serve as a sort of nontariff barrier to all but the very strong. Thus, when nine U.S. professional agricultural science societies (which encompassed 80,000 members) attempted in Congress in the late 1990s to oppose a much more elaborate — and costly — GMO regulation and certification process, the private sector broke ranks and supported the new legislation. This has led to a serious situation of overregulation of GMO products — and many would argue — spurious regulation.

Today, the documentation required by regulators for any single GMO "event" is likely to cost upward of US $1 million. This high cost becomes a barrier to all but the largest research organizations — and the most profitable GMO traits. It prevents smaller research laboratories from bringing products to market, as well as most public sector research efforts. Effectively, the regulations have created a large and growing divide between the R&D functions, with publicly funded organizations focusing on the "R" for an array of useful traits, but with little hope to engage actively in the "D" function of bringing research findings to fruition in the marketplace. This regulatory barrier has already been encountered with the so-called pro-Vitamin A "golden rice," and it is likely to affect most GMO research in orphan crops and the new work to expand micronutrient biofortification. A return to more reasonable regulations — based on the effect of the trait and not the transformation process itself — is urgently needed before the regulatory institutional structure becomes so formidable that changing it becomes almost an insurmountable quest.

3.5.2   Future of Public Research

As mentioned elsewhere, the private sector will only invest when and where a future stream of profits is envisioned. To the extent that private sector advances are to be captured for the benefit of the disadvantaged, public sector organizations must make the investments required to use, modify, or direct them to the crops, regions, and situations of the poor who largely operate outside of the standard market mechanisms. New types of public–private partnerships will be essential, ones that we are only beginning to explore and develop.

At the heart of the problem has been the growing separation between research and development in the public sector. While there is still considerable publicly funded research, the capacity of public sector institutions to deliver technology to farmers has atrophied considerably over the past 30 years. In the interest of efficiency, public sector production organizations (such as agricultural extension, seed and fertilizer suppliers, and grain marketing boards) have been greatly reduced or eliminated. The theory, often called the "Washington Consensus," calls for market-led development, and a shrinking role for the public sector in economic activities. International trade and the private sector have been identified as the engines of economic growth. While the logic seems compelling, market-led development has also led to a growing vacuum between publicly funded technology generation and the technology delivery dimension. National and international research organizations are producing research information and products with no obvious delivery mechanism to reach farmers and consumers.

Thus, new "contingency partnerships" with various organizations in both the private and public sectors are needed, along with the agricultural development roles that need to be played by governments and public goods research organizations. This is especially true in biotechnology. Public–private partnerships will be required to assemble the critical mass of talent, funding and access to technological components, plus the financial and technical resources to navigate the regulatory process, to address the difficult circumstances facing organizations with increasingly limited resources. Responsible public sector organizations will need to actively pursue such partnerships. One of the great challenges facing public goods research today is how to reconnect their scientific work with technology delivery, and particularly, how to take their research innovations to scale.

3.6 WHERE DO WE GO FROM HERE?

The world has the technology — either already available or well advanced in the research pipeline — to feed a population of 10 billion people. Fifty-percent increases in average global crop yields over the next 20 years are achievable, and in environmentally sustainable ways. The more pertinent question is whether governments and civil society will permit and encourage farmers and ranchers to employ new science and technology to expand world food production, and whether, in the case of the low-income, food-deficit nations, the large complementary investments in rural infrastructure — roads, power, and water resources — and education and health will be made to support dynamic smallholder commercial agricultural systems.

3.6.1 Rural Infrastructure Underpins All Development

Finding the way to provide effective and efficient infrastructure in remote and neglected areas that are typically drought-prone or mountainous or the inland areas typical of much of Sub-Saharan Africa underpins all other efforts to reduce poverty, improve health and education, and secure peace and prosperity.

Efficient transport is needed to facilitate production and enable farmers to bring their products to markets. Intensive agriculture is particularly dependent on vehicle access. But today, most agricultural production in Africa is generated along a vast network of footpaths, tracks, and community roads where the most common modes of transport are the "legs, heads, and backs of women." Indeed, the largest part of a household's time expenditure is for domestic transport.

Not only will improvements in transport systems and other rural infrastructure — especially potable water and electricity — greatly accelerate agricultural production and rural economic growth, it would reduce rural isolation, thus helping to break down tribal animosities and to establish rural schools and clinics in areas where teachers and health practitioners have been heretofore unwilling to settle. For instance, the conditions of local roads in Sub-Saharan Africa — both in rural areas and in urban centers — are especially poor and generally worsening as the volume and weight of traffic increases. Reversing this trend will require new thinking, new partnerships, and new institutional arrangements. Even if the public sector could manage transport effectively, the private sector will be increasingly asked to help finance it.

Today, experts are calling for new types of public–private sector partnerships to build and maintain integrated transport infrastructure — from farms to cities to ports. National governments, subregional associations, and large international donor organizations have important roles to play in the future planning, financing, and managing of integrated and coordinated systems. However, small- and medium-sized private entrepreneurs and local governments and communities must also be involved in the development and management process.

3.6.2 Agriculture and Peace

The first essential component of social justice is adequate food. And yet there are nearly 1 billion people who go to bed every night hungry. Particularly disheartening are the 150 million young children who go hungry each day, with this undernourishment often leading to irreversible damage to their bodies and minds.

Among developing countries with the lowest levels of hunger, only 8% were mired in conflict. In contrast, of those countries where more than half of the population was underfed, 56% were experiencing civil conflict. Since agriculture provides employment for the majority of people in low-income developing countries, it is not surprising that when this sector is allowed to falter, armed conflict often follows (Borlaug, 2000).

It is indeed troubling to see the persistence of large military budgets around the world, including in the United States. In total, about US$900 billion are spent annually on the military. The United States accounts for 50% of this total (about US$450 billion), and spends 40 times more on the military than it does on overseas development assistance. Indeed, trends in foreign assistance funding for agricultural and rural development have been declining, not only in the United States, but also in many other donor countries and institutions. The World Bank (2002) reported its lowest level of support to agriculture in its history in 2000–2001.

3.6.3 Mobilizing Political Will to Reduce Poverty and Hunger

Over the past two decades, the formerly low-income countries of Asia — led by China — have broken into global markets for manufactured products and a growing array of economic services. Extraordinary declines in poverty have accompanied this global integration and development. However, dozens of other poor countries, accounting for around 2 billion people, so far have been bypassed in the globalization process, and are in danger of becoming permanently marginalized. Many of these countries are found in Africa, although some are in Latin America and Western and Central Asia. For globalization to succeed, the global economy must become more inclusive. Past inequalities in education and access to resources must be redressed, and democratization of political systems will be critical to permit the poor to have a greater voice in government.

OECD agricultural subsidies — which exceed US$310 billion annually — are a barrier to increased agricultural trade for many developing countries and a disincentive to domestic production (FAO, 2003). OECD subsidies on cotton, sugar, fruits, and vegetables deny developing countries — most of which are located in tropical and subtropical latitudes — access to industrial country markets for commodities in which they have a comparative advantage. Subsidies on food crops lead to overproduction in temperate zones, which lowers world grain prices. This conspires against domestic food producers in developing countries by artificially lowering world food prices, and making it difficult to compete in global markets or even in local markets with locally produced food.

In 1970, the industrialized nations agreed to set a target of 0.7% of GNP for official foreign assistance. With the exception of some northern European nations and Canada, few OECD nations have honored that commitment. In 2000, the average OECD national contribution was 0.2% of GNP, with the United States at the bottom of the list, giving only 0.1%. Today OECD countries generate US$25 trillion in gross economic product per year. If they were to honor their previous foreign assistance pledge, there would be $175 billion in "north–south" income transfers annually, instead of the $57 billion that actually is provided in foreign assistance (UN Development Programme, 2003a). The additional transfers could make possible achievement of the Millennium Development Goals that were agreed on by 189 members of the United Nations in September 2000.

Poverty reduction and environmental protection, however, will not result from income transfers alone. Continuing, and possibly worsening, levels of corruption and greed in many low-income countries must be addressed. Ways to improve transparency and accountability must be found.

Foreign assistance programs will need to use both "carrots and sticks" more effectively in dealing with governments in developing countries. As a general rule, development assistance should become more "performance based" with poverty reduction an overriding criterion for sustaining (and increasing) support.

3.6.4 The Responsibility of Privileged Nations

Educated citizens in wealthy and democratic nations have valuable resources to help ensure a better life for all. A major resource consists of the large and respected publicly funded research organizations and universities (e.g., Ohio State University). Their teaching, research, extension and public service functions can help address and overcome the problems of poverty, hunger, poor health, and ignorance.

Vastly improved systems of animal health and management, particularly available to smallholders and pastoralists in developing countries, are needed. These target groups need to lower animal health-related losses and increase off-take so that herds and flocks provide dependable income generation and not just serve as ambulatory savings accounts.

Improved and more dependable systems of staple food production are also needed to provide better food security for the poor. Greater attention needs to be given to enhance broad-spectrum insect and disease resistance and control, and heat and drought tolerance.

Less expensive systems of irrigation are also required to produce off-season high-value crops for urban consumers, as are the programs and extension projects to deliver these systems to the rural areas.

As citizens privileged to speak our hearts and minds, we must become advocates for access of all people, both at home and abroad, to adequate health care and improved educational systems. We must also become more active environmentalists. It is truly a shame that by and large those of us highly trained in the biological and natural sciences have left the field of environmental advocacy to those who are much less well trained and frequently highly biased in their approach. Our silence has often contributed to unfortunate policies based on ignorance.

Finally, we can and must become spokespersons for "science in the service of humankind." The recent debates surrounding GMOs is but one of many contemporary issues, but it illustrates dramatically how ill-informed policymakers can directly harm the lives of poor people. We must defend good and useful science whenever and wherever it is challenged.

The responsibility of wealthy nations to help meet the Millennium Development Goals in poverty reduction, environment, health, and gender equality is a matter of enlightened self-interest. World peace and environmental protection will not be achieved with billions of people living at the edge of survival and marginalized in the world economy. If we shrink from our duty or fail to succeed, in the longer run, democracy, peace, and prosperity for all of us will be jeopardized.

REFERENCES

Borlaug, N.E. 2000. The Green Revolution revisted and the road ahead. Special 30th Anniversary Lecture, The Norwegian Nobel Institute, Oslo.

Conway, G. 1999. *The Doubly Green Revolution: Food for All in the 21st Century.* Cornell University Press, Ithaca, NY.

Delgado, C., M. Rosegrant, H. Steinfeld, S. Ehui, and C. Courbois. 1999. *Livestock to 2020: The Next Food Revolution.* International Food Policy Research Institute, Washington, DC.

Food and Agriculture Organization. 2003. *World Agriculture: Towards 2015/2020: An FAO Perspective.* FAO, Rome.

Gianessi, L. 2002. *Plant Biotechnology: Current and Potential Impact for Improving Pest Management in U.S. Agriculture.* National Center for Food and Agricultural Policy, Washington, DC.

International Food Policy Research Institute. 2002. *Reaching Sustainable Food Security for All by 2020: Getting the Priorities and Responsibilities Right.* International Food Policy Research Institute, Washington, DC.

Khush, G.S. 1995. Modern varieties — their real contribution to food supply and equity. *Geojournal,* 35:275–284.

Rajaram, S., and N.E. Borlaug. 1997. Approaches to Breeding for Wide Adaptation, Yield Potential, Rust Resistance and Drought Tolerance. Paper presented at First International Wheat Symposium, Obregón, Mexico, April 7–9.

Sen, A. 2000. *Democracy as Freedom*. Anchor Books, New York.

Smil, V. 1999. Long-range Perspectives on Inorganic Fertilizers in Global Agriculture. Travis P. Hignett Memorial Lecture, International Fertilizer Development Center, Muscle Shoals, AL.

U.N. Development Programme. 2003a. *Human Development Report. Millennium Development Goals: A Compact Among Nations to End Human Poverty*. U.N. Development Programme, New York.

U.N. Development Programme. 2003b. *Millennium Project. Halving Global Hunger — Background Paper of the Millennium Task Force on Hunger*. U.N. Development Programme, New York.

World Bank. 2002. *Reaching the Rural Poor: A Renewed Strategy for Rural Development*. World Bank: Washington DC.

World Meteorological Organization. 1997. *Comprehensive Assessment of the Freshwater Resources of the World*. World Meteorological Organization, Geneva.

4

Greenhouse Gases and Food Security in Low-Income Countries

ROY DARWIN, STACEY ROSEN, AND
SHAHLA SHAPOURI

CONTENTS

4.1 INTRODUCTION

Many people in the world are food insecure and are expected to remain so for some time to come (Rosegrant et al., 2001; Food and Agriculture Organization [FAO], 2003c; Shapouri and Rosen, 2003). Food insecurity is associated with low levels of agricultural productivity and/or inadequate purchasing power. Low levels of agricultural productivity may also result in inadequate purchasing power in regions in which sales of agricultural commodities are a major source of income. Emissions of greenhouse gases are expected to affect food security because of their potential impacts, both positive and negative, on agricultural productivity. Of particular concern, however, are the potentially damaging effects of rising temperatures associated with global climate change. Tropical and subtropical nations, which already experience temperatures that are higher than optimal for crop production, are especially vulnerable (Reilly et al. 1996; Gitay et al., 2001). Many of these nations are already food insecure as well.

Explicit estimates of the impacts of climate change on food security, such as the number of people at risk of hunger, are rare (Rosenzweig and Parry, 1994; Parry et al., 1999; Fischer et al., 2002). They are also limited for a variety of reasons. First, estimated impacts are based on changes in climate that are projected to occur in the relatively distant future, 2020 or later. Near-term estimates are not available. Second, the climate projections used in this research are derived from runs of general circulation models (GCMs). GCMs simulate the causal linkages between changing concentrations of greenhouse gases and changes in temperature, precipitation, and other climatic variables. They also strive to reflect regional and local differences caused by atmospheric circulation patterns and the Earth's topography. Nevertheless, uncertainty about projected changes in regional and local climate by GCMs is still relatively high (National Research Council, 2001).

The primary objective of this paper is to evaluate the extent to which global climate change may contribute to food insecurity in the relatively near term, such as 2012. Rather than relying on GCMs, climate change is simulated by extrapolating from long-term changes that occurred during the 20th century. Relying on the past to project the future is still subject to a certain level of uncertainty, but there is likely to be less uncertainty associated with this approach than with projections based on GCMs, especially when using a 10-year timeframe.

In the following, background information is provided on food security and the potential impacts that rising atmospheric concentrations of greenhouse gases might have on it. This includes some current estimates and short-term projections of food security. Next, impacts of climate change on growing-season length are estimated. They were an important determinant of agricultural productivity in some low-income countries during the 20th century. Potential implications of climate-induced changes on agricultural productivity and on food security in the short term are then explored. Finally, implications for management, policy, and research and development options for carbon sequestration in developing countries are discussed.

4.2 FOOD SECURITY

Global food production has grown faster than the world's population during the past 40 years. As a result, enough food is available worldwide so that countries with production shortfalls can import food as long as they have adequate financial resources. Yet many poor countries and millions of poor people do not share in this abundance. They suffer from food insecurity and hunger.

4.2.1 Agricultural Productivity

Most food-deficit countries allocate a large proportion of their domestic food production to food consumption, since a very small share of domestic consumption is provided by imports. Agricultural products are also a very important source of foreign exchange earnings to support the growing need for imports. Agricultural production growth depends on the availability and quality of resources. And in low-income, food-deficit countries, this means land and labor because of limited use of new technologies.

In many low-income countries, most increases in agricultural output have resulted from area expansion. In Sub-Saharan Africa, area expansion accounted for more than 80% of the region's grain output growth during the last two decades. This means that yield growth contributed to less than 20% of the growth. In Latin America, area expansion accounted for about two-thirds of the growth in grain production. In Asia, the reverse was true. Gains in yields contributed to nearly all growth experienced in the region's grain output.

The long-term prospects for acreage expansion are not good, because in most countries, much of the land that could be used for farming is unfit to cultivate without major investment. In Latin America and Sub-Saharan Africa, continued expansion of cropland means converting range and forestland to crop production, a process with high economic and environmental costs. The FAO estimates that about half of the land that could be used to produce food in Sub-Saharan Africa has poor quality soils (FAO, 1993). Sub-Saharan Africa has a vast and diverse land area, but the region faces a number of resource constraints, such as lack of water, that would ensure sustainable agricultural growth (Ingram and Frisvold, 1994). Some countries, such as Sudan and the Democratic Republic of Congo, have vast areas of rainfed land with crop potential, while other countries such as Kenya and Madagascar have already exhausted their high-potential land.

Demographic changes are placing increased pressure on land in the Sub-Saharan region. More than 20% of all vegetative land is degraded due to human causes. However, water and wind erosion still account for most of the degradation on affected ha. Much of this degraded area is in the Sahel, Sudan, Ethiopia, Somalia, Kenya, and southern Africa. Historically, farmers adjusted to resource constraints by following several years of planting with several fallow years. However, population pressures have reduced the practice of these sustainable agricultural techniques, leading to declines in land productivity.

4.2.2 Limited Use of New Technologies

The only option to sustain production growth is to increase yields. Within the developing world, average regional grain yields are the highest in North Africa, and the lowest in Sub-Saharan Africa. Yields are highly dependent on the use of additives, particularly fertilizers. The principal factor limiting yield response to fertilizer use is the inadequate supply of water during the growing season. Although water availability varies considerably across regions, it has been a serious problem in many countries. According to World Bank studies, depletion and degradation of water resources have become major problems facing many low-income countries (Crosson and Anderson, 1992; Cleaver and Schreiber, 1994). Within 10 years, if the population grows at projected rates, per capita water availability will decrease by an average 20% in developing countries, and by 34% in African countries. The agricultural sector consumes over half the annual freshwater withdrawals in most developing nations and could face greater competing demands from household and industrial uses in the future.

The sparse rainfall that characterizes much of Sub-Saharan Africa affects the response to and demand for fertilizer (Seckler et al., 1991; Harold et al., 1994). Farmers are reluctant to risk fertilizer use until rain begins to fall, because inadequate moisture will fail to dissolve fertilizer nutrients (especially nitrogen) and crops can "burn." Irrigation can make the use of fertilizer profitable and increase agricultural output. However, in Sub-Saharan Africa, only 4.3% of all arable land is irrigated. This is low, even when compared with other developing regions. In Latin America, 13% of the arable land is irrigated, and 38% is irrigated in Asia (FAO, 2003b). The world average is 19%. A potential exists to expand irrigated areas in Sub-Saharan Africa, but required investments will be costly. Increasing the use of fertilizer raises production costs. In many low-income countries, particularly in Sub-Saharan Africa and Latin America, almost all fertilizer is imported, and the lack of adequate foreign exchange constrains availability.

4.2.3   Assessing Food Security with the Food Security Assessment Model

The food security assessment (FSA) model, developed by the U.S. Department of Agriculture's (USDA) Economic Research Service (ERS), assesses food security at the country level and within country by income groups in order to account for both physical access (food availability) and economic access to food (see USDA, ERS, 2003 for details). The commodity coverage in this analysis includes grains, root crops, and "other" crops. Together, these three commodity groups account for 100% of all calories consumed in the study countries. In the above study, food consumption and food access were projected for 70 lower-income developing countries — 37 in Sub-Saharan Africa, 4 in North Africa, 11 in Latin America and the Caribbean, 10 in Asia, and 8 former republics of the Soviet Union.

The FSA model evaluates the food security of a country based on the gap between projected domestic food consumption, which is equivalent to all domestically produced food plus commercially imported food minus nonfood use, and a consumption requirement. Food gaps are calculated using two consumption targets: (1) maintaining base per capita consumption or status quo, which is the amount of food needed to support 1999–2001 levels of per capita consumption, and (2) meeting nutritional requirements, which is the gap between available food and food needed to support a minimum per capita nutritional standard. The minimum standard is a daily caloric intake of about 2100 calories per capita per day — depending on the region — recommended by the FAO. The caloric requirements are necessary to sustain life with minimum food-gathering activities. The FSA allows for the activity level of a refugee, but not for play, work, or any activity other than food gathering.

In combination, the two measures indicate two different aspects of food security: consumption stability and meeting a nutritional standard. They do not, however, account for food insecurity due to food distribution difficulties within a country. ERS attempted to capture this component of food security by allocating the projected level of food availability among different income groups. This represented an estimate of "nutrition distribution gap," which measures the quantity of food needed to raise food consumption of each income quintile to the minimum nutritional requirement. ERS also estimated the number of people who could not meet their nutritional requirements.

The FSA model estimated food security levels for 2002 (then current), 2007 (5 years out), and 2012 (10 years out). The 2002 estimates were calculated using contemporary data. Projections for 2007 and 2012 were calculated using historical commodity supply and use data for 1980 through 2001. This approach implicitly assumed that the historical trend in these key variables would continue into the future.

4.2.3.1 Food Security in 2002

An estimated 38% or roughly 1 billion people in 70 developing countries suffered from insufficient food intake in 2002 (Shapouri and Rosen, 2003). As indicated in Table 4.1, the nutrition gap for these study countries was nearly 18 Mt (million metric tons). Food insecurity occurred in 56 of the 70 countries evaluated. Sub-Saharan Africa, the most vulnerable region, was projected to account for almost 90% of the nutritional gap while accounting for only a quarter of the population. The distribution gap is estimated at 31 Mt.

Table 4.1 Food Availability and Food Gaps for 70 Low-Income Countries (thousands mt)

Year	Grain Production	Root Production (Grain Equivalent)	Commercial Imports (Grain Equivalent)	Food Aid Receipts (Grains)	Aggregate Availability of All Food
			—1000 tons—		
1993	404,514	58,988	45,251	6,145	604,451
1994	412,124	59,593	53,147	6,363	623,329
1995	411,629	61,063	57,882	6,568	670,279
1996	434,177	62,977	57,336	4,886	681,061
1997	424,980	65,053	60,754	5,042	683,394
1998	437,237	66,208	66,336	8,225	700,713
1999	457,515	70,880	69,246	6,526	728,445
2000	454,078	72,606	68,372	7,427	726,538
2001	464,281	73,128	69,879	7,218	749,660

				Food Gap		
		Projections		NR	NDR	
2002	452,265	74,880	72,073	**17,738**	**31,315**	729,785
2012	573,491	88,713	99,336	**16,928**	**25,318**	926,606

Notes: NDR = nutritional distribution requirements gap; NR = nutritional requirements.

Source: From U.S. Department of Agriculture, Economic Research Service. 2003. *Food Security Assessment.* Agriculture and Trade Report, GFA-14. U.S. Department of Agriculture, Washington, DC.

Data in Figure 4.1A indicate the distribution gap for all these countries except Cape Verde, which had a distribution gap equal to zero. There were 32 food insecure countries in Sub-Saharan Africa. Eight Sub-Saharan African countries had distribution gaps greater than 50 kg per person. Somalia, Zambia, and Zimbabwe had gaps greater than 100 kg per person. Zimbabwe's distribution gap was the largest at 176 kg per capita. Asia and Latin America each had ten food insecure countries. Only Afghanistan, however, had a distribution gap greater than 50 kg per person. Four countries among the former republics of the Soviet Union were food insecure.

For low-income countries, short-term production shocks intensify food security problems. The contribution of domestic production to consumption is large, often exceeding 90%, and imports are restricted because of the lack of foreign exchange. An examination of the degree of production instability of staple crops in these countries highlights the threat of production shocks. For example, annual grain production in 14 of the 70 countries was cut by more than half at least once during the last two decades. As shown in Figure 4.2, 53 of the 70 countries suffered production shortfalls of at least 20% at least once during the last 20 years, while 17 experienced such shortfalls more than five times. Successive years of drought caused grain production in Southern Africa to drop 20% in 2001 and 14% in 2002.

Frequent short-term events that negatively impact on domestic agricultural production and the lack of effective food safety net programs amplify the problem in Sub-Saharan Africa, thereby increasing the likelihood of famine. About half of the countries in the region had annual grain production shortfalls of more than a third during the last two decades. Thirteen of these countries experienced shortfalls of more than 20% once every 4 years, and per capita grain production growth was negative in 7 of these 13 countries between 1980 and 2001.

Policies that are not related to these frequent economic shocks also may affect food security. For example, in the early 1980s, Zimbabwe was a model of success in Sub-Saharan

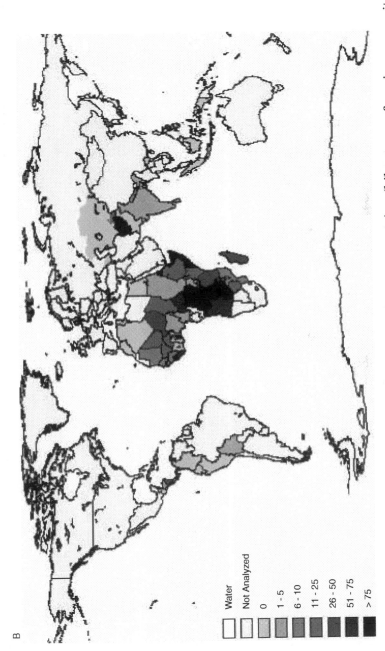

Figure 4.1 Food distribution gap in various low-income countries (kilograms of grain per capita). 1A. Estimates for 2002. 1B. Projections for 2012.

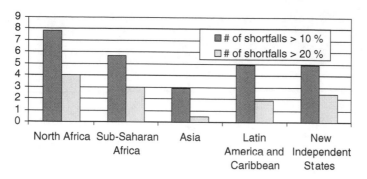

Figure 4.2 Frequency of production shortfalls from trend, 1980–2000. (From U.S. Department of Agriculture, Economic Research Service. 2003. *Food Security Assessment*. Agriculture and Trade Report, GFA-14. U.S. Department of Agriculture, Washington, DC.)

Africa because of the way it responded to the 1983–1984 drought, which reduced food production by half. However, two decades later, inappropriate policies and internal political problems led to a collapse of Zimbabwe's agricultural production, leaving the country with few resources to respond to the 2001–2002 drought. As a result, much of the population is food insecure and in some areas are in danger of experiencing famine.

4.2.3.2   Projections of Food Security in 2012

The number of food-insecure people is projected to decline to about 708 million by 2012 (Shapouri and Rosen, 2003). As shown in Table 4.1, the food gap to meet average nutritional requirements is projected to be 16.9 Mt, while the distribution gap is projected to be about 25.3 Mt. Estimated projections for 2012 by country are depicted in Figure 4.1B. The number of food-insecure countries is projected to decline from 56 in 2002 to 40 in 2012. Twenty-nine of these countries are in Sub-Saharan Africa. Somalia is projected to be the most food-insecure country with a distribution gap of about 100 kg per capita.

Chronic hunger in low—income countries is expected to grow at 1.5% per year, a rate lower than the population growth rate. Growth in agricultural productivity is necessary in order to improve food security in many developing countries. It means larger food supplies and lower prices for consumers as well as higher incomes for rural populations.

Food security in 37 nations of the Sub-Saharan Africa region is not expected to improve much during the next decade without a significant effort to address economic policies and to establish political stability. Based on all available indicators, the region will remain vulnerable to food insecurity unless a major commitment is made to improve the performance of the agricultural sector.

In general, food security is expected to improve over the next decade in Asia, Latin America, North Africa, and the former republics of the Soviet Union. Nevertheless, food security in several specific countries is not expected to improve much during the next decade. North Korea, Afghanistan, Haiti, Honduras, Nicaragua, and Tajikstan will continue to be chronically food insecure, both in terms of food availability and access to food by lower income groups. Even countries that are expected to become food secure by 2012 — based on our estimates — could face short-term problems because of production volatility. Countries such as Bangladesh and Cameroon are expected to barely pass the threshold of food security by 2012, but domestic production continues to contribute to more than 90% of their consumption. In this circumstance, any production shock can alter the situation significantly.

These projections are broadly consistent with projections of the number of malnourished children in 2020 (Rosegrant et al., 2001) and the incidence of undernourishment in 2015 and 2030 (FAO, 2003c). Rosegrant et al. (2001) estimates that the number of malnourished children will rise in Sub-Saharan Africa but fall in Western Asia/North Africa, Latin America, China, South Asia, and Southeastern Asia by 2020. The FAO (2003c) estimates that the incidence of undernourishment will rise in Sub-Saharan Africa and the Near East/North Africa, but fall in the Latin American/Caribbean, South Asia, and East Asia regions by 2015. The FAO also projects further declines in the incidence of malnourishment in the Latin American/Caribbean, South Asia, and East Asia regions as well as a decline (relative to 1997/1999) in Sub-Saharan Africa by 2030. The FAO estimates that the incidence of malnourishment in the Near East/North Africa will still be higher than 1997/1999 levels in 2030.

4.3 FOOD SECURITY AND GREENHOUSE GASES

Rising atmospheric concentrations of greenhouse gases are expected to affect food security through their impacts, both positive and negative, on agricultural productivity (FAO, 2003a). Impacts are generated through four mechanisms: CO_2 fertilization, climate change, extreme weather events, and rising sea level. Agricultural production, a requirement for food security, also emits greenhouse gases. Emissions associated with land-use change and land degradation are particularly important in low-income, food-insecure countries.

4.3.1 Impacts of Greenhouse Gas Emissions on Food Security

The direction of the various expected impacts on consumers and on farm income is summarized in Table 4.2. Farm income is defined as returns to land, labor, and capital. The net effect on food security depends on local conditions.

Table 4.2 General Impacts of Rising Concentrations of Greenhouse Gases on Factors Affecting Food Security

Greenhouse Gas Phenomenon	Impact on Food Availability		Impact on Local	
	Quantity	Price	Consumers	Farm Income
CO_2 fertilization	+	−	+	−
Climate change	?	?	?	?
Extreme weather events	−	+	−	−
Sea level rise	−	+	−	?

4.3.1.1 CO_2 Fertilization

Higher levels of CO_2 in the atmosphere tend to increase plant growth. However, increases in agricultural productivity due to this fertilization effect will vary considerably from one location to another. First, some crops benefit more than other crops. The response of maize, millet, sorghum, and sugarcane (known as C_4 crops) to increases in atmospheric CO_2, for example, is expected to be relatively small compared to other crops (Kirschbaum et al., 1996). Also, crop responses to atmospheric CO_2 are positively correlated with the level of nutrients available. If nutrients are limited, the response is negligible (Acock and Allen, 1985). Hence, productivity increases are likely to be smaller in low-income countries that rely heavily on C_4 crops and where the availability of fertilizers is a limiting factor in agricultural production.

CO_2 fertilization increases agricultural production and reduces food prices (Rosenzweig et al., 1993; Reilly and Hohmann, 1993; Tsigas et al., 1997; Darwin and Kennedy, 2000). This benefits consumers, but tends to reduce farm income (Darwin, 2003). Reductions in farm income are due to reduced revenues and the reallocation of resources (e.g., land, labor, and capital) from agriculture to other sectors that is expected after a decline in agricultural prices. The net effect on food security is not readily apparent. It will depend on the extent to which agriculture is a source of income in the country, and the ease with which alternative nonagricultural sources of income can be tapped by excess agricultural factors of production, particularly labor. Low-income countries that depend on agriculture for a large share of their income or that have few alternative sources of income are at greater risk of increasing food insecurity than other countries.

4.3.1.2   Climate Change

Climate change refers to average changes in meteorological patterns. Global mean temperature and precipitation are increasing. The distribution of changes, however, is far from uniform. Temperature and precipitation are declining in some areas (Folland et al., 2001). Climate change affects the quality

and quantity of land and water resources available for production in agriculture and other climate-dependent sectors
such as forestry and fisheries (Darwin et al., 1995, 1996;
Darwin, 1999; Fischer et al., 2001, 2002). Increases in agricultural production are accompanied by reductions in food
prices and vice versa. Lower or higher prices would respectively benefit or hurt consumers. They tend to have the opposite effect on farm income (Darwin, 1999, 2003).

4.3.1.3   Extreme Events

Some data suggest that extreme weather, such as drought,
heavy precipitation, and high winds, are increasing (Folland
et al., 2001). Because of their detrimental effects on food
production, storage, and delivery systems, extreme weather
events tend to reduce food quantities and raise food prices,
all of which, in turn, hurts consumers. Farm income is also
likely to decline for farmers impacted by them, because, in
addition to destroying crops or livestock, extreme events also
generate substantial losses in physical and natural capital.
Extreme events, for example, are often characterized by
increases in water and wind erosion. The increased incidence
of extreme events would also make it more difficult for farms
to remain economically viable, because savings generated in
good years may not be adequate to offset losses during bad
years.

4.3.1.4   Rising Sea Level

Sea levels are projected to rise by about 0.4 m during the 21st
century (Church et al., 2001). Rising sea levels are associated
with increased flooding, land loss, and salt water intrusions
into freshwater aquifers in coastal areas. This increases the
value of and competition for land and water resources in
noncoastal areas. And resulting decreases in food supplies
and increases in food prices hurt consumers. The impact on
farm income is less certain. Farm income falls for those whose
land is submerged or damaged. Farm income will tend to rise,
however, in adjacent areas.

4.3.2   Emissions of Greenhouse Gases Related to Food Security

Some agricultural practices that people in low-income countries use to maintain or improve their food security also emit greenhouse gases. In many low-income countries, particularly in Sub-Saharan Africa and Latin America, most increases in agricultural output have stemmed from area expansion as mentioned above. Shortened fallow periods and other practices also lead to land degradation. Both the conversion of natural vegetation to cropland or pasture and the degradation of land lead to the emission of formerly sequestered carbon into the atmosphere.

4.4   CLIMATE CHANGE AND FOOD SECURITY IN THE NEAR FUTURE

The remainder of this chapter focuses on the extent to which global climate change may contribute to food insecurity in the relatively near term, or 2012. The emission of greenhouse gases has already been associated with a rise in mean global temperature of from 0.3°C to 0.6°C since the late 19th century (Intergovernmental Panel on Climate Change, 2001). This has been accompanied by changes in precipitation in some locations as well.

Several major impacts of such climate changes on food security are evident. First, the impacts of climate change on growing-season length are estimated. Length of growing season was an important component of agricultural productivity in selected low-income countries during the 20th century. Percent changes in growing season for the 2002–2012 period are then extrapolated from the 20th century results. Using data from an earlier study, percent changes in growing season are transformed into percent changes in crop yield. Projections of food distribution gaps in 2012 (see Section 4.2.2.2) are then re-estimated with the climate-induced yield changes removed from the FSA's base yield projections. An analysis of these projections is presented below.

4.4.1   Climate Change and Length of Growing Season in 20th Century

Length of growing season is the length of time during the year that soil temperature and moisture are suitable for crop growth. It depends primarily on local temperature and precipitation. In this section the pattern of annual growing season lengths on current cropland during the 20th century are estimated and evaluated. Both long-run trends in length of growing season and their potential interaction on length of growing season during El Niño/Southern Oscillation (ENSO) cycles are assessed.

4.4.1.1   Length of Growing Season on All Land

Yearly series of growing season lengths for 1902 through 1997 were derived from a global data set of monthly surface temperature and precipitation extending from 1901 to 1997 over global land areas excluding Antarctica. The data set was prepared by the University of East Anglia's Climate Research Unit. Meteorological data are from an updated version of the data set described in New et al. (2000). Growing season is calculated with a soil temperature and moisture model developed by USDA's Natural Resource Conservation Service from monthly surface temperature and precipitation (Eswaran et al., 1995). The geographic resolution of the meteorological data and estimated length of growing season is 0.5° latitude-longitude grid boxes.

Estimated changes in growing season lengths are derived from regression models. (Details of the regression analysis are available from the authors upon request.) Independent variables include a trend term for average climate-induced changes in growing season length and terms for El Niño and La Niña, the two components of the ENSO cycle. El Niño and La Niña intermittently lengthen or shorten growing seasons in some regions. Two different ENSO indices — the Southern Oscillation Index (SOI) and the Cold Tongue Index (CTI) — are used in the analysis. The SOI, which is the traditional measure of ENSO cycles, is defined as the normalized Tahiti pressure minus the normalized Darwin pressure (Ternberth,

1984). The CTI was formed by subtracting the sea surface temperature (SST) averaged over all 2° latitude-longitude grid boxes within a rectangle defined by 6°N-6°S, 90°-180°W from the global mean SST (Mitchell and Wallace, 1996). Confidence levels of parameters were based on Student t or F statistics as appropriate. Parameters were considered statistically significant if the probability level is equal to or greater than 5% (two-tailed test). The location and direction of statistically significant trends are presented in Figure 4.3. Statistically significant parameters were used to compute the average change in length of growing season (in days) from 1902 to 1997 and the impact (in days) of an average ENSO event.

4.4.1.2   Length of Growing Season on Current Cropland

Cropland area by country and 0.5° latitude-longitude grid boxes was compiled from a 1-km–resolution, global land-cover characteristics database generated by U.S. Geological Survey, the University of Nebraska-Lincoln, and the European Commission's Joint Research Centre (U.S. Department of Interior, 2000). Cropland area by country was calibrated to data on arable land in 1997 (FAO, 2003b). Combining cropland location with growing season changes allowed for estimation of how much longer or shorter 1997 cropland growing seasons were relative to 1902 growing seasons, both in total and in areas simultaneously affected by the ENSO cycle. The latter provide information about changes in extreme events. Areas in which growing seasons were shortened by both climate change and ENSO, for example, become more susceptible to intermittent crop failure over time because the probability of reaching minimum growing season lengths declined.

Estimates of current cropland areas in low-income countries that were affected by climate-induced changes in growing season during the 20th century appear in Table 4.3. Estimates vary somewhat depending on which ENSO index was included in the regression models. The greatest difference occurred in Asia. For this region, SOI-based models indicated

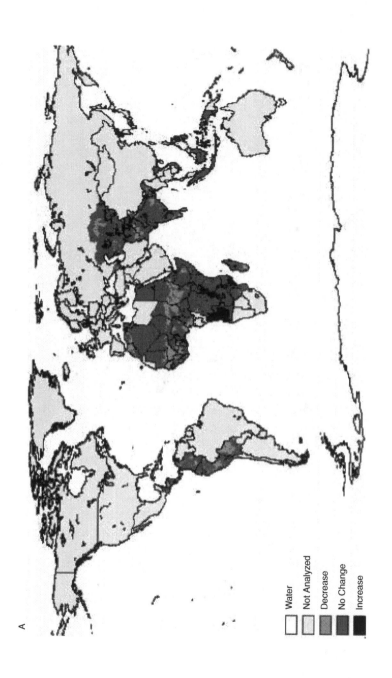

Figure 4.3 Impacts of climate change on growing seasons in selected low-income countries, 1902–1997: confidence and direction. 3A. Estimated from regression models using the Southern Oscillation Index for the El Niño/Southern Oscillation cycle. 3B. Estimated from regression models using the Cold Tongue Index for the El Niño/Southern Oscillation cycle.

Table 4.3 Current Cropland in Low-Income Countries Affected by Climate-Induced Changes in Growing Season, 1902 to 1997, by Region

Region	Length of Growing Season (Models with SOI for ENSO)[a] (1000 ha)		Length of Growing Season (Models with CTI for ENSO)[b] (1000 ha)	
	Longer	Shorter	Longer	Shorter
Sub-Saharan Africa	11,917	40,963	11,581	52,279
North Africa	1,597	7,931	1,820	8,171
Asia	38,347	27,828	21,146	51,688
Latin America and Caribbean	443	8,041	539	8,668
Former Soviet Republics	7,729	4,320	11,758	2,461
Total	60,033	89,093	46,844	123,267

Notes: CTI = Cold Tongue Index; ENSO = El Niño/Southern Oscillation; SOI = Southern Oscillation Index.
[a] Estimated from regression models with the Southern Oscillation Index for the El Niño/Southern Oscillation cycle.
[b] Estimated from regression models with the CTI for the ENSO cycle.
Source: From U.S. Department of Agriculture, Economic Research Service. 2003. *Food Security Assessment.* Agriculture and Trade Report, GFA-14. U.S. Department of Agriculture, Washington, DC.

that more cropland had longer growing seasons, while CTI-based models indicated that more cropland had shorter growing seasons. Estimates for the other regions were similar. In total, the SOI-based models estimated that length of growing season increased (relative to 1902) on 60 million ha of 1997 cropland, but decreased on 89 million ha of 1997 cropland. The CTI-based models estimated that length of growing season increased on 47 million ha of 1997 cropland, but decreased on 123 million ha of 1997 cropland.

Estimated impacts of climate change on the average length of growing seasons for 1997 cropland by country during the 1902–1997 period are depicted in Figure 4.4. The percent change in average length of growing season in a given country was equal to the average percent change in growing season on cropland affected by climate change multiplied by the share of cropland affected. Both the SOI- and CTI-based models estimate similar impacts. Except for India, both estimates for a country either have the same sign or one estimate is zero.

Using these criteria, climate change was associated with shorter growing seasons in 40 of the 69 countries evaluated. Regionally, climate change was associated with shorter growing seasons in 21 of 36 countries in Sub-Saharan Africa, 2 of 4 countries in North Africa, 5 of 10 countries in Asia, 10 of 11 countries in Latin America and the Caribbean, and 2 of 8 countries among the former Soviet republics. According to this model, climate change was associated with longer growing seasons in 24 countries, and had no impact on growing seasons in 4 countries. Impacts in one country, India, were uncertain because SOI-based models estimated longer growing seasons on average, while CTI-based models estimated shorter growing seasons on average.

Figures 4.5 and 4.6 indicate how the interaction of climate change with El Niño and La Niña affect average length of growing season on 1997 cropland by country during the 1902–1997 period. Percent change in average length of growing season in a given country was equal to the average percent change in growing season on cropland affected by both climate change and the ENSO cycle multiplied by the share of cropland affected. Estimated impacts from SOI- and CTI-based models were fairly similar to one another. For 49 countries, both estimates of climate change impacts either had the same sign or one estimate was zero. For 12 countries, both estimates of climate change impacts were zero. For eight countries estimates of climate change impacts had opposite signs.

The lighter shades indicate countries for which both climate change and the ENSO cycle were estimated to shorten growing seasons. El Niño and climate change, for example, shortened growing seasons in 16 countries in Sub-Saharan Africa, 5 countries in Asia, and 5 countries in Latin America (based on the assumption that both estimates of climate change impacts either had the same sign or that one estimate was zero). Of these, nine were shortened by more than 2%: Chad, Eritrea, Ivory Coast, Mauritania, Sudan, Zimbabwe, Sri Lanka, Ecuador, and Haiti. La Niña and climate change

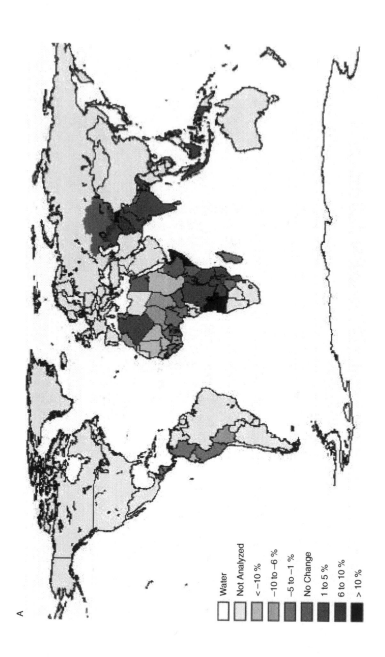

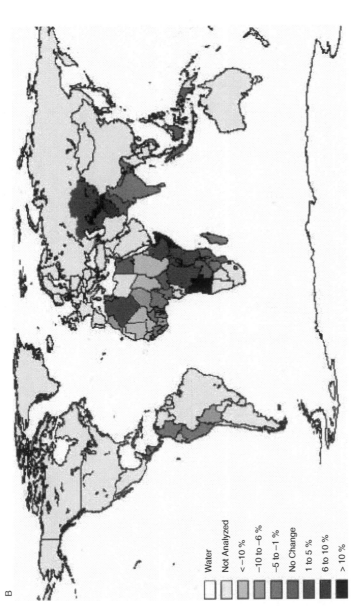

Figure 4.4 Impact of climate change on cropland growing seasons in selected low-income countries, 1902–1997 (average percent change, by country). 4A. Estimated from regression models using the Southern Oscillation Index for the El Niño/Southern Oscillation cycle. 4B. Estimated from regression models using the Cold Tongue Index for the El Niño/Southern Oscillation cycle.

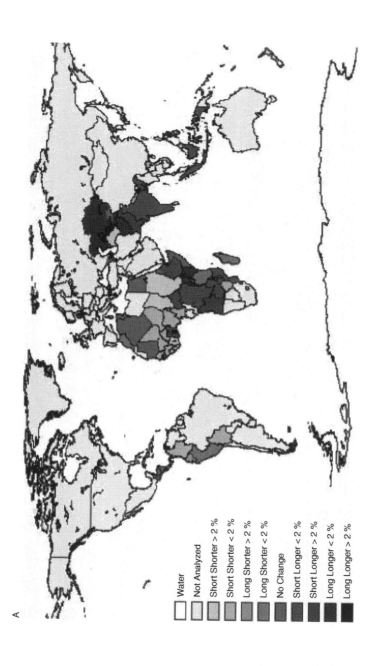

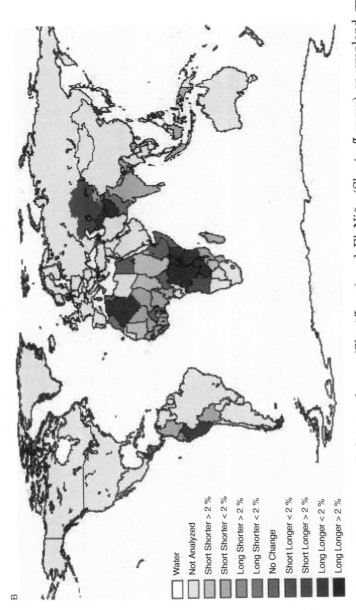

Water
Not Analyzed
Short Shorter > 2 %
Short Shorter < 2 %
Long Shorter > 2 %
Long Shorter < 2 %
No Change
Short Longer < 2 %
Short Longer > 2 %
Long Longer < 2 %
Long Longer > 2 %

Figure 4.5 Impact of climate change (Short/Long) and El Niño (Shorter/Longer) on cropland growing seasons in selected low-income countries, 1902–1997 (average percent change, by country). 5A. Estimated from regression models using the Southern Oscillation Index for the El Niño/Southern Oscillation cycle. 5B. Estimated from regression models using the Cold Tongue Index for the El Niño/Southern Oscillation cycle.

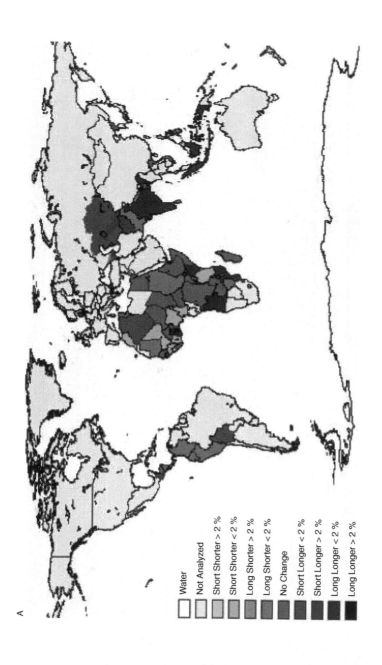

Water

Not Analyzed

Short Shorter > 2 %

Short Shorter < 2 %

Long Shorter > 2 %

Long Shorter < 2 %

No Change

Short Longer < 2 %

Short Longer > 2 %

Long Longer < 2 %

Long Longer > 2 %

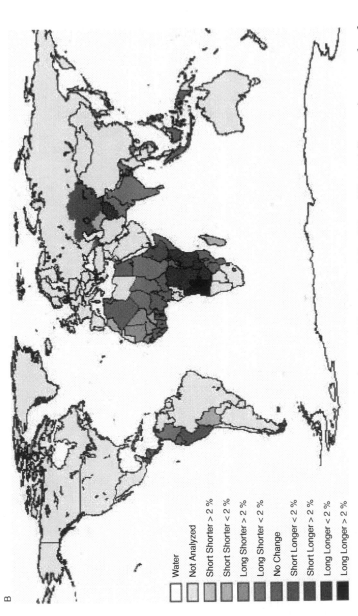

Figure 4.6 Impact of climate change (Short/Long) and La Niña (Shorter/Longer) on cropland growing seasons in selected low-income countries, 1902–1997 (average percent change, by country). 6A. Estimated from regression models using the Southern Oscillation Index for the El Niño/Southern Oscillation cycle. 6B. Estimated from regression models using the Cold Tongue Index for the El Niño/Southern Oscillation cycle.

Water
Not Analyzed
Short Shorter > 2 %
Short Shorter < 2 %
Long Shorter > 2 %
Long Shorter < 2 %
No Change
Short Longer < 2 %
Short Longer > 2 %
Long Longer < 2 %
Long Longer > 2 %

B

shortened the growing season in eight countries in Sub-Saharan Africa, and one country in North Africa.

4.4.2   Climate Change and Food Security
Projections for 2012

The food distribution gaps projected for 2012 (Section 4.2.2.2) implicitly include the effects of climate change because they were based in part on historical trends. The long-term impacts of climate were isolated from other factors by estimating total percent changes in growing season for the 2002–2012 period, quantifying the relationship between growing season lengths and crop yields, removing the climate-induced changes in yield from the base yield projections, and re-estimating the food distribution gaps in 2012.

Total percent changes in cropland growing season for the 2002–2012 period were estimated by simply extrapolating data from the estimated percent changes in growing season during the 1902–1997 period. The changes were in the same direction, but they were smaller in magnitude than the 1902–1997 impacts depicted in Figure 4.4. Length of growing season is an important component of agricultural productivity (FAO, 1996). Along with other factors, it determines what crops can be grown in a particular area. It also determines the number of crops that can be grown sequentially during the year.

In general, longer growing seasons mean greater agricultural productivity. Darwin et al. (1995), for example, estimated that in 1990 the average amount of crop produced on all cropland with short growing seasons of 100 days or less was 1.38 metric tons (mt) per hectare, while the average amount of crop produced on all cropland with long growing seasons of 301 days or more was 4.80 mt/ha. This is equivalent to 0.9% per day of additional crop produced (0.9% = 100 × [(4.80 mt/ha – 1.38 mt/ha)/1.38 mt/ha/(333 days – 50 days]). In this analysis, the link between growing seasons and yields is quantified by assuming that a 1% change in growing season is accompanied by a 1% change in yield. The climate-induced changes in yield are then removed from the base yield projections and food distribution gaps for 2012 are re-estimated. The difference between the original and revised estimates is attributable to climate change.

Results of the analysis are depicted in Figure 4.7. Estimated impacts from SOI- and CTI-based models were similar, as shown by the 0.975 correlation between them. Only in India are impacts predicted to differ. Except for Uzbekistan, the 27 countries that are food secure in 2012 are not affected by climate change. In Uzbekistan, a distribution gap is projected under the no-climate-change scenarios. This indicates that without a climate-induced increase in an average growing season, Uzbekistan would be food insecure in 2012. Distribution gaps in another five to seven food-insecure countries are also unaffected by climate change.

In 22 to 25 of the remaining countries, however, climate change is estimated to increase the distribution gap. In some countries, climate-induced decreases in average growing season are estimated to contribute to reductions in food security of greater than 5%. These include Cameroon, Mali, Nigeria, Niger, Togo, Chad, Senegal, Sudan, Guinea-Bissau, and Mauritania in Sub-Saharan Africa; Bangladesh in Asia; Bolivia and Honduras in Latin America; and Armenia among the former Soviet republics. In the 13 to 14 countries where climate-induced increases in average growing season are estimated to improve food security, the impacts tend to be more modest, or less than 5%. Uzbekistan and India are the exceptions, but the impacts in India are uncertain in that the SOI- and CTI-based models generate opposite impacts. The relative contribution of climate change to the total food distribution gap for the low-income countries evaluated in this analysis is less than 0.1%.

4.4.3 Discussion

The analyses presented in this chapter should be treated as exploratory. Nevertheless, they indicate that climate change probably did affect growing seasons in many low-income countries during the 20th century. They indicate that these changes probably affected food security in these countries as

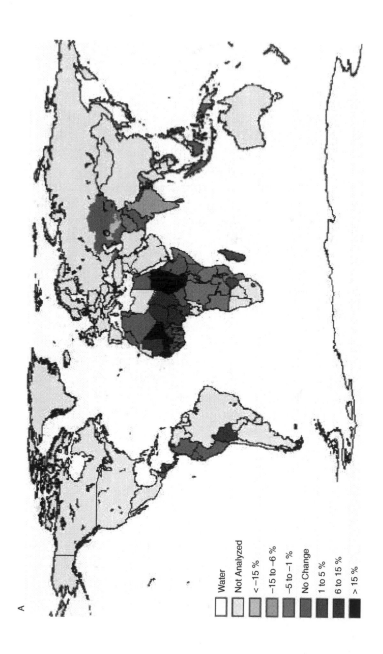

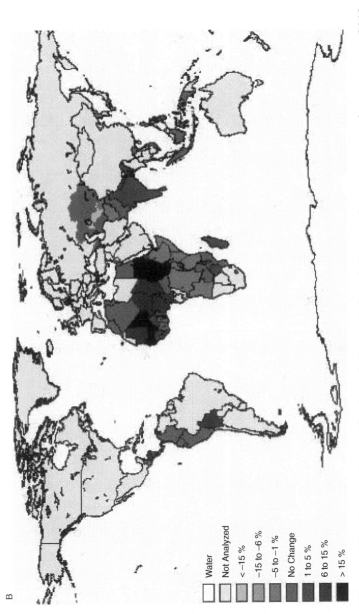

Water
Not Analyzed
< −15 %
−15 to −6 %
−5 to −1 %
No Change
1 to 5 %
6 to 15 %
> 15 %

Figure 4.7 Estimated contribution of climate-induced changes in average growing season to 2012 projections of food distribution gaps in selected low-income countries (percent). 7A. Estimated from regression models using the Southern Oscillation Index for the El Niño/Southern Oscillation cycle. 7B. Estimated from regression models using the Cold Tongue Index for the El Niño/Southern Oscillation cycle.

well. In most of the countries evaluated, climate change is estimated to reduce food security as early as 2012, and in some countries the impacts may be relatively large. The total contribution of climate change to the total food distribution gap for low-income countries, however, is on average probably very small. The analyses suggest that climate-induced changes in growing season probably interacted with the ENSO cycle as well. Some of these interactions would tend to exacerbate extreme events such as droughts in some countries. However, their potential impacts on food security are not explicitly quantified in this analysis.

Several limitations to this analysis merit attention. The scope of the analysis only considers climate change and a subset of extreme weather events. Other potential impacts of rising atmospheric concentrations of greenhouse gases, such as CO_2 fertilization and rising sea level, are ignored. In addition, climate change impacts are restricted to long-run trends in length of growing season and their potential interaction on length of growing season during ENSO cycles. Finally, only the relationship between average growing season length and food security was quantified.

Other limitations pertain to the methods used to simulate the agricultural and food security responses to global climate change. First, this analysis only focused on cropland, while climate-induced impacts on growing seasons on permanent pasture were ignored. Second, it was assumed that the relationship between length of growing season and agricultural productivity was the same in all countries included in the analysis. More precise results could be obtained with country-specific relationships. Third, the FSA model does not capture all the potential economic and policy responses available to countries that are vulnerable to global climate change. When climate change reduces agricultural productivity, for example, more land is typically devoted to agriculture than otherwise would be the case (Darwin, 2003). FSA does not take into account feedback loops that generate this behavior.

Projections based on historical trends appear to be promising alternatives for projections based on GCMs, at least for assessing impacts in the immediate future. They do not, however, reduce uncertainty to zero. As indicated in our results, models using different measures of the ENSO cycle generate slightly different results. And projections based on trends during the latter half of the 20th century instead of the whole century may yield different results. However, the prospects for resolving some of these issues are good and once addressed, researchers will have a more reliable method for simulating some important extreme weather-related events.

4.5 SUMMARY AND CONCLUSIONS

Many people in low-income countries are food insecure. Research suggests that rising concentrations of greenhouse gases are likely to affect food security even more in the future. Some effects of greenhouse gases, such as CO_2 fertilization, are likely to increase food security, while others, such as rising sea level, are likely to reduce food security. Of particular concern, however, are the potentially damaging effects of rising temperatures associated with global climate change. Countries in tropical and subtropical regions are especially vulnerable because they have temperatures that are already higher than optimal for crop production.

Results from this study suggest that climate change probably affected growing seasons in many low-income countries even during the 20th century. They also indicate that these changes probably affect food security in these countries as well. For example, climate change is estimated to reduce food security as early as 2012 in most of them, and in some, the impacts may be relatively large. However, the average total contribution of climate change to the total food distribution gap in low-income countries is probably very small. The data suggest that climate-induced changes in growing season probably interacted with the ENSO cycle as well. Some of these interactions would tend to exacerbate extreme events such as droughts in some countries.

The results of this study have implications for management options, policies, and research and development about carbon sequestration in developing countries. A decline in land quality usually leads to the conversion of nonagricultural land into agricultural uses, often in marginally productive areas. This typically triggers the release of carbon currently sequestered by other land covers into the atmosphere. Hence, management options that enhance food security by increasing agricultural productivity of existing agricultural land also may ensure continued carbon sequestration. Policies that are explicitly directed at carbon sequestration and that conflict with food security in developing countries will likely sequester less carbon than expected because of the ensuing expansion of agricultural land. In general, improving agricultural productivity tends to reduce land devoted to agricultural production (Ianchovichina et al., 2001). However, technological progress that substitutes capital for labor in labor-intensive agricultural systems can have the opposite effect, if the displaced labor is expelled to an agricultural frontier and begins clearing land (Angelsen and Kaimowitz, 2000, as reported in McNeely and Scherr, 2003).

Climate change is already occurring and is expected to continue throughout the 21st century. In addition to agricultural productivity, climate-induced changes in length of growing season also affect biological productivity in general. This in turn affects the accumulation and potential sequestration of terrestrial carbon. In general, longer growing seasons imply a greater ability to accumulate and sequester carbon and vice versa. But changing patterns of growing seasons over time at a given location may weaken the performance of management options or policies deemed suitable at their inception. This observation is particularly important when considering long-term carbon sequestration projects such as afforestation. One would want to plant trees, for example, that are compatible with the climate that is likely to occur over the time span of the project. Hence, consideration of the expected as well as the current climate, when evaluating management options, policies, and research and development designed to address carbon sequestration, will increase the probability of meeting program objectives.

ACKNOWLEDGMENTS

Gridded monthly weather data were supplied by the Climate Impacts LINK Project (United Kingdom Department of the Environment, contract EPG 1/1/16) on behalf of the Climate Research Unit, University of East Anglia. We would also like to thank Todd P. Mitchell of the University of Washington for help in obtaining and understanding the Cold Tongue Index, and directing us to the Southern Oscillation Index. We are also grateful to John Sullivan, Kevin Ingram, and three anonymous reviewers for their insights and suggestions. Any remaining errors, of course, are ours. The views expressed herein do not necessarily reflect those of the U.S. Department of Agriculture.

REFERENCES

Acock, B. and L.H. Allen Jr. 1985. Crop response to elevated carbon dioxide concentrations. In B.R. Strain and J.D. Cure, Eds. *Direct Effects of Increasing Carbon Dioxide on Vegetation*. U.S. Department of Energy, Washington, DC, pp. 53–97.

Angelsen, A. and D. Kaimowitz. 2000. When does technological change in agriculture promote deforestation? In D.R. Lee and C.B. Barretts, Eds. *Agricultural Intensification, Economic Development and the Environment,*. CAB International, Cambridge.

Church, J.A., J.M. Gregory, P. Huybrechts, M. Kuhn, K. Lambeck, M.T. Nhuan, D. Qin, and P.L. Woodworth. 2001. Changes in sea level. In Intergovernmental Panel on Climate Change. *Climate Change 2001: The Scientific Basis.* Contribution of Working Group I to the Third Assessment Report of the Intergovernmental Panel on Climate Change. Cambridge University Press, London; New York, pp. 639–693.

Cleaver, K. and G. Schreiber. 1994. *Reversing the Spiral: The Population, Agricultural, and Environment Nexus in Sub-Saharan Africa*. World Bank, Washington, DC.

Crosson, P. and J. Anderson. 1992. Resources and Global Food Prospects: Supply and Demand for Cereals. World Bank Technical Paper 184. World Bank, Washington, DC.

Darwin, R.F. 1999. A FARMer's view of the Ricardian approach to measuring effects of climatic change on agriculture. *Climatic Change,* 41:371–411.

Darwin, R.F. 2003. Impacts of rising concentrations greenhouse gases. In *Agricultural Resources and Environmental Indicators, 2001.* U.S. Department of Agriculture, Economic Research Service, Washington, DC, pp. 1–65.

Darwin, R.F. and D. Kennedy. 2000. Economic effects of CO_2 fertilization of crops: transforming changes in yield into changes in supply. *Environ. Modeling Assessment* 5:157–168.

Darwin, R.F., M. Tsigas, J. Lewandrowski, and A. Raneses. 1995. *World Agriculture and Climate Change: Economic Adaptations.* U.S. Department of Agriculture, Economic Research Service, Washington, DC.

Darwin, R.F., M. Tsigas, J. Lewandrowski, and A. Raneses. 1996. Land use and cover in ecological economics, *Ecological Econ.,* 17:157–181.

Eswaran, H., E. Van den Berg, P. Reich, R. Almaraz, B. Smallwood, and P. Zdruli. 1995. *Global Soil Moisture and Temperature Regimes.* U.S. Department of Agriculture, World Soil Resources Office, Natural Resources Conservation Service, Washington, DC.

Fischer, G., M. Shah, and H. van Velthuizen. 2002. *Climate Change and Agricultural Vulnerability.* International Institute for Applied Systems Analysis, Laxenburg, Austria.

Fischer, G., M. Shah, H. van Velthuizen, and F.O. Nachtergaele. 2001. *Global Agro-ecological Assessment for Agriculture in the 21st Century.* International Institute for Applied Systems Analysis, Laxenburg, Austria.

Folland, C.K., T.R. Karl, J.R. Christy, R.A. Clarke, G.V. Gruza, J. Jouzel, M.E. Mann, J. Oerlemans, M.J. Salinger, and S.-W. Wang. 2001. Observed climate variability and change. In Intergovernmental Panel on Climate Change. *Climate Change 2001: The Scientific Basis.* Contribution of Working Group I to the Third Assessment Report of the Intergovernmental Panel on Climate Change. Cambridge University Press, London; New York, pp. 99–181.

Food and Agriculture Organization. 1993. *Agriculture: Towards 2010*. Food and Agriculture Organization, Rome.

Food and Agriculture Organization. 1996. Agro-ecological zoning: guidelines. FAO Soils Bulletin 73. Food and Agriculture Organization, Rome.

Food and Agriculture Organization. 2003a. Climate change and agriculture: physical and human dimensions. In Bruinsma, Jelle, Ed. *World Agriculture: Toward 2015/2030: An FAO Perspective*. Earthscan, London, pp. 357–372.

Food and Agriculture Organization. 2003b. FAOSTAT. Food and Agriculture Organization Statistical Database. Available at: *http://apps.fao.org*.

Food and Agriculture Organization. 2003c. Prospects for food and nutrition. In J. Bruinsma, Ed. *World Agriculture: Toward 2015/2030: An FAO Perspective*, Earthscan, London, pp. 29–56.

Gitay, H., S. Brown, W. Easterling, B. Jallow, J. Antle, M. Apps, R. Beamish, T. Chapin, W. Cramer, J. Frangi, J. Laine, L. Erda, J. Magnuson, I. Noble, J. Price, T. Prowse, T. Root, E. Schulze, O. Sirotenko, B. Sohngen, and J. Soussana. 2001. Ecosystems and their goods and services. In Intergovernmental Panel on Climate Change. *Climate Change 2001: Impacts, Adaptation and Vulnerability*. Contribution of Working Group II to the Third Assessment Report of the Intergovernmental Panel on Climate Change. London; New York University Press, London; New York, pp. 235–342.

Harold, C., B. Larson, and L. Scott. 1994. *Fertilizer Consumption Remains Low*. International Agricultural and Trade Reports, Africa and Middle East Situation and Outlook Series, WRS-94-3. U.S. Department of Agriculture, Economic Research Service, Washington, DC.

Ianchovichina, E., R.F. Darwin, and R. Shoemaker. 2001. Resource use and technological progress in agriculture: a dynamic general equilibrium analysis. *Ecol. Econ.*, 38:275–291.

Ingram, K. and G. Frisvold. 1994. *Productivity in African Agriculture: Sources of Growth and Stagnation*. International Agricultural and Trade Reports, Africa and Middle East Situation and Outlook Series, WRS–94–3. U.S. Department of Agriculture, Economic Research Service, Washington, DC.

Intergovernmental Panel on Climate Change, Working Group I. 2001. Summary for policymakers. In *Climate Change 2001: The Scientific Basis.* Cambridge University Press, London; New York, pp. 1–20.

Kirschbaum, M.U.F., P. Bullock, J.R. Evans, K. Goulding, P.G. Jarvis, I.R. Noble, M. Rounsevell, and T.D. Sharkey. 1996. Ecophysiological, ecological, and soil processes in terrestrial ecosystems: a primer on general concepts and relationships. In Intergovernmental Panel on Climate Change, *Climate Change 1995. Impacts, Adaptations and Mitigation of Climate Change. Scientific-Technical Analyses.* Cambridge University Press, London; New York.

McNeely, J.A. and S.J. Scherr. 2003. *Ecoagriculture: Strategies to Feed the World and Save Biodiversity.* Island Press, Washington, DC.

Mitchell, T.P. and J.M. Wallace. 1996. ENSO seasonality: 1950–78 versus 1979–92. *J. Climate*, 9:3149–3161.

National Research Council. 2001. *Climate Change Science: An Analysis of Some Key Questions.* National Academy Press, Washington, DC.

New, M.G., M. Hulme and P.D. Jones. 2000. Representing twentieth-century space-time climate variability. Part II: Development of 1901–1996 monthly grids of terrestrial surface climate. *J. Climate*, 13:2217–2238.

Parry, M., C. Rosenzweig, A. Iglesias, G. Fischer, and M. Livermore. 1999. Climate change and food security: a new assessment. *Global Environ. Change,* 9(Suppl.):S51–S67.

Reilly, J., W. Baethgen, F.E. Chege, S.C. van de Geikn, Lin Erda, A. Iglesias, G. Kenny, D. Patterson, J. Rogasik, R. Rötter, C. Rosenzweig, W. Sombroek, and J. Westbrook. 1996. Agriculture in a changing climate: impacts and adaptation. In Intergovernmental Panel on Climate Change. *Climate Change 1995: Impacts, Adaptations and Mitigation of Climate Change. Scientific-Technical Analysis.* Cambridge University Press, London; New York, pp. 427–467.

Reilly, J. and N. Hohmann. 1993. Climate change and agriculture: the role of international trade, *Am. Econ. Rev.* 83:306–312.

Rosegrant, M.W., M.S. Paisner, S. Meijer, and J. Witcover. 2001. *Global Food Projections to 2020: Emerging Trends and Alternative Futures*. International Food Policy Research Institute, Washington, DC.

Rosenzweig, C. and M. Parry. 1994. Potential impact of climate change on world food supply. *Nature,* 367:133–138.

Rosenzweig, C., M. Parry, K. Frohberg, and G. Fisher. 1993. *Climate Change and World Food Supply*. Research Report No. 3. Environmental Change Unit, University of Oxford, Oxford.

Seckler, D., D. Gollin, and P. Antoine. 1991. Agricultural Potential of "Mid-Africa": A Technological Assessment. World Bank Discussion Paper 126. World Bank, Washington, DC.

Shapouri, S. and S. Rosen. 2003. Global food security: overview. In *Food Security Assessment*. Agriculture and Trade Report, GFA–14. U.S. Department of Agriculture, Economic Research Service, Washington, DC, pp. 1–8.

Ternberth, K.E. 1984. Signal versus noise in the southern oscillation. *Monthly Weather Rev.,* 112:326–332.

Tsigas, M.E., G.B. Frisvold, and B. Kuhn. 1997. Global climate change and agriculture. In T. W. Hertel, Ed. *Global Trade Analysis: Modeling and Applications*. Cambridge University Press, London; New York, pp. 280–304.

U.S. Department of Agriculture, Economic Research Service. 2003. *Food Security Assessment*. Agriculture and Trade Report, GFA-14. U.S. Department of Agriculture, Washington, DC.

U.S. Department of Interior. 2000. *Global Land Cover Characteristics Data Base Version 2.0*. U.S. Department of the Interior, Geological Survey. EROS Data Center, Washington, DC.

5

Climate Change, Soil Carbon Dynamics, and Global Food Security

RATTAN LAL

CONTENTS

Global climate change, global food security, and soil degradation are some of the most important challenges of the 21st century. The latter problem is exacerbated by the depletion of the pool of soil organic carbon (SOC) and an accompanying decline in soil quality. Global climate change, the greenhouse effect, and global warming are three interrelated processes. However, whereas global climate change and the greenhouse effect are natural processes, global warming is driven by anthropogenic activities.

Several natural processes affect the earth's climate, including its orbit around the sun, solar radiation, volcanic activities and their associated emissions, and meteorites. When a meteorite struck the Yucatan Peninsula 65 million years ago, it caused a drastic change in climate and subsequent mass extinction. Eighteen thousand years ago, during the most recent "ice age," a thick ice sheet covered a vast area of the northern hemisphere. The Labrador ice sheet covered the whole of eastern Canada and extended south to Cincinnati and New York. It was about 600 m (2000 ft) thick and covered 16 million km^2 (Coleman, 1926; Dawson, 1992). The Mount Pinatubo eruption in 1991 altered the climate of the Southeast Asian region, and increased photosynthesis in deciduous forests for the following 2 to 3 years.

The greenhouse effect is also a natural process. It has made the Earth inhabitable by raising its temperature from a frigid −18°C to a tolerable 15°C. The greenhouse effect is caused by the presence of various trace gases in the Earth's atmosphere, notably carbon dioxide (CO_2), methane (CH_4), and nitrous oxide (N_2O). They act as a glass does in a greenhouse — they permit shortwave radiation to enter Earth's atmosphere but impede the outward flow of infrared or longwave radiation, which causes a positive radiation balance; hence, the reference to "greenhouse gases" (GHGs). Because of this imbalance in radiation, the Earth — including its atmosphere and ocean — are warmer by about 33°C than they would be in the absence of this natural greenhouse effect. It is an extreme natural greenhouse effect that has caused the extremely high temperatures on Venus and extremely low temperatures on Mars. In contrast, the Earth's present temperature (15°C) is just right to support an incredible diversity of life.

Global warming is caused by acceleration of the natural greenhouse effect because of anthropogenic activities that enrich the atmospheric concentration of GHGs. In 1896, Swante Arrhenius warned that CO_2 emitted into the atmosphere from burning fossil fuel would act as a greenhouse gas and increase Earth's temperature. The accelerated greenhouse effect is called "global warming" if the rate of increase in global temperature exceeds 0.1°C per decade or 1°C per century. If the rate of increase in temperature is too rapid, the ecosystems cannot adjust. Each 1°C increase in temperature is likely to lead to a poleward shift of all cereal-growing and vegetation zones by several hundred kilometers (200 to 300 km) (Parry and Carter, 1988a; Kirschbaum et al., 1996). There will also be a poleward retreat of the permafrost boundary.

The observed climate change during the 20th century has already caused an increase in the Earth's temperature by 0.6 ±0.2°C, and the rate of increase in temperature since 1950 has been 0.17°C per decade for some regions (Intergovernmental Panel on Climate Change [IPCC], 2001). The projected global warming caused by a doubling of the atmospheric concentration of CO_2 ranges between 2°C and 4°C (Cheddadi et al., 2001; Prentice and Fung, 1990). There is growing concern that climate change may shift precipitation patterns; change the duration of the growing season; and affect regional agricultural productivity, the output of forests and fisheries, and global food security.

This chapter highlights the interactive effects of climate change, soil quality, and soil carbon dynamics on global food security. The chapter focuses on tropical ecosystems and on land use and soil management options that can enhance the SOC pool, thereby reducing the rate of enrichment of atmospheric CO_2 and improving food production by enhancing soil quality.

5.1 CLIMATE AND AGRICULTURE

Climate is the single most important determinant of agricultural productivity, primarily through its effects on temperature and water regimes (Oram, 1989). For example, the physiographic boundaries of principal biomes are determined by mean annual temperature (Mooney et al., 1993) and soil water regimes. Climate change is therefore expected to alter the biophysical environment of growing crops and to influence biomass productivity and agronomic yields (Rosenzweig and Hillel, 1998).

Positive effects may be associated with the fertilization effects of CO_2 enrichment, increases in the duration of growing seasons in higher latitudes and montane ecosystems, and possible increase in soil water availability in regions with an increase in annual precipitation. Each 1°C increase in temperature may lead to a 10-day increase in the growing season in northern Europe and Canada (Carter et al., 1991). The CO_2 fertilization effect is real (Allen, 1994; Wittwer, 1995; Allen and Amthor, 1995; Allen et al., 1996). However, the net positive effect may be moderated by other factors, such as the effective rooting depth and nutrient availability. Further, the productivity per unit of available water is expected to rise by 20% to 40% (van de Geijn and Goudriaan, 1996).

Negative effects of projected climate change on agriculture may be due to increases in respiration rate as temperature rises with attendant decreases in net primary productivity (NPP) (Abrol and Ingram, 1996); increases in the incidence of pests and diseases; shortening of the growing period in some areas; decrease in water availability as rainfall patterns change; poor vernalization; and increased risks of soil degradation caused by erosion and possible decline in SOC concentration. In contrast, the widespread incidence of drought in the United States in 1998 has been attributed to El Niño–related climate change (Bernard, 1993). The yield of rice has been estimated to decrease by 9% for each 1°C increase in temperature (Kropff et al., 1993). Phillips et al. (1996), using the explicit planetary isentropic coordinate (EPIC) model to examine the sensitivity of corn and soybean yields to climate change, projected a 3% decrease in both corn and soybean yields in response to a 2°C increase in temperature from a baseline precipitation level. However, a 10% precipitation increase balanced the negative effect of a 2°C temperature increase.

Rosenzweig et al. (1996) predicted the potential impact of climate change on citrus and potato production in the United

States. Their simulated treatments included combinations of three increased temperature regimes (+1.5, +2.5, and +5.0°C) and three levels of atmospheric CO_2 concentrations (440, 530, and 600 ppm). Citrus production may shift slightly northward in the southern states, but yields may decline in southern Florida and Texas. Fall potato production may be vulnerable to increased temperature in the northern states. Parry and Carter (1998) assessed the effects of increase in temperature on crop yields while maintaining the direct effects of CO_2 and precipitation held at current levels. They observed that average crop yields (wheat, rice, soybean, and maize) show a positive response (7% to 15%) to warming of +2°C and a negative response (−2% to −10%) to warming of +4°C. The effect is neutral in the range of 2°C to 4°C, whereas beyond this range, crop yields begin to decline.

These are global averages, however, and are unlikely to apply at regional or national levels. For example, a +2°C warming may have a strong positive effect in Canada or North America but a strong negative effect in Pakistan and India (Parry and Carter, 1998). In some sensitive ecoregions (Pakistan, Mexico, Egypt), crop yield declines could be 30% to 45% (Rosezweig and Parry, 1994). The effects of climate change on crop yields may be more negative at lower latitudes and generally positive at middle and high-middle latitudes. Further, crop growth is more affected by extremes of weather than by averages. The annual average changes in temperature or precipitation used in most predictive models do not reflect the short-term effects of so-called extreme events — droughts, floods, freezes, or heat waves.

5.2 GLOBAL FOOD DEMAND

Food security is defined as access to food to meet what people need biologically for a healthy and fulfilling life. It implies sustainable agricultural production, sufficient income so that people have access to food, and availability of a balanced diet to avoid hidden hunger. Swaminathan (2003) defines food security in terms of availability, access, and retention of food. According to the Food and Agriculture Organization (FAO)

(1996), "Food security exists when all people, at all times, have physical and economic access to sufficient, safe and nutritious food to meet their dietary needs and food preferences for an active and healthy life."

Food consumption as measured in terms of kilocalories per person per day is an important factor in the evaluation of regional or global food security. Global average food consumption increased from 2358 Kcal/person/day in 1965 to 2803 Kcal/person/day in 1998, and it is projected to increase to 2940 Kcal/person/day in 2015, and 3050 Kcal/person/day in 2030 (Table 5.1). The lowest levels of food consumption are observed in Sub-Saharan Africa and South Asia. Seven developing countries with a population of over 100 million are characterized by low food consumption. Among them, China, Indonesia, and Brazil have made notable progress, while India, Pakistan, and Nigeria are in a transitional phase, and Bangladesh is making progress in improving food consumption (Bruinsma, 2003). In general, developing countries still have food consumption levels under 2200 Kcal/person/day. This is not attributable only to supply deficits, however. Most food-insecure countries have been politically unstable, and ethnic conflicts have frequently exacerbated the problem of food insecurity, such as in Afghanistan, Angola, Bangladesh, Burundi, Cambodia, Central African Republics, Chad, Congo, Ethiopia, Guinea, Haiti, Kenya, North Korea, Liberia,

Table 5.1 Per Capita Food Consumption (kilocalories per capita per day)

Region	1965	1975	1985	1998	2015	2030
World	2358	2435	2655	2803	2940	3050
Sub-Saharan Africa	2058	2079	2057	2195	2360	2540
South Asia	2017	1986	2205	2403	2700	2900
Developing countries	2054	2152	2450	2681	2850	2980
Industrial countries	2947	3065	3206	3380	3440	3500

Source: Modified from Bruinsma, J., Ed. 2003. *World Agriculture: Towards 2015/2030.* Food and Agriculture Organization, Rome; Earthscan, London.

Table 5.2 Incidence of Undernourishment in
Developing Countries (millions)

Region	1991	1998	2015	2030
Sub-Saharan Africa	168	194	205	183
Near East/North Africa	25	32	37	34
Latin America, Caribbean	59	54	40	25
South Asia	289	303	195	119
East Asia	275	193	135	82
Total developing countries	816	776	612	443

Source: Modified from Bruinsma, J., Ed. 2003. *World Agriculture: Towards 2015/2030.* Food and Agriculture Organization, Rome; Earthscan, London.

Madagascar, Malawi, Mongolia, Mozambique, Namibia, Niger, Rwanda, Sierra Leone, Tanzania, Uganda, Yemen, and Zambia (Bruinsma, 2003).

The global incidence of undernourishment is summarized in Table 5.2. An estimated 816 million people were undernourished in 1991, and 776 million in 1998. This number is projected to decrease to 612 million in 2015, and 443 million in 2030 (Table 5.2). In any case, the demand for food will probably increase dramatically during the first half of the 21st century because of continuing population increases. Even though the rate of population growth is now declining, global population is projected to reach 7.2 billion by 2015 and 8.3 billion by 2030. An important concern about the growth in population is that most of the future increase (as much as 95%) will occur in developing countries where food security and malnutrition are already major issues (Table 5.3).

Population increases in food-insecure countries exacerbate food shortage by reducing per capita arable land area, which is already low (Lal, 2000). The world population in 1900 was 1.65 billion, while cropland area was 800 million ha, resulting in mean per capita cropland area of about 0.5 ha. Less than 8 million ha were under irrigation in 1900, and no chemical fertilizer was used. The world population in 2000 was 6 billion, and cropland area was 1.364 billion ha, resulting in mean per capita cropland area of 0.227 ha. Irrigated land

Table 5.3   World Population Growth (billions)

Region	1965	1975	1985	1998	2015	2030
World	3.3	4.1	4.8	5.9	7.2	8.3
Developing countries	2.3	2.9	3.6	4.6	5.8	6.9
Industrial countries	0.7	0.8	0.8	0.9	0.95	0.98
Transition countries	0.3	0.37	0.40	0.41	0.40	0.38

Source: Modified from Bruinsma, J., Ed. 2003. *World Agriculture: Towards 2015/2030.* Food and Agriculture Organization, Rome; Earthscan, London.

area in 2000 was 272 million ha (FAO, 2001), and total fertilizer use was 140 million metric tons (Mt) per year (International Fertilizer Development Center, 2001), which in part accounts for the increase in the production and productivity of cultivated area.

The cereal production required in developing countries to meet future food demands is given in Table 5.4 (Wild, 2003). These levels are estimated to be 778 Mt by 2025, and 1519 Mt by 2050. This projected increase in cereal production translates into an average increase of 30 Mt/year, or about 3.8% per annum. The required increase may come by cultivating more land, increasing yields, or both. If the projected increases were to come from increased area under cultivation alone, then the area under cereal production in developing countries, estimated at 467 million ha in 2000 would have to

Table 5.4   Projected Future Food Demands in Developing Countries

	2025	2050
Population (billions)	6.6	7.8
Required increase in cereal production (10^6 MT)	778	1519
Area required under cereal cultivation (million ha)	757	1032
Average yield of cereals (MT/ha/year)	4.4	6.0

Source: Modified from Wild, A. 2003. *Soils, Land and Food: Managing the Land During the 21st Century.* Cambridge University Press, Cambridge.

be increased to 757 million ha by 2025, and 1.032 billion ha by 2050. In most developing countries, however, there is little or no additional land available. Available land is agriculturally marginal, inaccessible, or located in ecologically sensitive regions. Thus, any major increases in food production must occur through increases in crop yields per unit of land and labor.

Wild (2003) estimated that average cereal yield in developing countries of 2.64 MT/ha in 2000 will have to be increased to 4.4 MT/ha/year by 2025, an increase of 67%, and to 6.0 MT/ha/year by 2050, an increase of 127% in 50 years. Several important technological options exist to increase crop yields, including crop varieties based on transgenics, increasing cropping intensity, enhancing fertilizer use and its efficiency by integrated nutrient management and precision farming, enhancing water use efficiency through soil–water conservation, and improved methods of irrigation and integrated pest management. Many studies have shown that agricultural production in the upper Midwest region of the United States (e.g., the heart of the U.S. Corn Belt) will at best remain stable under future climate change (Doering et al., 2002). However, there may be possible adverse effects of global climate change on crop production in ecologically sensitive regions where population growth rates are high and natural resources are already under great stress, such as Sub-Saharan Africa, North Africa, and South and Central Asia.

5.3 VARIATION IN CROP YIELDS AND CLIMATE CHANGE

Tremendous variation exists in average yields of major food crops among industrialized countries. Yields in these countries are higher by a factor of 7 to 8 than in developing countries. Data in Table 5.5 show that national average wheat yields range from a low of 0.8 MT/ha in Kazakhstan to a high of 7.8 MT/ha in the United Kingdom. Similarly, yields of maize range from a low of 1.3 MT/ha in Nigeria to a high of 9.4 MT/ha in Italy. In addition to these large yield variations, a large gap exists between actual yield and attainable yield,

Table 5.5 Average Yield of Maize and
Wheat in Countries That Produce 90%
of Global Output (MT/ha)

Country	Wheat Yield	Maize Yield
Argentina	2.4	5.0
Australia	2.0	—
Brazil	—	2.6
Canada	2.4	7.4
China	3.1	3.8
Denmark	7.1	—
Egypt	6.0	7.3
France	7.0	8.6
Germany	7.3	—
Hungary	3.9	6.0
India	2.6	1.7
Indonesia	—	2.6
Iran	1.6	—
Italy	3.2	9.4
Kazakhstan	0.8	—
Mexico	—	2.3
Nigeria	—	1.3
Pakistan	2.2	—
Philippines	—	1.6
Poland	3.4	—
Romania	2.6	3.0
Russia	1.4	—
South Africa	—	2.5
Spain	2.6	9.1
Thailand	—	3.5
Turkey	2.1	—
Ukraine	2.5	—
United Kingdom	7.8	—
United States	2.7	8.2
Yugoslavia	—	4.0

Source: Modified from Bruinsma, J., Ed. 2003.
World Agriculture: Towards 2015/2030. Food and
Agriculture Organization, Rome; Earthscan,
London.

and this gap is larger in developing than in industrialized
countries. As seen in Table 5.6, the gap in wheat yield is 4.2
MT/ha in Belarus; 4.0 MT/ha in Lithuania; 3.8 MT/ha in

Table 5.6 Actual and Attainable Wheat
Yield in 25 Countries (MT/ha)

Country	Attainable Yield	Actual Yield
Argentina	4.2	2.4
Australia	4.2	2.0
Belarus	6.7	2.5
Brazil	3.3	1.8
Canada	4.3	2.4
Ethiopia	4.0	1.2
France	6.6	7.1
Germany	7.6	7.3
Hungary	6.4	3.9
Italy	6.5	3.2
Japan	7.1	3.4
Latvia	6.2	2.5
Lithuania	6.8	2.8
Myanmar	2.5	0.9
Paraguay	3.4	1.4
Poland	7.1	3.4
Romania	6.3	2.5
Russia	4.4	1.4
Sweden	5.0	6.0
Tanzania	3.0	1.5
Turkey	4.8	2.1
Ukraine	6.2	2.5
United Kingdom	6.7	7.8
United States	5.8	2.7
Uruguay	5.3	2.3

Source: Modified from Bruinsma, J., Ed. 2003. *World Agriculture: Towards 2015/2030*. Food and Agriculture Organization, Rome; Earthscan, London.

Romania; 3.7 MT/ha in Latvia, Poland, and Ukraine; and 3.0 MT/ha in Uruguay. These estimates may, however, be affected by climate change, resulting in even lower yields and larger yield gaps.

Rosenzweig and Parry (1993) projected overall decrease in crop yields both with and without CO_2 effects. The projected decrease in crop yields without CO_2 fertilization effects ranged from −16% to −33% for wheat, −24% to −25% for rice, −20% to −31% for maize, and −19% to −57% for soybeans. In comparison, the projected decrease in crop yields with CO_2 effects ranged from -13% to 11% for wheat, -2% to -5% for rice, -15% to -24% for maize, and -33% to 16% for soybeans. Apparently, wheat and soybean yields may increase because of a CO_2 fertilization effect, provided that other factors such as soil quality remain the same.

There are numerous uncertainties in predicting the effects of CO_2 enrichment on plants. Short-term gains in photosynthesis due to CO_2 fertilization may be much greater than long-term effects, especially for C3 plants (Wolfe and Erickson, 1993). Relative crop responses to CO_2 fertilization also depend on ambient temperature (Idso et al., 1987), soil/water regime (Gifford, 1979), nutrient availability (Allen, 1991), and other factors such as weed competition and disease and insect pressure (Wolfe and Erickson, 1993). Consequently, projections must remain quite tentative for now.

Anticipated effects of climate change on maize yield in Zimbabwe are shown in Table 5.7. These three scenarios are under current climate, altered climate, and altered weather with CO_2 enrichment effect. Changes in maize yields were estimated using three separate models. Maize yields under all three conditions declined with climate change both with and without the CO_2 enrichment effect. The average decline in maize yield in comparison with a base yield of 3.7 MT/ha was 50% without any adaptive management option, 42% with additional fertilizer use, and 13% with additional fertilizer and irrigation use (Muchena and Iglesias, 1995; Parry and Carter, 1998).

Reilly (2002) reported an assessment of potential impacts of climate change on U.S. agriculture during the 21st century using various scenarios. A decline in the yield of wheat and sugarcane crops could be mostly alleviated by better management, but the latter would not significantly affect yield reduction in potatoes. The World Bank (1998) studied the impact of climate change on Indian agriculture. This report concluded that CO_2 fertilization could offset the harmful effects of climate change so that crop yields would be only marginally affected. According to Dinar and Mendelsohn (1998), adaptive management could mitigate any additional harmful effects,

Table 5.7 Effects of Projected Climate Change and
Adaptive Management on Maize Yield in Zimbabwe

Location	Global Circulation Model–Based Climatic Scenarios (MT/ha)		
	GISS	GFDL	UKMO
Blanket			
Base	4.8	4.8	4.8
Climate change	3.9	4.2	3.6
Climate change with direct CO_2 effect	4.1	4.4	3.7
Chisumbanje			
Base	2.8	2.8	2.8
Climate change	2.1	2.1	1.8
Climate change with direct CO_2 effect	2.2	2.2	1.9
Gweru			
Base	3.7	3.7	3.7
Climate change	2.6	2.7	1.8
Climate change with direct CO_2 effect	3.1	3.3	2.0

Soure: Modified from Parry, M. and T. Carter. 1998. *Climate Impact and Adaptation Assessment: A Guide to the IPCC Approach.* Earthscan, London. With permission. GFDL, Geophysical Fluid Dynamics Laboratory model; GISS, Goddard Institute for Space Studies model; UKMO, United Kingdom Meteorological Office model.

although this prediction is subject to the same uncertainty as predictions of negative consequences.

The adverse impacts of climate change on agriculture in general and on crop yields in particular under most scenarios are likely to be severe for the countries of Sub-Saharan Africa where institutional support to enable use of adaptive management is minimal, and where soil resources are already degraded because of long-term land misuse, soil mismanagement, and exploitative agricultural practices. The economic impact of extreme climate events may be more severe in semiarid regions of less developed countries than in cool temperate and cold regions (Parry and Carter, 1988a, 1988b). Indeed, some extreme climate scenarios predict regional disasters in food production even if global production is not affected. Furthermore, the impacts of climate change on crop yields have only been studied for principal crops such as wheat, rice, soybeans, and potatoes. Little if any research has been done on regional crops, especially those of importance to small landholders in Africa and elsewhere in the tropics, such as yam, cassava, and millet.

5.4 SOIL DEGRADATION AND CLIMATE CHANGE

Projected change in climate may influence several soil processes with a consequent adverse impact on soil quality (Brinkman and Sombroek, 1996). Important among these processes with an attendant adverse impact on soil quality are:

- Hydrolysis: the leaching of silica and basic cations
- Cheluviation: removal of Al and Fe by chelating organic acids
- Ferrolysis: transformation of clay by alternating oxidation and reduction processes, and reduction in cation exchange capacity
- Dissolution: of clay minerals by strong acids producing acid aluminum salts and amorphous silica
- Clay formation: reverse weathering leading to clay formation and transformation

Hydrolysis and cheluviation may accelerate with temperature increases, and ferrolysis may occur in soils subject to reduction and oxidation in high latitudes and monsoonal climates. These processes increase soil erodibility and decrease water- and nutrient-retention capacity. Two schools of thought exist with regard to the effects of projected climate change on soil quality.

The first school argues that climate change is likely to exacerbate global food insecurity, with increased risks of soil degradation. Accelerated soil erosion and other degrading processes already affect soil quality, especially in developing countries of the tropics and subtropics (Table 5.8). The land area affected by water erosion is estimated at 227 million ha in Africa, 441 million ha in Asia, 123 million ha in South America, and 46 million ha in Central America. Soil degradation by other

Table 5.8   Soil Degradation in Developing Countries (millions ha)

Region	Soil Erosion by Water	Soil Erosion by Wind	Chemical Degradation	Physical Degradation
Africa	227	186	62	19
Asia	441	222	74	12
South America	123	42	70	8
Central America	46	5	7	5
World total	1094	548	240	83

Source: Adapted from Oldeman, L.R. 1994. The global extent of soil degradation. In D.J. Greenland and I. Szabolcs, Eds. *Soil Resilience and Sustainable Land Use.* CAB International, Wallingford, United Kingdom, pp. 99–118.

processes is also a problem in developing areas. In South Asia, 25% of agricultural land is estimated to be affected by water erosion, 18% by wind erosion, 13% by fertility decline, 9% by salinization, 6% by lowering of the water table, and 2% by waterlogging (FAO, 1994). It sounds strange to say that fortunately these are overlapping categories.

Soil degradation is likely to be accelerated by projected climate change, especially in ecologically sensitive regions. Global hot spots of soil erosion include the Himalayan–Tibetan ecosystem, the unterraced slopes of China and Southeast Asia, tropical areas of Southeast Nigeria, the semiarid Sahelian region of West Africa, sloping lands of Central America, and the Andean valleys and cerrado region of South America (Scherr and Yadav, 1996).

Soil erosion rates are likely to change due to the erosive power of rainfall produced by more extreme precipitation events (IPCC, 1995). Since 1910 there has been a steady increase in the area of the United States affected by extreme events (>2" or >50.8 mm of rain in a 24-hour period) (Karl et al., 1996). Areas susceptible to wind erosion usually become drier and become more severely affected (Williams et al., 2002). In the U.S. Corn Belt, Lee et al. (1996) predicted that although a 2°C increase in temperature would decrease water erosion by 3% to 5%, wind erosion would increase by 15% to 18%. Thus, total erosion was predicted to increase as a consequence of increased temperature. Regions prone to salinization include the Indus, Nile, Tigris, and Euphrates river valleys; northeast Thailand; northern China; northern Mexico; and the Andean highlands (Scherr and Yadav, 1996; Norse et al., 2001). It is estimated that 20% of total irrigated area is already affected by salinization, and 12 million ha of irrigated land may have already gone out of production for this reason (Nelson and Mareida, 2001).

Higher temperatures due to projected climate change, especially in arid and semi-arid regions, may produce higher evaporative demand for water and exacerbate the drought that often follows the plow (Glantz, 1994). If the soil and water are adequate, as with irrigation, it turns out that an increase in evaporative demand may heighten the risks of salinization (Brinkman and Sombroek, 1996). However, under high atmospheric CO_2 conditions, there may be increased salt tolerance of crops (Bowman and Strain, 1987). Revegetation by overgrazing and other factors could exacerbate the problem of desertification (U.N. Environmental Programme, 1997), especially in Sub-Saharan Africa. Risks of overextraction of groundwater for irrigation in South Asia and also in the Near East/North Africa region are already recognized as serious (FAO, 1990).

Soil degradation is thus a major threat to global food security (Oldeman, 1998), and this threat may be increased with anticipated climate change. Soil degradation, especially that caused by accelerated erosion, characteristically involves depletion of soil organic matter. Most degraded soils contain an SOC pool that is below their potential set by ecological factors. Lee et al. (1996) observed that when the SOC pool decreased by 4.8 MT/ha in the U.S. Corn Belt, about 50% of this loss was due to accelerated soil erosion. Further, increased temperature and precipitation accelerate losses of SOC. These can exacerbate nutrient depletion in low-input agricultural systems that are already vulnerable to severe nutrient depletion, as is the case in Ghana (Rhodes, 1995) and elsewhere in Sub-Saharan Africa (Stoorvogel and Smaling, 1990). Sustainability of agriculture in the Sahel is already problematic (Reardon, 1995). It becomes even more difficult to attain with increased risks of soil degradation.

Alternatively, some scientists argue that soil quality may be improved, at least in some ecosystems, by the projected changes in climate (Brinkman and Sombroek, 1996). The CO_2 fertilization effect would increase biomass productivity with more litter and crop residues returned to the soil, and with higher root mass and greater root exudation. This could result in a gradual increase in soil fertility.

Part of this increase would be attributable to an increase in biological nitrogen fixation. A higher supply of N for plants would enhance growth and nutrient recycling. An increase in mycorrhizal activity due to fungal "infections" of plant roots would also enhance P uptake. Increases in root growth would enhance the SOC pool and the activity of soil biota such as earthworms and termites. Increases in the strength and quantity of stable soil aggregates that these organisms can create would improve soil structure, water infiltration rates, and available water capacity. Improvements in soil structure would decrease rates of soil erosion and reduce leaching losses of plant nutrients. Therefore, it is contended that the CO_2 fertilization effect over a long time may increase soil resilience and improve soil quality.

In this scenario, there would also be a positive effect on the weathering of parent material with a resulting increase in the rate of new soil formation, and thus in agroecosystem soil loss tolerance. The importance of increased soil microbial activity due to increases in temperature also should not be ignored. Greater microbial activity would strengthen elemental recycling mechanisms, increase the SOC pool, and enhance soil structure. CO_2 fertilization effects at high CO_2 concentrations may produce crop residues and litter with a higher C:N ratio, thereby decreasing the rate of mineralization and reducing the turnover rate.

Both schools of thought may be correct under specific ecosystems. Some of the projected climate change would certainly exacerbate soil degradation in some soils and ecosystems. But it could increase soil resilience and enhance soil quality in others. The net effect on soil quality would also depend on the adaptive options or the use of recommended management practices (RMPs) within a particular ecosystem. Adverse effects on soil quality may be more severe in ecologically sensitive regions populated by predominantly resource-poor farmers who cannot use or afford adoption of RMPs.

5.5 SOIL CARBON DYNAMICS

Several important carbon management strategies exist, including the adaptive and mitigative options found in Figure 5.1. Adaptive options are based on better management

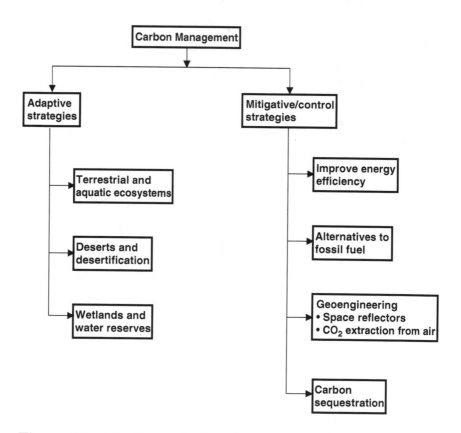

Figure 5.1 Adaptive and mitigative strategies of carbon management and sequestration to address global climate change.

of terrestrial and aquatic ecosystems, desert lands, and wetlands. Adopting RMPs such as fertilizer use and irrigation on croplands and grazing lands is an important C management option. Mitigative options include enhancement of energy-use efficiency, finding alternatives to fossil fuel, and using geoengineering techniques, such as space reflectors and CO_2 extraction from the atmosphere in order to influence the energy budget and the rate of enrichment of atmospheric concentration of CO_2. Carbon sequestration is a key mitigative strategy.

Carbon sequestration implies transferring CO_2 from a pool that has a short turnover time into a pool with a longer turnover time. Specifically, it involves the removal of CO_2 from the atmosphere and its storage in long-lived pools, such as soil, vegetation, wetlands, oceans, and geologic strata. There are two main ways to sequester C as illustrated in Figure 5.2. The biotic strategy involves conversion of CO_2 into carbohydrates, lignin, cellulose, and other forms of biomass through

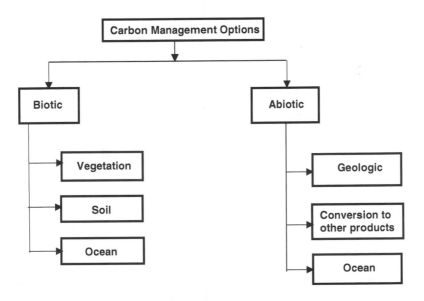

Figure 5.2 Categories of technological options for carbon sequestration through biotic and abiotic processes.

biotic processes such as photosynthesis. The biotic sequestration of CO_2 is relevant to the transfer of CO_2 into vegetation, soils, and the oceans. In contrast, the abiotic strategy involves a technical transfer of CO_2 from the atmosphere into geologic, oceanic, and other long-lived pools, and the conversion of CO_2 into other products such as $CaCO_3$.

Soil C sequestration is a biotic strategy, based on a transfer of atmospheric CO_2 into humic substances in soil. The terrestrial C pool is the third largest pool. As shown in Table 5.9, it holds at least 486 Pg (gigatons) C in biota and 2542 Pg C in the soil. Carbon sequestration in terrestrial

Table 5.9 Global Estimates of Land Area and Carbon Stocks in Plant Matter and Soil for Ecosystems

Ecosystem	Area (10^{12} m²)	Plant C (Pg[a])	Soil (Pg)	Total (Pg)
Forest, tropical	14.8	244.2	123	367
Forest, temperate and plantation	7.5	92.0	90	182
Forest, boreal	9.0	22.0	135	157
Woodland, temperate	2.0	16.0	24	40
Chaparral	2.5	8.0	30	38
Savanna, tropical	22.5	65.9	263	329
Grassland, temperate	12.5	9.0	295	304
Tundra, arctic and alpine	9.5	6.0	121	127
Desert and semidesert, scrub	21.0	6.9	168	175
Desert, extreme	9.0	0.3	23	23
Perpetual ice	15.5	0.0	0	0
Lake and stream	2.0	0.0	0	0
Wetland	2.8	12.0	202	214
Peatland, northern	3.4	0.0	455	455
Cultivated and permanent crop	14.8	3.0	117	120
Human area	2.0	1.0	10	11
Total	150.8	486.4	2056	2542

[a] Pg, 10^{15} grams, or 1 gigaton.

Source: From Amthor, J.S., M.A. Huston, et al., 1998. Terrestrial ecosystems responses to global change: a research strategy. ORNL/TM-1998/27. Oak Ridge National Laboratory, Oak Ridge, TN; and U.S. Department of Energy. 1999. *Carbon Sequestration: Research and Development.* National Technical Information Service, Springfield, VA.

Table 5.10 Terrestrial Carbon Sequestration
Potential

Biome	C Sequestration Potential (Pg[a] C/year)
Agricultural lands	0.85–0.90
Biomass croplands	0.5–0.8
Grasslands	0.5
Rangelands	1.2
Forest lands	1–3
Deserts and degraded lands	0.8–1.3
Terrestrial sediments	0.7–1.7
Boreal peatlands and wetlands	0.1–0.7
Total	5.7–10.1

[a] Pg, 10^{15} grams, or 1 gigaton.
Source: From U.S. Department of Energy. 1999. *Carbon Sequestration: Research and Development.* National Technical Information Service, Springfield, VA.

ecosystems can be achieved by enhancing the C pool in living plant matter, roots, and soil. Soil C storage involves both organic and inorganic C pools. Formation of secondary carbonates is one of the mechanisms of soil C sequestration.

There are numerous estimates of the potential for terrestrial C sequestration. The estimate of 5.7 to 10.1 Pg C/year shown in Table 5.10 is quite large (U.S. Department of Energy, 1999). However, the attainable potential of terrestrial C sequestration may only represent 10% to 20% of this amount through the use of adaptive strategies. IPCC (2000) estimated the potential of terrestrial C sequestration in agricultural and forestry ecosystems at 2.5 Pg C/year over the next 40 years. As indicated in Table 5.11, such a level of sequestration will yield a net increase of only 0.5 to 0.7 Pg C/year in atmospheric concentration, assuming the same rate of increase as observed in the 1990s.

Soil-specific research on adaptive and mitigative strategies aimed at soil C sequestration and enhancement of the SOC pool is needed. This would be a "no-regret approach" (Wittwer, 1995). Soil C sequestration would yield benefits with or without future climate change as it improves soil quality

Table 5.11 Carbon Sequestration Potential in Managed Terrestrial Ecosystems

Scenario	Area (million ha)	Average Adoption/Conversion (% of area) 2010	2040	Average Carbon Sequestration (Mg C/ha/year)	Potential (Tg[a] C year) 2010	2040
Improved Management						
Cropland	1289	29	59	0.33	125	258
Rice paddies	153	51	81	0.10	8	13
Agroforestry	400	22	40	0.28	26	45
Grazing land	3401	10	20	0.70	237	474
Forest land	4051	10	39	0.41	170	703
Urban land	100	5	15	0.3	2	4
Land Use Change						
Agroforestry	630	20	30	3.1	391	586
Restoring degraded soils	277	5	10	0.25	4	8
Grassland	1457	3	7	0.8	38	82
Wetland restoration	230	5	15	0.4	4	14
Forest Products	—	—	—	—	300	300
Totals					1302	2485

[a] Tg= teragram, or 10^{12} grams, or 1 megaton.

Note: Average adoption rate and average carbon sequestration rate are calculated as weighted mean averages.

Source: Modified from Inter-Government Panel on Climate Change 2000. Land Use, Land Use Change and Forestry. Cambridge University Press, London, 181–281.

and subsequent productivity. Adaptive strategies involve: (1) the use of transgenic plants that are more resilient to environmental stresses such as drought and heat and with pest resistance in crops, trees, and livestock; (2) conservation-effective measures to improve soil quality and reduce the risks of soil degradation; and (3) improved energy-use efficiency and development of fossil-fuel offsets.

5.6 SOIL CARBON SEQUESTRATION AND GLOBAL FOOD SECURITY

Soil C sequestration can be made a win–win proposition. It is an important objective of both adaptive and mitigative strategies to minimize or reverse the effects of global climate change, although they approach this aim differently. The adaptive strategy by increasing the SOC pool would enhance soil microbial activity and would accentuate soil physical, chemical, and biological quality. Soil quality improvements could negate some or most of the adverse effects of climate change on crop growth and yield. Technological options related to enhancement of the SOC pool through adaptive strategies include conversion from plow tillage to no till, incorporating cover crops in the rotation cycle, using compost and other biosolids, adoption of agroforestry practices, and growing adapted species that tolerate climatic extremes. Mitigative strategies focus on SOC because this has a strong potential to reduce the rate of enrichment of atmospheric concentration of CO_2. Technological options related to enhancement of the SOC pool using mitigative strategies include restoration of degraded soils and ecosystems, desertification control, conversion of marginal agricultural soils from crop and pastures to forests or any other perennial vegetation, and adoption of RMPs on agricultural lands.

In this regard, the impact of agricultural intensification in developing countries in Sub-Saharan Africa, South and Central Asia, and Central America and the Caribbean becomes central. This strategy would include growing improved varieties, returning crop residues and other biomass to the soil, improving soil fertility through a judicious combination of fertilizers and manures that enhance microbial processes in soil, and adoption of conservation-effective measures that improve water and nutrient-use efficiency.

5.7 CONCLUSIONS

An increase in atmospheric concentration of CO_2 and other GHGs is likely to cause regional and global climate change. Temperature increases are likely to be more accentuated at higher than at lower latitudes and to have an evident impact on precipitation and its distribution. The frequency of extreme events may increase with global warming, and projected climate change is likely to affect soil quality. In some cases, water and wind erosion of soils and salinization may be accelerated. In other situations, there may be improvement in soil biological activity, improvement in the SOC pool, and strengthening of the elemental cycling mechanisms. Crop yields may also be affected. Modest temperature increases of up to +2°C may have positive impacts on crop yields in middle and high-middle latitudes, but negative impacts on crop yields in lower latitudes and in semi-arid climates. In contrast, a temperature increase of +4°C may have clear adverse effects on all crop yields, especially in developing countries.

Improving soil quality through soil C sequestration is a win–win strategy for both adaptive and mitigative options. Estimates of the potential of SOC sequestration range from 1 Pg C/year to 2.5 Pg C/y. The attainable potential may be modest in developing countries with many resource-poor farmers. Yet, it appears that some and maybe many of the adverse impacts of projected climate change on food security can be kept in check by enhancing the quality and quantity of the SOC pool by restoring degraded soils and ecosystems and adopting RMPs on agricultural soils.

REFERENCES

Abrol, Y.P. and K.T. Ingram. 1996. Effects of higher day and night temperatures on growth and yields of some crop plants. In F. Bazzaz and W.G. Sombroek, Eds. *Global Climate Change and Agricultural Production*. John Wiley & Sons, Chichester, United Kingdom, pp. 123–140.

Allen, L.H. Jr. 1991. *Carbon Dioxide Effects on Growth, Photosynthesis and Evapotranspiration of Rice at Three Nitrogen Fertilizer Levels: Response of Vegetation to CO₂*. Report 62. Plant Stress and Protection Unit, U.S. Department of Agriculture-Agriculture Research Service, Gainesville, FL.

Allen, L.H. Jr. 1994. Carbon dioxide increase: direct impacts on crops and indirect effects mediated through anticipated climate change. In K.J. Boote, J.M. Bennett, T.R. Sinclair, and G.M. Paulsen, Eds. *Physiology and Determination of Crop Yield*. Agronomy Society of America, Madison, WI, pp. 425–459.

Allen, L.H. Jr., and J.S. Amthor. 1995. Plant physiological response to elevated CO_2, temperature, air pollution and UV-B radiation. In G.M. Woodwell and F.T. Mackenzie, Eds. *Biotic Feedbacks in the Global Climatic System: Will the Warming Feed the Warming?* Oxford University Press, New York, pp. 51–84.

Allen, L.H. Jr., J.T. Baker, and K.J. Boote. 1996. The CO_2 fertilization effect: higher carbohydrate production and retention as biomass and seed yield. In F. Bazzaz and W.G. Sombroek, Eds. *Global Climate Change and Agricultural Production*. John Wiley & Sons, Chichester, United Kingdom, pp. 65–100.

Amthor, J.S., M.A. Huston, et al., 1998. *Terrestrial Ecosystems Responses to Global Change: A Research Strategy*. ORNL/TM-1998/27. Oak Ridge National Laboratory, Oak Ridge, TN.

Bernard, H.W. Jr. 1993. *Global Warming Unchecked: Signs to Watch For*. Indiana University Press, Bloomington.

Bowman, W.D. and B.R. Strain. 1987. Interaction between CO_2 enrichment and salinity stress in the C4 non-halophyte *Andropogan glomeratus* (Walter) BSP. *Plant Cell Environ.*, 10:267–270.

Brinkman, R. and W.G. Sombroek. 1996. The effects of global change on soil conditions in relation to plant growth and food production. In F. Bazzaz and W.G. Sombroek, Eds. *Global Climate Change and Agricultural Production*. John Wiley & Sons, Chichester, United Kingdom, pp. 49–63.

Bruinsma, J., Ed. 2003. *World Agriculture: Towards 2015/2030*. Food and Agriculture Organization, Rome; Earthscan, London.

Carter, T.R., J.H. Porter, and M.L. Parry. 1991. Climate warming and crop potential in Europe: prospects and uncertainties. *Global Environ. Change,* 1:291–312.

Cheddadi, R., J. Guiot, and D. Jolly. 2001. The Mediterranean vegetation: what if the atmospheric CO_2 increased? *Landscape Ecol.,* 16:667–675.

Coleman, A.P. 1926. *Ice Ages: Recent and Ancient*. Macmillan, New York.

Dawson, A.G. 1992. *Ice Age Earth: Late Quaternary Geology and Climate*. Routledge, London.

Dinar, A. and R. Mendelsohn. 1998. Overview: Climate Change and Crop Yield. World Bank Technical Paper 402. World Bank, Washington, DC.

Doering, O.C., J.C. Randolph, J. Southworth, and R.A. Pfeiffer. 2002. Conclusions. In O.C. Doering, III, J.C. Randolph, J. Southworth, and R.A. Pfeiffer, Eds. *Effects of Climate Change and Variability on Agricultural Production Systems*. Kluwer, Dordrecht, pp. 265–271.

Food and Agriculture Organization. 1990. *Water and Sustainable Agricultural Development: A Strategy for the Implementation of the Mar del Plata Action Plan for the 1990s*. FAO, Rome.

Food and Agriculture Organization/UN Development Programme/ UN Environmental Programme. 1994. *Land Degradation in South Asia: Its Severity, Causes and Effects upon the People*. World Soil Resources Report 78. FAO, Rome.

Food and Agriculture Organization. 1996. *Rome Declaration on World Food Security and World Food Summit Plan of Action*. World Food Summit, FAO, Rome.

Food and Agriculture Organization. 2001. *Production Yearbook*. FAO, Rome.

Gifford, R.M. 1979. Growth and yield of CO_2-enriched wheat under water-limited conditions. *Aust. J. Plant Physiol.*, 6:367–378.

Glantz, M.H., Ed. 1994. *Drought Follows the Plow.* Cambridge University Press, London; New York.

Idso, S.B., B.A. Kimball, M.G. Anderson, and J.R. Mauney. 1987. Effects of atmospheric enrichment on plant growth: the interactive role of air temperature. *Agric. Ecosys. Environ.*, 20:1–10.

Intergovernmental Panel on Climate Change. 1995. *Second Assessment Synthesis of Scientific-Technical Information Relevant to Interpreting Article 2 of the UN Framework Convention on Climate Change.* Intergovernmental Panel on Climate Change, Geneva, Switzerland.

Intergovernmental Panel on Climate Change. 2000. *Land Use, Land Use Change, and Forestry.* Cambridge University Press, London; New York, pp. 181–281.

Intergovernmental Panel on Climate Change. 2001. *Climate Change 2001: The Scientific Basis.* Cambridge University Press, London; New York.

International Fertilizer Development Center. 2001. *Global and Regional Data on Fertilizer and Consumption 1961/62–2000/01.* International Fertilizer Development Center, Muscle Shoals, AL.

Karl, T.R., R.W. Knight, D.R. Easterling, and R.G. Quayle. 1996. Indices of climate change for the United States. *Bull. Am. Meteorolog. Soc.*, 77:279–292.

Kirschbaum, M.U.F., A. Fischlin, M.G.R. Cannell, R.V.O. Cruz, W. Galinski, and W.P. Cramer. 1996. Climate change impacts on forests. In *Climate Change 1995: Impacts, Adaptation and Mitigation of Climate Change. Scientific-Technical Analyses.* Contributions of Working Group II to the Second Assessment Report of the Intergovernmental Panel on Climate Change. Cambridge University Press, London; New York, pp. 95–129.

Kropff, M.J., G. Centeno, D. Bachelet, et al. 1993. *Predicting the Impact of CO_2 and Temperature on Rice Production.* IRRI Seminar Series on Climate Change and Rice. International Rice Research Institute, Los Baños, Philippines.

Lal, R. 2000. Soil management in developing countries. *Soil Sci.*, 165:57–72.

Lee, J.J., D.L. Phillips, and R.F. Dodson. 1996. Sensitivity of the U.S. Corn Belt to climate change and elevated CO_2. II. Soil erosion and organic carbon. *Agric. Syst.*, 52:503–521.

Mooney, H.A., E.R. Fuentes, and B.I. Kronberg, Eds. 1993. *Earth System Responses to Global Change: Contrasts between North and South America*. Academic Press, San Diego, CA.

Muchena, P. and A. Iglesias. 1995. Vulnerability of maize yields to climate change in different farming sectors in Zimbabwe. In *Climate Change and Agriculture: Analysis of Potential International Impacts*. ASA Publication 59. American Society of Agronomy, Madison, WI, pp. 229–240.

Nelson, M. and M. Mareida. 2001. Environmental Impacts of the CGIAR: An Assessment. Document SDR/TAC:IAR/01/11. Presented at mid-term meeting of the Consultative Group for International Agricultural Research, Durban, South Africa, May 21–25.

Norse, D., J. Li, and Z. Zhang. 2001. *Environmental Costs of Rice Production in China: Lessons from Hunan and Hubei*. Aileen International Press, Bethesda, MD.

Oldeman, L.R. 1994. The global extent of soil degradation. In D.J. Greenland and I. Szabolcs, Eds. *Soil Resilience and Sustainable Land Use*. CAB International, Wallingford, United Kingdom, pp. 99–118.

Oldeman, L.R. 1998. *Soil Degradation: A Threat to Food Security*. ISRIC Report 98/01. International Soil and Reference Information Centre, Wageningen, Netherlands.

Oram, P.A. 1989. Views on the new global context for agricultural research: implications for policy. In *Climate and Food Security*. International Rice Research Institute, Manila.

Parry, M. and T. Carter. 1988a. The assessment of effects of climate variations on agriculture: aims, methods and summary of results. In M.L. Parry, T.R. Carter, and N.T. Koniju, Eds. *The Impact of Climate Variations on Agriculture. Vol. 1: Assessments in Cool Temperate and Cold Regions*. Kluwer, Dordrecht, pp. 11–95.

Parry, M. and T. Carter. 1988b. The assessment of effects of climatic variations on agriculture: a summary of results from studies in semi-arid regions. In: M.L. Parry and T.R. Carter, Eds. *The Impact of Climatic Variations on Agriculture. Vol. 2: Assessment in Semi-Arid Regions*. Kluwer, Dordrecht, pp. 9–60.

Parry, M. and T. Carter. 1998. *Climate Impact and Adaptation Assessment: A Guide to the IPCC Approach*. Earthscan, London.

Phillips, D.L., J.L. Lee, and R.F. Dodson. 1996. Sensitivity of the U.S. Corn Belt to climate change and elevated CO_2. I: Corn and soybean yields. *Agric. Syst.,* 52:481–502.

Prentice, K.C. and I.Y. Fung. 1990. The sensitivity of terrestrial carbon storage to climate change. *Nature,* 346:48–50.

Reardon, T. 1995. Sustainability issues for agricultural research strategies in the semi-arid tropics: focus on the Sahel. *Agric. Syst.,* 48:345–359.

Reilly, J.M. 2002. *Agriculture: The Potential Consequences of Climate Variability and Change for the United States*. Cambridge University Press, London; New York.

Rhodes, E.R. 1995. Nutrient depletion by food crops in Ghana and soil organic nitrogen management. *Agric. Syst.,* 48:101–118.

Rosenzweig, C. and M.L. Parry. 1993. Potential impacts of climate change on world food supply: a summary of recent international study. In H.M. Kaiser and T.E. Drennen, Eds. *Agricultural Dimensions of Global Climate Change*. St. Lucie Press, Delray Beach, FL, pp. 87–116.

Rosenzweig, C. and M.L. Parry. 1994. Potential impacts of climate change on world food supply. *Nature,* 367:133–138.

Rosenzweig, C., J. Phillips, R. Goldberg, J. Carroll, and T. Hodges. 1996. Potential impact of climate change on citrus and potato production in the U.S. *Agric. Syst.,* 52:455–479.

Rosenzweig, C. and D. Hillel. 1998. *Climate Change and the Global Harvest: Potential Impacts of the Greenhouse Effect*. Oxford University Press, New York.

Scherr, S. and S. Yadav. 1996. Land degradation in the developing world: implications for food, agriculture and the environment to 2020. Food, Agriculture and the Environment Discussion Paper 14. International Food Policy Research Institute, Washington, DC.

Stoorvogel, J.J. and E.M.A. Smaling. 1990. Assessment of soil nutrient depletion in sub-Saharan Africa. The Winand Staring Center for Integrated Soil and Water Research, Wageningen, Netherlands.

Swaminathan, M.S. 2003. The century of hope. In R. Lal, D.O. Hansen, N. Uphoff, and S. Slack, Eds. *Food Security and Environmental Quality in the Developing World*. Lewis Publishers, Boca Raton, FL, pp. 1–11.

U.N. Environmental Programme. 1997. *Global Environment Outlook*. U.N. Environmental Programme and Oxford University Press, Oxford.

U.S. Department of Energy. 1999. *Carbon Sequestration: Research and Development*. National Technical Information Service, Springfield, VA.

van de Geijn, S.C. and J. Goudriaan. 1996. The effects of elevated CO_2 and temperature change on transpiration and crop water use. In F. Bazzaz and W.G. Sombroek, Eds. *Global Climate Change and Agricultural Production*. John Wiley & Sons, Chichester, United Kingdom, pp. 101–121.

Wild, A. 2003. *Soils, Land and Food: Managing the Land During the 21st Century*. Cambridge University Press, London; New York.

Williams, A., F.F. Pruski, and M.A. Nearing. 2002. Indirect impacts of climate change that affect agricultural production: soil erosion. In O.C. Doering, III, J.C. Randolph, J. Southworth, and R.A. Pfeiffer, Eds. *Effects of Climate Change and Variability on Agricultural Production Systems*. Kluwer, Dordrecht, pp. 249–264.

Wittwer, S.H. 1995. *Food, Climate and Carbon Dioxide: The Global Environment and World Food Production*. CRC Press/Lewis Publishers, Boca Raton, FL.

Wolfe, D.W. and J.D. Erickson. 1993. Carbon dioxide effects on plants: uncertainties and implications for modeling crop response to climate change. In H.M. Kaiser and T.E. Drennen, Eds. *Agricultural Dimensions of Global Climate Change*. St. Lucie Press, Delray Beach, FL, pp. 153–178.

World Bank. 1998. *Measuring the Impact of Climate Change on Indian Agriculture*. World Bank Technical Paper 402. Washington, D.C. pp. 266.

Section II

Climate Change and Net Primary
Productivity

6

Climate Change Effects on the Water Supply in Some Major River Basins

RANJAN S. MUTTIAH AND RALPH A. WURBS

CONTENTS

6.1 INTRODUCTION

While the Green Revolution during the latter part of the 20th century may have been facilitated by higher-yield grain varieties, the impact of increased water harvesting techniques (dams, irrigation systems) on agricultural production cannot be ignored. The promotion of agriculture to sequester carbon will require the careful evaluation of future water availability. The following are widely thought to impact the water cycle in a future climate: (1) greenhouse gases (GHGs) such as CO_2, CH_4, and N_2O, are expected to increase from human related activities such as fuel emission and fertilizer application; (2) air and sea surface temperatures (SST) will rise due to GHGs; (3) the number of extreme events (flooding, drought, tornados due to SST-related El Niño/Southern Oscillation [ENSO] events) precipitation intensity may increase, that is, the wet periods will get wetter and the dry periods will get drier; (4) the quality of arable land may decline due to increased salinization, erosion, and poor management; and (5) urban population growth will continue at or above current rates. Iraq serves as an example of point 4. While about 3.5 million ha are potentially cultivable in irrigated agriculture, roughly half, 1.94 million ha, are actually cultivated due to water logging and salinization problems (Food and Agriculture Organization [FAO], 1997). Recent extreme events from the late 1990s to the present — such as the flooding of the Elbe in Central Europe in August 2002, the 1998 flooding of the Yangzte in China, the 2000 and 2002 droughts in monsoon-dependent India, and the highest recorded tornado activity in the United States in 2003 — are visible signs of potential trends in extreme events. During 2003, the World Meteorological Organization took the unprecedented step of announcing likely changes in extreme events in its reporting. Historical analysis has traced changes in civilization from changes in the Holocene climate (DeMenocal, 2001).

This chapter examines the likely consequences of climate change on the water supply in Texas and in ten major basins of the world. The scope of this chapter is to investigate how climate change may affect water supply systems in Texas for a highly urbanized watershed (San Jacinto with drainage area 7300 km²), a large basin (Brazos with drainage area 118,000 km²), and in ten other basins worldwide (Figures 6.1 and 6.2). Although the focus of this chapter is on evaluation of water resources, the potential of irrigated crops to sequester carbon,

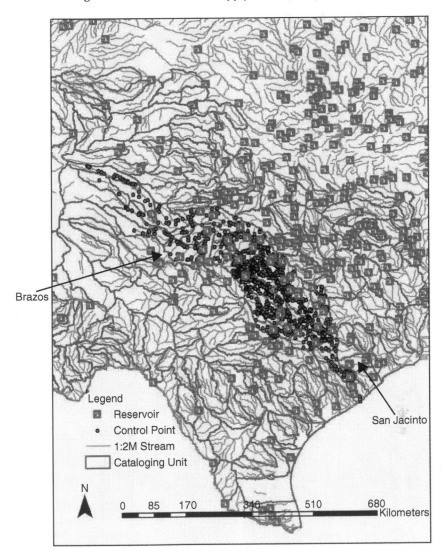

Figure 6.1   Brazos and San Jacinto Basins in Texas.

and of the water supply in these basins to meet future water demands are also discussed. Assessment of water resources should consider both water supply and demand. In many natural and human systems such as irrigated agriculture,

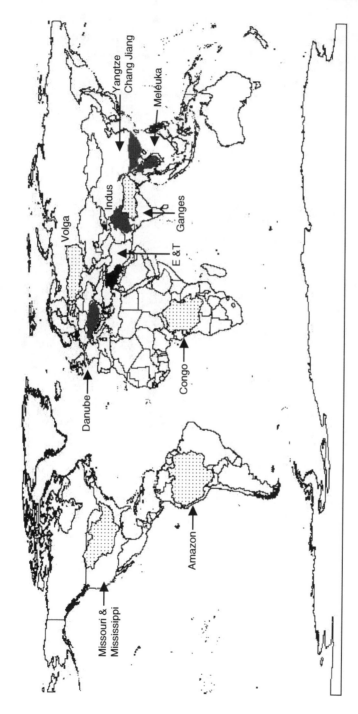

Figure 6.2 The ten basins of the world. (Map reproduced with permission of Aaron T. Wolf, Department of Geosciences, Oregon State University.)

water supply and demand are in some form of equilibrium through an evolutionary process for natural systems or trial and error experimentation for human systems. When one side of this equation is changed, there is bound to be a temporary imbalance before onset of another equilibrium state.

The sources of consumptive water are streams, reservoirs (or storage systems), and groundwater wells. While aquifer groundwater supply is sensitive to climate recharge, we examine surface water supplies only. A comprehensive assessment of water supply requires water rights and flow databases. The databases contain hydrologic information by control points (CPs). A CP is a point of water transfer or storage (reservoir) in a stream network. Hydrologic information consists of historical stream flows, water diversion amounts, reservoir storage, hydropower generation, and priority of water rights. Due to lack of intensive data, water supply analysis on a worldwide basis is currently not possible. Compilation of water supply and demand data for all major river basins in a comprehensive database is therefore highly desirable. The Texas examples highlight the importance of comprehensive water allocation databases for effective estimation of likely changes to water supply under climate change. To date, water resource assessments during climate change have ignored the influence of storage systems.

Hydrologic assessments depend on global circulation million models (GCMs) for downscaled weather data. The GCMs range from models that consider the atmosphere only (AGCMs) such as the Goddard Institute for Space Studies GISS Models I and II (Hansen et al., 1983), to coupling between oceans and atmosphere with terrestrial biosphere feed back such as the Canadian Climate Center model, CCCma (Flato et al., 2000) and the U.K. Meteorological Office Hadley Centre models HadCM2 (Johns et al., 1997) and HadCM3 (Gordon et al., 2000). Since the GCMs capture physical processes of atmospheric circulation from the surface boundary layers to the upper layers involving atmospheric chemistry and radiation physics, model results are generated at coarse grid resolutions of about 2 to 3 degrees at the equator. The Japanese Earth Simulator (called AGCM/AFES), run jointly by the National

Space Development Agency (NASDA), Japanese Atomic
Energy Research Institute (JAERI), and Japanese Marine Science and Technology Center (JAMSTEC) has the ambitious
goal of simulating Earth's climate on 10-km grids (see
www.es.jamstec.go/jp/esc/eng/index.html) (Shingu et al.,
2002; Ohfuchi, 2003). Depending on the size of the hydrologic
basin, downscaling techniques range from direct use of GCM
output (Arora and Boer, 2001), interpolation between grids
(Jones and Thornton, 1999), regional circulation models forced
with GCM boundary conditions (Giorgi et al., 1994), and use
of surrogate variables such as GCM atmospheric pressures to
estimate precipitation (Burlando and Rosso, 1991). Whether
any one downscaling technique is superior to another is unresolved at the moment.

6.2 METHODOLOGY

Our methodology for the Texas examples consisted of obtaining
CCCma (model CGCM1) daily precipitation and temperatures
for the Brazos and San Jacinto regions between 2040 and 2060
(2050). The climate was supplied to a watershed model called
the Soil and Water Assessment Tool (SWAT) (Arnold et al.,
1993, 1998) to generate naturalized flows in watersheds under
historical climate conditions (see Figure 6.3). Naturalized
flows are defined as stream flows obtained after subtracting
flow influences due to manufactured structures. The SWAT
flows were calibrated to measured flows using observed historical climate from 1960 to 1989. The SWAT model was run
with future (2040 to 2060) weather generated by the CCCma
(at about $2.5° \times 2.5°$) model for a GHG increase of 1% per year
plus aerosols. A separate SWAT control run was made with
2040–2060 weather from CCCma with GHGs set at 1995 levels. Monthly flow multiplication factors were generated by
ratio of SWAT flows with and without GHG change (Muttiah
and Wurbs, 2002; Wurbs et al., 2003). The flow multipliers
were then multiplied against historical naturalized flows
(monthly flows, 1940 to 1996) in the Water Rights and Analysis
Package (TAMU-WRAP). The TAMU-WRAP program accounts
for water allocation by control points (CPs) in a river network

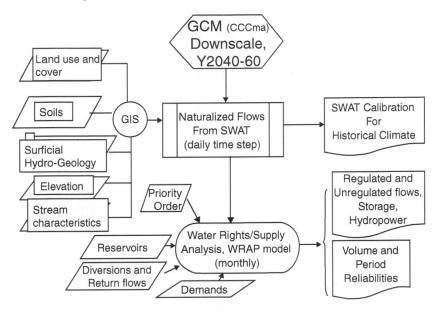

Figure 6.3 Flow chart describing the linkage between the Soil and Water Assessment Tool (SWAT) model for generating naturalized flows, and the water rights and analysis package (WRAP) model.

(Wurbs, 2001). Increased water abstraction due to population changes were based on Texas Water Development Board (TWDB) projections. The volume reliability for flow diversions were expressed as (v/V) 100, where v is the water volume supplied, and V is the amount demanded by the water right holder; equivalently, the period of reliability was defined as (n/N) 100, where n was the period (in months) during which the demand target was fully met, and N the total number of months in the simulation.

Flow changes in the ten basins of the world were based on literature review, especially the work of Arora and Boer (2001) who modeled flows using CCCma, and Arnell (1999) who used weather generated from Hadley Centre circulation models. Arnell generated six different scenarios with five scenarios coming from the Hadley HadCM2 and one from HadCM3. In the HadCM3, flux adjustments are made automatically. The HadCM2 scenarios differed in the model initial conditions. To discern the likely impact of regulations on stream flows, the dams in the basins were classified into major and minor. The lower limit capacity of major dams was set at 200 million m^3. While the International Commission on Large Dams (ICOLD) based in Paris has a lower limit of 1 million m^3 or 15 m in height; the higher 200 million m^3 was selected, since the FAO country descriptions on which we relied had many instances of the 200 million m^3 capacity as the lower limit for a "major" dam. For the Danube, Mekong, and Mississippi and Missouri (MMR) basins we obtained dam data from previous basin-wide studies, respectively, from the Danube River Basin Pollution Reduction Program (International Commission for the Protection of the Danube River, 1999; Zinke, 2003), the International Water Management Institute in Sri Lanka (Kite, 2000), and the U.S. Army Corp of Engineers. The dam estimates for the Volga River Basin were obtained from Volga Ltd. Consulting Engineers (Galant, 2003). Since dam capacity data were not available for all the dams in the Amazon–Tocantins Basin, the hydropower generation potential was prorated against known capacities using Fearnside (1995). In final consideration, the dam estimates in our opinion are not very reliable at present due to different definitions, and lack of a common dam database. The major and minor dams in our analysis include main stem and tributary dams; in brief, we did not distinguish between main stems and tributaries for the worldwide basins.

The irrigated areas and crops within basins were obtained from several sources including the FAO-AQUASTAT (data from the mid-1990s), World Resources Institute (WRI) basin assessments (*http://earthtrends.wri.org/pdf_library*), and the 1998 special irrigated area census done by the National Agricultural Statistics Service (NASS) of the U.S. Department of Agriculture for the MMR. The irrigated areas were also checked against those supplied by Mark Rosegrant (2003) from the International Food Policy Research Institute. The WRI estimates were based on satellite (AVHRR) estimates, while the AQUASTAT (countrywide) estimates were based on country surveys. In the case of large discrepancies, we selected the FAO survey estimates. Table 6.1 summarizes characteristics of the ten basins and dams.

The carbon uptake potential of irrigated areas was determined by selecting the dominant crops from the basins, and using dry matter estimates from ambient and above-ambient CO_2 open-top chambers (OTC), and free-air CO_2 enrichment (FACE) experiments for rice (De Costa et al., 2003; Kim et al., 2001), CO_2 OTC for wheat (Hakala, 1998), soil plant atmosphere units (SPAR) for cotton (Reddy et al., 1998), and OTC experiments for barley (Fangmeier et al., 2000). Table 6.2 gives the biomass and new soil organic carbon additions from experiments. The corn biomass in the MMR was estimated from reported yields for the 1997 census year from NASS (at harvest index 0.5). The change to corn biomass under doubled CO_2 was estimated by assuming a 3% increase based on experiments reported by Ziska and Bunce (1997). A conversion factor of 0.42 was used to estimate amount of organic carbon in biomass (Izzauralde, 2003). The change in cumulative water use *through a growing season* due to increase in CO_2 is not well documented. While water use efficiency (biomass fixed per unit of water use by plants) has been observed to significantly increase (upward of 20%) under CO_2 fertilization, the seasonal cumulative water use of wheat in FACE experiments has been observed to be significantly ($P > F = 0.3$) lower by only 4% for well-fertilized (350 kg N/ha) and watered conditions (Hunsaker et al., 2000). The new soil organic carbon (SOC) input to the soil at the time of crop harvest was determined from the lower limit given in Leavitt et al. (2001) for wheat, and the mean value given in Torbert et al. (1997) for soybeans and grain sorghum (as surrogate for C_4 crops corn and maize). Since no SOC data were found for crops such as rice and barley, the C_3 wheat estimates were used. Our SOC estimates differ from those of Sperow et al. (2003) and Lal and Bruce (1999), since we did not account for soil C savings from reduced erosion and management practices (no-till, reclamation, set-asides). Our carbon uptake estimates also assume cropping conditions similar to the CO_2 experiments.

Table 6.1 Characteristics of World's Ten Major River Basins and Dams

Name	Region	DA (x 1000 km³)	Length (km)	Sectors/Value	Estimated n Large Dams (km³)	Minor/ Major[a] Dams	Irrigated Area (millions ha)	Irrigated Croplands mm/year (km³/year)	Mean Flow (km³/ year)[b]	Δ Runoff (%)
Amazon-Tocantins	Latin America	7180	6771	Navigation, industrial	4(70)	U	<0.25	Ignored[c]	5676	–10/–34
Zaire (Congo)	Central Africa	3822	4700	Navigation	6(U)	U	<0.1	Ignored[c]	1325	+10/x
Indus	Pakistan/ India/ China	1060	3180	Navigation, industrial, irrigation	2(26)	26	14	940(132)	121	–12/–30
Euphrates & Tigris	Middle East	808	4540	Navigation, shipping, municipal, irrigation	50(112)	40	4.1	580(24)	27	–20/x
Mekong	Southeast Asia	—	—	Irrigation, municipal	16(20)	3	3.7	1630(60)	325	–30/x
Yangtze (Chang Jiang)	China	1808	5800	Irrigation, flood control, municipal	17(106)	7	30	805(241)	950	+22/–32
Ganges/ Brahmaputra/ Meghna	South Asia	1073	2700	Irrigation, municipal	6(U)	U	10.8	622(67)	473	x/+5
Volga	Russia	1360	3688	Industrial, irrigation, municipal	24(>100)	35	1.7	U	230	+5/+13

Missouri & Mississippi	United States	3221	6109	Industrial, navigation, flood control	27(122)	3	6.8	375(25)	553	+27/+5
Danube	Central Europe	817	2780	Municipal, industrial irrigation	12(15)	33	<0.1	Ignored[c]	203	−20/−16

[a] "Major" dam capacity >0.2 km^3.
[b] Flow estimates at mouth of river.
[c] Ignored due to smallness of area.

Notes: DA = drainage area; irrigated croplands = average water use by irrigated crops in units of millimeters per year cubic kilometers per year; Δrunoff = percent change in runoff; U = unknown.
Source: Change in runoff data from Arnell, N.W. 1999. *Global Environ. Change,* 9:S31–S49; and Arora, V.K., and G.J. Boer. 2001. *J. Geophys. Res.,* 106:3335–3348.

Table 6.2 Source of CO_2 Experiments Used in Biomass Fixation and Soil Carbon Sequestration Estimates

Crop	Site	Type	$-CO_2$ Concentration (μmol/mol)	Biomass (metric tons ha^{-1})		Increase in SOC		Reference
				Ambient	$-CO_2$	Ambient	$-CO_2$	
Rice	Batalagoda, Sri Lanka	OTC	570	11.8	16.07	NA	NA	De Costa et al. (2003)
	Northern Japan	FACE	665–755	14.95 Average 13.37	17.24 (medium N) Average 16.63	NA	NA	Kim et al. (2001)
Wheat	Southern Finland	OTC	700	12.22	14.33	NA	NA	Hakala (1998)
Cotton	Maricopa, Arizona	FACE	700			29–41 g C m^{-2} year^{-1}	29–41 g C m^{-2} year^{-1}	Leavitt et al. (2001)
Cotton	Mississippi State	SPAR units	700	19	26.5	NA	NA	Reddy et al. (1998)
Barley	Giessen, Germany	OTC	650	18.76	25.9	NA	NA	Fangmeier et al. (2000)
Sorghum	Auburn, Alabama	OTC	705–732	6.05	8.4	3.26%	14.8% (of biomass)	Torbert et al. (1997)
Soybeans	Auburn, Alabama	OTC	705–732	6.2	9.25	27.9%	8.36% (of biomass)	Torbert et al. (1997)

Notes: FACE = free-air CO_2 enrichment; NA = not available; OTC = open-top chambers; SPAR = soil plant atmosphere units.

6.3 RESULTS

6.3.1 Two Basins in Texas

The San Jacinto Basin in Texas is highly urbanized due to the presence of the Houston metroplex (the 1990 population was 2.7 million). About 95% of water usage in the basin is due to municipal and industrial demand. Figure 6.4 shows the naturalized flow multiplication factors obtained for the 2040–2060 climate from CCCma relative to that of the historical climate (1960–1989) for the "west" cataloging units (subwatershed) in the San Jacinto Basin (Muttiah and Wurbs, 2002). The flow changes were typical of the anticipated changes in many parts of the world during the fall/winter and spring/summer months (Arora and Boer, 2001), although the amplitude and phase changes to flow worldwide are more pronounced in areas dependent on snow melt. Tables 6.3 and 6.4 show the water balance and reservoir storages in the upstream Lake Conroe (capacity of 531 million m^3), and the

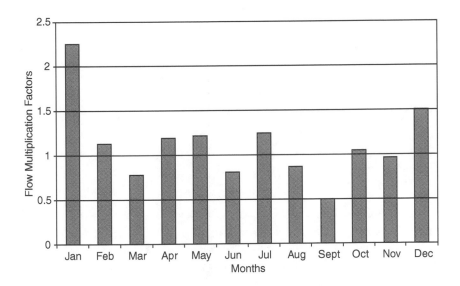

Figure 6.4 Flow multiplier for the west cataloging unit in San Jacinto Basin.

Table 6.3   Summary of 57-Year Simulation Results for Lake Conroe

Water Use Climate	2000 Historical	2000 2050	2050 Historical	2050 2050
Water Balance				
Stream inflow to reservoir (m³/s)	7.28	10.26	7.33	10.27
Outflow to river (m³/s)	2.80	5.91	0.99	2.21
Water supply diversions (m³/s)	3.88	3.91	6.48	8.07
Evaporation-precipitation (m³/s)	0.69	0.47	0.16	0.24
Change in reservoir storage (m³/s)	0.09	0.03	0.30	0.25
Mean Storage Over 57 Years				
Million m³	409	471	64	166
Percent of capacity	77.1	88.7	12.0	31.2

Table 6.4   Summary of 57-Year Simulation Results for Lake Houston

Water Use Climate	2000 Historical	2000 2050	2050 Historical	2050 2050
Water Balance				
Stream inflow to reservoir (m³/s)	52.16	65.61	51.46	62.84
Outflow to river (m³/s)	43.21	56.55	31.75	32.21
Water supply diversions (m³/s)	8.49	8.72	19.31	19.54
Evaporation-precipitation (m³/s)	0.46	0.33	0.41	0.09
Change in reservoir storage (m³/s)	0.00	0.00	0.01	0.00
Mean Storage Over 57 Years				
Million m³	192	196	131	155
Percent of capacity	97.3	99.1	66.6	78.8

downstream (closest to Houston) Lake Houston (capacity 197 million m³). Due to increased municipal demand of the Houston metropolitan area, Lake Conroe is forced into higher regulation (change in reservoir storage) relative to that of Lake Houston, and there is markedly less water in Lake Conroe compared to Lake Houston in a future (2050) climate and water use scenario.

In the Brazos Basin, municipal and industrial demand accounts for 78% of water withdrawal, while irrigation accounts for most of the rest. The 12 major reservoirs with total capacity of 3437 million m^3, which accounts for 63% of all surface water storage, are operated by the Brazos River Authority (BRA) essentially as a unit. There are 578 other smaller reservoirs with combined capacity of 5428 million m^3. Segmentation and trend analysis of the naturalized flows showed a significant increase of between 168% to 180% above normal flow due to ENSO forcing (Wurbs et al., 2003). Table 6.5 shows the flow multipliers and net evaporation (evaporation – precipitation) at four sites in the basin. As in San Jacinto, the summer months have reduced flows relative to the autumn months. The increased evaporation in the spring/summer months is driven by increased temperatures, while negative net evaporation is due to increased precipitation in the autumn months. Table 6.6 shows the unappropriated flows at the four sites in the basin. Generally, there is increased appropriation, and about 25% less exceedance of flows for the 2050 climate. Table 6.7 compares the storages (as percent of capacity) of the 12 major reservoirs operated by BRA, and the 578 reservoirs operated by small stakeholders. There is a larger difference in storages in the smaller reservoirs, than the BRA reservoirs, suggesting reduced regulation and overspill from the smaller reservoirs. Since the BRA operates the reservoirs as one system or unit, storage depletion tends to be balanced between the BRA reservoirs.

6.3.2 Ten Major Basins of the World

Table 6.1 lists the drainage areas (DA), sectors or value of the river basin to each region, number of estimated major dams (dams, each with capacity greater than 200 million m^3), the ratio of minor (1 to 200 million m^3 capacity) to major dams, irrigated areas in millions of hectares, mean outflow at mouth of basin, and estimated changes in mean naturalized flows (change in runoff) in a future 2050 climate from Arnell (1999) for the Hadley Centre models, and Arora and Boer (2001) for the CCCma model. The total irrigated areas in the ten basins

Table 6.5 Adjustment Factors for WRAP Input

Month	Naturalized Streamflow Multipliers				Net Reservoir Evaporation Added (mm)			
	Aquilla Gauge	Waco Gauge	Cameron Gauge	Hempstead Gauge	Aquilla Gauge	Waco Gauge	Cameron Gauge	Hempstead Gauge
Jan	0.79	1.10	0.95	0.76	1.6	2.0	3.2	4.2
Feb	0.71	0.90	1.08	0.66	5.8	5.8	7.5	9.3
Mar	0.61	0.43	0.41	0.42	24.5	23.4	27.3	31.4
Apr	0.59	0.66	0.72	0.51	22.1	21.6	26.1	30.8
May	0.37	0.54	0.57	0.38	19.5	19.1	23.7	28.6
Jun	0.30	0.39	0.34	0.27	22.4	22.4	26.6	30.0
Jul	0.35	0.68	0.94	0.53	13.8	15.5	18.9	19.1
Aug	0.34	1.00	0.88	0.42	3.8	6.3	8.6	7.4
Sep	1.42	2.66	2.84	1.58	−27.4	−22.7	−22.8	−27.2
Oct	1.56	2.18	2.27	1.88	−18.7	−15.3	−15.0	−17.6
Nov	1.06	0.96	1.43	1.29	6.6	7.0	9.1	10.7
Dec	0.94	1.65	1.77	1.13	−7.7	−6.4	−6.1	−6.5
Average	0.75	1.10	1.18	0.82	5.5	6.6	8.9	10.0

Table 6.6 Comparison of Unappropriated Streamflows with Historical and 2050 Climate for the Brazos

	Aquilla Gauge			Waco Gauge			Cameron Gauge			Hempstead Gauge		
	Hist (m³/s)	2050 (m³/s)	2050/H (%)	Hist (m³/s)	2050 (m³/s)	2050/H (%)	Hist (m³/s)	2050 (m³/s)	2050/H (%)	Hist (m³/s)	2050 (m³/s)	2050/H (%)
Mean	20.3	7.3	35.9	11.0	8.6	78.4	42.3	36.3	86.0	153.7	77.6	50.5
Stand. Dev.	78.4	37.0	47.1	24.6	24.9	101.2	74.6	80.9	108.4	249.6	152.4	61.1
Exceedance Frequency												
90%	0.0	0.0	–	0.0	0.0	–	0.0	0.0	–	0.0	0.0	–
75%	0.0	0.0	–	0.0	0.0	–	5.8	0.0	–	12.7	0.0	–
50%	0.0	0.0	–	0.0	0.0	–	12.5	10.1	–	50.7	23.6	46.6
25%	2.9	0.0	0.0	10.9	7.5	69.0	46.6	37.2	79.8	198.7	86.7	43.6
10%	56.8	4.9	8.6	33.9	24.7	72.8	114.3	99.0	86.6	437.3	212.8	48.6

Table 6.7 Comparison of Reservoir Storage with Historical and 2050 Climate

	Total 590 Reservoirs		12 BRA Reservoirs		578 Other Reservoirs	
	Historical	2050	Historical	2050	Historical	2050
Capacity (Mm³)	5,428	5,428	3,437	3,437	1,990	1,990
Mean (Mm³)	4,810	4,185	3,011	2,711	1,799	1,474
Exceedance Frequency	Storage as a Percent of Storage Capacity					
100%	58.4	33.2	51.1	23.4	71.0	47.4
98%	63.6	37.1	57.1	31.1	74.8	50.2
90%	75.2	55.4	71.3	52.0	82.0	61.3
75%	83.4	69.9	81.2	71.5	87.3	67.0
50%	90.4	80.2	89.8	82.6	91.3	76.1
25%	96.3	87.1	97.4	91.1	94.5	80.1
10%	99.0	93.6	99.9	97.6	97.3	86.6

amount to about 30% of irrigated lands worldwide (FAO, 1995). Assuming a uniform flow rate in the basin to that of the outflow at the mouth, it is apparent that the Indus and the Euphrates and Tigris (E&T) Basins use nearly all river flow to meet agricultural crop water demand. If not for upstream storage in reservoirs, the outflow from E&T could be significantly more if naturalized flows were to dominate the basin. The Tarbela and Mangla dams in Pakistan provide a limited buffer against flow shortages in the upstream areas of the Indus. The Mekong has very high irrigation water demands because of two rice crops per year. Using the changes to naturalized flows predicted by the climate models, the Amazon–Tocantins, Indus, E&T, Mekong, and Danube are the most affected in terms of shortages. The higher anticipated flows could potentially be buffered by storage in reservoirs. For example, the Volga Basin has considerable storage capacity in major reservoirs, and flooding is less of a concern there than in the Missouri–Mississippi, given current and anticipated changes to flow rates. The Danube at present is not that dependent on streams for agricultural water supply due to adequate rainfall. But, reduced naturalized flows in the

2050 climate reflect reduced rainfall amounts. Therefore, basins such as the Danube and Amazon–Tocantins may become more reliant on stream water supply for agricultural production needs in a future climate.

Table 6.8 gives the estimated carbon fixed as biomass of irrigated crops in the basin, and the organic carbon in soil due to crop production. The Yangtze basin stands out due to its large irrigated crop area (30 million ha). Taking an upper limit for terrestrial sinks of 2.5 Gt (gigatons) C/year (Sarmiento and Gruber, 2002), the total carbon biomass fixed under ambient CO_2 by the irrigated crops is about 17% of worldwide total sinks. The basin carbon estimates are an upper limit, since CO_2 release from crop and food processing, and CO_2 respiration from soil biota were ignored. For CO_2 concentration, increases of between 550 to 750 µmol mol^{-1}, the carbon fixed as biomass for a roughly doubled CO_2 climate increases by 23% from 0.425 Gt C/year to 0.522 Gt C/year. The carbon sequestered in the soil due to the growing crop (ignoring all management options) is estimated to increase by 24.5% from 24.5 Mt of C per year to 30.4 Mt of C per year. Even though soybeans show a reduction in soil carbon under increased CO_2 due to a differential rate of decomposition (Torbert et al., 1997), worldwide the soil organic carbon seems to increase as a simple function of biomass.

6.4 DISCUSSION

The two Texas examples suggested that the larger water permit holders such as urban centers, river authorities, and large water suppliers will regulate their reservoirs in order to maintain stable water capacities in their reservoirs. Generalizing from the Texas examples to the rest of the developed and developing worlds (Table 6.1), increased regulation by the larger water rights holders can be expected in the MMR and the Mekong. On the other hand, due to reduced flows under climate change, larger reservoirs in the E&T, Indus, and Danube basins will become more dependent on releases from the smaller (tributary) reservoirs to maintain their capacities. While increased production can be anticipated in basins with

Table 6.8 Potential Carbon Input from Biomass and Soil Carbon Sequestration in Irrigated Areas of Basins

Basin	Dominant Crops	Crop Ratio	Carbon Fixed as Biomass (Mt C/year)		Soil Sequestered (Mt C)	
			Ambient	↑ [CO_2]	Ambient	↑ [CO_2]
Amazon-Tocantins		Negative				
Zaire (Congo)		Negative				
Indus	Wheat: rice: cotton	3.0:1.0:1.2	0.082	0.103	4.2	4.2
Euphrates & Tigris	Wheat: barley: cotton	2.5:1:1	0.027	0.034	1.23	1.23
Mekong	Rice	1.0	0.041	0.051	1.11	1.11
Yangtze	Rice: wheat: maize: soybean: cotton	50:7:2.3:2.23:1.0	0.165	0.204	11.65	11.69
Ganges-Brahmaputra-Meghna	Rice: wheat: cotton	7.4:4.3:1.0	0.061	0.075	3.24	3.24
Volga	Assumed wheat: barley	1:1	0.011	0.014	Negative	0.42
Missouri & Mississippi	Corn: soybeans: wheat: cotton	5.7:3.5:3.5:1.0	0.037	0.041	3.14	8.48
Danube		Negative				
Totals			0.425 Gt C/year	0.522 Gt C/year	24.5 Mt C/year	30.4 Mt C/year

Notes: Inputs were determined by selecting dominant crops within basins from the Food and Agriculture Organization's AQUASTAT database, and ambient vs. elevated CO_2 experiments referenced in the methodology section. Two growing seasons per year for rice was assumed.

irrigated agriculture due to CO_2 stimulation, this may come at no overall water benefit over a growing season. With increased population pressures, irrigated areas may get less water due to diversion to meet municipal demand. Since the Hadley Centre and CCCma models predict different signs for the Yangtze mean flows, it was difficult to make predictions about consequences of climate change on the Yangtze water supply. While increased naturalized mean flows by 22% may be intercepted by the larger dams (such as the Three Gorges), and thus reduce downstream flooding, reduced flows may lead to less irrigation in this high production region of China.

Since many of the CO_2 fertilization experiments in the field, or through OTC are usually performed on non-water stressed plants, our estimates of carbon fixed in biomass during the growing season may be realistic. While irrigated areas of the world are about a 20% fraction of all arable lands, they may be an extremely important and significant part of the worldwide carbon sequestration budget. Since irrigated lands are generally well managed relative to rainfed agriculture that is prone to rainfall-runoff-erosion soil loss, the carbon storage pool (per year) may be stable and eventually predictable in irrigated lands. A more accurate assessment must involve accounting for yield export from place of production, CO_2 emission from processing of crops, CO_2 respiration loss from the soil substrate, and additional efflux of GHG such as CH_4.

There has been considerable debate about the benefits of surface water storage as highlighted by the World Commission on Dams case studies for the developed and developing world (Asianics Agro-Development International, 2000). Our analysis suggests that in addition to examination of dams and their viability, a river-basin wide assessment at the systems level is required to determine the vulnerability of water supply and potential benefits of irrigated agriculture to sequester carbon. None of the ten basins examined here is immune to the vagaries of extreme events or changes in means, as shown on Table 6.1. Due to high water demand on the Indus and the E&T Basins and reduced naturalized flows under future climate conditions, these basins may be particularly vulnerable.

Since irrigation application efficiencies in many parts of the developing world are relatively low (30% or less), improved application methods such as buried drip lines and low leakage canal conveyance systems may be required. Additional agronomic practices could include changes in the crop mix to drought-tolerant crops, or adoption of drought-tolerant varieties. Innovative funding mechanisms through aid agencies and donor banks will be required to bring the vital capital investments to these areas. Basins such as the Amazon-Tocantins that have geared their water supply systems to hydropower generation may have to slightly alter water management practices to account for agriculture and municipal demand. Our analysis did not take into account the planned construction of dams. Basins such as Yangtze and Danube may be at sufficient storage-outflow balance with existing dam capacities, and there may be limited additional storage possible on the Mekong without impacting sustainable future flows.

6.5 CONCLUSIONS

We have presented a method to study the water supply systems of river basins using a watershed model (SWAT) to predict naturalized flows, and a water rights analysis program (TAMU-WRAP) to assess the impact of changing naturalized flows and populations on future water supply. Development of comprehensive water storage, flows, and water rights and permits databases for all major river basins of the world would yield more realistic estimates of consequences of climate change on the surface water resources of the world. Water planners and policymakers in turn would find these estimates useful. We estimated the biomass carbon fixed by the irrigated crops in ten basins to cover both the developing and developed parts of the world. With sufficient development of databases and biophysical modeling technology, it may be feasible to both assess water and carbon budgets of the basins of the world in the not too distant future. Regional water planners should develop mitigation plans now in order to deal with the vagaries of river flows due to climate change.

ACKNOWLEDGMENTS

We are grateful to Cesar Izaurralde from the Pacific Northwest Laboratories, Maryland, and Hyrum Johnson, U.S. Department of Agriculture, Agricultural Research Service, in Temple, Texas.

REFERENCES

Arnell, N.W. 1999. Climate change on global water resources. *Global Environ. Change,* 9:S31–S49.

Arnold, J.G., P.M. Allen, and G. Bernhardt. 1993. A comprehensive surface-groundwater flow model. *J. Hydrol.,* 142:55–77.

Arnold, J.G., P.R. Srinivasan, R.S. Muttiah, and J.R. Williams. 1998. Large area hydrologic modeling and assessment, part I: model development. *J. Am. Water Res. Assoc.,* 34:73–89.

Arora, V.K. and G. J. Boer. 2001. Effects of simulated climate change on the hydrology of major river basins. *J. Geophys. Res.,* 106:3335–3348.

Asianics Agro-Development International, Ltd. 2000. Tarbela Dam and Related Aspects of the Indus River Basin, Pakistan. A WCD Case Study Prepared as an Input to the World Commission on Dams. World Commission on Dams, Capetown, South Africa. Available at: *www.dams.org/Kbase/studies/PK*

Burlando P. and R. Rosso. 1991. Extreme storm rainfall and climate change. *Atmospheric Res.,* 27:169–189.

De Costa W.A.J.M., W.M.W. Weerakoon, H.M.L.K. Herath, and R.M.I. Abeywardena. 2003. Response of growth and yield of rice (Oryza sativa) to elevated atmospheric carbon dioxide in the sub-humid zone of Sri Lanka. *J. Agron. Crop Sci.,* 189:83–95.

DeMenocal, P.B. 2001. Cultural responses to climate change during the late Holocene. *Science* 292:667–673.

Fangmeier A., B. Chrost, P. Hogy, and K. Krupinska. 2000. CO_2 enrichment enhances flag leaf senescence in barley due to greater grain nitrogen sink capacity. *Environ. Exp. Bot.,* 44:151–164.

Fearnside, P.M. 1995. Hydroelectric dams in the Brazilian Amazon as sources of "greenhouse" gases. *Environ. Conserv.,* 22:7–19.

Flato, G.M., G.J. Boer, W.G. Lee, N.A. McFarlane, D. Ramsden, M.C. Reader, and A.J. Weaver. 2000. The Candian Centre for Climate Modeling and Analysis global coupled models and its climate. *Climate Dynamics,* 16:451–467.

Food and Agriculture Organization. 1995. *Production Yearbook.* Food and Agriculture Organization, Rome.

Food and Agriculture Organization. 1997. AQUASTAT Iraq report. Available at: *www.fao.org / waicent / faoinfo / agricult / agl / aglw / aquastat / countries / iraq / index.stm.* Accessed September 24, 2003.

Galant, M. 2003. Personal communication, Volta Ltd. Consulting Engineers, Moscow, Russia. Available at: *www.volgaltd.ru.* Accessed September 24, 2003.

Giorgi, F., C.S. Brodeur, and G.T. Bates. 1994. Regional climate change scenarios over the United States produced with a nested regional climate model. *J. Climate,* 7:375–399.

Gordon, C., C. Cooper, C.A. Senior, H. Banks, J.M. Gregory, T.C. Johns, J.F.B Mitchell, and R.A. Wood. 2000. The simulation of SST, sea ice extents and ocean heat transports in aversion of the Hadley Centre Coupled Model without flux adjustments. *Climate Dynamics,* 16:147–168.

Hakala, K. 1998. Growth and yield potential of spring wheat in a simulated changed climate with increased CO_2 and higher temperature. *Eur. J. Agron.,* 9:41–52.

Hansen, J., G. Russell, D. Rind, P. Stone, A. Lacis, S. Lebedeff, R. Ruedy, and L. Travis. 1983. Efficient three-dimensional global models for climate studies: Models I & II. *Monthly Weather Rev.,* 111:609–662.

Hunsaker, D.J., B.A. Kimball, P.J. Pinter, G.W. Wall, R.L. LaMorte, F.J. Adamsen, S.W. Leavitt, T.L. Thompson, A.D. Matthias, T.J. Brooks. 2000. CO_2 enrichment and soil nitrogen effects on wheat evapotranspiration and water use efficiency. *Agric. For. Meteorol.,* 104:85–105.

International Commission for the Protection of the Danube River. 1999. Danube River Basin Pollution Reduction Program Report UNDP/GEF, 1999. International Commission for the Protection of the Danube River, Vienna International Center, Vienna, Austria.

Izzauralde, C. 2003. Personal communication, Joint Global Change Research Institute, College Park, Maryland.

Johns, T.C., R.E. Carnell, J.F. Crossley, J.M. Gregory, J.F.B. Mitchell, C.A. Senior, S.F.B. Tett, and R.A. Wood. 1997. The second Hadley Centre coupled ocean-atmosphere GCM: model description, spinup and validation. *Climate Dynamics,* 13:103–134.

Jones, P.G. and P.K. Thornton. 1999. Fitting a third-order Markov rainfall model to interpolated climate surfaces. *Agric. For. Meteorol.,* 63:1–19.

Kim, H.Y., M. Lieffering, S. Miurra, K. Kobayashi, and M. Okada. 2001. Growth and nitrogen uptake of CO_2-enriched rice under field conditions. *New Phytologist,* 150:223–229.

Kite, G. 2000. Developing a Hydrological Model for the Mekong Basin: Impacts of Basin Development on Fisheries Production. Working Paper 2. International Water Management Institute, Colombo, Sri Lanka.

Lal, R. and J.P. Bruce. 1999. The potential of world cropland soils to sequester C and mitigate the greenhouse effect. *Environ. Sci. Policy,* 2:177–185.

Leavitt, S.W., E. Pendall, E.A. Paul, T. Brooks, B.A. Kimball, P.J. Pinter, H.B. Johnson, A. Matthias, G.W. Wall, and R.L. LaMotte. 2001. Stable-carbon isotopes and soil organic carbon in wheat under CO_2 enrichment. *New Phytologist,* 150:305–314.

Muttiah, R.S. and R.A. Wurbs. 2002. Modeling the impacts of climate change on water supply reliabilities. *Water Intl.,* 27:407–419.

Ohfuchi, W. 2003. Personal communication, Japanese Marine Science and Technology Center, Tokyo, Japan.

Reddy K.R., R.R. Robana, H.F. Hodges, X.J. Liu, and J.M. KcKinion. 1998. Interactions of CO_2 enrichment and temperature on cotton growth and leaf characteristics. *Environ. Exp. Bot.,* 39:117–129.

Rosegrant, M. 2003. Personal communcation, International Food Policy Research Institute, Washington, DC.

Sarmiento, J.L. and N. Gruber. 2002. Sinks for anthropogenic carbon. *Physics Today,* August, pp. 30–36.

Shingu, S., H. Takahara, H. Fuchigami, M. Yamada, Y. Tsuda, W. Ohfuchi, Y. Sasaki, K. Kobayashi, T. Hagiwara, S. Habata, M. Yokokawa, H. Itoh, and K. Otsuka. 2002. A 26.58 Tflops Global Atmospheric Simulation with the Spectral Transform Method on the Earth Simulator. Proceedings Super Computing 2002, November 16–22, Baltimore, Maryland. IEEE Computer Society, Los Alamitos, CA.

Sperow, M., M. Eve, and K. Paustian. 2003. Potential soil C sequestration on U.S. agricultural soils. *Climate Change,* 57:319–339.

Torbert, H.A., H.H. Rogers, S.A. Prior, W.H. Schlesinger, and G.B. Runion. 1997. Effects of elevated atmospheric CO_2 in agroecosystems on soil carbon storage. *Global Change Biol.,* 3:513–521.

Wurbs, R.A. 2001. Assessing water availability under a water rights priority system. *J. Water Resour. Plann. Manage.,* 127:235–243.

Wurbs, R.A., R.S. Muttiah, F. Felden. 2003. Incorporation of climate change in water availability modeling. *ASCE, Hydrological J.*

Zinke, A. 2003. Personal communication, Zinke Environment Consulting, Vienna, Austria. Available at: *www.zinke.at.* Accessed September 24, 2003.

Ziska, L.H. and J.A. Bunce. 1997. Influence of increasing carbon dioxide concentration on the photosynthetic and growth stimulation of selected C4 crops and weeds. *Photosynthesis Res.,* 54:199–208.

7

Climate Change and Terrestrial Ecosystem Production

WILFRED M. POST AND ANTHONY W. KING

CONTENTS

7.1 INTRODUCTION

Photosynthesis and the incorporation of photosynthate into plant biomass (net primary productivity, NPP) varies spatially and through time at individual locations as a response to climate. In particular, the amount and seasonal distribution of precipitation and the length and warmth of the growing season are the primary determinants of biomass production in natural ecosystems and also for agroecosystems. Over the past century regional meteorological measurements have shown that global climate is changing (Easterling et al., 1997). Some of this change has been the result of rising atmospheric concentration of CO_2 from fossil fuel burning and land-use change (Folland et al., 2001). As CO_2 concentrations in the air continue to rise, additional solar radiation is trapped in the atmosphere and warms the planet. Over the next century, CO_2 concentrations will continue to rise and additional changes in climate are expected.

Global climate models project significant changes in temperature regimes and precipitation patterns over the next 100 years when they are forced with expected scenarios of atmospheric CO_2 concentration changes from increasing fossil fuel burning. The response of the terrestrial ecosystems to this CO_2 increase and climate change may be modeled based on information from field studies that examine the response of ecosystems to interannual climate variation and from experiments where temperature, moisture, or CO_2 concentration has been manipulated.

We have developed and employed a terrestrial biogeochemistry model that uses fundamental processes of plant and soil carbon dynamics to estimate NPP of terrestrial ecosystems over the past century and then into the future for the next 100 years. We completed two simulations with this model to examine the response of terrestrial ecosystems to climate change and rising CO_2 concentration. In both simulations, a changing climate was used to generate historical and projected estimates of NPP. In the first simulation terrestrial ecosystems were stimulated by rising atmospheric CO_2 concentrations in accordance with experimental findings. In a second simulation, we eliminated this CO_2 stimulation by running the model with a CO_2 concentration fixed at the 1930 amount. From the difference in response we are able to infer the relative effects of rising CO_2 compared to those of climate change. Terrestrial ecosystem NPP increases with rising atmospheric CO_2 concentration, and decreases with recent historical and future projected climate. The relative strength of these two opposing trends indicates whether global change will result in an increase or decrease in terrestrial ecosystem NPP over the next century.

7.2 METHODS

Ecosystem manipulations, including studies with elevated CO_2, temperature, and water or nitrogen additions provide insights into physiological, ecological, and biogeochemical processes underlying carbon storage. Long-term monitoring of ecosystem carbon fluxes and how they vary annually and seasonally with climate and between ecosystems with different environmental conditions also provide valuable information. This information must be extrapolated to regional and global scales.

The terrestrial biosphere is heterogeneous enough that simple extrapolation of a reasonable number of experiments and measurements is not sufficient. The processes that regulate carbon storage and fluxes are uniform enough that mechanistic models can be used. Global simulations are presented to indicate how this approach may be used to quantify some of the processes involved in terrestrial carbon uptake, their global magnitude, and their spatial distribution.

7.3 MODEL DESCRIPTION

Many direct and indirect interactions among interacting ecosystem processes result in responses to changing environmental conditions at many different scales of time and space. Large-scale responses over several years pose the greatest challenges to elucidate and quantify. We used the global terrestrial ecosystem carbon (GTEC) model to analyze terrestrial carbon storage and exchange with the atmosphere over the 1930–2100 period. In this model, the carbon dynamics of each 1° latitude by 1° longitude vegetated land cell are described by a mechanistic soil-plant-atmosphere model of ecosystem carbon cycling and exchange.

The GTEC model has implemented a big-leaf version of the so-called Farquhar model for C3 plants (Farquhar et al., 1980) and a similar model for C4 plants (Collatz et al., 1992). Photosynthesis is coupled to a description of the dependence of stomatal conductance on assimilation rate, temperature, and available soil moisture to form a leaf productivity model. Autotrophic maintenance respiration is a function of tissue nitrogen concentration and temperature, while growth respiration is proportional to the change in biomass.

Soil moisture is calculated using a multilayer soil with simple piston flow dynamics. Canopy photosynthesis and maintenance respiration are calculated hourly, while carbon allocation, growth, and growth respiration and soil water balance are calculated daily. Carbon in dead organic matter is partitioned as in the Rothamsted (RothC) model (Jenkinson, 1990) with litter inputs assigned to decomposable and resistant plant material compartments. The model is thus capable of responding to interactions among climate, rising atmospheric CO_2 concentration, soil moisture, and solar radiation. This detailed physiological model is considerably more sensitive to rising atmospheric CO_2 concentration than most biogeochemical terrestrial ecosystem models.

The model requires many inputs and parameters for a simulation. Some inputs remain constant during the simulations and also between different simulations. These include vegetation type for each grid cell (Table 7.1), and associated photosynthesis parameters, respiration coefficients, partitioning and turnover parameters for plant tissue allocation and longevity, and litter quality characteristics. Soil type and associated hydrological parameters are also required.

Initial carbon pools for plant and soil C compartments must also be supplied. We supplied this information by spinning the model up using climate data for the past 70 years. The model was run from very small pool sizes using yearly climate data chosen at random from the historical record until all the model compartments remained more or less constant without a strong time-dependent trend. Using the

Table 7.1 Ecosystem Types in Global Terrestrial Ecosystem Carbon Model

Terrestrial Ecosystem Type	Global Area (10^6 ha)
Broadleaf evergreen forest	1342
Broadleaf deciduous forest and woodland	330
Mixed coniferous and broadleaf deciduous forest and woodland	660
Coniferous forest and woodland	1298
High-latitude deciduous forest and woodland	575
C4 Wooded grassland	1710
C4 Grassland	893
Shrubs and bare ground	1100
Tundra	710
Desert, bare ground	1687
Cultivation	1328
C3 Wooded grassland	460
C3 Grassland	1152

Notes: Each grid square is assigned to an ecosystem type. The ecosystem types are used to assign photosynthesis, respiration, allocation, and tissue turnover parameters. Ecosystem type and associated areas may be used to sum geographically extensive fluxes and pools by report to ecosystem type as is done in Figures 7.1, 7.2, and 7.3.

simulated initial conditions, the model was then run using inputs that varied during the time-dependent simulations that included atmospheric CO_2 concentration and the climate variables temperature, precipitation, solar radiation, and relative humidity.

7.4 HISTORICAL AND FUTURE CLIMATE AND CO_2 CONCENTRATION

Monthly historical climate data from 1930 to 1996, compiled and interpolated to a regular geographic grid by New et al. (2000), were obtained from data sets maintained by the National Aeronautics and Space Administration's Distributed Active Archive Center at Oak Ridge National Laboratory (*http://daac.ornl.gov/CLIMATE/climate_collections.html*).

Projected future climate was generated by the coupled parallel climate model (PCM) (Washington and Weatherly, 1997). PCM model output for the stabilization scenario was obtained from the National Energy Research Scientific Computing Center at Lawrence Berkeley National Laboratory (*www.nersc.gov/projects/gcm_data/*). The stabilization scenario prescribes a future time-varying atmospheric greenhouse gas forcing that saturates at the equivalent of 550 ppm CO_2 by the year 2150 (Dai et al., 2001b). The PCM for the stabilization scenario simulation projects a 2°C increase in temperature and a global increase in precipitation from 3.07 to 3.17 mm per day. Compared to other coupled atmosphere–ocean climate models, these responses are at the low end of projected changes.

Monthly data for each grid cell were interpolated to daily information using algorithms outlined by Richardson and Wright (1984). The atmospheric CO_2 concentrations used in the model simulations are based on observations for 1930 to 1996. For 1997 to 2100, the atmospheric CO_2 concentrations are the same as those used in the PCM climate simulations, and were estimated using an energy economics model driven by regionally-specific assumptions regarding population growth, economic growth, energy use per capita, technology development, and so on. Details are provided in Dai et al. (2001a).

7.5 RESULTS

Figure 7.1 gives the estimated annual NPP of the most geographically extensive terrestrial ecosystems (Table 7.1) as calculated by the GTEC model using historical climate data for 1930 to 1995 and projected climate data for 1995 to 2100. The CO_2 concentration of the atmosphere used in the simulation follows observed concentrations for the historical period that are rising exponentially to 385 ppm by 1995. The rate of atmospheric CO_2 increase slows thereafter according to the stabilization scenario and reaches equilibrium of 550 ppm by 2150.

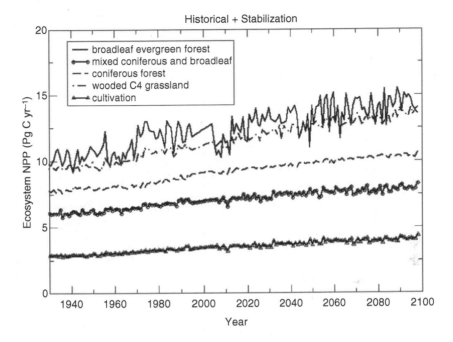

Figure 7.1   Model estimate of terrestrial net primary productivity for ecosystem types that cover large areas. Simulations are forced with historical climate and CO_2 concentration and projected climate change and CO_2. Projected climate change is output from parallel climate model using a stabilization CO_2 emission scenario.

During the historical period and into the future, even after CO_2 concentration stabilizes, the model simulations indicate an increase in terrestrial ecosystem NPP. This increase in NPP is greatest in the tropical ecosystems (broadleaf evergreen forest and wooded C4 grassland), but holds for all ecosystems including cultivation. Cultivated ecosystems are expected to show even greater increases in NPP than indicated here due to expected improvements in management (irrigation, fertilization, genetic selection and engineering, etc.) that are not included in the simulations. Tropical ecosystems show a greater interannual variation in NPP. This is the result of interannual climate fluctuations from the El

Niño–Southern Oscillation affecting the balance of ecosystem production and vegetation respiration.

Most of the increase in ecosystem NPP may be attributed to CO_2 fertilization effects. Both the direct influence on photosynthesis and indirect effect on water balance contribute to the simulated response (DeLucia et al., 2005, this volume). We can demonstrate the influence of CO_2 on the simulation results by observing what happens in simulations in which we effectively turn off the CO_2 effects. Figure 7.2 shows simulation results with historical and projected climate change, but with CO_2 held constant at the 1930 concentration. In this simulation, which shows the influence of climate alone, terrestrial ecosystem NPP remains largely constant for the period 1930 to 1990 and then begins to decline. The decline in terrestrial ecosystem NPP, due to climate alone, is most noticeable for the tropical ecosystems with a stronger decrease in NPP after 2000.

The global total NPP over all ecosystems, including those with small land areas not presented in Figures 7.1 and 7.2, shows the opposing effects of CO_2 fertilization and climate change more dramatically. With CO_2 fertilization included, global NPP increases more or less linearly, although the simulation starts to level off toward the end of the next century after the CO_2 concentration stabilizes. The influence of rising CO_2 concentration is stronger than indicated since climate change impact on NPP has counteracted some of the potential increase in NPP. This is indicated by the climate change only simulation results that show a decrease in NPP from 47 to 40 Pg C year^{-1} (Pg = petagram = 1 gigaton). With a CO_2 effect included, NPP increased to 63 Pg C year^{-1}. If we add the estimated decrease of 7 Pg C year^{-1} due to climate change, we estimate that the effect of CO_2 alone would have increased NPP to nearly 70 Pg C year^{-1} by 2100. The effect of CO_2 therefore was to counteract a 15% decrease in global NPP due to climate change and result in a net 34% increase in NPP by the end of the 21st century.

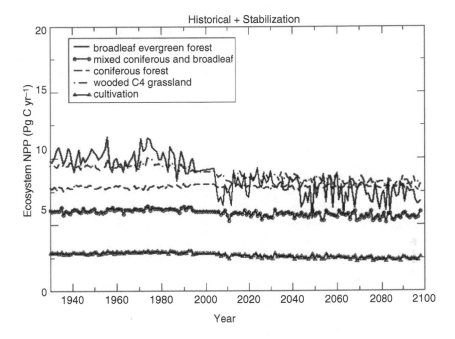

Figure 7.2 Model estimate of terrestrial net primary productivity forced with historical climate and projected climate estimated using output from the parallel climate model as in Figure 7.1 but without a CO_2 fertilization effect. The effect of rising atmospheric CO_2 concentration on photosynthesis, in this simulation, is effectively turned off by holding CO_2 constant at the 1930 concentration.

7.6 SUMMARY

Terrestrial NPP response to expected global changes has two opposing trends: (1) NPP increases with rising atmospheric CO_2; and (2) NPP decreases with recent and projected climate change. The relative balance between these two trends determines the long-term response of the global carbon cycle to global changes. This balance also determines the physiological response of ecosystems to altered environmental factors that influence productivity. This applies also for agricultural ecosystems from which we derive food and biomass for fuels, and

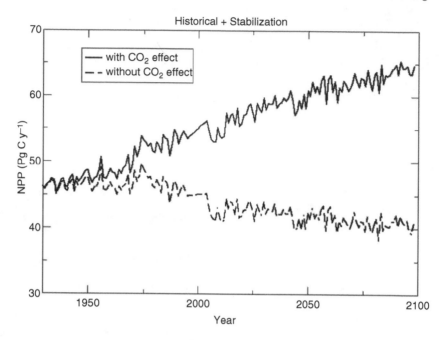

Figure 7.3 Global total estimate net primary production using the Oak Ridge National Laboratory global terrestrial ecosystem carbon (ORNL-GTEC) model. Two scenarios are presented. Both are forced with historical and projected climate change. One run used a constant CO_2 concentration, and the other includes a response to historical and projected future CO_2 concentration.

for forest ecosystems from which we derive wood products, fiber, and some energy. Ecosystems with greatest NPP have the greatest expected change with respect to CO_2 and climate, both for increases with rising CO_2, and decreases with climate change.

Other factors to be considered when projecting future global NPP that are not considered in the analysis here include historical and current land-use patterns, ecosystem N cycling and impact of atmospheric N-deposition, changes in hydrology of wetland ecosystems, and potential changes in fire frequency, insect outbreaks, and other disturbances that could be associated with climate changes. We have presented only the main biochemical and biophysical responses of terrestrial ecosystems to recent historical and future anticipated global change. The analysis is thus not yet comprehensive.

In our simulations, the changes in NPP were driven by historical and prescribed global changes in atmospheric CO_2 concentration and associated climate change. The simulated changes in NPP are large enough to influence the global carbon cycle, and therefore potentially to feed back and change atmospheric CO_2 and alter climate further. The simulated CO_2 fertilization response would result in a negative feedback with simulated NPP resulting in the potential to increase carbon sequestration in terrestrial ecosystems reducing atmospheric CO_2. On the other hand, the simulated response to climate change alone suggests the possibility of a positive feedback where additional warming leads to greater decreases in NPP that result in higher CO_2 concentrations. The relative strengths of positive and negative feedbacks will determine the relative contribution of terrestrial ecosystem ecophysiology to enhancement or mitigation of climate change.

Other climate models are likely to project stronger decreases in NPP, and other ecosystem biogeochemistry models (Pan et al., 1998) are known to have more subdued responses to increased CO_2. In combination, responses to future global change could range from the negative or stabilizing response indicated with the detailed processes models used in this study, to positive or destabilizing responses. Unstable positive feedback where climate change reductions in NPP become larger and the increase in NPP by CO_2 fertilization become saturated is, however, possible (Cox et al., 2000). Our simulations indicate that this outcome is less likely than stabilizing negative feedback, as suggested by the coupled carbon cycle-climate change simulations by Freidlingstein et al. (2001).

ACKNOWLEDGMENTS

This research used resources of the Center for Computational Sciences at Oak Ridge National Laboratory, and was supported by the U.S. Department of Energy, Office of Science. Oak Ridge National Laboratory is managed by University of Tennessee–Battelle, LLC, for the U.S. Department of Energy under contract DE-AC05-00OR22725.

REFERENCES

Collatz, G.J., M. Ribas-Carbo, and J.A. Berry. 1992. Coupled photosynthesis-stomatal conductance model for leaves of C_4 plants. *Aust. J. Plant Physiol.,* 19:519–538.

Cox, P.M., R.A. Betts, C.D. Jones, S.A. Spall, and I.J. Totterdell. 2000. Acceleration of global warming due to carbon-cycle feedbacks in a coupled climate model. *Nature,* 408:184–187.

Dai, A., T.M.L. Wigley, B.A. Boville, J.T. Kiehl, and L.E. Buja. 2001a. Climates of the twentieth and twenty-first centuries simulated by the NCAR Climate System Model. *J. Climate,* 14:484–519.

Dai, A., G.A. Meehl, W.M. Washington, T.M.L Wigley, and J.A. Arblaster. 2001b: Ensemble simulation of 21st-century climate changes: business as usual vs. CO_2 stabilization. *Bull. Am. Meteorol. Soc.,* 82:2377–2388.

DeLucia, E.H., D.J. Moore, J.G. Hamilton, R.B. Thomas, C.J. Springer, and R.J. Norby. 2005. Chapter 8, this volume.

Easterling, D.R., B. Horton, P.D. Jones, T.C. Peterson, T.R. Karl, D.E. Parker, M.J. Salinger, V. Razuvayev, N. Plummer, P. Jamason, and C.K. Folland. 1997. Maximum and minimum temperature trends for the globe. *Science,* 277:364–367.

Farquhar, G.D., S. von Cammerer, and J.A. Berry. 1980. A biochemical model of photosynthetic CO_2 assimilation in leaves of C_3 species. *Planta,* 149:78–90.

Folland, C.K., T.R. Karl, J.R. Christy, et al. 2001. Observed climate variability and change. In J.T. Houghton, et al., Eds. *Climate Change 2001: The Scientific Basis.* Cambridge University Press, London; New York, pp. 99–181.

Friedlingstein, P., L. Bopp, P. Ciais, J.-L. Dufresne, L. Fairhead, H. LeTreut, P. Monfray, and J. Orr. 2001. Positive feedback of the carbon cycle on future climate change, *Geophys. Res. Lett.,* 28:1543–1546.

Jenkinson, D.S. 1990. The turnover of organic-carbon and nitrogen in soil. *Philos. Trans. R. Soc. London B Biol. Sci.,* 329:361–368.

New, M.G., M. Hulme, and P.D. Jones. 2000. Representing twentieth century space-time climate variability. Part II: Development of a 1901–1996 monthly terrestrial climate field. *J. Climate,* 13:2217–2238.

Richardson, W.W. and D.A. Wright. 1984. *WGEN: A Model for Generating Daily Weather Variables.* Publication ARS–8. Available from U.S. Department of Agriculture, Agricultural Research Service, Temple, TX.

Pan, Y.D., J.M. Melillo, A.D. McGuire, D.W. Kicklighter, L.F. Pitelka, K. Hibbard, L.L. Pierce, S.W. Running, D.S. Ojima, W.J. Parton, and D.S. Schimel. 1998. Modeled responses of terrestrial ecosystems to elevated atmospheric CO2: a comparison of simulations by the biogeochemistry models of the Vegetation/Ecosystem Modeling and Analysis Project (VEMAP). *Oecologia,* 114:389–404.

Washington, W.M. and J.W. Weatherly. 1997. Simulations with a climate model with high-resolution ocean and sea ice. In *Polar Processes and Global Climate: Draft Summary Report from the Conference on Polar Processes and Global Climate.* WMO/ICSU/IOC World Climate Research Programme. International Arctic Climate System Study (ACYSYS) Project Office, Oslo, Norway, pp. 250–252.

8

The Changing Role of Forests in the Global Carbon Cycle: Responding to Elevated Carbon Dioxide in the Atmosphere

EVAN H. DeLUCIA, DAVID J. MOORE,
JASON G. HAMILTON, RICHARD B. THOMAS,
CLINT J. SPRINGER, AND RICHARD J. NORBY

CONTENTS

8.1 INTRODUCTION

Worldwide, forests have an enormous impact on the global C cycle. Of the 760 gigatons (10^{15} g, Gt) of C in the atmosphere, photosynthesis by terrestrial vegetation removes approximately 120 Gt, almost 16% of the atmospheric pool each year, and about half of this amount (56 Gt) is returned annually by plant respiration (Figure 8.1). The difference between gross canopy photosynthesis and plant respiration (see below) is defined as net primary production (NPP), and represents the annual production of organic matter that is available to consumers. Although estimates vary considerably, forests make up almost half of the global NPP, and approximately 80% of the terrestrial NPP (Figure 8.2). Thus, small changes in the capacity of forests to remove C from the atmosphere by photosynthesis, or return it to the atmosphere by respiration, or store it in wood and soils greatly affect the distribution of C between the terrestrial and atmospheric pool. Because trees use the C_3 pathway of photosynthesis, they are very responsive to increases in atmospheric CO_2, and it has been hypothesized that a stimulation of photosynthesis and growth of trees may reduce the rate of accumulation of C in the atmosphere derived from fossil fuels. Mounting evidence suggests that a significant portion of the imbalance in the global C cycle, the 2.8 Gt year^{-1} that is unaccounted for when all known sinks are subtracted from known sources (Figure 8.1), may be explained by additional C uptake in temperate forests (Fan et al., 1998; Pacala et al., 2001; Janssens et al., 2003). How much of this sink is derived from land use change vs. growth enhancement of trees by elevated CO_2, nitrogen deposition, and changes in climate remains uncertain.

The combustion of fossil fuels and other human activities, including deforestation and other changes in land use, is driving an imbalance in the global C cycle. Prior to the Industrial Revolution, the concentration of CO_2 in the atmosphere was approximately 280 μl l^{-1}, and it was at this level for at least the previous 1000 years (Houghton et al., 1996). The injection of CO_2 into the atmosphere by the widespread combustion of fossil fuels currently adds approximately 6.4 Gt C to the

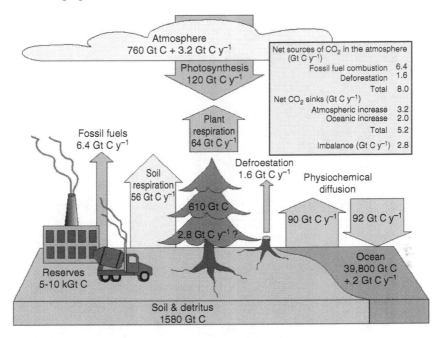

Figure 8.1 The global carbon cycle. Anthropogenic emissions are causing the amount of C in the atmosphere to increase by approximately 3.2 Gt year^{-1}. The cement plant and truck represent anthropogenic fluxes of C to the atmosphere caused by the combustion of fossil fuels during cement production, and the tree stump represents the contribution of changes in land use, primarily deforestation. The mass balance indicates a missing C sink of approximately 2.8 Gt. (Values compiled by K.L. Griffin from Field [2001], Prentice et al. [2001], and Schimel et al. [2001]. Drawing courtesy of K.L. Griffin, Lamont-Doherty Earth Observatory of Columbia University. With permission.)

atmosphere each year, and deforestation contributes another 1.6 Gt (Figure 8.1). About half of this anthropogenic CO_2 remains in the atmosphere. Carbon dioxide is a potent greenhouse gas; along with water vapor, methane and other gases, it maintains the habitable temperatures on Earth, but its further accumulation in the atmosphere is the primary driver of global warming. During 2002, the CO_2 concentration in the atmosphere was ~373 µl l^{-1}, and it is expected to double from

DeLucia et al.

Field - 1998   Whittaker - 1975

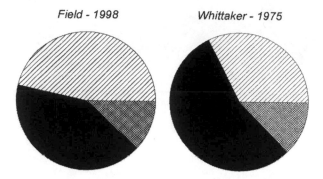

Figure 8.2 Two estimates of the contribution of forest ecosystems to global net primary production (NPP); black represents forest NPP, fine cross-hatch represents NPP from other terrestrial ecosystems and diagonal hatch represents ocean NPP. Field et al. (1998) combined estimates of radiation-use efficiency with satellite estimates of foliage cover to calculate NPP (global value: 105 Pg C year^{-1}). Whittaker (1975) derived NPP by census and biometric methods (Lieth and Whittaker, 1975) (global value: 85 Pg C year^{-1}, assuming C = 0.5 dry biomass). Tropical savannas were included as forest in both estimates.

its pre-Industrial level to ~560 μl l^{-1} during the 21st century. Recent estimates suggest that a doubling of atmospheric CO_2 will force a 1.4°C to 5.8°C increase in global mean temperature (Houghton et al., 2001); this magnitude of warming is similar to the increase that occurred from the peak of the last ice age, approximately 15,000 years ago. But current climate change is happening over a much shorter time scale, a mere 50 to 100 years, and is far too rapid for many biological and ecological systems to adapt to the change.

The objectives of this chapter are to describe the major components of the terrestrial C cycle with an emphasis on current uncertainties in estimating these components, particularly for forest ecosystems; and, to compare the responses of two contrasting forest ecosystems to an experimental increase in atmospheric CO_2.

8.2 CARBON FLOW IN FOREST ECOSYSTEMS: POOLS AND FLUXES

Carbon is transferred through and stored in ecosystems by a myriad of physiological, ecological, and geochemical processes (Schlesinger, 1997; Clark et al., 2001) that may respond independently to the different facets of global change. Microbial respiration in the soil, for example, is extraordinarily sensitive to temperature, whereas photosynthesis responds strongly to changes in both atmospheric CO_2 and temperature. Predicting the effects of changing climatic conditions and atmospheric chemistry on the C cycle requires a clear understanding of these different processes. Broadly defined, C cycles comprise pools and fluxes, where a pool is a C reservoir lasting 1 year or longer (Hamilton et al., 2002), and fluxes represent the rates of C transfer from one pool to another. C pools typically are expressed per unit land area; fluxes are expressed per unit land area per annum. Although the major processes (fluxes) and pools in ecosystem C cycles were identified over 30 years ago (Whittaker, 1975), quantification of many of these processes remains clouded with uncertainty, and few studies have attempted to "close" the C budget for individual forest stands. The following discussion presents the currently held definitions for the major process regulating the flow of C through forest ecosystems and highlights a few of the most important uncertainties in their quantification.

Carbon enters ecosystems from the atmosphere by net photosynthesis, and is expressed as gross primary production (GPP) (Figure 8.1). Here photosynthesis is defined as the net reduction of CO_2 by plant canopies to sugars and structural materials. This net value represents C fixed by the primary carboxylating enzyme in chloroplasts, rubisco (rubisco: ribulose bisphosphate carboxylase-oxygenase [EC 4.1.1.39]), minus the amount immediately returned to the atmosphere by photorespiration, but does not include oxidative respiration in mitochondria. GPP cannot be directly measured because of the need to separate C uptake by photosynthesis and losses by mitochondrial respiration. At the stand or ecosystem level,

GPP typically is estimated by scaling appropriate leaf-specific measurements in space and time (Wofsy et al., 1993), by process-based models (Aber et al., 1996; Luo et al., 2001), or calculated from stand-level estimates of transpiration (Schäfer et al., 2003). Alternatively, the sum of all subordinate C increments and losses can be used to calculate the rate of GPP necessary to meet these demands (Ryan et al., 1996; Hamilton et al., 2002).

Particularly in monospecific stands where obtaining representative physiological parameters and a physical description of the canopy are relatively straightforward, process-based models provide an effective means of calculating GPP. In many models, the strength of this approach stems from the rigorous and mature theory relating the biochemistry of carbon fixation to leaf biochemical properties and environmental conditions (von Caemmerer, 2000).

Once CO_2 becomes chemically reduced by photosynthesis, carbohydrates are expended to meet metabolic needs of plants and the corresponding rates of autotrophic respiration (R_a) result in a substantial return of C back to the atmosphere, globally amounting to ~64 Gt yr^{-1} (Figure 8.1). Carbohydrates consumed in oxidative respiration by mitochondria are used to support the maintenance and construction of plant tissues in roots, stems, and foliage. Assumed to be 50% of GPP (Waring et al., 1998), R_a can exceed 70% of GPP in some forests (Hamilton et al., 2002) and varies with forest age (Makela and Valentine, 2001). Because of the shear magnitude of R_a and its strong sensitivity to temperature, it is an important process defining the capacity of forests to store atmospheric C (Valentini et al., 2000).

The calculations of respiratory fluxes at the ecosystem level are made by scaling tissue specific rates, usually as a function of tissue nitrogen content and temperature (Ryan, 1995; Ryan et al., 1996; Hamilton et al., 2002; Meir and Grace, 2002), or measured directly at the stand level by micrometeorological methods (Baldocchi et al., 1988; Baldocchi, 2003). Unfortunately, the current understanding of environmental and physiological regulation of R_a is far from complete.

In contrast to photosynthesis, there is no theory for predicting variation in respiration rates for individual tissues. Respiration used to maintain cellular integrity, build ion gradients, and transport materials is highly sensitive to temperature, and this sensitivity results in a strong relationship between short-term variation in temperature and the rates of CO_2 evolution by plant tissues (Atkin and Tjoelker, 2003). The temperature dependence of R_a provides a convenient tool for extrapolating tissue-specific rates to the stand level. However, the rate of oxidative respiration also varies with the supply of carbohydrates, resulting in a linkage to the rate of photosynthesis and to the utilization of carbohydrates by sink tissues (Dewar et al., 1999; Atkin and Tjoelker, 2003). Under this form of control, respiration rates should vary with photosynthesis, as is implied by the observed correlation between R_a and GPP (Waring et al., 1998). The absence of a clear understanding of when seasonally and when during ecosystem development that R_a is under temperature or substrate control is a major impediment to estimating its response to current and future conditions.

The origin of respired CO_2 is not always clear, which contributes additional uncertainty to our estimates of ecosystem R_a. The fraction of soil respiration derived from plant roots vs. soil microbes (Andrews et al., 1999), the capacity of C to move in solution in the transpiration stream through trees (Teskey and McGuire, 2002), and "contaminate" estimates of bole respiration, and the activity of mitochondrial respiration during photosynthesis (Loreto et al., 2001; Wang et al., 2001) are just a few of the uncertainties that undermine quantitative estimates of R_a. An improved understanding of the origins and regulation of plant respiration under field conditions will greatly enhance the accuracy of forest carbon budgets.

Net primary production (NPP) is the difference between GPP and R_a; in addition to providing energy in the form of reduced C compounds for nonphotosynthetic organisms, NPP contributes to the accumulation of C in ecosystems. In contrast to GPP and R_a, the annual increment of woody tissue, a major component of NPP in forest ecosystems, can be estimated directly from measurements of diameter growth and allometric relationships. However, a number of processes, some small and some large, often are not included and their absence may contribute to substantial underestimates of NPP (Clark et al., 2001). NPP includes the annual production of foliage as well as coarse and fine roots. Estimating fine root production and turnover is inherently difficult (Nadelhoffer and Raich, 1992; Publicover and Vogt, 1993; Hertel and Leuschner, 2002), which is compounded by high spatial variability within forests. The contribution of fine root production to NPP range from 33% to 67% (Jackson et al., 1997; Grier et al., 1981; Santantonio and Grace, 1987), and recent evidence that root longevities may have been substantially underestimated (Matamala et al., 2003) will likely alter these previous estimates. The production of short-lived materials (less than 1 year), including losses of dissolved organic carbon, and herbivory and volatilized organic compounds should be included (Clark et al., 2001), although with the exception of leaf litter, these other elements tend to be relatively small proportion of NPP. Herbivory, fine root mortality, and losses of volatile and nonvolatile organic compounds contributed to less than 10% of NPP in a rapidly growing loblolly pine plantation (Hamilton et al., 2002).

Globally, heterotrophic respiration (R_h) returns approximately 56 Gt C to the atmosphere each year; similar to the amount from R_a and almost ten times more than the amount of C injected into the atmosphere annually from the combustion of fossil fuels (Figure 8.1). Most of this flux is derived from rhizosphere and soil organisms including bacteria, fungi, and soil invertebrates. Estimates of R_h suffer from many of the same scaling issues as R_a, and present an additional formidable challenge — separating R_h from root-derived R_a. Carbon dioxide evolved from soils is derived from a combination of autotrophic respiration from plant roots and soil microorganisms, and assigning this C to the proper source is critical for determining NPP and NEP. Mycorrhizal fungi and bacteria living in the sphere of influence of fine roots, the rhizosphere, raise an additional problem. Should they be considered part of the root or part of the soil C budget? Wiant (1967) argued that root respiration should include all processes oxidizing plant-derived organic compounds in and on the surface of fine roots, including mycorrhizal fungi and microorganisms oxidizing root exudates.

Hanson et al. (2000) identified three general approaches to quantify the contribution of R_a and R_h to soil CO_2 efflux. The first approach, "component integration," involves physically isolating and measuring the individual fluxes of different components of the plant–soil system and then adding them up. Hamilton et al. (2002) and George et al. (2003) used a variation of this approach to quantify the belowground C budget for a pine forest and a sweetgum forest exposed to elevated CO_2; R_h was calculated as the difference between total soil CO_2 efflux and the respiration rate of unearthed fine roots. The potential effects of disturbing the plant-soil system are the primary limitations of this approach. The second category includes various methods of "excluding roots," by inserting barriers or digging trenches in the soil, physically removing roots, or measuring soil CO_2 fluxes from soil under large canopy gaps where the influence of plant roots presumably is minimal. Disturbance effects and large increases in CO_2 efflux derived from rapid decomposition of newly severed roots are potential limitations of these methods. The third approach includes various methods of "isotope labeling," where intrinsic variation in soil and root C isotopic composition or the introduction of a label is used to identify the source of respired CO_2. Andrews et al. (1999) used the sudden exposure of an intact pine forest to ^{13}C-depleted air to determine that roots contributed 55% of total soil respiration at the surface late in the growing season.

Estimates of the contribution of R_a derived from roots to total soil CO_2 efflux vary considerably. In a literature survey, Hanson et al. (2000) reported that the median value for the contribution of R_a to total soil CO_2 efflux was 50% to 60%. However, ~31% of the studies in this survey reported a percent contribution of fine roots to soil CO_2 efflux of at least 60%, and ~20% of the studies reported a proportional root contribution of at least 40%. A recent study of a rapidly growing pine plantation indicated that R_h was ~22% of total soil CO_2

efflux, but increased sharply in plots exposed to elevated levels of atmospheric CO_2 (Hamilton et al., 2002).

Net ecosystem production (NEP) is of great importance to our understanding of the transfers of C between the atmosphere and terrestrial ecosystems as it represents the net accumulation of C in ecosystems, and thus provides a measure of C sequestration. As defined by Woodwell and Whittaker (1968), NEP is the difference between NPP and heterotrophic respiration (R_h). This definition is incomplete in that it does not include a number of nonrespiratory fluxes, such as C losses as volatile organic carbon and methane (Randerson et al., 2002). While the absolute magnitude of these nonrespiratory losses often is small (e.g., Hamilton et al., 2002), the impact of these molecules can be quite large. For example, the volatile organic compound isoprene is a precursor to tropospheric ozone, a potent oxidant with enormous potential to reduce ecosystem productivity. A more inclusive definition of NEP, therefore, is simply the change in total C stocks in a given ecosystem over time (Randerson et al., 2002). Net biome production (NBP) (Schulze and Heimann, 1998) is functionally equivalent to NEP, but applies to regional C increments and losses from fire, harvest, and other episodic disturbances. Although mathematically simple, quantifying NEP represents a formidable challenge.

Setting aside modeling approaches (Aber et al., 1996; Kicklighter et al., 1999; Waring and McDowell, 2002) and inversion methods based on static gas sampling of the atmosphere (Pacala et al., 2001), estimation of net C storage in ecosystems takes one of three approaches: (1) direct measurement of C fluxes, either by chamber or micrometeorological methods; (2) physiological scaling in combination with biometric measurements; or (3) direct measurement of C stocks over time, either within a site or along a chronosequence. The chamber and micrometeorological methods require a continuous record of C fluxes for a given ecosystem over an entire year that is either measured or backfilled with a model. The annual integral of net ecosystem exchange by these methods is equivalent to NEP.

Chamber-based measurements are restricted to low-stature communities (e.g., Drake et al., 1996; Shaver et al., 1998; Dore et al., 2003; Obrist et al., 2003) and require intermittent sampling, while the eddy flux method provides a continuous record of CO_2 fluxes over relatively large land areas (Baldocchi et al., 1988; Curtis et al., 2002; Baldocchi, 2003). For forest ecosystems, this latter micrometeorological approach is not amenable to comparative studies involving relatively small plots. In free-air CO_2 enrichment (FACE) experiments, for example, where the plot size exposed to ambient or elevated levels of atmospheric CO_2 is less than 1/10 ha (DeLucia et al., 1999; Norby et al., 2002), the scaling of physiological fluxes combined with biometric estimates of standing biomass (O'Connell et al., 2003) or direct measurement of changes in C stocks (Boone et al., 1988; Lichter, 1998) are more appropriate for estimating NEP. In accounting for changes in C stocks through time, particular attention is paid to changes in live and dead vegetation, the major components of NEP, with smaller changes in forest floor and soil C (Boone et al., 1988; Hamilton et al., 2002). Approaches to estimating NEP from physiological scaling suffer from the same uncertainties employed in estimating R_a and R_h discussed above. Changes in most C stocks can readily be measured on annual or greater time steps, but often it is difficult to detect change in soil C pools over relatively short periods.

8.3 ECOSYSTEM RESPONSES OF PINE AND SWEETGUM FORESTS TO ELEVATED CO_2

Although there is a rich understanding of the response of potted plants and small trees to elevated CO_2, until recently experiments had not been conducted at an appropriate spatial and temporal scale to examine the effect of elevated CO_2 on ecosystem processes regulating the C cycle. With the development of FACE technology (Hendrey et al., 1999; McLeod and Long, 1999; Miglietta et al., 2001; Okada et al., 2001), it became possible to elevate atmospheric CO_2 in large plots in intact ecosystems without altering other microclimatic variables and without restricting the movement of animals, including important herbivores. Initially employed in agricultural systems (Hendrey and Kimball, 1994; Kimball et al., 2002), approximately 24 FACE experiments currently are underway in nonagricultural ecosystems, ranging from deserts and grasslands to large-stature forests (Nowak et al., 2004). Two of the longest running forest experiments, a loblolly pine (*Pinus taeda* L.) plantation (DeLucia et al., 1999; Naidu and DeLucia, 1999) and a sweetgum (*Liquidambar styraciflua* L.) plantation (Norby et al., 2001), provide a unique opportunity to examine the responses of contrasting evergreen and deciduous forest ecosystems, respectively, to elevated atmospheric CO_2.

Loblolly pine and sweetgum trees are both early successional species of southeastern forests in North America and often compete with one another following agricultural abandonment, with sweetgum favoring moister soils (Keever, 1950). Although these species share similar life history characteristics, the difference in leaf and fine root longevity may directly alter the retention and cycling of C in these different forests, and may further affect the C cycle indirectly by altering the rate of ecosystem nitrogen transformations. Foliage of loblolly pine, an evergreen species, lives for approximately 18 months, whereas sweetgum is a deciduous species, and the leaves live for 6 months or less. Similarly, longevity of loblolly pine fine roots is about 3.4 times longer than that of sweetgum fine roots (Matamala et al., 2003).

In the Duke Forest FACE experiment, 30-m diameter plots in a loblolly pine forest have been exposed to ambient plus 200 μl l^{-1} CO_2 almost continuously since late 1996. This forest, located near Chapel Hill, North Carolina (35° 58' N, 79° 05' W), is on heavily weathered clay-rich Alfisol soils with relatively low nitrogen and phosphorus availability (Schlesinger and Lichter, 2001; Hamilton et al., 2002). Trees were 13 years old when fumigation was initiated. The experimental sweetgum plantation located on the Oak Ridge National Environmental Research Park in Roane County, Tennessee (35° 54' N, 84° 20' W) was established on moderately well-drained, silty-clay-loam soils classified as an Aquatic Hapludult; these soils are somewhat richer in nutrients than the pine forest in North Carolina (Norby et al., 2001; George et al., 2003). The sweetgum trees were 10 years old at the initiation of fumigation, slightly younger than the pine forest, and daytime CO_2 concentration in the experimental plots has averaged approximately 550 µl l^{-1} during the growing season since April 1998. These CO_2 treatments were chosen for the two experiments because, based on current projections, this CO_2 level is anticipated by 2050 (Houghton et al., 2001). Both experiments use the same FACE technology (Hendrey et al., 1999) and include fully instrumented control plots.

Although these experiments employed similar technology to comparably sized forest stands at similar developmental stages, there are important differences between them, and direct comparisons of the results, particularly of the absolute values of various C pools and fluxes, must be treated cautiously. In addition to having a less diverse community of plants in the understory, the sweetgum experiment experiences cooler temperatures and is established on more nutrient rich soils than the pine experiment (Zak et al., 2003). Moreover, estimates of the major ecosystem pools and fluxes of C were made by different investigators who used, in some cases, different scaling approaches and measurements. With these caveats in mind, these contrasting forests provide the most direct and comprehensive comparison of the response of different forests types (evergreen and deciduous) to elevated CO_2 currently available.

Exposure to elevated CO_2 substantially increased C storage and cycling in these forests (Table 8.1), but the magnitude of stimulation for different components of the carbon budget varied considerably. Gross primary production was stimulated by elevated CO_2 to a similar amount (18% to 23%) in the pine and sweetgum plantations. Neither experiment included direct measurement of GPP. In the pine experiment, GPP was estimated as NEP plus ecosystem respiration, which included the sum of the major respiratory C losses from plants and microbes ($R_e = R_{soil} + R_{wood} + R_{canopy} + \text{herbivory}_{aboveground} + $ dissolved inorganic carbon [DIC]); for sweetgum, it was calculated as NPP plus plant respiration ($R_a = R_{wood} + R_{canopy} + R_{fine\ root}$). Carbon losses by herbivory$_{aboveground}$ and DIC in the

Table 8.1 Carbon Budgets (g C m^{-2} year^{-1}) for Loblolly Pine and Sweetgum Forests Under Ambient and Elevated Atmospheric CO_2

	Pine			Sweetgum		
	Ambient	Elevated	% Δ	Ambient	Elevated	% Δ
GPP	2371	2805	18	1952	2377	23
R_a	1704	1604	−6	970	1240	32
NPP	705	897	27	972	1137	17
R_h	216	574	166	622	680	9
NEP	428	602	41	401	512	28

Note: GPP = gross primary production, Ra = plant respiration, NPP = net primary production, R_h = heterotroph respiration and NEP = net ecosystem production. Each value represents an average of three (pine) or two (sweetgum) experimental plots. The percent difference between ambient and elevated CO_2 plots is indicated by % Δ. The budget was calculated for plots exposed to elevated CO_2 for 2 years for pine and 3 years for sweetgum.

pine plantation were small and can be ignored (Hamilton et al., 2002). Although a quantitative analysis of the sources of error in these estimates for either forest is not available, it is important to note that a number of simplifying and perhaps imprecise assumptions were employed in scaling the respiratory fluxes measured for individual tissues at a given instant in time to annual values for the entire ecosystem.

Plant respiration (R_a) includes C losses from wood (R_{wood}; stems, branches, and coarse roots); foliage (R_{canopy}), and fine roots ($R_{fine\ root}$). The proportion of GPP lost by R_a appeared greater in the pine (57% to 72%) than in the sweetgum forest (34% to 52%; Table 8.1), but was stimulated by elevated CO_2 only in the sweetgum stand. With the exception of foliage, annual respiratory losses were greater from pine wood and fine roots than for sweetgum. Greater tissue specific rates of leaf respiration for sweetgum (Tissue et al., 2002) than for pine (Hamilton et al., 2001) contributed to slightly higher R_{canopy} in the former (560 to 570 g C m^{-2} year^{-1}) than in the latter (463 to 492 g C m^{-2} year^{-1}), even though peak canopy mass was approximately twice as large in the pine (1054 to 1105 g dry matter [DM] m^{-2}) (DeLucia et al., 2002) than in the sweetgum (486 to 553 g DM m^{-2}) (Norby et al., 2003) forest. Respiration from pine stems, branches, and coarse roots (ambient plots: 488 g C m^{-2} $year^{-1}$; elevated plots: 519 g C m^{-2} $year^{-1}$) (Hamilton et al., 2002), although unaffected by CO_2 was considerably greater than for sweetgum (ambient plots: 150 g C m^{-2} $year^{-1}$; elevated plots: 230 g C m^{-2} y^{-1}) (R.J. Norby, 2004, unpublished results, based on Edwards et al., 2002). As with stems, C losses by $R_{fine\ root}$ were higher in the pine forest than the sweetgum forest. Although maintenance respiration per unit root mass was slightly greater in sweet-gum than for pine, the average annual standing biomass of fine roots was two- to three-fold greater in the pine forest (George et al., 2003).

Although it is becoming increasingly evident that short-duration changes in atmospheric CO_2 do not affect tissue-specific respiration rates (Hamilton et al., 2001; Davey et al., 2003), substantial increases in R_{wood} and $R_{fine\ root}$ for trees grown under elevated CO_2 contributed to an increase in R_a in the sweetgum forest (Table 8.1). Increased biomass increment and substrate levels under elevated CO_2 caused a 23% increase in growth respiration and a 48% increase in main-tenance respiration, respectively, for sweetgum stems (Edwards et al., 2002). This stimulation was driven in part by greater wood production under elevated CO_2. Given that wood production in the pine forest also was stimulated, it is curious that this forest did not exhibit an increase in R_{wood}. The answer may be found in the different assumptions about the relative respiration rates of branches vs. boles in these two forests.

Although absolute respiratory losses by fine roots appear lower in the sweetgum forest than the pine forest, elevated CO_2 caused a substantial increase in $R_{fine\ root}$ only in the former (ambient plots: 245 g C m^{-2} $year^{-1}$; elevated plots: 455 g C m^{-2} $year^{-1}$) (George et al., 2003). Tissue-specific rates of respira-tion for sweetgum were unaffected, but a 73% increase in the standing mass of fine roots contributed to the large stimula-tion of $R_{fine\ root}$ for this species exposed to elevated CO_2.

Greater respiratory losses may have contributed to lower NPP in the pine forest relative to sweetgum, but NPP was substantially increased by elevated CO_2 in both forests (Table 8.1). Values of NPP have been calculated somewhat differently at both sites, but these differences have relatively little effect on the absolute values and the magnitude of the treatment effect. For the pine forest, NPP was calculated as the sum of biomass increments (I_{wood} + I_{leaf} + $I_{coarse\ root}$ + $I_{fine\ root}$), plus the major inputs to detritus, litterfall, and fine root turnover ($D_{litterfall}$ + $D_{fine\ root}$), plus losses as dissolved organic carbon (DOC) in the soil (Hamilton et al., 2002). For the deciduous sweetgum trees, I_{leaf} is 0, and root production was calculated directly from minirhizotron analysis rather than from $I_{fine\ root}$ plus $D_{fine\ root}$. DOC was not measured. Given these differences and that these forests experience different edaphic factors and climatic regimes, it is not possible to conclude that elevated CO_2 caused a greater stimulation of NPP for the pine forest (27%) than for the sweetgum forest (17%). In 2000, when the C budget for sweetgum was calculated, R_a was stimulated by elevated CO_2, which provides a plausible explanation for why this forest may have experienced a lesser stimulation of NPP for that year. However, the percent stimulation for sweetgum varied between 16% and 38% depending on the year (Figure 8.3). Calculating long-term averages of the behavior of the more labile C pools will strengthen the direct comparisons of the response of these forests to elevated CO_2.

Perhaps more important than the potential differences in the magnitude of the response between these contrasting forest types is the observation that the distribution of the response to elevated CO_2 among various C stocks was quite different. Enhanced wood production was the primary factor increasing NPP for the pine forest exposed to elevated CO_2 (DeLucia et al., 1999; Hamilton et al., 2002), while the treatment caused a substantial shift in C allocation in sweetgum (Norby et al., 2002). After the first year of exposure to elevated CO_2, the stimulation of wood production in sweetgum abated and was replaced by an equivalent increase in fine root production. If these differences are sustained they have important implications for the forest products industry as well as the future role of forests in the global C cycle. A stimulation of

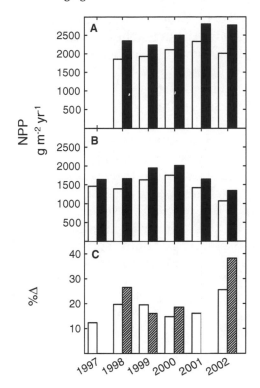

Figure 8.3 Net primary production (NPP; g DM m^{-2} y^{-1}) for experimental plots in a loblolly pine forest (A) and sweetgum forest (B) exposed to ambient (~370 μl l^{-1}, open bars) and elevated (~570 μl l^{-1}, closed bars) levels of atmospheric CO_2. The percent stimulation of NPP (pine, open bars; sweetgum, hatched bars) is illustrated in (C). NPP was calculated as the sum of woody biomass increment and annual litterfall. In the pine forest, the treatment was initiated in August 1996, and some of the 1997 litter was formed before the initiation of the treatment. (Data from D. Moore, E. DeLucia, and R. Norby, unpublished, 2004.)

harvestable wood production, particularly in young pine stands, will be beneficial to the forestry industry (Groninger et al., 1999), and provided that this wood is used in durable products, will contribute to a net removal of C from the atmosphere. Allocation of extra C to highly labile fine roots in sweetgum, however, may contribute far less to C sequestration in biomass, as the mean residence time of C in sweetgum fine roots is just 1.2 years (Matamala et al., 2003). Although the potential for C sequestration in biomass is less, more C is cycled into the soil in the sweetgum forest, where there is a potential for some of it to be sequestered in soil organic matter.

In both forests, the increases in NPP with elevated CO_2 were driven by greater rates of biomass accumulation associated with a stimulation of photosynthesis rather than increases in the capacity of the forest canopy to capture light energy. The canopies of both forests at the time these C budgets were calculated were at their maxima, with leaf area indices (LAI) of ~4 and ~6 for the pine and sweetgum forest, respectively (DeLucia et al., 2002; Norby et al., 2003). The stimulation of biomass increment without corresponding increases in LAI and light absorption resulted in 23% to 27% stimulation in radiation-use efficiency (ε), defined as biomass increment per unit absorbed photosynthetically active radiation. Values of ε for the sweetgum forest (2001 ambient plot: 2.01 g MJ^{-1} and elevated plot: 2.48 g MJ^{-1}) (Norby et al., 2003) were considerably greater than for the pine forest (ambient plot: 0.49 g MJ^{-1}; elevated plot: 0.62 g MJ^{-1}) (DeLucia et al., 2002). Most of the difference in the absolute magnitude of ε between these forests is likely to stem from the year-round light absorption in pine without corresponding growth during the winter. In fact, the ratio of NPP/LAI, a proxy for ε, is remarkably similar between forests (pine: 176; sweetgum: 170). Current evidence suggests that LAI and light absorption of forests is not likely to be affected by increasing CO_2.

Calculation of microbial respiration from the soil (R_h) continues to be problematic, as it requires differentiating C derived from plant roots from C derived from soil microorganisms (Kelting et al., 1998; Edwards and Norby, 1999), yet quantifying this variable is important for estimating NEP from NPP. The pine and sweetgum experiments used different approaches to solve this problem. In the pine experiment R_h was estimated as the difference between R_{soil} and $R_{fine\ root}$; where R_{soil} was measured as CO_2 efflux from the soil surface (Andrews and Schlesinger, 2001), and $R_{fine\ root}$ was calculated as the product of standing root biomass and temperature-adjusted respiration rates measured on unearthed but attached roots (Hamilton et al., 2002; George et al., 2003). For the sweetgum forest, R_h was calculated as the product of R_{soil} and the ratio of fine root-to-microbial respiration ($R_{fine\ root}/R_h$), where the ratio $R_{fine\ root}/R_h$ was based on an analysis of the isotopic composition of C evolved from the soil, as in Andrews et al. (1999). In sharp contrast to R_a, estimates of R_h revealed a strong stimulation by elevated CO_2 in the pine forest (166%) compared to the sweetgum forest (9%) (Table 8.1).

A number of factors may have contributed to the differential responsiveness of R_h to elevated CO_2 in these forests, including broad differences in the composition of the soil microbial communities. Pine roots, for example, are associated with ectomycorrhizal fungi, while sweetgum roots are associated with vesicular–arbuscular mycorrhizal fungi. In addition, elevated CO_2 disproportionately stimulated litter inputs of C to the soil in the pine relative to the sweetgum forest, thereby providing more substrate for soil microbial populations. Leaf litter represents a highly labile C source, and the amount of litter was ~19% greater in the elevated CO_2 plots in the pine forest (Finzi et al., 2001), but only ~10% greater in the elevated CO_2 plots in the sweetgum forest (Norby et al., 2003). The nutrient contents of pine and hardwood leaf litter in the pine forest were unaffected by growth under elevated CO_2 (Finzi et al., 2001); microbial populations in this forest should therefore respond solely to the increased input of litter C. Nitrogen concentration is significantly lower in the CO_2-enriched sweetgum litter, but since the litter decomposes so quickly, potential effects of litter quality on decomposition are minimal (Johnson et al., 2004).

NEP was stimulated by elevated CO_2 and the absolute values were comparable between these forests (Table 8.1). For the pine experiment, NEP was calculated as the sum of biomass increments plus the increase in forest floor biomass (I_{wood} + I_{leaf} + $I_{coarse\ root}$ + $I_{fine\ root}$ + $I_{forest\ floor}$); in the sweetgum forest, NEP was calculated as NPP − R_h. That NEP was calculated independently from the respiratory fluxes in the pine forests revealed an interesting and potentially important inconsistency. While this estimate of NEP is consistent with the value calculated as NPP – R_h for the ambient plots, there is a large discrepancy for the elevated CO_2 plots, where NEP calculated by subtraction is only ~54% of the value presented in Table 8.1. This discrepancy suggests that the estimate of $R_{fine\ root}$ for this forest is too small, or the value of R_h is too large or some combination of the both. Andrews et al. (1999) estimated that the root contribution to R_{soil} was 55% under elevated CO_2, similar to the 48% in this analysis, suggesting that many small errors may have contributed to this discrepancy. A similar inability to close the C budget for this pine forest under elevated CO_2 was recently reported by Schäfer et al. (2003).

Ignoring human-induced changes in land cover, NEP represents C sequestration in ecosystems (International Geosphere-Biosphere Programme, 1998). There is great uncertainty about the potential for forest ecosystems to abate the accumulation of anthropogenic C in the atmosphere and the contribution of the CO_2-fertilization effect to the observed residual terrestrial C "sink" of ~2.8 Gt year^{-1}. Recent evidence suggests that reforestation and afforestation in eastern North America and Western Europe contribute substantially to this sink (Fan et al., 1998; Pacala et al., 2001; Janssens et al., 2003). How much of this sink is derived from changes in land use relative to stimulations in tree growth caused by elevated CO_2, nitrogen deposition and climate remains controversial. Schimel et al. (2000) estimate that as much as one third of additional C stored in forest ecosystems in North America is derived from a combined stimulation of tree growth by CO_2 and climate, whereas Caspersen et al. (2000) estimate that approximately 2% but with an upper limit of 7% of the observed increase in aboveground net ecosystem production was caused by a CO_2-stimulation of growth. Although their approach has been criticized for not being sufficiently robust to estimate small changes in growth (Joos et al., 2002), this upper limit is consistent with experimental data.

Based on the observed stimulation of NEP in these forests and assuming that the response of NEP to CO_2 is linear, the ~55 µl l^{-1} increase in CO_2 between 1930 and 1995, the approximate interval examined by Caspersen et al. (2000), should have contributed to an 8% to 12% stimulation in C sequestration. Although young forests have considerable capacity to respond to increases in atmospheric CO_2, the magnitude of the responses observed for these forests suggests that the effect of changes in land use on C sequestration are greater than the effect of a CO_2-induced growth stimulation. Detecting a response to CO_2 that is independent of other environmental influences, stand developmental history, and regional-scale land-use patterns remains a problem.

The Duke and the Oak Ridge FACE experiments provide novel insights into the response of forest ecosystems to an increase in atmospheric CO_2, but the picture they paint is incomplete. Both experiments exposed trees to a step change in CO_2 — one day the experimental plots experienced ambient CO_2 and the next day and from then on it was elevated to the level expected in the year 2050. Extrapolations from perturbation experiments such as these are difficult because ecosystem C sequestration rates are projected to respond differently to gradual vs. step increases in atmospheric CO_2 (Luo et al., 2003). Respiration is generally proportional to the sizes of various C pools, and increases in pool sizes are cumulative. The difference between a step increase in GPP and a gradual increase in respiration translates to a transient response in C sequestration rate (Luo et al., 2003). Hence, we cannot assume that the effect of CO_2 enrichment on stimulation of NEP in these experiments will persist. By compiling several data sets including growth measurements of trees growing next to natural CO_2 springs, Idso (1999) concluded that the growth stimulation caused by elevated CO_2 attenuates strongly with time. A second but no less important limitation of these experiments is that trees are exposed to elevated CO_2, but without the intimately related increase in air temperature that is expected, leaving the question unresolved of how elevated CO_2 and temperature interact to affect C cycling. The recent construction of a forest nitrogen budget for the Duke loblolly pine experiment and an analysis of interannual variation in the growth response to elevated CO_2 for this forest provide tentative answers to these questions.

8.4 INTERANNUAL VARIATION IN THE GROWTH RESPONSE OF A PINE FOREST TO ELEVATED CO_2

One might expect that on nutrient deficient soils, the stimulation of tree growth should abate as forests outpace the capacity of soils to provide N and other nutrients. Oren et al. (2001) and Finzi et al. (2002) have demonstrated that at least early in the loblolly pine FACE experiment, the growth of pine trees is co-limited by CO_2 and N. The N and C cycles are tightly coupled, and it has been hypothesized that elevated CO_2, by increasing carbon and decreasing the N concentration of foliage, will reduce N mineralization rates from decomposing litter, thereby retarding the growth stimulation by elevated CO_2 (Zak et al., 1993; Finzi et al., 2002). Evidence to date tacitly supports this hypothesis, but it is far from conclusive.

The annual N requirement for pine grown under elevated CO_2 increased by 16% (Finzi et al., 2002), but there is no evidence yet that elevated CO_2 has altered microbial N cycling in this forest or in the sweetgum forest (Zak et al., 2003). Greater litter production in these forests when exposed to elevated CO_2 has not yet altered the supply of microbial N for tree growth. Thus, it appears that the N demand under elevated CO_2 is outpacing supply, but this potential imbalance has not yet reduced the growth stimulation. Johnson et al. (2004) concluded that in the sweetgum experiment, increased demand for N is small relative to its availability, and an N limitation is not likely to constrain the growth response to elevated CO_2 in the foreseeable future. It is reasonable to anticipate that the stimulation in growth without a corresponding increase in the supply of N will lead to a reduction in the response to elevated CO_2, but given the potentially large storage of N in tree stems and soils, the relatively high spatial variation of N in various components of these ecosystems and low experimental replication, it may take several years for an N limitation to become evident.

At just over 6 years of exposure to elevated CO_2, the Duke experiments provide a unique opportunity to examine the strength of the growth stimulation with time as well as its interaction with changing environmental conditions. The reduction in stomatal conductance often observed for plants grown under elevated CO_2 (Curtis, 1996; Medlyn et al., 2001) leads to the hypothesis that the growth enhancement should be disproportionately greater in drought years (Strain and Bazzaz, 1983). And, because photorespiration becomes a larger drain on carbon assimilation as temperature increases, it also has been suggested that the stimulation of photosynthesis, and perhaps growth, will be greatest at high temperatures (Long, 1991; Drake et al., 1997). Although manipulative experiments to directly test these hypotheses at the scale of an intact forest ecosystem are not yet possible, an examination of the interannual variation in the response of NPP to CO_2 may provide an indirect test.

From the first year of the treatment in 1997 for the pine forest and in 1998 for the sweetgum forest, elevated CO_2 caused a substantial and sustained increase in NPP (~12% to ~38%) (Figure 8.3). In addition to being responsive to CO_2 and soil N availability, regression analyses revealed that NPP in this pine forest was highly responsive to precipitation during the growing season. Precipitation during the growing season varied from approximately 500 to 800 mm over the 6 years of this experiment, and this variation caused a ~27% increase in NPP. In contrast to one of the hypotheses posed above, the pine forest plots exposed to ambient and elevated CO_2 responded similarly to increasing precipitation (e.g., there was no trend in the percent stimulation of NPP with rainfall, Figure 8.4). The absence of an interaction between NPP and rainfall stems from the observation that unlike many angiosperms, the stomata of loblolly pine needles are relatively insensitive to growth under elevated CO_2 (Ellsworth, 1999). Because more litter accumulated on the forest floor, thereby retarding soil evaporation, soil moisture was somewhat greater in plots exposed to elevated CO_2 (Schäfer et al., 2002), but this had a negligible effect on NPP. The proportional response to elevated CO_2 was, however, greatest in warm years, as predicted by the kinetic properties of the primary carboyxlating enzyme in C_3 photosynthesis (Drake et al., 1997).

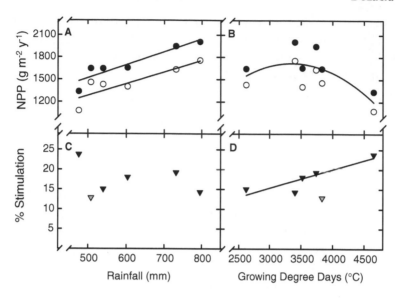

Figure 8.4 Net primary production (NPP; g DM m^{-2} year^{-1}) for loblolly pine forest plots exposed to ambient (open symbols, ~370 µl l^{-1}) and elevated atmospheric CO_2 (~570 µl l^{-1}, closed symbols), and its percent stimulation, plotted as a function of total rainfall during the growing season and growing degree days. The shaded symbol represents data collected in 1997; foliage in this year developed before the treatment. Data are for the Duke free-air CO_2 enrichment experiment, and each point represents a mean value for a given year (1997–2002). The coefficients of determination and *p* values for all regressions were >0.5 and <0.05, respectively, and where a single line is shown, the regressions for the control and treatment plots did not differ. (Data are from D. Moore and E. DeLucia, unpublished, 2004.)

As inferred from interannual variation in the growth response, elevated temperature stimulated NPP in cool years but caused NPP to decrease in warm years (Figure 8.4). Respiration consumes a major portion of the carbon fixed by GPP (57% to 71%) (Table 8.1), and is profoundly temperature dependent (Atkin and Tjoelker, 2003), thus explaining the decline in NPP in warm years. Unlike the response to precipitation, pine forest plots exposed to elevated CO_2 responded to temperature differently from those under ambient CO_2; the percent stimulation caused by elevated CO_2 increased with increasing temperature. Although this relationship must be interpreted with caution, as it was derived from a correlation, it is consistent with the theoretical prediction for photosynthesis and GPP.

Photosynthesis in C_3 plants is responsive to CO_2 because rubisco, the primary carboxylating enzyme, is not saturated at current concentrations, and because the reaction catalyzed by this enzyme is competitively inhibited by O_2 (Zelitch, 1973). Moreover, the specificity of rubisco, its relative affinity for CO_2 vs. O_2, is temperature dependent, decreasing strongly with rising temperature (Long and Drake, 1992). A consequence of this decline in specificity is that the stimulation of photosynthesis by elevated CO_2 progressively increases with increasing temperature (Long, 1991; Long and Drake, 1992). A greater percentage stimulation of photosynthesis by elevated CO_2 has been confirmed for loblolly pine (Myers et al., 1999), and it is therefore likely that this disproportionate increase in the stimulation of photosynthesis and GPP explain the observed increase in the percent stimulation of NPP in warm years. Further affirmation of this mechanism stems from the use of process-based models. Application of the PnET-II model (Aber et al., 1995, 1996) to this forest produced the same pattern of increasing percent stimulation of NPP by CO_2 in warm years, as observed in Figure 8.4 (C.J. Springer and R.B. Thomas, unpublished data, 2004).

8.5 CONCLUSIONS

At least early in stand development, loblolly pine and sweetgum forests on nutrient-deficient soils and experiencing the full suite of biological interactions and variation in the environment have considerable capacity to respond to changes in atmospheric CO_2 derived from the combustion of fossil fuels. The experimental simulation of plus 200 µl l^{-1} CO_2 caused an additional 174 g C m^{-2} to be stored in the pine forest, representing a 41% stimulation of NEP, and an additional 111 g C m^{-2} to be stored in the sweetgum forest, representing a 28%

stimulation of NEP (Table 8.1). Although there is no evidence yet that the observed stimulation of forest productivity is systematically declining with time, there is considerable interannual variation in its absolute magnitude and enhancement. Net primary production in the pine forest was greater in years with more precipitation than in dry years, and this response to precipitation was not altered by elevated CO_2. In contrast, elevated CO_2 lessened somewhat the reduction in NPP observed at elevated temperature (Figure 8.4). Forest productivity is likely to be stimulated by increasing atmospheric CO_2, at least early in stand development, and for pine forests this increase will be greater in warm than in cool years.

Respiratory fluxes return large quantities of C to the atmosphere and are important determinants of the total C sequestration in ecosystems, yet the magnitude and regulation of these fluxes remains poorly understood. In the pine and sweetgum stands, R_a alone returned 50% to 72% of GPP to the atmosphere and greater R_a in the sweetgum forest than in the pine forest may explain its lower NPP (Table 8.1). Further research on the different components of respiration, with an emphasis on understanding seasonal variation in its temperature dependence and the interaction of temperature dependence with the rate of substrate supply, will greatly enhance our understanding the forest C cycles.

Differences in how forests respond to elevated CO_2 will alter their capacity to store additional C. In the pine forest exposed to elevated CO_2 additional C was allocated to boles and branches, whereas the sweetgum forest responded with a disproportional increase in fine root production. These tissues have profoundly different mean residence times potentially altering the duration of C storage. The residence time of C is longer in wood than in fine roots suggesting that the pine forest offers a longer-term storage of atmospheric C than the sweetgum forest. This statement must be tempered by our lack of understanding of the fate of C derived from decomposition of sweetgum roots. Insofar as this C becomes incorporated in recalcitrant soil organic matter, its residence time in the soil can be greatly extended.

The imbalance between the rate of N supply and utilization in the pine forest (Finzi et al., 2002) suggests that the stimulation of productivity by elevated CO_2 would be short-lived; 7 years or less, the duration of these experiments, admittedly is a small fraction of the "life" of these forests and an abatement in the growth response may appear in the future. The stimulation of productivity observed in these experiments may provide a short-term benefit to the forest products industry and slow the rate of increase of CO_2 in the atmosphere; however, faced with such an enormous injection of C into the atmosphere, even if sustained, these CO_2-induced stimulations in productivity are far from sufficient to reverse the accumulation of C in the atmosphere (DeLucia et al., 1999; Hamilton et al., 2002).

REFERENCES

Aber, J.D., P.B. Reich, and M.L. Goulden. 1996. Extrapolating leaf CO_2 exchange to the canopy: a generalized model of forest photosynthesis validated by eddy correlation. *Oecologia,* 106:257–265.

Aber, J.D., S.V. Ollinger, C.A. Federer, et al. 1995. Predicting the effects of climate change on water yield and forest production in the northeastern United States. *Climate Res.,* 5:207–222.

Andrews, J.A. and W.H. Schlesinger. 2001. Soil CO_2 dynamics, acidification and chemical weathering in a temperate forest with experimental CO_2 enrichment. *Global Biogeochem. Cycles,* 15:149–162.

Andrews, J.A., K.G. Harrison, R. Matamala, and W.H. Schlesinger. 1999. Separation of root respiration from total soil respiration using carbon-13 labeling during free-air carbon dioxide enrichment (FACE). *Soil Sci. Soc. Am. J.,* 63:1429–1435.

Atkin, O.K. and M.G. Tjoelker. 2003. Thermal acclimation and dynamic response of plant respiration to temperature. *Trends Plant Sci.,* 8:343–351.

Baldocchi, D.D. 2003. Assessing the eddy covariance technique for evaluating carbon dioxide exchange rates of ecosystems: past, present and future. *Global Change Biol.,* 9:479–492.

Baldocchi, D.D., B.B. Hicks, and T.P. Meyer. 1988. Measuring biosphere-atmosphere exchanges of biologically related gases with micrometeorological methods. *Ecology,* 69:1331–1340.

Boone, R.D., P. Sollins, and K. Cromack Jr. 1988. Stand and soil changes along a mountain hemlock death and regrowth sequence. *Ecology,* 69:714–722.

Caspersen, J.P., S.W. Pacala, J.C. Jenkins, G.C. Hurtt, P.R. Moorcroft, and R.A. Birdsey. 2000. Contribution of land-use history to carbon accumulation in U.S. forests. *Science,* 290:1148–1151.

Clark, D.A., S. Brown, D.W. Kicklighter, J.Q. Chambers, J.R. Thomlinson, and J. Ni. 2001. Measuring net primary production in forests: concepts and field methods. *Ecol. Appl.,* 11:356–370.

Curtis, P.S. 1996. A meta-analysis of leaf gas exchange and nitrogen in trees grown under elevated carbon dioxide. *Plant Cell Environ.,* 19:127–137.

Curtis, P.S., P.J. Hanson, P. Bolstad, C. Barford, J.C. Randolph, H.P. Schmid, and K.B. Wilson. 2002. Biometric and eddy-covaraince based estimates of annual carbon sorage in five eastern North American deciduous forests. *Agric. For. Meteorol.,* 113:3–19.

Davey, P.A., S. Hunt, G.J. Hymus, E.H. DeLucia, B.G. Drake, D.F. Karnosky, and S.P. Long. 2003. Respiratory oxygen uptake is not decreased by an instantaneous elevated of [CO_2], but is increased with long-term growth in the field at elevated [CO_2]. *Plant Physiol.,*

DeLucia, E.H., J.G. Hamilton, S.L. Naidu, R.B. Thomas, J.A. Andrews, A. Finzi, M. Lavine, R. Matamala, J.E. Mohan, G.R. Hendrey, and W.H. Schlesinger. 1999. Net primary production of a forest ecosystem under experimental CO_2 enrichment. *Science,* 284:1177–1179.

DeLucia, E.H., K. George, and J.G. Hamilton. 2002. Radiation-use efficiency of a forest exposed to elevated concentrations of atmospheric carbon dioxide. *Tree Physiol.,* 22:1003–1010.

Dewar, R.C., B.E. Medlyn, and R.E. McMurtrey. 1999. Acclimation of the respiration/photosynthesis ratio to temperature: insights from a model. *Global Change Biol.,* 5:615–622.

Dore, S., G.J. Hymus, D.P. Johnson, C.R. Hinkle, R. Valentini, and B.G. Drake. 2003. Cross validation of open-top chamber and eddy covariance measurements of ecosystem CO_2 exchange in a Florida scrub-oak ecosystem. *Global Change Biol.,* 9:84–95.

Drake, B.G., M.A. Gonzalez-Meler, and S.P. Long. 1997. More efficient plants: a consequence of rising atmospheric CO_2. *Ann. Rev. Plant Physiol. Plant Mol. Biol.,* 48:609–639.

Drake, G.B., M.S. Muehe, G. Peresta, M.A. Gonzalez-Meler, and R. Matamala. 1996. Acclimation of photosynthesis, respiration and ecosystem carbon flux of a wetland on Chesapeake Bay, Maryland to elevated atmospheric CO_2 concentration. *Plant Soil,* 187:111–118.

Edwards, N.T. and R.J. Norby. 1999. Below-ground respiratory responses of sugar maple and red maple saplings to atmospheric CO_2 enrichment and elevated air temperature. *Plant Soil,* 206:85–97.

Edwards, N.T., T.J. Tschaplinski, and R.J. Norby. 2002. Stem respiration increases in CO_2- enriched sweetgum trees. *New Phytologist,* 155:239–248.

Ellsworth, D.S. 1999. CO_2 enrichment in a maturing pine forest: are CO_2 exchange and water status in the canopy affected? *Plant Cell Environ.,* 22:461–472.

Fan, S., M. Gloor, J. Mahlman, S. Pacala, J. Sarmiento, T. Takahashi, and P. Tans. 1998. A large terrestrial carbon sink in North America implied by atmospheric and oceanic carbon dioxide data and models. Science, 282:442–446.

Field, C.B. 2001. Plant physiology of the "missing" carbon sink. *Plant Physiol.,* 125:25–28.

Field, C.B., M.J. Behrenfeld, J.T. Anderson, and P. Falkowski. 1998. Primary production of the biosphere: integrating terrestrial and oceanic components. *Science,* 281:237–240.

Finzi, A.C., A.S. Allen, E.H. DeLucia, D.S. Ellsworth, and W.H. Schlesinger. 2001. Forest litter production, chemistry, and decomposition following two years of free-air CO_2 enrichment. *Ecology,* 82:470–484.

Finzi, A.C., E.H. DeLucia, J.G. Hamilton, D.D. Richter, and W.H. Schlesinger. 2002. The nitrogen budget of a pine forest under free air CO_2 enrichment. *Oecologia,* 132:567–578.

George, K., R.J. Norby, J.G. Hamilton, and E.H. DeLucia. 2003. Fine-root respiration in a loblolly pine and sweetgum forest growing in elevated CO_2. *New Phytologist,* 160:511–522.

Grier, C.C., K.A. Vogt, M.R. Keyes, and R.L. Edmonds. 1981. Biomass distribution and above- and below-ground production in young and mature *Abies amabilis* zone ecosystems of the Washington Cascades. *Can. J. For. Res.,* 11:155–167.

Groninger, J.W., K.H. Johnsen, J.R. Seiler, R.E. Will, D.S. Ellsworth, and C.A. Maier. 1999. Implications of elevated atmospheric carbon dioxide for loblolly pine plantation management and productivity. *J. For.,* 97:4–10.

Hamilton, J.G., E.H. DeLucia, K. George, S.L. Naidu, A.C. Finzi, and W.H. Schlesinger. 2002. Forest carbon balance under elevated CO_2. *Oecologia,* 131:250–260.

Hamilton, J.G., R.B. Thomas, and E.H. DeLucia. 2001. Direct and indirect effects of elevated CO_2 on leaf respiration in a forest ecosystem. *Plant Cell Environ.,* 24:975–982.

Hanson, P.J., N.T. Edwards, C.T. Garten, and J.A. Andrews. 2000. Separating root and soil microbial contributions to soil respiration: a review of methods and observations. *Biogeochemistry,* 48:115–146.

Hendrey, G., D. Ellsworth, K. Lewn, and J. Nagy. 1999. A free-air enrichment system for exposing tall forest vegetation to elevated atmospheric CO_2. *Global Change Biol.,* 5:293–309.

Hendrey, G.R. and B.A. Kimball. 1994. The FACE program. *Agric. For. Meteorol.,* 70:3–14.

Hertel, D. and C. Leuschner. 2002. A comparison of four different fine root production estimates with ecosystem carbon balance data in a Fagus-Quercus mixed forest. *Plant Soil,* 239:237–251.

Houghton, J.T., L.G. Meira Filho, J. Bruce, et al., Eds. 1996. *Climate Change 1995: Impacts, Adaptations and Mitigation of Climate Change. Scientific-Technical Analyses.* Cambridge University Press, London; New York.

Houghton, J.T., Y. Ding, D.J. Griggs, et al., Eds. 2001. *Climate Change 2001: The Scientific Basis.* Intergovernmental Panel on Climate Change, Third Assessment Report. Cambridge University Press, London; New York.

Idso, S.B. 1999. The long-term response of trees to atmospheric CO_2 enrichment. *Global Change Biol.,* 5:493–495.

International Geosphere-Biosphere Programme Terrestrial Carbon Working Group. 1998. The terrestrial carbon cycle: implications for the Kyoto Protocol. *Science,* 280:1393–1394.

Jackson, R.B., H.A. Mooney, and E.-D. Schulze. 1997. A global budget for fine root biomass, surface area, and nutrient contents. *Proc. Natl. Acad. Sci. U. S. A.,* 94:7362–7366.

Janssens, I.A., A. Freibauer, P. Ciais, et al. 2003. Europe's terrestrial biosphere absorbs 7 to 12% of European anthropogenic CO_2 emissions. *Science,* 300:1538–1542.

Johnson, D.W., W. Cheng, J.D. Joslin, R.J. Norby, N.T. Edwards, and D.E. Todd Jr. 2004. Effects of elevated CO_2 on nutrient cycling in a sweetgum plantation. *Biogeochemistry,* 69:379–403.

Joos, F., I.C. Prentice, and J.I. House. 2002. Growth enhancement due to global atmospheric change as predicted by terrestrial ecosystem models: consistent with US forest inventory data. *Global Change Biol.,* 8:299–303.

Keever, C. 1950. Causes of succession on old fields of the Piedmont, North Carolina. *Ecol. Monogr.,* 20:231–250.

Kelting, D.L., J.A. Burger, and G.S. Edwards. 1998. Estimating root respiration, microbial respiration in the rhizosphere, and root-free soil respiration in forest soils. *Soil Biol. Biochem.,* 30:961–968.

Kicklighter, D.W., M. Bruno, S. Doenges, et al. 1999. A first order analysis of the potential of CO_2 fertilization to affect the global carbon budget: a comparison of four terrestrial biosphere models. *Tellus,* 51:343–366.

Kimball, B.A., K. Kobayashi, and M. Bindi. 2002. Responses of agricultural crops to free-air CO_2 enrichment. *Adv. Agron.,* 77:293–368.

Lichter, J. 1998. Primary succession and forest development on coastal Lake Michigan sand dunes. *Ecol. Monogr.,* 4:487–510.

Lieth, H. and R.H. Whittaker, Eds. 1975. *The Primary Productivity of the Biosphere.* Ecological Studies 14. Springer, New York.

Long, S.P. and B.G. Drake. 1992. Photosynthetic CO_2 assimilation and rising atmospheric CO_2 concentrations. In N.R. Baker, and H. Thomas, Eds. *Topics in Photosynthesis. Crop Photosynthesis: Spatial and Temporal Determinants, vol. 2.* Amsterdam, Elsevier, pp. 69–107.

Long, S.P. 1991. Modification of the response of photosynthetic productivity to rising temperature by atmospheric CO_2 concentrations: has its importance been underestimated? *Plant Cell Environ.,* 14:729–739.

Loreto, F., V. Velikova, and G. Di Marco. 2001. Respiration in the light measured by $^{12}CO_2$ emission in $^{13}CO_2$ atmosphere in maize leaves. *Aust. J. Plant Physiol.,* 28:1103–1108.

Luo, Y., B. Medlyn, D. Hui, D. Ellsworth, J.F. Reynolds, and G. Katul. 2001. Gross primary productivity in the Duke Forest: Modeling synthesis of the free-air CO_2 enrichment experiment and eddy-covariance measurements. *Ecol. Appl.,* 11:239–252.

Luo, Y., L.W. White, J.G. Canadell, E.H. DeLucia, D.S. Ellsworth, A. Finzi, J. Lichter, and W.H. Schlesinger. 2003. Sustainability of terrestrial carbon sequestration: a case study in Duke Forest with inversion approach. *Global Biogeochem. Cycles,* 17:1021.

Makela, A. and H.T. Valentine. 2001. The ratio of NPP and GPP: evidence of change over the course of stand development. *Tree Physiol.,* 21:1015–1030.

Matamala, R., M.A. Gonzalez-Meler, J.D. Jastrow, R.J. Norby, and W.H. Schlesinger. 2003. Impacts of fine root turnover on forest NPP and soil C sequestration potential. *Science,* 302:1385–1387.

McLeod, A.R. and S.P. Long. 1999. Free-air carbon dioxide enrichment (FACE) in global change research: a review. *Adv. Ecol. Res.,* 28:1–55.

Medlyn, B.E., C.V.M. Barton, M.S.J. Broadmeado, et al. 2001. Stomatal conductance of forest species after long-term exposure to elevated CO_2 concentration: a synthesis. *New Phytologist,* 149:247–264.

Meir, P. and J. Grace. 2002. Scaling relationships for woody tissue respiration in two tropical rain forests. *Plant Cell Environ.,* 25:963–973.

Miglietta, F., M.R. Hoosbeek, J. Foot, et al. 2001. Spatial and temporal performance of a miniFACE (free air CO_2 enrichment) system on bog ecosystems in northern and central Europe. *Environ. Monitoring Assessment,* 66:107–127.

Moore, D., E. DeLucia, and R. Norby. 2004. Unpublished data.

Myers, D.A., R.B. Thomas, and E.H. DeLucia. 1999. Photosynthetic capacity of loblolly pine (Pinus taeda L.) trees during the first year of carbon dioxide enrichment in a forest ecosystem. *Plant Cell Environ.,* 22:473–481.

Nadelhoffer, K.J. and J.W. Raich. 1992. Fine root production estimates and belowground carbon allocation in forest ecosystems. *Ecology,* 73:1139–1147.

Naidu, S.L. and E.H. DeLucia. 1999. First-year growth response of trees in an intact forest exposed to elevated CO_2. *Global Change Biol.,* 5:609–614.

Norby, R.J., P.J. Hanson, E.G. O'Neill, et al. 2002. Net primary productivity of a CO_2-enriched deciduous forest and the implications for carbon storage. *Ecol. Appl.,* 12:1261–1266.

Norby, R.J., D.E. Todd, J. Fults, and D.W. Johnson. 2001. Allometric determination of tree growth in a CO_2-enriched sweetgum stand. *New Phytologist,* 150:477–487.

Norby, R.J., J.D. Sholtis, C.A. Gunderson, and S.S. Jawdey. 2003. Leaf dynamics of a deciduous forest canopy: no response to elevated CO_2. *Oecologia,* 136:574–584.

Nowak, R.S., D.S. Ellsworth, and S.D. Smith. 2004. Tansley Review: Functional responses of plants to elevated atmospheric CO2: Do photosynthetic and productivity data from FACE experiments support early predictions? *New Phytologist,* 162:253–280.

O'Connell, K.E.B., S.T. Gower, and J.M. Norman. 2003. Net ecosystem production of two contrasting boreal black spruce forest communities. *Ecosystems,* 6:248–260.

Obrist, D., E.H. DeLucia, and J.A. Arnone III. 2003. Consequences of wildfire on ecosystem CO_2 and water vapour fluxes in the Great Basin. *Global Change Biol.,* 9:563–574.

Okada M., M. Lieffering, H. Nakamura, M. Yoshimoto, H.Y. Kim, and K. Kobayashi. 2001. Free-air CO_2 enrichment (FACE) using pure CO_2 injection: system description. *New Phytologist,* 150:251–260.

Oren R., D.E. Ellsworth, K.H. Johnsen, et al. 2001. Soil fertility limits carbon sequestration by forest ecosystems in a CO_2-enriched atmosphere. *Nature,* 411:469–472.

Pacala, S.W., G.C. Hurtt, D. Baker, et al. 2001. Consistent land- and atmosphere-based U.S. carbon sink estimates. *Science,* 292:2316–2320.

Prentice, I.C., et al. 2001. The carbon cycle and atmospheric carbon dioxide. In *Climate Change 2001: The Scientific Basis.* Contribution of Working Group I to the Third Assessment Report of the Intergovernmental Panel on Climate Change, Cambridge University Press, London; New York, pp. 183–238.

Publicover, D.A. and K.A. Vogt. 1993. A comparison of methods for estimating forest fine root production with respect to sources of error. *Can. J. For. Res.,* 23:1179–1186.

Randerson, J.T., F.S. Chapin III, J.W. Harden, J.C. Neff, and M.E. Harmon. 2002. Net ecosystem production: a comprehensive measure of net carbon accumulation by ecosystems. *Ecol. Appl.,* 12:937–947.

Ryan, M.G. 1995. Foliar maintenance respiration of subalpine and boreal trees and shrubs in relation to nitrogen content. *Plant Cell Environ.,* 18:765–772.

Ryan, M.G., R.M. Hubbard, S. Pongracic, R.J. Raison, and R.E. McMurtrie. 1996. Foliage, fine-root, woody-tissue and stand respiration in *Pinus radiatia* in relation to nitrogen status. *Tree Physiology,* 16:333–343.

Santantonio, D. and J.C. Grace. 1987. Estimating fine-root production and turnover from biomas and decomposition data — a compartment flow model. *Can. J. For. Res.,* 17:900–908.

Schäfer, K.V.R., R. Oren, D.S. Ellsworth, C-T. Lai, J.D. Herrick, A.C. Finzi, D.D. Richter, and G. Katul. 2003. Exposure to an enriched CO_2 atmosphere alters carbon assimilation and allocation in a pine forest ecosystem. *Global Change Biol.,* 9:1378–1400.

Schäfer, K.V.R., R. Oren, C.-T. Lai, and G.G. Katul. 2002. Hydrologic balance in an intact temperate forest ecosystem under ambient and elevated atmospheric CO_2 concentration. *Global Change Biol.*, 8:895–911.

Schimel, D.S., J. Melillo, H. Tan, et al. 2000. Contribution of increasing CO_2 and climate to carbon storage by ecosystems in the United States. *Science*, 287:2004–2006.

Schimel, D.S., J.I. House, K.A. Hibbard, et al. 2001. Recent patterns and mechanisms of carbon exchange by terrestrial ecosystems. *Nature*, 414:169–172.

Schlesinger, W.H. 1997. *Biogeochemistry: An Analysis of Global Change*. Academic Press, New York.

Schlesinger, W.H. and J. Lichter. 2001. Limited carbon storage in soil and litter of experimental forest plots under increased atmospheric CO_2. *Nature*, 411:66–469.

Schulze, E.-D. and M. Heimann. 1988. Carbon and water exchange of terrestrial ecosystems. In J. Galloway and J.M. Melillo, Eds. *Asian Change in the Context of Global Change*. Cambridge University Press, London; New York, pp. 145–161.

Shaver, G.R., L.C. Johnson, D.H. Cades, G. Murray, J.A. Laundre, E.B. Rastetter, K.J. Nadelhoffer, and A.E. Giblin. 1998. Biomass and CO_2 flux in wet sedge tundras: responses to nutrients, temperature and light. *Ecol. Monogr.*, 68:75–97.

Springer, C.J. and Thomas, R.B. 2004. Unpublished data.

Strain, B.R. and F.A. Bazzaz. 1983. Terrestrial plant communities. In E.R. Lemon, Ed. *CO_2 and Plants. The Response of Plants to Rising Levels of Atmospheric Carbon Dioxide*. Westview Press, Boulder, CO.

Teskey, R.O. and M.A. McGuire. 2002. Carbon dioxide transport in xylem causes errors in estimation of rates of respiration in stems and branches of trees. *Plant Cell Environ.*, 25:1571–1577.

Tissue, D.T., J.D. Lewis, S.D. Wullschleger, J.S. Amthor, K.L. Griffen, and R. Anderson. 2002. Leaf respiration at different canopy positions in sweetgum (*Liquidambar styraciflua*) grown in ambient and elevated concentrations of carbon dioxide in the field. *Tree Physiology*, 22:1157–1166.

Valentini, R., G. Matteucci, A.J. Dolman, E-D. Schulze, C. Rebmann, E.J. Moors, et al. 2000. Respiration as the main determinant of carbon balance in European forests. *Nature,* 404:861–865.

von Caemmerer, S. 2000. *Biochemical Models of Leaf Photosynthesis.* Techniques in Plant Sciences 2. Commonwealth Scientific and Industrial Research Organisation, Collingwood, VIC, Australia.

Wang, X.Z., J.D. Lewis, D.T. Tissue, J.R. Seemann, and K.L. Griffin. 2001. Effects of elevated atmospheric CO_2 concentration on leaf dark respiration of *Xanthium strumarium* in light and in darkness. *Proc. Natl. Acad. Sci. U. S. A.,* 98:2479–2484.

Waring, R.H. and N. McDowell. 2002. Use of a physiological process model with forestry yield tables to set limits on annual carbon balances. *Tree Physiology,* 22:179–188.

Waring, R.H., J.J. Landsberg, and M. Williams. 1998. Net primary production of forests: a constant fraction of gross primary production? *Tree Physiology,* 18:129–134.

Whittaker, R.H. 1975. *Communities and Ecosystems,* 2nd ed. Macmillan, New York.

Wiant, H.V. 1967. Has the contribution of liter decay to forest soil respiration been overestimated? *J. For.,* 65:408–409.

Wofsy, S.C., M.L. Goulden, J.W. Munger, S.M. Fan, P.S. Bakwin, B.C. Daube, S.L. Bassow, and F.A. Bazzaz. 1993. Net exchange of CO_2 in a midlatitude forest. *Science,* 260:1314–1317.

Woodwell, G.M. and R.H. Whittaker. 1968. Primary production in terrestrial ecosystems. *Am. Zoologist,* 8:19–30.

Zak, D.R., K.S. Pregitzer, P.S. Curtis, J.A. Teeri, R. Fogel, and D.L. Randlett. 1993. Elevated atmospheric CO_2 and feedback between carbon and nitrogen cycles. *Plant Soil,* 151:105–117.

Zak, D.R., Holmes, W.E., Finzi, A.C., Norby, R.J., and Schlesinger, W.H. 2003. Soil nitrogen cycling under elevated CO_2: a synthesis of forest FACE experiments. *Ecol. Appl.,* 13:1508–1514.

Zelitch, I. 1973. Plant productivity and the control of photorespiration. *Proc. Natl. Acad. Sci. U.S.A.* 70:579–84.

9

Impact of Climate Change on Soil Organic Matter Status in Cattle Pastures in Western Brazilian Amazon

CARLOS C. CERRI, MARTIAL BERNOUX,
CARLOS E. P. CERRI, AND KEITH PAUSTIAN

CONTENTS

9.1 INTRODUCTION

Global climate changes caused by increased greenhouse gas emissions to the atmosphere from anthropogenic activities have direct influence on natural and agrosystem functioning (Lal, 2002). Modifications in hydrologic regimes and atmosphere temperature due to anthropogenic greenhouse effects provoke variations in plant productivity, and therefore affect food production (Intergovernmental Panel on Climate Change [IPCC], 1998).

Crop simulation models driven by future climate scenarios from global circulation models suggest that reduction in agricultural production would be more severe in tropical regions (IPCC, 2001), where there is still a shortage of food production. Brazil, located almost entirely in the tropical zone, is not an exception to this rule, and therefore is susceptible to reductions in agricultural and livestock production. Moreover, agriculture comprises the largest single sector of the Brazilian economy, representing 29% of gross domestic product in 2002, and about 47.5% of export revenue in 2003. Consequently, understanding the possible impacts of climate change on Brazilian agriculture is crucial to government decision makers.

Research about the impact of climate change on Brazilian agriculture is scanty, and has focused on grain production (Siqueira et al., 1994, 2001). Simulations of grain production are usually done by coupling a crop growth model with a climate change scenario and projected increases in CO_2 from a future emission scenario, using historical climate data and current CO_2 levels as a base scenario. Siqueira et al. (1994, 2001) presented results of wheat (*Triticum vulgare Vill*), maize (*Zea mays L.*), and soybean (*Glycine max L. Merr*) production simulations with the crop growth model CERES and SOYGRO for 13 situations under climate change scenarios generated by the Goddard Institute for Space Studies, Geophysical Fluid Dynamic Laboratory, and United Kingdom Meteorological Office general circulation models (GCMs) run with 330 and 555 ppm CO_2.

Siqueira et al. (2001) reported that simulations show an increase in the mean air temperature between 3°C to 5°C and an increase of about 11% in the mean precipitation for the Center-South region throughout the year 2050. This scenario would cause a reduction of 30% and 16% of wheat and corn production, respectively; and an increase of about 21% in soybean production. These figures correspond to a reduction of 1 million metric tons of wheat and 2.8 million metric tons of corn and an increase of 3.5 metric tons of soybeans. The major problems resultant from additional rainfall are related to higher probability of disease incidence, greater difficulty in cultivation management, and higher risks of soil water erosion (Siqueira et al., 1994).

Most studies concerning the impact of climate change on food security deal with grain production only. But beef production, with a herd of 175 million cattle in 2001, represents a large component of Brazilian agriculture. About 30% of the total is in the Amazon region where pasturage is typically extensively managed on low-fertility soils. The sustainability of these fragile ranching systems can be evaluated through soil organic matter (SOM) status. A changing climate can induce losses of SOM that upset the input–output nutrient balance and provoke losses in plant grass productivity, and subsequently sustainability of the overall system.

The main objective of this chapter is to simulate changes induced by potential climate change on SOM stocks in extensive pasturage of the Brazilian Amazon region.

9.2   MATERIALS AND METHODS

The northern region of Brazil, largely comprised of the Amazon forest and deforested areas managed as cattle pasturage, has an important role in the global climate change issue. Small changes in temperature and precipitation can markedly influence soil organic carbon dynamics that will contribute to global climate change responses.

In order to illustrate this scenario, we applied the Century ecosystem model (Century 4.0) using Tyndall Center climatic predictions to simulate soil carbon stocks and fluxes in a chrono-sequence of forest to pasture located within the Nova Vida Ranch, located in Rondônia State, in the western part of the Brazilian Amazon.

9.2.1 Study Area

Nova Vida Ranch (Figure 9.1) covers an area of 22,000 ha, and is a mixture of native forest and well-managed pastures of various ages. The climate of the region is humid tropical, with a dry season from May to September. Annual rainfall is 2200 mm. Annual mean temperature is 25.6°C. Mean temperature for the warmest and coolest months varies by less than 5°C, and mean annual relative humidity is 89% (Bastos and Diniz, 1982). Soils are classified as Podzólicos Vermelho-Amarelo (Red-Yellow Podzolic) in the Brazilian classification scheme, and as Ultisols (kandiuldults) in the U.S. soil taxonomy (Moraes et al., 1995). This soil type covers about 35% of the Brazilian Amazon Basin (Bernoux et al., 1998a, 1998b; Cerri et al., 1999).

The native forest vegetation of the ranch is classified as "open humid tropical forest," with large numbers of palms, mainly *Orbignya barbosiana, Oenocarpus spp., Jessenia bataua, Euterpe precatoria*, and *Maximiliana regia* (Pires and

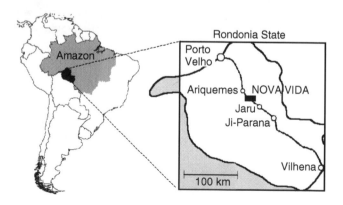

Figure 9.1 Location of Nova Vida Ranch, Rondônia State, Amazon region, Brazil.

Prance, 1986). Selective logging removed three or four economically valuable trees per hectare in the forest sites (Piccolo et al., 1996). The pasture sites were developed by a slash-and-burn technique used to clear the original forest, and then establish the grass species (Graça et al., 1999). This was done by cutting brush in March, followed by tree harvest in June and July. The remaining trees and bush were burned at the beginning of the next rainy season in September or early October, followed by seeding of pasture grasses. All pastures were created in a similar manner, and after grass establishment pasture areas have been well managed, that is, adequate grazing control (maximum of two animals per hectare) and frequent control of weed invasions. The pastures cleared in 1911, 1972, 1987, and 1989 are dominated by brachiarão (*Brachiaria brizantha*), and the pastures cleared in 1951, 1979, and 1983 are dominated by colonião (*Panicum maximum*). Mechanized agricultural practices or chemical fertilizers were not used on any of the pastures (Steudler et al., 1996).

9.2.2 Century Ecosystem Model

Century is a model of SOM and nutrient dynamics that emphasizes the decomposition of SOM, and the flux of C and N within and between different components (Parton et al., 1987). The grassland/crop and forest systems have various plant production submodels that are linked to a common SOM and nutrient cycling submodel (Figure 9.2) that has been fully described before (Metherell et al., 1993; Parton et al., 1994; Paustian et al., 1997; Kirschbaum and Paul, 2002).

In brief, the model includes two fractions of litter (metabolic and structural) and three SOM pools (active, slow, and passive), which differ in their potential decomposition rates (Figure 9.2). In addition, there are residue pools representing different size fractions of woody debris. The active pool represents microbial biomass and metabolites that turn over relatively rapidly (annual time scales), and the slow pool consists of partially stabilized SOM constituents with an intermediate turnover time (on the order of decades), while

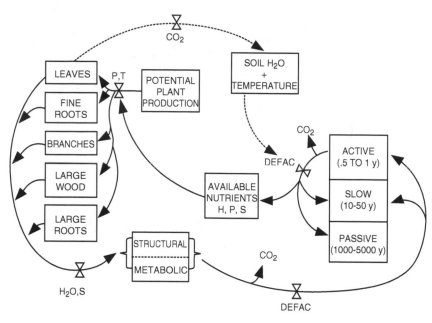

Figure 9.2 Structure of the Century ecosystem model.

the passive pool represents recalcitrant materials that turn over on time scales of centuries. Separate pools for surface vs. soil locations are maintained for the two litter fractions and the active pool, while the slow and passive pools are represented only within the soil. Various environmental factors (e.g., temperature, moisture), litter quality, soil texture, and management activities affect the parameters controlling decomposition rates and coefficients governing the flow of C and N between the SOM pools.

9.2.3 Climate Change Scenarios

For our soil C stock simulations, we modified the current climatic data (Table 9.1) measured at the study area, using Tyndall Center predictions (TYN CY 3.0 data set for Brazil available at *www.cru.uea.ac.uk/~timm/grid/table.html*). Details on the TYN SC 3.0 data set are fully reported by Mitchell et al. (2003). Briefly, the TYN CY 3.0 data set

Table 9.1 Actual and Predicted Weather Data for Nova Vida Ranch

Month	Actual Data			HadCM3-A1F1			DOE PCM-B1		
	Precipitation (cm)	Maximum Temperature (°C)	Minimum Temperature (°C)	Precipitation (cm)	Maximum Temperature (°C)	Minimum Temperature (°C)	Precipitation (cm)	Maximum Temperature (°C)	Minimum Temperature (°C)
January	38.5	34.0	18.7	33.0	40.1	12.6	38.4	35.1	17.6
February	39.2	27.6	15.3	35.5	33.2	9.7	39.1	28.6	14.3
March	33.3	27.0	15.4	28.3	33.0	9.4	33.2	28.1	14.3
April	17.4	31.4	23.6	14.5	38.0	17.0	17.3	32.7	22.3
May	17.7	32.4	19.6	14.9	39.6	12.4	17.7	33.7	18.3
June	5.8	32.9	18.9	3.2	40.2	11.6	5.8	34.3	17.5
July	3.4	33.9	25.3	1.8	41.3	17.9	3.3	35.2	24.0
August	3.1	32.9	24.1	1.5	39.9	17.1	3.1	34.4	22.6
September	11.1	32.2	23.5	8.3	39.0	16.7	11.0	33.5	22.2
October	16.2	28.0	18.3	11.3	35.0	11.3	16.1	29.5	16.8
November	29.7	31.9	20.7	23.3	38.8	13.8	29.6	33.2	19.4
December	33.0	34.3	24.3	26.8	40.9	17.7	32.9	35.5	23.1

Note: Actual data derive from local meteorological station, and predicted data derive from Tyndall Center calculations.

comprises predicted climate for the 2001–2100 period. There are five climatic variables available: cloud cover, diurnal temperature range, precipitation, temperature, and vapor pressure. There are 16 climate change scenarios that comprise all permutations of four GCM models (GCMM2, CSIRO mk 2, DOE PCM, and HadCM3, used by the IPCC [2001]) with four contrasting emissions scenarios (A1F1, A2, B2, B1) used by the IPCC Special Report on Emission Scenarios (SRES) (IPCC, 2001). This study is based on results from the HadCM3 model combined with the A1F1 SRES scenario (HadCM3-A1F1), and DOE PCM model with the B1 scenario (DOE PCM-B1), covering the maximum range in temperature and precipitation changes applied to actual data registered for the Nova Vida region.

Data from the Tyndall Center predicted that scenario HadCM3-A1F1 would cause an increase of 6.7°C in annual mean temperature and a decrease of 461 mm in annual mean precipitation. Scenario DOE PCM-B1 indicated smaller variations in annual temperature and precipitation compared to the former scenario, that is, an average annual mean temperature increase of only 1.3°C and a 50 mm increase in annual mean precipitation (Mitchell et al., 2003). It is important to emphasize that modifications made on actual data (Table 9.1) were performed using specific simulated results of temperature and precipitation for each month of the year (i.e., we did not spread simulated climatic differences uniformly throughout the year). Preserving the monthly differences in temperature and precipitation is important in areas where there are marked differences between wet and dry seasons, which is the case for our study area.

9.3 RESULTS AND DISCUSSION

The effects of the conversion of tropical forest to pasture on total soil C, using current weather data (Table 9.1), was analyzed in detail by Cerri et al. (2003), using the Century model and chrono-sequence data collected from the Nova Vida Ranch. First, the model was applied to estimate equilibrium organic matter levels, plant productivity, and residual C inputs under native forest conditions. Then Century was set to simulate the deforestation following slash and burn. SOM dynamics were simulated for pastures established in 1911, 1951, 1972, 1979, 1983, 1987, and 1989.

Using input data from Nova Vida Ranch, the Century model predicted that forest clearance and conversion to pasture would cause an initial fall in soil C stocks, followed by a slow rise to levels exceeding those under native forest (Figure 9.3). The model predicted longer-term changes in soil C under pasture close to those inferred from the pasture chrono-sequence. Mean differences between simulated and observed values were about 9.32 g m^{-2} for total soil C (data not shown). After approximately 80 years under pasture cultivation, simulated results showed that soils of the Nova Vida Ranch chrono-sequence sequestered about 1.7 kg C m^{-2} in comparison to the levels presented by the soil under native forest (Figure 9.3). Bearing in mind that this figure is model derived, it is interesting to observe that soil C stocks increase, even for a period longer than 100 years. On the other hand, many studies related to soil C sequestration suggest that there is probably a limit for nutrient storage in a particular soil type under specific climatic and management conditions (Schlesinger, 2000; Schuman et al., 2002; Lal, 2003).

Figure 9.4 shows the simulated effects of climate change scenarios suggested by the Tyndall Center (Mitchell et al., 2003) on soil C stock dynamics in the Nova Vida forest-to-pasture chrono-sequence. Note that the same parameterization procedures were adopted in the three simulated conditions (current climate, HadCM3-A1F1, and DOE PCM-B1), except for the climatic variables temperature and precipitation, which were modified (Table 9.1) according to criteria discussed above.

Simulated results gave similar curve shapes for all three modeled situations, that is, an initial decline in soil C stock in the first 2 to 3 years following conversion from forest to pasture, and then a steady increase during pasture establishment. Small differences of simulated soil C dynamics between DOE PCM-B1 and HadCM3-A1F1 scenarios can be observed. For instance, in the first 16 to 17 years after deforestation,

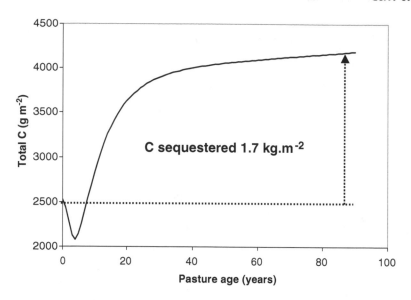

Figure 9.3 Simulated result of soil C content in the 0- to 20-cm layer at the forest to pasture chrono-sequence, Nova Vida Ranch, Rondônia State, Amazon region.

the scenario DOE PCM-B1 presented slightly higher soil C stock results compared to the HadCM3-A1F1 scenario. Around year 20, the difference between those two modeled conditions disappeared. Moreover, after about 80 years of pasture cultivation, simulated results showed an inversion of the pattern presented in the early period, that is, soil C stock results were approximately 2% higher for the HadCM3-A1F1 scenario compared to the DOE PCM-B1 scenario (Figure 9.4).

According to Century model predictions, Nova Vida chrono-sequence soils under current climate conditions would store much more C in the 0 to 20 cm layer than the other two considered scenarios. Actually, simulated results applying weather data measured at the study area indicated that soil would sequester about 4160 g C m^{-2} after 80 years of continuous well-managed pasture cultivation, which is approximately 400 g C m^{-2} and 465 g C m^{-2} more than the HadCM3-A1F1 and DOE PCM-B1 scenarios, respectively (Figure 9.4).

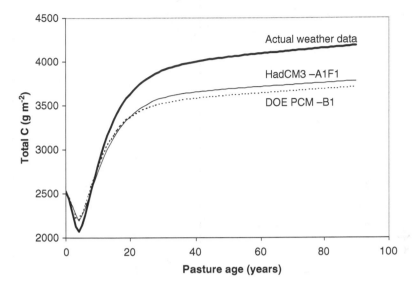

Figure 9.4 Century-simulated scenarios of soil C stock dynamics at Nova Vida Ranch forest to pasturage chrono-sequence, applying Tyndall Center predictions.

Despite the enhancement in annual mean temperature of 1.3°C or 6.7°C (scenarios DOE PCM-B1 and HadCM3-A1F1, respectively), simulated results for those scenarios did not reflect in an increase of soil C stocks compared to levels in the actual weather data scenario. A plausible reason for this condition may be directly related to the Century model decomposition structure and concept. In the Century model, average monthly soil temperature near the soil surface is calculated using equations developed by Parton et al. (1987). These equations calculate maximum soil temperature as a function of the maximum air temperature and the canopy biomass (lower for high biomass), while minimum soil temperature is a function of the minimum air temperature and canopy biomass (higher for high biomass). The actual soil temperature used for decomposition and plant growth rate functions is the average of the minimum and maximum soil temperature (Metherell et al., 1993). Therefore, increasing temperature by 1.3°C or 6.7°C simulated decomposition rates, reducing the storage rates of C into the surface soil layer (Figure 9.4).

Moreover, it is interesting to observe that the former inference of decomposition levels would probably occur more intensively in the slow C pool, which is responsible for about 68% of the total C in the first 20 cm below the surface (Figure 9.5). As expected, independent of the soil C pool (active, slow, or passive) considered, simulated results followed the same pattern presented in Figure 9.4, that is, the highest soil C content showed for simulations applying actual weather data, and the lowest for simulations using climatic predictions from the DOE PCM-B1 scenario.

The decomposition of SOM in the Century model is assumed to be mediated with an associated loss of CO_2 as a result of microbial respiration. The potential decomposition rate is reduced by multiplicative functions of soil moisture and soil temperature. Decomposition products flow into one

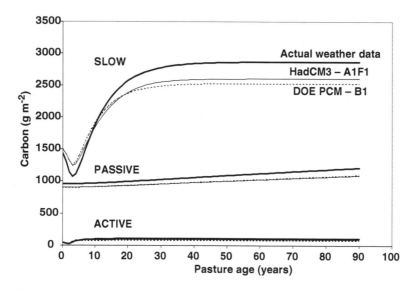

Figure 9.5   Simulated soil C pools in the 0- to 20-cm layer, using actual weather data, HadCM3-A1F1 and DOE PCM-B1 scenarios, for forest-to-pasturage chrono-sequence conditions.

of the three SOM pools, each characterized by different maximum decomposition rates (Metherell et al., 1993). The active pool (Figure 9.5) represents soil microbes and microbial products, and has a turnover time of months to a few years, depending on the environment. Soil texture influences the turnover rate of the active SOM (higher rates for sandy soils) and the efficiency of stabilizing active into slow SOM. The slow pool in Figure 9.5 includes resistant plant material derived from the structural pool and soil-stabilized microbial products derived from the active and surface microbe pools. It has a turnover time of 20 to 50 years. Finally, the passive pool is very resistant to decomposition and includes physically and chemically stabilized SOM, and has a turnover time of 400 to 2000 years.

From the standpoint of soil C sequestration, the ideal situation is to store C in the passive pool, due to its stabilization state and long turnover time. Analyzing simulated results presented in Figure 9.5, it is possible to verify that about 29% of the total C content is in the passive pool. Moreover, simulated values are increasing steadily (the change appears linear because of the slow rate of change) throughout the simulation period, and are not maintaining a constant level as in the active pool.

The climate change scenarios impact other soil chemical, physical, and biological properties that we could not directly validate with measured data from the Nova Vida Ranch chrono-sequence. Another important aspect related to soil C dynamics in the Amazon region that we have not dealt with here is related to pasture management. Fearnside and Barbosa (1998) showed that trends in soil carbon were strongly influenced by pasture management. Sites that were judged to have been under bad management generally lost soil C, whereas sites under improved management had gained carbon. Trumbore et al. (1995) reported soil C losses in overgrazed pasture, but soil C gains from fertilized pasture in the Amazon region. Neill et al. (1997) suggested that degraded pastures with little grass cover probably will be less likely to accumulate soil C because inputs to soil organic C from pasture roots will be diminished, but that might not be true in

Table 9.2 Simulated Changes in Plant
Productivity of Pasture at Nova Vida Ranch
Using Current (Actual Data), HadCM3-A1F1
and DOE PCM-B1 Scenarios

Scenario	Simulated Plant Productivity	
	Above Ground	Below Ground
	$(g\ m^{-2})$	
Current (actual data)	520	411
DOE PCM-B1	500	392
HadCM3-A1F1	416	333

more vigorous secondary forest regrowth. Greater grazing
intensity and soil damage from poor management would in
all likelihood cause soil C and N losses. Similar processes that
influence magnitude of annual SOM inputs also regulate the
accumulation of C in soils of North American grasslands (Conant et al., 2001).

We have also simulated changes in above- and below-
ground plant productivity for pastures at Nova Vida Ranch,
using weather data from current (actual data) HadCM3-A1F1
and DOE PCM-B1 scenarios (Table 9.2). Modeled results
showed a decrease in above- and below-ground productivity
of 4% using DOE PCM-B1 data, and about 20% reduction
using data from the HadCM3-A1F1 scenario compared to the
plant productivity simulated for the current scenario
(Table 9.2). Those simulated results suggest that climate
change would cause a reduction in cattle stock rate (animals
per hectare) for pastures in the Nova Vida Ranch.

Finally, as shown in the present study, modeling provides
a flexible and powerful way to assess how different scenarios
of climate and land use changes can affect soil C dynamics.

9.5 CONCLUSIONS

The research literature emphasizes that conversion of forest
to pasture is an important source of greenhouse gas emissions,
notably CO_2, to the atmosphere. Globally, land use changes
are responsible for about 14% of total emissions. Research on

SOM dynamics under well-managed pastures has shown that soil organic carbon stocks progressively increase with time after pasture cultivation. This means that if considered in isolation, soils under well-managed pastures can be considered as a CO_2 sink. However, when the whole system is evaluated, including the slash-and-burn process, it acts as a source of CO_2 to the atmosphere. The disequilibria between inputs and outputs of carbon can be aggravated if we consider the simulations performed in the present study. The scenarios used here indicated that there is a negative feedback in the soil carbon stocks due to increased temperatures caused by climate change. Under this new climatic condition, the soil carbon accumulation rate under Amazonian pastures tends to decrease, which reduces its atmospheric CO_2 mitigation effect.

ACKNOWLEDGMENTS

This work was supported by the Fundação de Amparo à Pesquisa do Estado de São Paulo (FAPESP-99/07103-0), Coordenação de Aperfeiçoeamento de Pessoal de Nivel Superior (CAPES-1240/01-3), National Science Foundation (IBN-9987996), Consortium for Agricultural Mitigation of Greenhouse Gases (CASMGS), Ecosystem Center (Woods Hole Oceanographic Institution, Woods Hole, MA), and Global Environment Facility (GFL/2740-02-4381). We are grateful to João Arantes Jr., who kindly permitted us to work at Nova Vida Ranch and assisted with logistical support.

REFERENCES

Bastos, T.X. and A.S. Diniz. 1982. Avaliação do Clima do Estado de Rondônia para Desenvolvimento Agrícola. Boletim de Pesquisa 44. Empresa Brasileira de Pesquisa-/Centro de Pesquisa Agropecuária do Trópico Úmido (CPATU), Belém, Amazonas, Brazil.

Bernoux, M., D. Arrouays, C.C. Cerri, B. Volkoff, and C. Jolivet. 1998a. Bulk density of Brazilian Amazon soils related to other soil properties. *Soil Sci. Soc. Am. J.,* 62:743–749.

Bernoux, M., D. Arrouays, C.C. Cerri, P.M. de A. Graça, B. Volkoff, and J. Trichet. 1998b. Estimation des stocks de carbone des sols du Rondônia (Amazonie brésilienne). *Études et gestion des sols,* 5:31–42.

Cerri, C.C., M. Bernoux, D. Arrouays, B.J. Feig, and M.C. Piccolo. 1999. Carbon stocks in soils of the Brazilian Amazon. In R. Lal, J. Kimble, R. Follet, and B.A. Stewart, Eds. *Global Climate Change and Tropical Ecosystems. Advances in Soil Science.* CRC Press, Boca Raton, FL, pp. 33–50.

Cerri, C.E.P., K. Paustian, M. Bernoux, R.L. Victoria, J.M. Mellilo, and C.C. Cerri. 2004. Modeling changes in soil organic matter in Amazon forest to pasture conversion, using the Century model. *Global Change Biol.,* 10:815–832.

Conant, R.T., K. Paustian, and E.T. Elliott. 2001. Grassland management and conversion into grassland: effects on soil carbon. *Ecol. Appl.,* 11:343–355.

Fearnside, P.M. and R.I. Barbosa. 1998. Soil carbon changes from conversion of forest to pasture in Brazilian Amazon. *For. Ecol. Manage.,* 108:147–166.

Graça, P.M.L.A., P.M. Fearnside, and C.C. Cerri. 1999. Burning of Amazon forest in Ariquemes, Rondônia, Brazil: biomass, charcoal formation, and burning efficiency. *For. Ecol. Manage.,* 120:179–191.

Intergovernmental Panel on Climate Change. 1998. *The Regional Impacts of Climate Change: An Assessment of Vulnerability.* Cambridge University Press, London; New York.

Intergovernmental Panel on Climate Change. 2001. *Climate Change 2001: The Scientific Basis. Third Assessment Report. www.ipcc.ch/press/pr.htm.*

Kirschbaum, M.U.F., and K.I. Paul. 2002. Modelling C and N dynamics in forest soils with a modified version of the Century model. *Soil Biol. Biochem.,* 34:341–354.

Lal, R. 2002. Soil carbon dynamic in cropland and rangeland. *Environ. Pollution,* 116:353–362.

Lal, R. 2003. Global potential of soil carbon sequestration to mitigate the greenhouse effect. *Crit. Rev. Plant Sci.,* 22:151–184.

Metherell, A.K., L.A. Harding, C.V. Cole, and W.J. Parton. 1993. CENTURY Soil Organic Matter Model Environment. Technical documentation. Agroecosystem version 4.0. Great Plains System Research Unit Technical Report 4. U.S. Department of Agriculture, Fort Collins, CO.

Mitchell, T.D., T.R. Carter, P.D. Jones, M. Hulme, and M. New. 2003. A comprehensive set of high-resolution grids of monthly climate for Europe and the globe: the observed record (1901–2000) and 16 scenarios (2001–2100). *Climatic Change,* 60:217–242.

Moraes, J.F.L., C.C. Cerri, J.M. Melillo, D. Kicklighter, C. Neill, D. Skole, and P.A. Steudler. 1995. Soil carbon stocks of the Brazilian Amazon basin. *Soil Sci. Soc. Am. J.,* 59:244–247.

Neill, C., C.C. Cerri, J. Melillo, B.J. Feigl, P.A. Steudler, J.F.L. Moraes, and M.C. Piccolo. 1997. Stocks and dynamics of soil carbon following deforestation for pasture in Rondonia. In R. Lal, J.M. Kimble, R.F. Follett, and B.A. Stewart, Eds. *Soil Processes and the Carbon Cycle.* CRC Press, Boca Raton, FL, pp. 9–28.

Parton, W.J., D.S. Schimel, C.V. Cole, and D.S. Ojima. 1987. Analysis of factors controlling soil organic matter levels in Great Plains grasslands. *Soil Sci. Soc. of Am. J.,* 51:1173–1179.

Parton, W.J., D.S. Schimel, D.S. Ojima, and C.V. Cole. 1994. A general model for soil organic matter dynamics: sensitivity to litter chemistry, texture, and management. In R.B. Bryant and R.W. Arnold, Eds. *Quantitative Modeling of Soil Farming Processes.* SSSA Special Publication 39. Soil Science Society of America, Madison, WI, pp. 147–167.

Paustian, K, E. Levine, W.M. Post, and I.M. Ryzhova. 1997. The use of models to integrate information and understanding of soil C at the regional scale. *Geoderma,* 79:227–260.

Piccolo, M.C., C. Neill, J. Melillo, C.C. Cerri, and P.A. Steudler. 1996. [15]N natural abundance in forest and pasture soils of the Brazilian Amazon Basin. *Plant Soil,* 182:249–258.

Pires, J.M. and G.T. Prance. 1986. The vegetation types of the Brazilian Amazon. In G.T. Prance and T.M. Lovejoy, Eds. *Amazonia.* Pergamon Press, Oxford, pp. 109–115.

Schlesinger, W.H. 2000. Carbon sequestration in soils: some cautions amidst optimism. *Agric. Ecosys. Environ.,* 82:121–127.

Schuman, G.E., H.H. Janzen, and J.E. Herrick. 2002. Soil carbon dynamics and potential carbon sequestration by rangelands. *Environ. Pollution,* 116:391–396.

Siqueira, O.J.F., J.R.B Farias, and L.M.A. Sans. 1994. Potential effects of global climate change for Brazilian agriculture and adaptative strategies for wheat, maize and soybean. *Revista Brasileira de Agrometeorologia,* 2:115–129.

Siqueira, O.J.W., S. Steinmetz, L.A.B. Salles, and J.M. Fernandes. 2001. Efeitos potenciais das mudanças climáticas na agricultura brasileira e estratégias adaptativas para algumas culturas. In Lima, M.A., O.M.R. Carbral, and J.D.G. Miguez, Eds. *Mudanças Climáticas Globais e a Agropecuária Brasileira.* Embrapa Meio Ambiente, Jaguariúna, São Paulo, pp. 33–63.

Steudler, P.A., J.M. Melillo, B.J. Feigl, C. Neill, M.C. Piccolo, and C.C. Cerri. 1996. Consequence of forest-to-pasture conversion on CH_4 fluxes in the Brazilian Amazon Basin. *J. Geophys. Res.,* 101:18547–18554.

Trumbore, S.E., E.A. Davidson, P.B. Camargo, D.C. Nepstad, and L.A. Martinelli. 1995. Below-ground cycling of carbon in forest and pastures of eastern Amazonia. *Global Biogeochem. Cycles,* 9:515–528.

Section III

Climate Change and Agronomic Production

10

Climate Change, Agriculture, and Sustainability

CYNTHIA ROSENZWEIG AND DANIEL HILLEL

CONTENTS

The first global climate model experiments projecting the atmospheric responses of increasing carbon dioxide (CO_2) and other greenhouse gases were published in the early 1980s. Soon after, research began on the agricultural implications of the changing atmospheric composition and its projected climate shifts. As the primary land-based human activity most intimately connected with climate and as the very foundation for human nutrition and indeed survival, agriculture naturally became a key focus for early climate change impact studies.

Through the ensuing two decades, scientists have employed a variety of analytic approaches in a multitude of studies to answer such research questions as: What might be the major effects of climate changes in the 21st century? Are some regions likely to gain, while others lose? What response measures are indicated? How climate change affects agriculture and how agriculture responds to a changing climate will invariably shape the sustainability of this vital sector.

Research in the area of climate change impacts on agriculture has involved field experiments, regression analyses, and modeling studies. The fields concerned have included agronomy, resource economics, and geography. Climate change and agriculture studies continue, with broad-brush explorations giving way to more detailed studies of biophysical processes and social responses. In this chapter, we review some of the main lessons learned from two decades of research on climate change and agriculture, and then delineate several pathways for continuing research that will help to elucidate further the interactions of climate change and agricultural sustainability.

10.1 CLIMATE CHANGE

Climate change projections are fraught with much uncertainty in regard to both the rate and magnitude of temperature and precipitation alterations in the coming decades. This uncertainty derives from a lack of precise knowledge of how climate system processes will change and of how population growth, economic and technological development, and land use will proceed in the coming century (Intergovernmental Panel on Climate Change [IPCC], 2000, 2001).

Nevertheless, three points regarding climate change can be made with some certainty (Figure 10.1). First, greenhouse gas concentrations have increased progressively since the beginning of the Industrial Revolution. Second, the natural presence of greenhouse gases is known to affect the planetary energy balance, causing the planet to be warmer than it would be otherwise. Thus, any increases in greenhouse gases will tend to enhance the natural "greenhouse effect." Third, the planet has indeed been warming over the last century, especially in the most recent two and a half decades.

The IPCC has attributed the observed warming over the last century to anthropogenic emissions of greenhouse gases, especially carbon dioxide (CO_2), methane (CH_4), and nitrous oxide (N_2O) (IPCC, 2001). Thus, anthropogenic emissions of greenhouse gases appear to be altering our planetary energy balance and to be manifested in a large-scale warming of the planet. If warming continues at the global scale, the association among greenhouse gas emissions, greenhouse effect, and surface warming will trend toward greater and greater certainty. The ultimate significance of the climate change issue is related to its planetary scale.

A

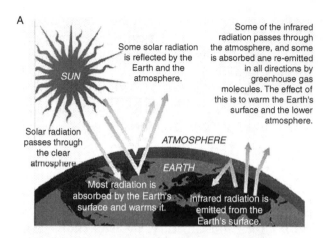

B

Carbon Dioxide Concentrations

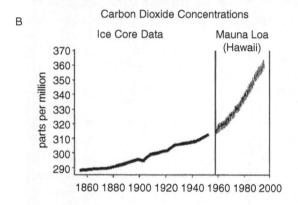

C

Global Average Temperature

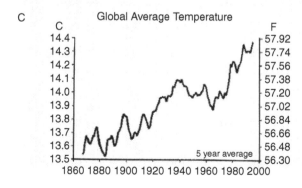

Figure 10.1 (opposite page) The three certainties of global climate change: (A) the greenhouse effect and planetary energy balance; (B) atmospheric concentrations of greenhouse gases, 1860 to present; and (C) mean global surface temperature, 1860 to present. (From OSTP. 1997. *Climate Change: State of Knowledge.* Office of Science and Technology Policy, Washington, DC.)

10.2 GOALS OF IMPACT STUDIES

Many of the climate change studies done to date, particularly the early ones, were undertaken to aid national policymakers to assess the significance of global climate change and its implications for broad regions as well as for whole countries. These studies are thus "policy relevant" in the sense that they may contribute to national decisions on whether and how to participate in the U.N. Framework Convention on Climate Change (UNFCCC) and the Kyoto Protocol. Questions here revolve around how serious the ultimate warming may be; who may be the "winners and losers"; and what the potential is for adaptation in broad-brush terms.

Recently, attention has been turning to how to respond to global climate change, including more detailed explorations of adaptation strategies and adaptive capacities at finer spatial scales — even down to individual villages. Many of these adaptation studies are focused on defining vulnerability and represent a link between the experience of current climate extremes, disaster management, and potential decadal-to-century warming. A further shift in focus involves the potential role of carbon sequestration in climate change mitigation, and to what extent this can reduce the anthropogenic build-up of greenhouse gases in the atmosphere.

10.3 AGRO-ECOSYSTEM PROCESSES

Determining what the net effect of a changing climate may be on an agro-ecosystem is complicated due to the interactions of several simultaneous biophysical processes. In some cases, changes in climate may be beneficial, while in others they may be detrimental (Figure 10.2). On the beneficial side,

Possible benefits

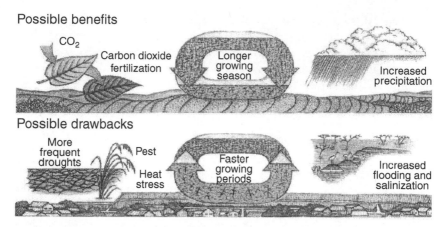

Possible drawbacks

Figure 10.2 Agro-ecosystem processes and a changing climate. (Redrawn from Bongaarts, J. 1994. *Sci. Am.*, 270:36–42.)

increasing levels of atmospheric CO_2 have been shown to increase photosynthesis rates and to increase stomatal resistance in crops, leading to overall increased water-use efficiency (Kimball et al., 2002). These processes have been called "CO_2 fertilization."

Another beneficial impact would be the prolongation of crop growing seasons in areas where they are now limited by cold temperature, that is, at high latitudes and high elevations. A further benefit for crops may accrue in some semi-arid locations from increased precipitation, since a warmer atmosphere can hold more water vapor. However, the location and extent of any such regions of enhanced precipitation is not known precisely, due primarily to the difficulty of simulating the regional-scale hydrological cycle in global climate models.

A warmer and more variable climate is likely to have negative as well as positive effects on agricultural regions around the world. Potential negative effects include more frequent droughts and floods, heat stress, increased outbreaks of diseases and pests, shortening of crop growing periods, and — in coastal regions — increased flooding and salination due to sea-level rise and impeded drainage. While the absolute magnitude of precipitation change in any one region or decade is not predictable, global climate models project that hydrological regimes are likely to become more intense as well as more variable (IPCC, 2001). Episodes of heat stress are known to be detrimental to crops, especially during critical growth stages, and such episodes are likely to be more frequent and prolonged in the future.

An important, albeit counterintuitive, negative effect that warming has on crops is the shortening of their growing period (not their overall growing season). Warmer temperatures speed crops through their growing cycle, especially the grain-filling stage. Total yield is a product of the rate and duration of grain filling, which is determined by accumulated temperature. Since higher temperatures shorten the duration of grain filling, higher temperature tends to exert a negative pressure on the yield of most annual crops.

Finally, in agricultural regions close to the ocean, sea-level rise and associated saltwater intrusion and flooding can harm crops through impeded soil aeration and salination. This is likely to be most serious in countries such as Egypt and Bangladesh, which have major crop-growing areas in low-lying coastal regions.

10.4 WHAT WE HAVE LEARNED

10.4.1 Agriculture Regions Will Experience Change over Time

Due to all the agro-ecosystem processes described above, it is fairly certain that agricultural regions will experience some changes, and that these changes will evolve continuously through the coming decades. Shifts in crop zonation are likely to occur, with some crop types expanding their ranges and others contracting. Given the range of projected temperature and precipitation changes from global climate models, and the unknown degree of manifestation of direct CO_2 effects on crops growing in farmers' fields, however, the magnitudes and rates of these changes are uncertain.

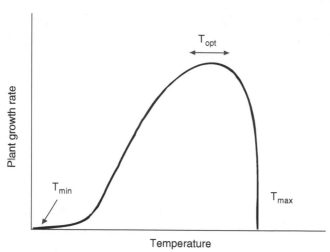

Figure 10.3 Temperature response curve for biological processes.
(From Rosenzweig, C., and D. Hillel. 1998. *Climate Change and the
Global Harvest: Potential Impacts of the Greenhouse Effect on Agri-
culture*. Oxford University Press, New York. With permission.)

The interactions between beneficial and detrimental
agro-ecosystem processes are likely to change over time for
several reasons. First, as biophysical responses move through
their temperature–response curves, responses to change in
temperature may shift from positive to neutral, and then to
negative (Figure 10.3). Another reason that climate change
effects are likely to be transformed over time is the potential
for decadal shifts in the hydrological cycle. While it is difficult
to predict the direction of change in any specific agricultural
region, global climate models do show increased decadal variability in hydrological regimes. Finally, as crop breeding and
pest species evolve in the coming decades under changing
climate conditions, new agro-ecosystem weeds, insects, and
diseases are likely to emerge, and the adjustment to these
may be costly.

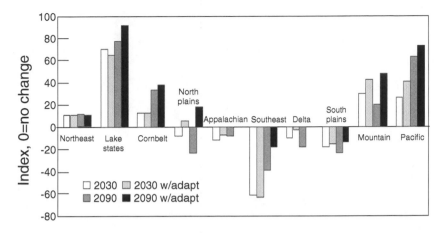

Figure 10.4 Simulated percentage changes in U.S. regional agricultural production, with adaptation, under the Canadian Climate Center scenario. (From Reilly, J., F. Tubiello, B. McCarl, et al. 2003. *Climatic Change,* 57:43–69. With permission.)

10.4.2 Effects on Agricultural Production Systems Will Be Heterogeneous

Global studies done to date show that negative and positive effects will occur both within countries and across the world. In large countries such as the United States, Russia, Brazil, and Australia, agricultural regions will likely be affected quite differently (Figure 10.4). Some regions will experience increases in production and some declines (e.g., Reilly et al., 2003). At the international level, this implies possible shifts in comparative advantage for export crop production. This also implies that adaptive responses to climate change will necessarily be complex and varied.

10.4.3 Agricultural Production in Many Developing Countries Is Especially Vulnerable

Despite general uncertainties about the rate and magnitude of climate change and about consequent hydrological changes, regional and global studies have consistently shown that

Table 10.1 Effects of Climate Change on Global Food Production
Under Various Scenarios (Percent Change in Yield)

Scenario	HadCM3 2080s							HadCM2 2080s	
	A1F1	A2a	A2b	A2c	B1a	B2a	B2b	S550	S750
CO_2 ppm	810	709	709	709	527	561	561	498	577
World	−5	0	0	−1	−3	−1	−2	−1	1
Developed countries	3	8	6	7	3	6	5	5	7
Developing countries	−7	−2	−2	−3	−4	−3	−5	−2	−1
Developed countries–developing countries differences	10.4	9.8	8.4	10.2	7.0	8.7	9.3	6.6	7.7

agricultural production systems in the mid and high latitudes
are more likely to benefit in the near term (to mid-century),
while production systems in the low latitudes are more likely
to decline (IPCC, 2001). In biophysical terms, rising temperatures will likely push many crops beyond their limits of optimal
growth and yield. Higher temperatures will create more
atmospheric water demand leading to greater water stress,
especially in semi-arid regions. Since most of the developing
countries are located in lower-latitude regions, while most developed countries are located in the mid- to high-latitude regions,
this finding suggests a divergence in vulnerability between
these groups of nations, with far-reaching implications for
future world food security (Table 10.1) (Parry et al., 2004).

Furthermore, developing countries often have fewer
resources with which to devise appropriate adaptation measures to meet changing agricultural conditions. The combination of potentially greater climate stresses and low adaptive
capacity in developing countries creates different degrees of
vulnerability as rich and poor nations confront global warming. This differential is due in part to the potentially greater
detrimental impacts of a changing climate in areas that are
already warm, and in part to the generally lower levels of
adaptive capacity in developing countries. The latter tend to
be food-recipient countries in times of food crises, while developed countries are more often donors.

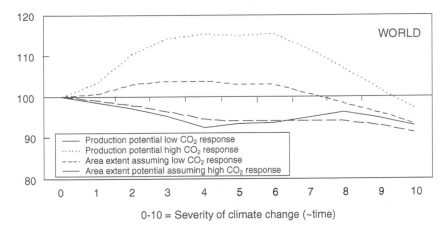

0-10 = Severity of climate change (~time)

Figure 10.5 Generalized projection of world agricultural production potential and areal extent under low and high CO_2 responses for increasing severity of climate change. (Note: severity of climate change may be taken as a proxy for decadal-to-century timeframe.) (From Fischer, G., and F. Tubiello. Personal communication, 2003.)

10.4.4 Long-Term Effects on Agriculture Are Negative

If the effects of climate change are not abated, even production in the mid- and high-latitude regions is likely to decline in the long term (end of 21st century) (Figure 10.5). These results are consistent over a range of temperature, precipitation, and direct CO_2 effects tested, and are due primarily to the detrimental effects of heat and water stress as temperatures rise. While the beneficial effects of CO_2 will eventually level out, the detrimental effects of warmer temperatures and greater water stress will be progressive in all regions.

10.4.5 Agricultural Systems Can Adapt, but Not Completely

Adaptation is integral to the study of climate change impacts on agriculture. Social scientists have made a significant contribution to the field of climate change impacts by bringing forward this important point (IPCC, 2001; Smith et al., 2003).

The task now is to integrate the findings and insights of economists, sociologists, political scientists, anthropologists, and psychologists in providing guidance to decision makers, so as to promote sectoral and international cooperation in minimizing the potential negative impacts and maximizing the opportunities for adjustment to climate change.

10.5 KEY INTERACTIONS

We need to understand how interactions within a changing climate will affect agriculture. (1) How will large-scale climate variability systems such as the El Niño–Southern Oscillation change? (2) How will supply and demand for water resources be affected? (3) How will pests of crops and livestock — including weeds, insects, and diseases — evolve?

10.5.1   El Niño-Southern Oscillation

The El Niño–Southern Oscillation (ENSO) is a large-scale ocean–atmosphere, quasi-regular interaction in the Pacific Ocean, which has reverberations in the climate system worldwide. These climate "teleconnections" bring droughts and floods to many agricultural regions, especially in the tropics, but to some degree in mid-latitudes as well. During an El Niño event, droughts tend to occur in Northeast Brazil, Australia, Indonesia, and southern Africa, among other locations, while floods tend to occur in southeastern South America, the west coast of North America, and the Horn of Africa.

In the La Niña phase, the reverse tends to occur. Figure 10.6 shows the normalized difference vegetation index for southeastern South America under the plentiful rainfall conditions of the El Niño in February 1998, and the severe drought brought on by the opposite La Niña conditions in February 2000. While it is uncertain exactly how a warming climate will affect this major variability system, there is potential for more frequent El Niño-like conditions that may affect agricultural regions around the world (IPCC, 2001).

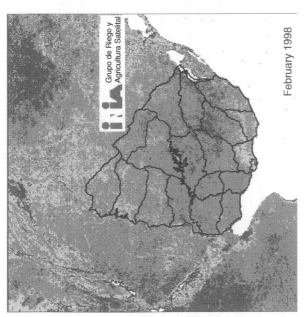

Figure 10.6 Effects of El Niño–Southern Oscillation on vegetation index in southeast South America: El Niño, February 1998; La Niña, February 2000. (From W. Baethgen, personal communication, 2003.)

10.5.2   Water Resources

Both supply of and demand for water resources are likely to change in a warming climate. As population increases and urbanization proceeds apace, there is also likely to be greater demand for water from competing domestic and commercial users. Studies show that increased water requirements for agriculture in many regions are likely under warming conditions, and that there is potential for decadal "surprises" in the reliability and percentage of water demand that can be met (Figure 10.7a) (Strzepek et al., 1999; Doll, 2002).

Research on water resources for agriculture in temperate areas has shown that changes in seasonality such as earlier snowmelt will likely change the filling and use of reservoirs and hence of water availability for irrigation (Figure 10.7b) (Strzepek et al., 1999). Current utilization plans for such facilities will need to be adjusted and readjusted as the decades proceed. Whereas early work on climate change impacts on agriculture tended to focus on the effects of more frequent droughts, recent work has emphasized the important role of damage from floods and excess soil moisture as well (Figure 10.7c) (Rosenzweig et al., 2002). Early decades in the century may tend to be wetter and the later ones drier, due to the greater effect of rising temperatures on evaporative demand later in the century.

10.5.3   Agricultural Pests

Pest–crop interactions play a crucial role in agro-ecosystems. Pest problems are very likely to be exacerbated under changing climate conditions since pests tend to thrive in warmer conditions (Rosenzweig et al., 2001). This is due to the lengthening of the frost-free seasons, allowing for more generations of pests; the extension of overwintering ranges with warmer winters; and the potential for new pests to emerge and spread, such as has occurred with the soybean cyst nematode and the soybean sudden death syndrome caused by *Fusarium solani* f. sp. alycines (Figure 10.8). Even in the current climate, warmer temperature and increases in rainfall tend to increase average per acre pesticide usage costs for many

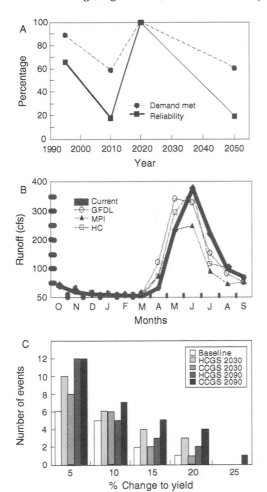

Figure 10.7 (A) Demand met (monthly average percentage of water demand met) and reliability (percentage of years in which water demands are met) for the Lower Missouri Water Region for the present and for the Max Planck (MPI) climate change scenario for the 2010s, 2020s, and 2050s. (B) Runoff in the water regions supplying the U.S. Cornbelt for current climate, and the Geophysical Fluid Dynamics Laboratory, MPI, and Hadley Center (HC) climate change scenarios. (C) Number of events causing damage to maize yields due to excess soil moisture conditions, averaged over all study sites, under current baseline (1951–1998), and the HC and Canadian Climate Centre climate change scenarios. Events causing a 20% simulated yield damage are comparable to the 1993 U.S. Midwest floods. (From Strzepek, K.M., D.C. Major, C. Rosenzweig, A. Iglesias, D.N. Yates, A. Holt, and D. Hillel. 1999. *J. Am. Water Resour. Assoc.*, 35:1639–1655; and Rosenzweig, C., F.N. Tubiello, R. Goldberg, E. Mills, and J. Bloomfield. 2002. *Global Environ. Change*, 12:197–202. With permission.)

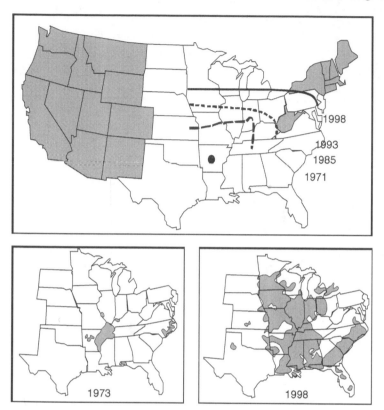

Figure 10.8 Spread of agricultural pests under current climate conditions: (top) Spread of soybean cyst nematode, 1971-1998; (bottom) Spread of soybean sudden death syndrome. (From Rosenzweig, C., A. Iglesias, X.B. Yang, P.R. Epstein, and E. Chivian. 2001. *Global Change Hum. Health,* 2:90–104. With permission.)

crops (Chen and McCarl, 2001). The emergence of new pests could produce situations for which agricultural systems may not be prepared.

10.6 MITIGATION AND ADAPTATION RESPONSES

After nearly two decades of research on the potential impacts of climate change on agriculture, attention is now turning to the assessment of appropriate responses. A distinction can be made between two types of responses: mitigation and adaptation. *Mitigation* is action to check the rising atmospheric concentration of greenhouse gases, thereby moderating their effects. *Adaptation* refers to actions that reduce the negative effects or enhance the beneficial effects of climate changes that are already occurring or are projected to occur in the future. Adaptations may be either autonomous or planned (IPCC, 2001). Research on these two types of responses have proceeded on parallel tracks. Here we suggest that they be considered conjunctively.

10.6.1 Mitigation

The practice of agriculture plays a major role in the global carbon cycle (Figure 10.9) (Rosenzweig and Tubiello, 2004; Rosenzweig and Hillel, 2000). On a global scale, the process of photosynthesis by agricultural crops fixes about 2 Gt C year^{-1}, with about 1 Gt C year^{-1} providing sustenance for the world's population that is respired back to the atmosphere as it is consumed; about 1 Gt C is returned to the soil annually as plant residues. Some of the latter carbon, however, subsequently is returned to the atmosphere by soil microbial activity, and some is stored in the soil matrix. Furthermore, the fossil fuel that powers the machinery to sow, irrigate, harvest, and dry crops worldwide is responsible for atmospheric emissions of about 150 MT (million metric tons) C year^{-1}. Large amounts of fossil fuel energy are used to produce fertilizers, especially nitrogenous compounds. Rice cultivation, livestock production, and soil processes are also responsible for considerable methane and nitrous oxide emissions (Rosenzweig and Hillel, 1998).

The agricultural carbon cycle offers several entry points for mitigation of greenhouse gas accumulation in the atmosphere. An important one is the potential for agricultural soils to store carbon, particularly to the extent that its "active" carbon stores had been depleted by past soil management practices (Rosenzweig and Hillel, 2000). Other ways that agriculture may help to mitigate the enhanced greenhouse effect

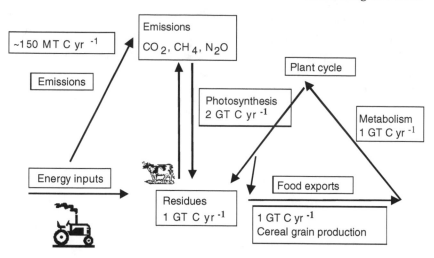

Figure 10.9 The agricultural carbon cycle. (From Rosenzweig, C., and F. Tubiello. 2005. Accepted. Mitigation and adaptation in agriculture: an interactive approach. *Mitigation and Adaptation Strategies for Global Change*. With permission.)

are through the production of biofuels, the development of more efficient rice and livestock production systems, and the reduction of fossil fuel use by farm machinery.

10.6.2 Adaptation

Farmers have always had to adapt to the vagaries of weather, whether on weekly, seasonal, or annual timescales. They will undoubtedly continue to adapt to the changing climate in the coming decades, applying a variety of agronomic techniques, such as adjusting the timing of planting and harvesting operations, substituting cultivars, and ultimately changing the entire cropping system.

However, it is important to remember that farming systems have never been completely adapted even to the current climate (witness the recurrent effects of droughts and floods on various agricultural regions around the world). Hence, it seems unreasonable to expect perfect adaptation in the future to changing climate conditions. Some adaptations will likely be successful (e.g., change in planting dates to avoid heat stress), while other attempted adaptations (e.g., changing cultivars) may not always be effective in avoiding the negative effects of droughts or floods on crop and livestock production (Figure 10.10). There are numerous social constraints to adaptation, as

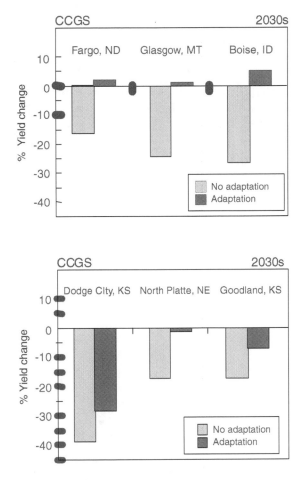

Figure 10.10 Percent yield changes with and without adaptation under the Canadian Climate Centre climate change scenario in the 2030s. Spring wheat with change of planting date (top). Winter wheat with change of cultivar (bottom). (From Tubiello, F. Personal communication, 2003.)

well, some of which have been highlighted recently by social
scientists (Smith et al., 2003).

10.7 INTERACTIONS

The joint consideration of agricultural mitigation and adaptation is needed for several reasons. Our research shows that
the soil carbon sequestration potential of agricultural soils
varies under changing climate conditions (Figure 10.11).
Thus, a changing climate clearly will affect the mitigation
potential of agricultural practices. If changing climate is not
taken into consideration, calculations such as those pertain-
ing to carbon to be sequestered may be in serious error.

On the other hand, mitigation practices can also affect
the adaptation potential of agricultural systems. For example,
by enhancing the ability of soils to hold moisture, carbon
sequestration in agricultural soils helps crops withstand

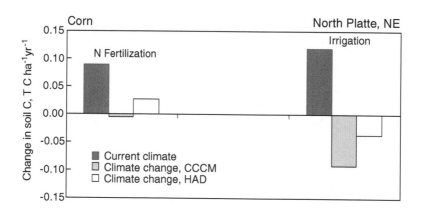

Figure 10.11 Change in soil carbon in corn production under
nitrogen fertilization and irrigation under current climate and
under the Canadian Climate Centre and Hadley Center climate
change scenarios (From Rosenzweig, C. and F. Tubiello. 2005.
Accepted. Mitigation and adaptation in agriculture: an interactive
approach. *Mitigation and Adaptation Strategies for Global Change.*
With permission.)

droughts and/or floods, both of which are projected to increase in frequency and severity in a warmer climate. Additionally, sequestering carbon in soil supports larger and more diverse populations of microbes and other organisms that provide services to plants and indirectly to animals, such as producing root growth–promoting hormones. All these functions can contribute substantially to the sustainability of agricultural systems. Adaptation practices may in turn affect the mitigation potential. For example, irrigation and nitrogen fertilization may greatly enhance the ability of soils in semi-arid regions to sequester carbon.

Finally, since it is likely that efforts to reduce the enhanced greenhouse effect (such as the Kyoto Protocol) will not be completely effective, farmers and others in the agricultural sector will be faced with the dual tasks of reducing carbon dioxide and other greenhouse gas emissions, while having to cope with an already changing climate.

10.7.1 Research Pathways

To better address the interactions between climate change and sustainability of food and fiber production, we suggest the following areas for future research attention.

10.7.1.1 Climate Variability and Change

Another bifurcation in the field of climate impacts has occurred between research on responses to major systems of climate variability, such as between the El Niño–Southern Oscillation and long-term global warming. The insights that have been gained from studies of agriculture in regard to these two timescales — seasonal to interannual vs. decadal to century — need to be reconciled.

The work on seasonal-to-interannual climate forecasts has tended to focused on short-term decision making in regard to predictions of El Niño and La Niña events, which are manifested in terms of climate extremes. The role of local stakeholders is crucial at these timescales, and responses are focused on adaptation.

The work on the decadal-to-century timescale, on the other hand, has focused primarily on responses to mean changes and long-term decisions. The stakeholders for climate change impact studies have often been national policymakers. The goal here has usually been to provide information needed to help these decision makers to decide long-term strategies in regard to the climate change issue, in terms of both mitigation and adaptation.

New theoretical constructs are needed to link climate–agriculture interactions on the two timescales, as are new ways to use analytic tools such as dynamic crop growth models and statistical analyses. We need to move beyond the more tractable projections of crop responses to mean changes, and tackle the more difficult, yet more relevant, issue (to farmers and agricultural planners) of how crops may respond to altered climate variability, such as changes in the frequency and intensity of extreme events.

10.7.1.2 Observed Effects of Warming Trends

Analysis of temperature records from around the world shows that many regions are already experiencing a warming trend, especially from the 1970s to the present (Figure 10.12). Warmer-than-normal springs have been documented in western North America since the late 1970s (Cayan et al., 2001). In some areas of the world, there have also been recent episodic increases in floods (e.g., North America) and droughts (e.g., Sahel) (IPCC, 2001), with likely but as yet mostly undocumented effects on food production. The responses of agricultural systems to such changes need to be monitored and documented. Have farmers indeed switched to earlier planting dates? Have they changed cultivars? And are there any trends in yields that can be discerned in conjunction with the climate trends?

Such questions are difficult to answer because other factors besides climate, such as land-use change and pollution, have been occurring simultaneously. But they are important for furthering our understanding of agricultural adaptation to climate, and for validating the many simulation studies

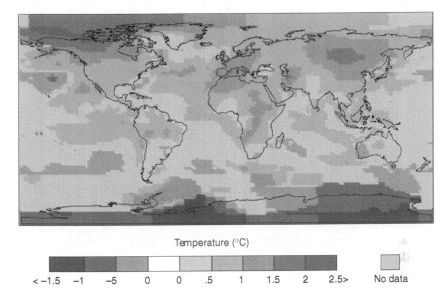

Temperature (°C)

< −1.5   −1   −5   0   0   .5   1   1.5   2   2.5>   No data

Figure 10.12 Observed temperature trends, 1970–2000. (From National Aeronautics and Space Administration/Goddard Institute for Space Studies. http:///www.giss.nasa.gov/data/update/gistemp/maps/)

done on potential climate change impacts in the future. These analyses will contribute to the IPCC Fourth Assessment now under way.

10.7.1.3 Global and Local Scales

A final bifurcation that needs to be resolved is the reconciling of global and local/regional scales. Recent work has emphasized the importance of scale in estimating the impacts of climate variability and change on agriculture (Mearns, 2003). In order to understand how a changing climate will affect agriculture, we must find new ways to bring detailed knowledge at local and regional scales to bear on global analyses. If we do not, our analyses may be in error.

This is because agriculture in any one region is linked to other agricultural regions, and indeed to the world food system, both through trade and the food donor system. As a changing climate shifts the comparative advantage in one or more regions, other regions will inevitably be affected. Thus, in our research on agriculture and climate change, we need to link regional "place-based" studies of vulnerability and adaptation, as well as mitigation, into a global synthesis.

10.8 CONCLUSION

Improving responses to climate variability and change must be a crucial requirement for future agricultural sustainability. The challenge for the field of climate change impacts on agriculture is to integrate insights from the physical, biophysical, and social sciences into a comprehensive understanding of climate–agriculture interactions at seasonal-to-interannual and decadal-to-century timescales, as well as at regional and global spatial scales. The final challenge is to disseminate and apply this knowledge to "real-world" agricultural practices and planning worldwide, so that long-term sustainability may be truly enhanced.

REFERENCES

Baethgen, W. International Research Institute for Climate Prediction. Palisades, New York. Personal communication, 2003.

Bongaarts, J. 1994. Can the growing human population feed itself? *Sci. Am.,* 270:36–42.

Cayan, D.R., S.A. Kammerdiener, M.C. Dettinger, J.M. Caprio, and D.H. Peterson. 2001. Changes in the onset of spring in the western United States. *Bull. Am. Meteorol. Soc.,* 82:399–415.

Chen, C.-C. and B.A. McCarl. 2001. An investigation of the relationship between pesticide usage and climate change. *Climatic Change,* 50:475–487.

Doll, P. 2002. Impact of climate change and variability on irrigation requirements: a global perspective. *Climatic Change,* 54:269–293.

Fischer, G., M. Shah, H. Velthuizen, and F.O. Nachtergael. 2001. Global Agro-Ecological Assessment for Agriculture in the 21st Century. International Institute for Applied Systems Analysis, Laxenburg, Austria.

Fischer, G. and F. Tubiello. International Institute for Applied Systems Analysis, Laxenburg, Austria and Goddard Institute for Space Studies, New York, New York. Personal communication, 2003.

Intergovernmental Panel on Climate Change. 2001. *Climate Change 2001: Impacts, Adaptation, and Vulnerability.* Contribution of Working Group II to the Third Assessment Report of the Intergovernmental Panel on Climate Change. Cambridge University Press, London; New York.

Kimball, B., K. Kobayashi, and M. Bindi. 2002. Responses of agricultural crops to free-air CO_2 enrichment. *Adv. Agron.,* 77:293–368.

Mearns, L.O., Ed. 2003. *Issues in the Impacts of Climate Variability and Change on Agriculture: Applications to the Southeastern United States.* Kluwer, Dordrecht.

Nakienovi, N., J. Alcamo, G. Davis, et al., Eds. 2000. *Special Report on Emissions Scenarios (SRES).*Working Group III, Intergovernmental Panel on Climate Change. Cambridge University Press, London; New York. Available at: *www.grida.no / climate / ipcc / emission / index.htm.*

Office of Science and Technology Policy. 1997. *Climate Change: State of Knowledge.* Office of Science and Technology Policy, Washington, DC.

Parry, M.L., C. Rosenzweig, A. Iglesias, M. Livermore, and G. Fisher. 2004. Assessing the effects of climate change on global food production under socio-economic scenarios. *Global Environ. Change,* 14(1):53–67.

Reilly, J., F. Tubiello, B. McCarl, et al. 2003. U.S. agriculture and climate change: new results. *Climatic Change,* 57:43–69.

Rosenzweig, C. and D. Hillel. 1998. *Climate Change and the Global Harvest: Potential Impacts of the Greenhouse Effect on Agriculture.* Oxford University Press, New York.

Rosenzweig, C. and D. Hillel. 2000. Soils and global climate change: challenges and opportunities. *Soil Sci.,* 165:47–56.

Rosenzweig, C., A. Iglesias, X.B. Yang, P.R. Epstein, and E. Chivian. 2001. Climate change and extreme weather events: implications for food production, plant diseases, and pests. *Global Change Hum. Health,* 2:90–104.

Rosenzweig, C., F.N. Tubiello, R. Goldberg, E. Mills, and J. Bloomfield. 2002. Increased crop damage in the US from excess precipitation under climate change. *Global Environ. Change,* 12:197–202.

Rosenzweig, C. and F. Tubiello. 2005. Mitigation and adaptation in agriculture: an interactive approach. *Mitigation and Adaptation Strategies for Global Change* (accepted).

Smith, J.B., R.J.T. Klein, and S. Huq. 2003. *Climate Change, Adaptive Capacity, and Development.* Imperial College Press, London.

Strzepek, K.M., D.C. Major, C. Rosenzweig, A. Iglesias, D.N. Yates, A. Holt, and D. Hillel. 1999. New methods of modeling water availability for agriculture under climate change: the U.S. Cornbelt. *J. Am. Water Resour. Assoc.,* 35:1639–1655.

Tubiello, F. Goddard Institute for Space Studies, New York, New York. Personal communication, 2003.

11

Assessing the Consequences of Climate Change for Food Security: A View from the Intergovernmental Panel on Climate Change

WILLIAM EASTERLING

CONTENTS

11.1 INTRODUCTION

Considerable progress in understanding how global climate change is likely to affect agricultural production and forest resources has been made in recent years. In the recent Third Assessment Report (TAR) of the Intergovernmental Panel on Climate Change's (IPCC) Working Group (WG) II (IPCC, 2001), agriculture was combined in a chapter with other, less managed terrestrial and aquatic ecosystems in order to assess its response to climate change. This was intended to facilitate comparison of the impacts of climate change on basic biological and ecological processes across ecosystems (including agroecosystems and forest ecosystems) in a consistent manner. However, agroecosystems are fundamentally different from less managed ecosystems such as wetlands, tundra, and savannas: they produce economically valuable goods and services within a system of clear and enforceable property rights. Such greatly complicates understanding of their response to climate change because of intense human intervention into climate–ecosystem interactions and because responses of these ecosystems to climate change can have direct and immediate economic impacts. Hence, I focus on agriculture in this chapter.

The aim of this chapter is to distill important insights into the vulnerability, potential impacts, and adaptation prospects of agriculture in response to climate change from the IPCC's WG II TAR. Most of the discussion here is drawn from material generated for chapter 5 and reported in Easterling and App (in press). The purpose of that chapter was to review and assess scientific progress in understanding of how ecosystems (including agroecosystems) and their coupled social systems may respond and adapt to climate change, and to provide a global perspective on possible agricultural outcomes. We recognize the importance of understanding small-scale regional variation in such outcomes, but leave detailed discussion of such for others to consider.

We follow a modified version of the state-pressure-response model (to include adaptation) of the Organization for Economic Cooperation and Development (OECD) to report key findings. *State* refers to the status or condition and future trends in food, fiber, fuel, and fodder systems under current climate conditions. *Pressure* refers to environmental and social stresses, including those arising from climate change, on such systems. *Response* refers to the induced changes in these systems arising from the imposed pressures (including climate change). *Adaptation* refers to the managed changes in ecosystems and deliberate human actions aimed at adjusting to climate change. This model disciplines our synthesis of the large amount of research on the consequences of climate change by focusing the review on the following questions: What is the current state of the Earth's agricultural ecosystems, and how effectively are we meeting the demands for their goods and services? What major challenges confront the world's food and fiber sectors over the next several decades, whether the climate changes or not? What are the likely biophysical and socioeconomic effects of climate change? What are the prospects for successful adaptation by agricultural systems to those effects? How vulnerable will those systems be after accounting for the potential for adaptation to alleviate stress or take advantage of opportunity?

11.2   STATE OF THE GLOBAL AGRICULTURE SECTOR

Agriculture and forests account for approximately 41% of the Earth's land covers (Houghton, 1990). According to the United Nation's Food and Agriculture Organization (FAO)

(*www.fao.org/DOCREP/003/X7470E/X7470E00.HTM*), global exports of their commodities and services were valued at
$440 billion in 1999. As noted above, unlike less managed
ecosystems, the products of agriculture and forests are traded
as commodities on world markets. Those products possess
critical life-giving properties and are part of the Earth's life
support system. There is a consensus that the global food and
fiber enterprise will be challenged over the coming decades
to expand capacity in step with anticipated expansion in global demand (World Bank, 1993; Alexandratos, 1995; Rosegrant et al., 1995; Antle et al., 1999; Johnson, 1999).
Furthermore, the most severe challenge to the ability of global
agricultural capacity to expand space with demand, with or
without climate change, will come in the next 25 to 40 years,
with the challenge abating after that, as population growth
is projected to slow and global income elasticity of food
demand is projected to decline. That is, the real story of
climate change impacts on global agriculture is likely to be
played out over the next 25 to 40 years, with the rest of the
century being anticlimactic.

11.2.1   State of Agriculture

Agricultural production in the latter half of the 20th century
increased, with the global food supply outstripping the
increase in global demand for food; this was accomplished in
spite of increases in global population and incomes. As a
result, prices for most major crops declined when adjusted for
inflation. Wheat and feed corn declined at an annual average
rate of 1% to 3% over the period (Johnson, 1999; Antle et al.,
1999). In the absence of climate change, several analysts (e.g.,
World Bank, 1993; Rosegrant et al., 1995; Johnson, 1999)
expect inflation-adjusted food prices to remain stable or slowly
to decline over the next two decades. Confidence in this outcome is high over the next two decades.

Declining food prices will likely ease but not fully erase
problems of food security, particularly in low-income countries
where lack of access to food, political instability, and inadequate physical and financial resources will remain major

challenges. In some instances, especially in the small number of nations with little immediate prospect for a successful transition from agricultural economies to manufacturing or service-based economies, lower global food prices could be stressful. However, the anticipated spread of technology and science-based production practices even to the poorest agricultural economies will likely reduce costs of production to help farmers cope with lower prices. Agricultural trade policies tend to decrease the efficiency of production both in high- and low-income countries. In high-income countries, policies tend to subsidize production in order to protect the agriculture sector, while in low-income countries policies tend to tax and discourage production (Schiff and Valdez, 1996).

Much of the optimism for future growth in agricultural production hinges on anticipated technological progress that increases crop yields. Rosegrant and Ringler (1997) argue that considerable unexploited capacity to raise crop yields exists in current crop varieties. Other analysts (Pingali, 1994; Tweeten, 1998) argue that the declining supply of new agricultural land combined with large-scale degradation of soil and water resources will slow the increase in global agricultural output, which may slow or negate the expected decline in real food prices. Approximately 50% of cereal production in developing countries is irrigated and, although it accounts only for 16% of the world's crop land, irrigated land produces 40% of the world's food. It appears that the rate of expansion of irrigation is slowing, and 10% to 15% of irrigated land is degraded to some extent by waterlogging and salinization (Alexandratos, 1995). It is questionable whether irrigation water supplies necessary to meet future irrigation demands will be available. The two conflicting views represented in this paragraph make the future trend in prices beyond the first third of the century highly uncertain.

11.3 ENVIRONMENTAL PRESSURES ON AGRICULTURE

The degradation of environmental assets, especially soils, air, and water, severely challenges the productivity of agriculture and forest resources (Pinstrup-Andersen and Pandya-Lorch, 1998; Price et al., 1999a, 1999b). In the post–World War II period, approximately 23% of the world's agricultural and forest lands were classified as degraded by the U.N. Environment Programme (Oldeman et al., 1991). Irrigated land is particularly vulnerable, although the expansion of irrigation is slowing.

Although the economic impacts of the long-term environmental degradation of forest and agricultural systems are difficult to determine, the general consensus is that they will eventually begin to undermine the necessary expansion of food and fiber production if allowed to increase at current rates.

11.4   RESPONSE OF AGRICULTURE TO RISING ATMOSPHERIC CO$_2$ AND CLIMATE CHANGE

11.4.1   Direct Effects of Rising Atmospheric CO$_2$

11.4.1.1   CO$_2$ Effects on Crops

Results from experimental studies have established that it is no longer realistic to examine the effects of climate change on crop and forage plants without also accounting for the direct effects of rising atmospheric CO$_2$ at the same time. The short-term responses to elevated CO$_2$ of isolated plants grown in artificial conditions remain difficult to extrapolate to crops in the field (Körner, 1995). Even the most realistic free-air carbon dioxide enrichment (FACE) experiments yet undertaken impose an abrupt change in CO$_2$ concentration and create a modified area (Kimball et al., 1993) analogous to a single irrigated field in a dry environment. Natural ecosystems and croplands are experiencing a gradual increase in atmospheric CO$_2$. Nonetheless, a cotton crop exposed to FACE increased biomass and harvestable yields by 37% and 48% in elevated (550 ppm) CO$_2$ (Mauney et al., 1994). At the same CO$_2$ increase, spring wheat yields increased by 8% to 10% when water was nonlimiting (Pinter et al., 1996). A simple linear extrapolation of spring wheat FACE results to a doubling of CO_2 produces a 28% increase in yields.

Several important breakthroughs in the understanding of how direct effects on crops via CO_2 were accomplished since the Second Assessment Report (SAR). Most concern improvements in the understanding of how climate interacts with the physiology of CO_2 direct effects. Horie et al. (2000) found that moderate temperatures accompanied by a doubling of CO_2 increases the seed yield of rice by 30%. However, with each 1°C increase in temperature above 26°C, rice yields declined by 10% because of shortened growth period and increased spikelet sterility. This raises concerns that CO_2 benefits may decline quickly as temperatures warm (established but incomplete). On the positive side, crop plant growth may benefit more from CO_2 enrichment in drought conditions than in wet soil because photosynthesis would be operating in a more sensitive region of the CO_2 response curve (Samarakoon and Gifford, 1995). Significantly, this effect was observed in C_4 photosynthesis. The most likely explanation for this thus far is that drought-induced stomatal closure causes CO_2 to become limiting in the absence of CO_2 enrichment (established but incomplete). It is not clear how much this effect is likely to offset the overall effect of drought on crop yield. Also, the notable dearth of testing of tropical crops and suboptimal growth conditions (nutrient deficits, weed competition, pathogens) continues from the SAR as a major research gap.

11.4.2   Impacts of Climate Change

11.4.2.1   Impacts on Crops and Livestock

Major advances were made since the SAR in the understanding of how changes in climate elements such as temperature, precipitation, and humidity, are likely to affect crop plants and livestock; CO_2 direct effects were included in much of this new crop research. A review of 43 crop modeling studies performed since the SAR revealed important geographic differences in the predicted impact of climate change on yields (Gitay et al., 2001, Table 5.3). The studies incorporated a wide range of climate change scenarios, including several different

general circulation model experiments, historical climate fluctuations, and simple sensitivity experiments. While change
in climate variability, defined as change in the higher
moments of climate elements, was not explicitly examined in
this part of our review, it is incorporated in many of the
scenarios used in the crop modeling experiments. The modeling studies were separated into tropical and temperate
regions for comparison. Predicted percentage changes in
yields (relative to current yields) in response to climate
change from each study were plotted against local temperature increases; the crops were rice and corn in the tropics,
and corn and wheat in the temperate regions. All studies
accounted for CO_2 direct effects but not for adaptation. Only
study results based on local precipitation increase were
selected for this comparison. We focused on cases of precipitation increase for three reasons: (1) to permit evaluation of
the response of crops to the least stressful expected conditions
as a conservative estimate of crop sensitivity; (2) to be able
to report an acceptable number of studies performed with
comparable climate change characteristics — there are more
modeling studies in important agricultural regions based on
positive precipitation change than negative; and (following
from 2) (3) among the more discernable patterns of agreement
among climate model projections reported by the TAR-WG II
(Carter and LaRovere et al., 2001) are increases in summer
precipitation in high northern hemisphere latitudes, tropical
southern Africa, and south Asia, with little change in southeast Asia. (Continental drying can be expected even when
warming is accompanied by increased precipitation due to the
effects of higher evapotranspiration.) The distribution of raw
modeled yields vs. temperature change was converted to a log
normal distribution in order to damp the distorting effect of
outlier yield estimates. The logged yield values were then
were averaged across studies at each degree of temperature
change — that is, yield estimates for all studies reported at,
for example, +1°C were averaged to create a mean value for
+1°C, +2°C, and so on out to +4°C. The mean log yields were
then converted back to their original units (MT^{-ha}) and plotted
to produce the line graphs shown in Figures 11.1 and 11.2.

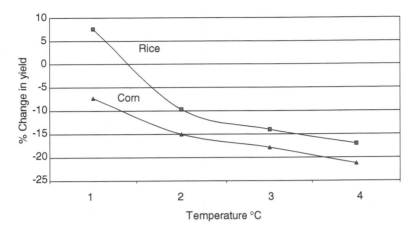

Figure 11.1   Corn and rice yields vs. temperature increase in the tropics averaged across 13 crop modeling studies. All studies assumed a positive change in precipitation. CO_2 direct effects were included in all studies.

Comparison of Figures 11.1 and 11.2 demonstrates the relatively greater sensitivity of tropical crops to climate warming than temperate crops. In the tropics, although rice yields increase by approximately 7% above current yields with 1°C of warming, they decline sharply beginning at 2°C of warming, falling to 17% below current yields at the maximum of 4°C of warming. The initial positive response of rice was heavily skewed by a preponderance of studies at the northern edges of the tropics. Rice yields everywhere else in the tropics declined with the initial 1°C of warming. Tropical corn yields decline by nearly 7% with the initial 1°C of warming, by more than 20% with 4°C of warming. This will pose a challenge to adequate food production in a majority of the world's least developed nations.

In temperate regions, corn was slightly benefited by warming of up to 2°C of warming before slipping below current yields at +3°C (Figure 11.2). Wheat yields tended to be less resilient in response to the climate change, slipping below current yields at +2°C, and declining to 25% below current yields at +4°C.

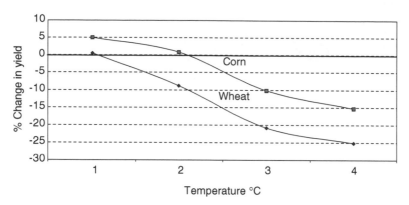

Figure 11.2 Corn and wheat yields vs. temperature increase in the temperate zone averaged across 30 crop modeling studies. All studies assumed a positive change in precipitation. CO_2 direct effects were included in all studies.

The greater sensitivity of tropical crops to warming is partly explained by the fact that crops there are grown under normal temperatures that approach theoretical optima for photosynthesis, and any additional warming is deleterious, even when accompanied by increased precipitation. Temperate crops are normally cold temperature limited, and the early stages of warming, accompanied by increasing precipitation, undoubtedly stimulate higher productivity — for a while. However, as temperate warming proceeds, so does evapotranspiration. At temperature increases of +3°C or greater, evapotranspiration appears to overcome the benefits of warming and increased precipitation, leading to increasing aridity and decreasing yields. Hence, all major planetary granaries are likely to require adaptive measures by +2°C to 3°C of warming no matter what happens to precipitation. It would be reasonable to expect adaptive measures to become necessary at lesser amounts of warming in those regions experiencing precipitation decreases with the warming.

Recent research on the impact of climate change directly on livestock supports the major conclusions of the SAR. Farm animals experience climate change directly by altered physiology and indirectly by changes in feed supplies. A dearth of physiological models that relate climate to animal physiology limits confidence in predictions of impacts, although model building is underway (Hahn, 1995; Klinedinst et al., 1993). However, there is general consensus from experimental results that climate warming likely will alter heat exchanges between animals and their environment such that mortality, growth, reproduction, and milk and wool production would be affected.

Livestock managers routinely cope with weather and climate stresses on their animals, using techniques such as strategic shading and use of sprinklers. This bodes well for adapting to climate change.

11.4.2.1.1 Accounting for Climate Variability

Natural climate variability and its changes with mean warming regulate the frequency of extreme events such as drought, excessive moisture, heat waves, and the like, which are critical determinants of crop and livestock production. Carter and LaRovere et al. (2001) list several *likely* to *very likely* changes in extreme events of importance to agriculture including, for example, higher maximum temperatures over nearly all land areas and increased summer drying over most mid-latitude continental interiors (even in cases of increased precipitation due to increased evapotranspiration). Research has only begun to consider the effect of change in frequency of extremes on agricultural production explicitly. Some analysts find that increased interannual climate variability accompanying mean climate changes disrupts crop yields more than mean climate changes alone (Mearns et al., 1995; Rosenzweig et al., 2000). Stochastic simulations of wheat growth indicated that a greater interannual variation of temperature reduces average grain yield more than a simple change in mean temperature. The potential of a change in extreme events with climate change to amplify the impact of climate change on crop productivity (both positively and negatively) is established but research is incomplete.

Analysts argue that it is important that the effect of change in climate variability on crops be distinguished from that of the change in mean climate conditions as a basis for distinguishing the impacts of natural swings in climate variability from those of climate change. Hulme et al. (1999) found it difficult to distinguish the impact on modeled wheat yield of simulated natural climate variability from that of simulated changed variability due to climate change. Hulme et al. (1999) compared wheat yields simulated with a multi-century modeled control climate containing realistic natural climate variability with those simulated with a multi-century climate change containing a change in climate variability. They found that yields under the control climate were indistinguishable from yields under climate change in a majority of the modeling sites. Such simulation results emphasize the need for greater efforts to distinguish the "noise" of natural climate variability from the "signal" of climate change (Semenov et al., 1996).

11.4.3   Response and Adaptation to Impacts on Crops, Livestock, and Forest Resources

11.4.3.1   Response and Adaptation of Crops and Livestock

The impacts of climate change will induce responses from farmers and ranchers aimed at adapting. Initial responses likely will be autonomous adjustments to crop and livestock management such as changes in agronomic practices (e.g., earlier planting, cultivar switching) or microclimate modification to cool animals' environment. They require little government intervention and are likely to be made within the existing policy and technological regimes. Methodologically, there has been little progress since the SAR in modeling agronomic adaptations. On the positive side, the adaptation strategies being modeled are limited to a small sample of the many possibilities open to farmers, which may underestimate adaptive capacity. On the negative side, the adaptations tend unrealistically to be implemented as though farmers possess perfect knowledge about evolving climate changes, which may overstate their effectiveness (Schneider et al., 2000). The preponderance of studies finds agronomic adaptation to be most effective in mid-latitude developed regions and least effective in low-latitude developing regions (Parry et al., 1999; Rosenzweig and Iglesias, 1998). However, differences in assumptions and modeling methodology among studies often lead to conflicting conclusions in specific regions. For example, in two studies using the same climate change scenarios, Matthews et al. (1997) simulate large increases while Winters et al. (1999) simulate large decreases in rice yield with adaptation across several countries in Asia. This lowers confidence in these simulations.

Like crop producers, livestock managers are likely to implement routinized adaptive techniques that were developed to deal with short-term climate variability during the initial stages of warming. For example, Hahn and Mader (1997) suggest several proactive management countermeasures that can be taken during heat waves (e.g., shades and/or sprinklers) to reduce excessive heat loads. The success livestock producers have had in the past with such countermeasures gives optimism for dealing with future climate change. However, coping can entail significant dislocation costs for certain producers. Confidence in the ability of livestock producers to adapt their herds to the physiological stresses of climate change is difficult to judge. As noted above, the absence of physiologically based animal models with well-developed climate components suggests a major methodological void.

11.4.3.1.1 Economic Costs of Agricultural Adaptation

The agricultural cost (both to producers and consumers) of responding to climate change will mostly be for the implementation of measures to adapt. (See referred Gitay et al., 2001, Table 5.3 for details of climate scenarios used in model simulations reported in this section.) At the individual farm or ranch level, these costs will reflect changes in revenues, while at national and global levels they will reflect changes in prices paid by the consumer. Crop and livestock producers who possess adequate levels of capital and technology should be capable of adapting to climate change, although changes in types of crops and animals that are grown may be required. Two different studies in the U.S. Midwest demonstrate this point. Doering et al. (1997) used a crop–livestock linear programming model linked to the Century biogeochemistry model to show that climate change may cause substantial shifts in the mix of crops grown in the upper Midwest, with much less land planted to a corn–soybean rotation and more land devoted to wheat than now observed. Earlier planting of corn increased returns, hence a more frost-resistant corn variety was found to be important to farm-level adaptation. Antle et al. (1999) used an econometric process simulation model of the dryland grain production system in Montana also linked to the Century model (as reported in Paustian et al., 1999) to assess the economic impacts of climate change in that region. Simulations were conducted for baseline and doubled CO_2 (Canadian climate model) with the observed production technology, with and without land-use adaptation and with and without CO_2 fertilization. With climate change, CO_2 fertilization and adaptation, mean returns change by −11% to +6% relative to the base climate and variability in returns increases by +7% to +25%, whereas without adaptation mean returns change by −8% to −31% and variability increases by 25% to 83%.

There will be important regional variation in the success of adaptation to climate change. It appears that developed countries will be less challenged than developing countries and countries in transition, especially in the tropics and subtropics. Winters et al. (1999) examined the impacts of climate change on Africa, Asia, and Latin America using a computable general equilibrium model. They focus on the most vulnerable groups in poor countries: poor farmers and urban poor consumers. The results show that impacts on incomes of these vulnerable groups after adaptation would tend to be negative and in the range of 0% to −10%, as compared to the impacts on consumer and producer groups predicted for the United States by Adams et al. (1998), which ranged from −0.1% to

+1%. Darwin (1999) reports results disaggregated by region, and also concludes that the developing regions are likely to have welfare effects that are less positive or more negative than the more developed regions. These findings provide evidence to support the hypothesis advanced in the SAR that climate change is likely to have its greatest adverse impacts on areas where resource endowments are the poorest and the ability of farmers to respond and adapt is most limited.

At the global level, adaptation is expected to result in small percentage changes in income. These changes are expected to be generally positive for small to moderate amounts of warming, account taken for CO_2 effects. The price of agricultural commodities is a good all-around quantity to reflect the net consequences of climate change for the regional or global supply–demand balance and on food security. A global economic model used by Darwin et al. (1995) and a U.S. model developed by Adams et al. (1998) predict that, with the rate of average warming expected by the IPCC scenarios over the next century, agricultural production and prices are likely to continue to follow the downward path observed in the 20th century. As a result, the impact on aggregate welfare comprises a small percentage of gross domestic product, and tends to be positive, especially when the effects of CO_2 fertilization are incorporated. The only study that predicts real price increases with only modest amounts of climate change is Parry et al. (1999).

Is there a threshold of climate change below which the global food production system is unimpaired, but above which is clearly impaired? The question can only be answered with very low confidence at this time. Response of prices to climate change provides insight into the question because prices determine the accessibility of a majority of the world's population to an adequate diet. Two of three recent global economic studies project real agricultural output prices to decline with a mean global temperature increase of up to 2.5°C, especially if accompanied by modest increase in precipitation (Darwin et al., 1995; Adams et al., 1998). Another study (Parry et al., 1999) projects output prices to rise with or without climate change and even a global mean temperature increase of −1°C (projected by 2020) causes prices to rise relative to the case of no climate change. When studies from the SAR are combined with the recent ones, there is general agreement that a mean global temperature rise of more than 2.5°C could increase prices (Reilly et al., 1996; Adams et al., 1998; Parry et al., 1999), with one exception (Darwin et al., 1995). Thus, with very low confidence, it is concluded from these studies that a global temperature rise of more than 2.5°C will exceed the capacity of the global food production system to adapt without price increases.

11.4.3.1.2   A Note on Environmental Damage from Adaptation to Climate Change

Degradation of the natural resource base for agriculture, especially soil and water quality, is one of the major future challenges for global food security. Those processes are likely to be intensified by adverse changes in temperature and precipitation. Land use and management have been shown to have a greater impact on soil conditions than the direct effects of climate change, thus adaptation has the potential to significantly mitigate but may, in some cases, intensify degradation. Such environmental damage may raise the costs of adaptation. Lewandrowski and Schilmmelpfenning (1999) suggest that the increased demand for irrigation predicted by a suite of studies of land and water resources, wild species, and natural ecosystems likely will increase the opportunity cost of water and possibly reduce water availability for wildlife and natural ecosystems. Strzepek et al. (1999) show that some scenarios of climate change may reduce irrigation system reliability in the lower Missouri River in the U.S. Corn Belt, which may induce in-stream environmental stress. In many developing countries, current irrigation efficiencies are very low by developed country standards. Irrigation efficiency in the Philippines in 1990 was 18% compared to the global average of 43% (Asian Development Bank [ADB], 1998). Some 3480 to 5000 liters of water are currently used to produce 1.0 kg rough rice (equivalent to 640 g milled rice) in the Philippines (Baradas, 1999) and some neighboring countries. At those irrigation efficiencies, increased irrigation demand caused by climate change would strain irrigation supplies. Hence, one adaptation strategy is to increase irrigation efficiency.

11.5 SUMMARY AND CONCLUSIONS

A clearer picture of the manner in which climate change likely will affect agriculture and forestry has taken shape over the past few years. Some important generalizations are now possible. Levels of uncertainty indicated in the text above are not repeated in the list that follows, although we urge the reader to keep these in mind. We state these as major findings from the TAR under the headings of the state-pressure-response-adaptation framework. They include the following:

11.5.1 State

Constant or declining food prices are expected for at least the next 25 years, although food security problems will persist in many developing countries as those countries deal with population increases, political crisis, poor resource endowments, and steady environmental degradation. Most economic model projections suggest that low relative food prices will extend beyond the next 25 years, although our confidence in these projections erodes farther out into the 21st century.

11.5.2 Pressure

According to U.N. estimates, approximately 23% of all forest and agricultural lands were classified as degraded over the period since World War II.

11.5.3 Response

The most realistic experiments to date — free air experiments in an irrigated environment — indicate that C3 agricultural crops particularly respond favorably to the gradual rise of atmospheric CO_2 concentrations over current levels (e.g., wheat yield increases by an average of 28%), although extrapolation of experimental results to real-world production where several factors (e.g., nutrients, temperature, precipitation, and others) are likely to be limiting at one time or another remains problematic. Moreover, little is known of crop response to elevated CO_2 in the tropics, as most of the research has been conducted in the mid-latitudes.

Research suggests that for some crops, for example rice, CO_2 benefits from a doubling of atmospheric CO_2 concentrations over current concentrations may decline quickly as temperatures warm beyond optimum photosynthetic levels. However, crop plant growth may benefit relatively more from CO_2 enrichment in drought conditions than in wet conditions.

Modeling studies suggest that any warming above current temperatures will diminish crop yields in the tropics while up to 2°C to 3°C of warming in the mid-latitudes may be tolerated by crops, especially if accompanied by increasing precipitation. Many developing countries are located in or near the tropics; this finding does not bode well for food production in those countries.

Recent advances in modeling of vegetation response suggest that transient effects associated with dynamically responding ecosystems to climate change will increasingly dominate over the next century, and that during these changes the global forest resource is likely to be adversely affected.

11.5.4 Adaptation

The ability of livestock producers to adapt their herds to the physiological stress of climate change appears encouraging due to a variety of techniques for dealing with climate stress, but this issue is not well constrained, in part because of the general lack of experimentation and simulations of livestock adaptation to climate change.

Crop and livestock farmers who have sufficient access to capital and technologies should be able to adapt their farming systems to magnitudes of climate change common in the agricultural literature. Substantial changes in their mix of crops and livestock production may be necessary, however, as considerable costs could be involved in this process because of investments in learning and gaining experience with different crops or irrigation.

Impacts of climate change on agriculture after adaptation are estimated to result in small percentage changes in overall global income. Nations with large resource endowments (i.e., developed countries) will fare better in adapting to climate change than those with poor resource endowments (i.e., developing countries and countries in transition, especially in the tropics and subtropics) which will fare worse. This, in turn, could worsen income disparities between developed and developing countries.

Global agricultural vulnerability to climate change is assessed by the anticipated effects such change will have on food prices. Based on the accumulated evidence of modeling studies, a global temperature rise of greater than 2.5°C is likely to reverse the trend of falling real food prices. This would greatly stress food security in many developing countries.

ACKNOWLEDGMENTS

I was a convening lead author on chapter 5 (Easterling and App, in press) from which much of the material for this chapter was drawn. Co-authors of chapter 5 include J. Antle, S. Brown, H. Bugmann, L. Erda, R. Fleming, L. Hahn, E. Schulze, O. Sirotenko, B. Sohngen, J. Soussana, G. Takle, J. van Minnen, and T. Williamson.

REFERENCES

Adams, R.M., B.H. Hurd, S. Lenhart, and N. Leary. 1998. Effects of global climate change on agriculture: an interpretive review. *Climate Res.*, 11:19–30.

Alexandratos, N., Ed. 1995. *World Agriculture: Towards 2010. An FAO Study.* John Wiley & Sons, Chichester, United Kingdom.

Antle, J.M., S.M. Capalbo, and J. Hewitt. 1999. Testing Hypotheses in Integrated Impact Assessments: Climate Variability and Economic Adaptation in Great Plains Agriculture. FY 1998/99 Annual Report. National Institute for Global Environmental Change. Available at: *http://nesen.unl.edu/nigec/facts/projects/Antle98.html*.

Asian Development Bank. 1998. *The Bank's Policy on Water*. Working paper. Asian Development Bank, Manila, Philippines.

Baradas, M.W. 1999. Using PVC pipes to double the irrigated area in the Philippines. In Proceedings, Philippine Society of Agricultural Engineers National Convention, General Santas City, Philippines, April, 26–29.

Carter, T.R. and E.L. LaRovere, et al. 2001. Developing and applying scenarios. In Intergovernmental Panel on Climate Change. *Climate Change 2001: Impacts, Adaptation, and Vulnerability*. Contribution of Working Group II to the Third Assessment Report of the Intergovernmental Panel on Climate Change. Cambridge University Press, London; New York, pp. 145–190.

Darwin, R.F., M. Tsigas, J. Lewandrowski, and A. Raneses. 1995. World Agriculture and Climate Change: Economic Adaptations. Agricultural Economic Report 703. U.S. Department of Agriculture, Economic Research Service, Washington, DC.

Darwin, R. 1999. A farmer's view of the Richardian approach to measuring agricultural effects of climatic change. *Climatic Change*, 41:371–411.

Doering, O.C., M. Habeck, J. Lowenberg-DeBoer, J.C. Randolph, J.J. Johnston, B.Z. Littlefield, M.A. Mazzocco, and R. Pfeifer. 1997. Mitigation strategies and unforseen consequences: a systematic assessment of the adaptation of upper midwest agriculture to future climate change. *World Resour. Rev.*, 9:447–459.

Easterling, W. and M. App. In press. Assessing the consequences of climate change for food and forest resources: a view from the IPCC. In J. Salinger (Ed.) *Increasing Climate Variability and Change: Reducing the Vulnerability of Agriculture and Forestry*, to be published in the special issue of *Climatic Change* 70:1–2.

Gitay, H., S. Brown, W.E. Easterling, et al. 2001. Ecosystems and their services. In Intergovernmental Panel on Climate Change. *Climate Change 2001: Impacts, Adaptation, and Vulnerability.* Contribution of Working Group II to the Third Assessment Report of the Intergovernmental Panel on Climate Change. Cambridge University Press, London; New York, pp. 235–342.

Hahn, G.L. 1995. Environmental influences on feed intake and performance, health, and well-being of livestock. *Jpn. J. Livestock Manage.,* 30:113–127.

Hahn, G.L. and T.L. Mader. 1997. Heat waves in relation to thermoregulation, feeding behavior, and mortality of feedlot cattle. In Proceedings, 5th International Livestock Environment Symposium, Minneapolis, MN, pp. 563–571.

Horie, T., J.T. Baker, H. Nakagawa, and T. Matsui. 2000. Crop ecosystem responses to climatic change: rice. In Reddy, K.R., and H.F. Hodges, Eds. *Climate Change and Global Crop Productivity.* CAB International, Wallingford, United Kingdom, pp. 81–106.

Houghton, R.A. 1990. Carbon. In Turner, B.L., W.C. Clark, R.W. Kates, J.F. Richards, J.T. Matthews, and W.B. Meyer, Eds. *The Earth as Transformed by Human Action*, Cambridge University Press, London; New York, pp. 393–408.

Hulme, M., E.M. Barrow, N.W. Arnell, P.A. Harrisson, T.C. Johns, and T.E. Downing. 1999. Relative impacts of human-induced climate change and natural climate variability. *Nature,* 397:688–691.

Intergovernmental Panel on Climate Change. 2001. *Climate Change 2001: Impacts, Adaptation, and Vulnerability.* Contribution of Working Group II to the Third Assessment Report of the Intergovernmental Panel on Climate Change. Cambridge University Press, London; New York.

Johnson, D.G. 1999. Food security and world trade prospects. *Am. J. Agric. Econ.,* 80:941–947.

Kimball, B.A., F.S. Mauney, F.S. Nakayama, and S.B. Idso. 1993. Effects of elevated CO_2 and climate variables on plants. *J. Soil Water Conserv.,* 48:9–14.

Klinedinst, P.L., D.A. Wilhite, G.L. Hahn, and K.G. Hubbard. 1993. The potential effects of climate change on summer season dairy cattle milk production and reproduction. *Climatic Change*, 23:21–36.

Körner, C.H. 1995. Towards a better experimental basis for upscaling plant responses to elevated CO_2 and climate warming. *Plant Cell Environ.*, 18:1101–1110.

Lewandrowski, J. and D. Schilmmelpfenning. 1999. Economic implications of climate change for U.S. agriculture: assessing recent evidence. *Land Econ.*, 75:39–57.

Matthews, R.B., M.J. Kropff, and D. Bachelet. 1997. Simulating the impact of climate change on rice production in Asia and evaluating options for adaptation. *Agric. Syst.*, 54:399–425.

Mauney, J.R., B.A. Kimball, P.J. Pinter, R.L. Lamortne, K.F. Lewin, J. Nagy, and G.R. Hendrey. 1994. Growth and yield of cotton in response to a free-air carbon dioxide enrichment (FACE) environment. *Agric. For. Meteorol.*, 70:49–67.

Mearns, L.O., F. Giorgi, L. McDaniel, and C. Shields. 1995. Analysis of climate variability and diurnal temperature in a nested regional climate model: comparison with observations and doubled CO_2 results. *Climate Dynamics*, 11:193–209.

Oldeman, R.L., T.A. Hakkeling, and W.G. Sombroek. 1991: *World Map of the Status of Human-Induced Soil Degradation*, 2nd rev. ed. International Soil Reference and Information Centre, Wageningen, Netherlands.

Parry, M., C. Fischer, M. Livermore, C. Rosenzweig, and A. Iglesias. 1999. Climate change and world food security: a new assessment. *Global Environ. Change*, 9, S51–S67.

Paustian, K., E.T. Elliott, and L. Hahn. 1999. *Agroecosystem Boundaries and C Dynamics with Global Change in the Central United States*. FY 1998/1999 Progress Report. National Institute for Global Environmental Change. Available at: *http://nigec.ucdavis.edu*.

Pingali, P. 1994. Technological prospects for reversing the declining trend in Asia's rice productivity. J.R. Anderson, Ed. In *Agricultural Technology: Policy Issues for the International Community*. CAB International and World Bank, Wallingford, United Kingdom, pp. 384–401.

Pinstrup-Andersen, P., and R. Pandya-Lorch. 1998. Food security and sustainable use of natural resources: a 2020 vision. *Ecol. Econ.*, 26:1–10.

Pinter, P.J. Jr., B.A. Kimball, R.L. Garcia, G.W. Wall, D.J. Hunsaker, and R.L. LaMorte. 1996. Free-air CO_2 enrichment: responses of cotton and wheat crops. In Koch, G.W. and H.A. Mooney, Eds. *Carbon Dioxide and Terrestrial Ecosystems,* Academic Press, San Diego, CA, 215–249.

Pittock, A.B. 1999. Coral reefs and environmental change: adaptation to what? *Am. Zool.*, 39:10–29.

Price, D.T., D.H. Halliwell, M.J. Apps, and C.H. Peng. 1999a. Adapting a patch model to simulate the sensitivity of Central-Canadian boreal ecosystems to climate variability. *J. Biogeogr.*, 26:1101–1113.

Price, D.T., C.H. Peng, M.J. Apps, and D.H. Halliwell. 1999b. Simulating effects of climate change on boreal ecosystem and carbon pools in central Canada. *J. Biogeogr.*, 26:1237–1248.

Reilly, J., et al. 1996. Agriculture in a changing climate: impacts and adaptations. In J.T. Houghton, L.G. Meiro Filho, B.A. Callander, N. Harris, A. Kattenberg, and K. Maskell, Eds. *Climate Change 1995: The Science of Climate Change. Contribution of Working Group I to the Second Assessment of the Intergovernmental Panel on Climate Change.* Cambridge University Press, London; New York.

Rosegrant, M.W., M. Agcaoili-Sombilla, and N.D. Perez. 1995. *Global Food Projections to 2020: Implications for Investment.* 2020 Vision for Food, Agriculture, and the Environment. Discussion Paper 5, International Food Policy Research Institute, Washington, DC.

Rosegrant, M.W. and C. Ringler. 1997. World food markets into the 21st century: environmental and resource constraints and policies. *Aust. J. Agric. Resour. Econ.*, 41:401–428.

Rosenzweig, C. and A. Iglesias. 1998. The use of crop models for international climate change impact assessment. In G.Y. Tsuji, G. Hoogrnboom, and P.K. Thorton, Eds. *Understanding Options for Agriculture Production.* Kluwer, Dordrecht, pp. 267–292.

Rosenzweig, C., A. Iglesias, X. Yang, P. Epstein, and E. Chivian. 2000. *Climate Change and U.S. Agriculture: The Impacts of Warming and Extreme Weather Events on Productivity, Plant Diseases, and Pests.* Center for Health and the Global Environment, Harvard Medical School, Boston.

Samarakoon, A.B. and Gifford, R.M. 1995. Soil water content under plants at high CO_2 concentration and interactions with the direct CO_2 effects: a species comparison. *J. Biogeogr.*, 22:193–202.

Schiff, M. and A. Valdez. 1996. Agricultural incentives and growth in developing countries: a cross-country perspective. In J.M. Antle and D.A. Sumner, Eds. *The Economics of Agriculture.* Vol. 2 of Papers in Honor of D. Gale Johnson. University of Chicago Press, Chicago, pp. 386–399.

Schneider, S.H., W.E. Easterling, and L.O. Mearns. 2000: Adaptation: sensitivity to natural variability, agent assumptions, and dynamic climate changes. *Climatic Change*, 45:203–221.

Semenov, M.A. and J.R. Porter. 1995. Climate variability and the modeling of crop yields. *Agric. Forest Meterol.*, 38:127–145.

Semenov, M.A., J. Wolf, L.G. Evans, H. Eckersten, and A. Eglesias, 1996. Comparison of wheat simulation models under climate change, 2: application of climate change scenarios. *Climate Res.*, 7:271–281.

Strzepek, K.M., C.D. Major, C. Rosenzweig, A. Iglesias, D. Yates, A. Holt, and D. Hillel. 1999. New methods of modeling water availability for agriculture under climate change: the U.S. Corn Belt, *J. Am. Water Resour. Assoc.*, 35:1639–1655.

Tweeten, L. 1998. Dodging a Malthusian bullet in the 21st century. *Agribusiness*, 14:15–32.

Winters, P., R. Murgai, A. de Janvry, E. Sadoulet, and G. Frisvold. 1999. Climate change and agriculture: effects on developing countries. In G. Frisvold and B. Kuhn, Eds. *Global Environmental Change and Agriculture*. Edward Elgar Publishers, Cheltenham, United Kingdom.

World Bank. 1993. *Water Resources Management.* A World Bank Policy Study. World Bank, Washington, DC.

12

Climate Change and Tropical Agriculture: Implications for Social Vulnerability and Food Security

HALLIE EAKIN

CONTENTS

12.1 INTRODUCTION

A review of relevant data would lead one to be optimistic about the world's future capacity to feed itself under projected scenarios of climate change. On the whole, food production has kept up with population growth, the prices of cereals have been gradually declining, and trade models suggest that markets can move food effectively between countries and continents to compensate for deficits (Food and Agriculture Organization [FAO], 2003). Assuming that farmers can and will make adjustments in crop varieties, planting dates, irrigation applications, and the area they plant, research has shown that anticipated negative impacts of climate change on yields can be partially or perhaps even largely mitigated at the global scale (Helms et al., 1996; Rosenzweig and Parry, 1993).

However, when analyses are disaggregated at a regional level, these conclusions are not so optimistic. For example, maize is a crop that is central to the food security of much of Africa and Latin America. Changes in its yields are likely to be negative, particularly in lower latitude regions where it is more important to food security than in higher latitude regions (Rosenzweig and Parry, 1993). The latest Intergovernmental Panel on Climate Change (IPCC) report gives similar findings from modeling research of rice, millet, and wheat production under climate change scenarios, although the uncertainty in the models' outputs may prohibit any conclusive statements (IPCC, 2001).

Although the IPCC is also cautious about studies of future agricultural price changes (IPCC, 2001), several studies have also suggested that future cereal prices may rise anywhere between 10% and 200% with global warming greater than 2.5°C (Reilly, 1995). The precise numbers related to yield and price changes are uncertain; yet there is some concern that the stronger adaptive capacities of countries at higher latitudes will facilitate continued market dominance in cereal production, thus fostering increased dependence on food imports in many developing countries (Reilly and Schilmmelpfenning, 1999).

Such dependence is not necessarily a concern if prices continue on their negative trend. However, if cereal prices were to increase or become more volatile under progressive global warming, import-dependent countries would be negatively affected (IPCC, 2001). Countries already largely dependent on food imports would feel such impacts first (Parry, 1990; Reilly and Schilmmelpfenning, 1999). With rising food prices, the number of people at risk from hunger in developing nations would also rise by tens to perhaps hundreds of millions (Chen and Katz, 1994; Downing, 1996; Parry et al., 2001; Rosenzweig and Parry, 1993). Thus, while aggregate world food production may be able to keep up with population growth, the social consequences of climate change impacts in areas already struggling with high rates of poverty and malnutrition are cause for considerable concern.

The following section reviews the biophysical and social features that make tropical agricultural systems particularly vulnerable to climatic change, recognizing that it is not the physical region of the tropics that defines its vulnerability, but rather a complex combination of environmental, economic, and social characteristics at different scales that may have relatively little to do with latitude. This is particularly true when one focuses on the implications of climate change for regional food security, given that food security is essentially a social problem that typically involves historical inequities in resource distribution, the politics of resource access, and the differential capacities of population groups to exert an effective command over their basic necessities (Sen, 1981).

The analysis of maize and climate change in Mexico shows that while climatic impacts are not a determinant of food insecurity in the country, overall vulnerability is exacerbated when such impacts coincide with adverse trends in agriculture policy and rural welfare. Farmers, the government, and civic associations in Mexico have recently mobilized around the issue of food sovereignty or the power of the country's population to determine the content and quality of their diet according to their preferences whether through food imports or domestic production. There is some concern that climate change may not only contribute to a loss in productive capacity in agriculture, but also to a more intangible but troubling loss of control and command over the country's preferred food staple, maize.

12.2 SENSITIVITY OF TROPICAL AGRICULTURE TO CLIMATIC CHANGE

The tropics is a geographic region, defined by 23.5 degrees north and 23.5 degrees south latitude (Agnew, 1998). The tropics are home to some of the world's poorest populations. Although considerable variability exists within the tropics, their climates are generally distinguished by the persistence of high year-round temperatures and the convective and highly seasonal nature of rainfall. And, although precipitation is one of the more difficult variables to simulate accurately in climate models, rainfall is the principal biophysical determinant of tropical vegetation and crop production (Agnew, 1998).

The variability of precipitation in the tropics exposes countries of this region to the full range of extreme events — floods, droughts, cyclones, hurricanes, hail storms, and, in the areas of higher elevation, frost (Gommes, 1993). Among other driving forces, the influence of the El Niño–Southern Oscillation (ENSO) tends to be stronger in the tropics (Cane, 2001) and in recent years it has contributed to the high frequency of disasters that have exacted such a high toll on developing economies (Dilley and Heyman, 1995).

These characteristics pose particular challenges to agricultural production. As described by Rosenzweig and Liverman (1992), agricultural productivity in the humid tropics is limited by fragile soils with low organic matter content and serious problems with soil leaching and erosion. In the subhumid and semi-arid tropical regions, yields are often affected by problems of soil water retention, limited irrigation potential due to poor geographic distribution of water resources and infrastructure costs and soil salinity. The year-round warm conditions that generate the rich biodiversity for which the tropics are well known also facilitate the proliferation of crop diseases, pests, and weeds that continually threaten agricultural yields (Rosenzweig and Liverman, 1992).

Although, as was already mentioned, the tropics exhibit a diversity of climates and regional characteristics, research has shown that these limitations are more likely to be exacerbated rather than mitigated under climate change. Small changes in mean climatic conditions are likely to mean disproportionately larger changes in climatic variability, and consequently an increase in the frequency of extreme events (Burton, 1997; IPCC, 2001; Parry and Carter, 1998). For example, should the ENSO change in temporal duration and frequency — causing several subsequent years of poor production conditions — the impact on food security and economic stability of highly sensitive regions, such as northeastern Brazil and southern Africa, would be severe.

Even changes in mean conditions are unlikely to be helpful. Production in the tropics already occurs near the higher temperature limits of many crops. Although in some areas precipitation is expected to increase, simultaneous increases in temperatures also mean increased evapotranspiration and thus perhaps a decrease in moisture available for plant growth. The prevalence of C4 crops in the tropics, coupled with nitrogen limitations, indicates that yields are unlikely to improve as much for C3 crops in temperate zones from increases in atmospheric carbon levels (IPCC, 2001).

In semi-arid regions, capturing any increase in rainfall for human use is also difficult, particularly if it arrives in concentrated storms as many models anticipate. In these cases, increases in temperature will do little to improve yields and, depending on the timing and frequency of heat spells during the crop growth cycle, may be quite damaging to annual tropical crops (IPCC, 2001; Rosenzweig and Liverman, 1992).

12.3 SOCIAL VULNERABILITY AND FOOD SECURITY

Agriculture in the tropics is more often than not constrained by widespread poverty in rural areas, poor levels of investment in agricultural research, and limited access to markets and improved production technologies (IPCC, 2001; Reilly and Schilmmelpfenning, 1999; Ribot et al., 1996). In many countries, the distribution of fertile land is highly inequitable. Large-scale industrialized producers coexist along with subsistence farmers or semicommercial smallholders, who may contribute little to a country's gross national product, but more often than not represent the majority of a country's farmers. The percent of gross national product coming from agricultural activities in the tropics tends to be far higher than in Organization for Economic Cooperation and Development (OECD) countries (Table 12.1). This necessarily increases the sensitivity of the national economy to environmental and/or economic shocks in the agricultural sector.

This is perhaps nowhere more obvious than in Sub-Saharan Africa, where regional drought often severely constrains economic growth. The 1992 drought in Zimbabwe, for

Table 12.1 Agricultural GDP and Labor Force, Selected Countries (1993)

	Percent of GDP from Agriculture	Percent of Labor Force in Agriculture
Botswana	5.67	60.67
Costa Rica	15.29	22.01
Cote d'Ivoire	37.42	52.69
Ecuador	12.11	28.06
Ethiopia	60.45	72.65
Gambia	27.53	80.19
Guatemala	25.16	49.46
Honduras	19.73	53.24
Kenya	28.92	75.65
Mexico	8.57	28.10
Zimbabwe	15.18	66.65
Organization for Economic Cooperation and Development Countries	2.00	2.00

Source: Data Distribution Center, Intergovernmental Panel on Climate Change. Available at: *http://ipcc-ddc.cru.uea.ac.uk*.

example, reportedly resulted in a 5.8% drop in gross domestic product (GDP) (Benson and Clay, 1996). The same drought in Malawi was viewed as the driving force behind a 6% rise in the fiscal deficit, a 25% drop in agricultural output, and an 8% contraction in the country's economy (World Bank, 1996). By some accounts, the 1992 drought cost the southern Africa region over $4 billion, and had lasting implications for national debt and economic growth.

More significant, however, is the percentage of national populations dependent on agricultural production as a central livelihood activity. In Sub-Saharan Africa and Latin America, the population involved in agricultural activities varies from 25% to as much as 80% of the national labor force (Table 12.1). These large rural populations are also often the most impoverished, the least likely to have reliable access to health care and educational services, and the most likely to be isolated from new sources of information and technology. Climate impacts thus can have a significant impact on national economic growth by affecting the capacity of a population to meet its basic needs. They also directly affect the livelihoods of millions of people through impacts on agriculture yields and productivity.

Discouragingly, during the 1980s and 1990s, income disparities between rural farming populations and urban groups in many developing nations did not significantly improve, and in some cases they actually worsened (Dixon et al., 2001). Although some countries, such as China, have made remarkable improvements in reducing poverty levels since the early 1980s, most other problematic regions have remained so. A report commissioned by the FAO and the World Bank claims that GDP per capita in Sub-Saharan Africa, where an estimated 61% of the total population is dedicated to agricultural activities, was lower at the end of the 1990s compared to the 1970s (Dixon et al., 2001). The same report indicated that 90% of the poor in Sub-Saharan Africa are from rural areas.

Between 1980 and 1998, the percentage of Latin American rural households classified as indigent increased from 28% to 31% (CEPAL, 2000). The World Bank reports persistent income inequality across Latin America, with many countries reporting GINI coefficients of 50% or higher. (The GINI coefficient measures the extent to which income is less than perfectly distributed across a population. A coefficient of 100 is perfectly inequitable; a coefficient of 0 is perfectly equitable. The United States has a GINI coefficient of 40.1, France 32.7, and Germany 30.0, according to the World Bank's 2002 development indicators. In comparison, the GINI coefficient for Guatemala, Honduras, Mexico, Bolivia, and Paraguay are 55.8, 56.3, 53.1, 44.7, and 57.7, respectively.) Poverty in Central America by many accounts has worsened, particularly since the devastation wreaked by Hurricane Mitch (FAO, 2001).

These statistics suggest that calculations of potential future food production in a country, or the impact of climate on simulated future crop yields, may say relatively little about future food security. Any discussion of food security is by definition a discussion of public policy, resource distribution, access, and power (Sen, 1981, 1990). Food insecurity and famine risk are driven not simply by climatic variability and hazards, but also are dictated by political instability, armed conflict, changes in land tenure and/or food pricing policy, changes in income and purchasing power, and progressive degradation of resources available for production (Blaikie et al., 1994; Dreze and Sen, 1989; Sen, 1990). As Reilly and Schimmelpfennig (1999) note, while climate change studies have tended to focus on food supply, food security is generally a question of food access.

12.4 AGRICULTURE, ECONOMIC POLICY CHANGE, AND FOOD SECURITY

The interaction of future climatic variability and change with economic and political uncertainty has received little attention in climate change research (Leichenko and O'Brien, 2002; O'Brien and Leichenko, 2000). Greater openness of global markets has reportedly improved consumer choice and the efficiency of food distribution (FAO, 2003). Markets have opened to foreign investment and trade, and agricultural technologies and information are likely to be more available now to farmers than in the past. However, some populations, often those in developing countries who were the beneficiaries of former protectionist policies and public intervention, have not always felt the full benefits of market liberalization (FAO, 2003). These populations are often the same ones singled out as potentially vulnerable to the negative impacts of climatic change (Leichenko and O'Brien, 2002).

The impacts of market liberalization vary among sectors and countries, making generalization difficult. However, the process of market liberalization in world agriculture has been particularly hard for smallholder and peasant producers who lack the resources to participate actively in export markets. The continued protection of particular agricultural commodities (e.g., cotton) in industrialized nations has also prohibited the competition of many commercial farmers in developing nations. Many farmers in the tropics have faced declining prices for their harvests, rising costs of imported inputs, reduced access to private and public finance, and greater market volatility (Buttel, 1997; Dixon et al., 2001; Gledhill, 1995; Leichenko and O'Brien, 2002; McMichael, 1994).

These farmers enter the 21st century at a considerable disadvantage. They tend to have been subject to considerable past state intervention and protectionism. In many countries, smallholders were also unable to benefit directly from the technological advances of the Green Revolution of the 1960s and 1970s, and since that period have often been ignored in national agricultural research and development programs in favor of more "modern" farming subsectors.

Open markets can mean lower food prices for net food buyers, but they can also mean increased dependence on food imports, and thus greater sensitivity to market volatility and the politics of international agricultural trade. As commodity chains become increasingly integrated and controlled by a handful of large transnational agribusinesses (FAO, 2003), this dependence is also subject to the policies and performance of a limited number of large economic enterprises. (The FAO report, *World Agriculture Towards 2015/2030*, estimates that only three transnational companies now control 80% of U.S. maize exports. This means that the performance of these three companies has important implications for countries such as

Mexico, which derives 98% of its maize imports from the United States.)

There is also a concern that, as agriculture in developing nations is restructured to meet global consumer preferences and demands, production will be increasingly disconnected from local consumption (FAO, 2003; Friedmann, 1994). As part of this process, major food staples and the places in which they are produced that have no place in global markets, such as millet, cassava, and preferred varieties of maize or rice, can be devalued (Marsden, 1997). With food supplies determined by the preferences of wealthier consumers at the global scale, and food-deficit countries increasingly import dependent, populations in developing regions have progressively less control over the price, quality, and content of the foods they import, and when and where food is available (Raynolds, 1997). For many developing countries, the impact of climate change will thus depend largely on how their agricultural sectors respond to both old challenges such as persistent hunger, population growth, and degraded resources, and new challenges, such as market globalization, and access to technology and information, in addition to climatic variability and change (Reilly and Schilmmelpfenning, 1999).

On a more optimistic note, many nations considered to be vulnerable to hunger and climatic hazards may also be a valuable source of knowledge and experience about adaptation to risk and survival under climatic uncertainty. Farmers in Latin America have a long history of adaptation to climatic and other biophysical constraints (Denevan, 1980; Whitmore and Turner, 1992). A wide diversity of climatic-risk mitigation strategies, including soil moisture conservation techniques, tillage practices, seed selection for stress tolerance, intercropping, and other forms of microclimate adjustments, continue to be practiced in many traditional agricultural systems in Mexico (see, for example, Altieri and Trujillo, 1987; Bellon, 1991; Brush et al., 1988; Doolittle, 1989; Trujillo, 1990; Wilken, 1982, 1987).

The extent of and possibilities for adaptation have also been illustrated in research on climate forecast applications in Sub-Saharan Africa, Mexico, and South America. Small farmers are acutely aware of climatic variability and its consequences for production as well as a wide range of possible management techniques for mitigating anticipated impacts (Eakin, 1999, 2000; O'Brien and Vogel, 2003; Phillips et al., 1998). While it would be erroneous to suggest that these methods are always effective, they represent a wealth of knowledge and experience that should be used as a point of departure for research on more resilient and adaptive agricultural systems.

However, to be adaptive a system needs to be flexible, and in order to respond proactively, it also needs to have a certain amount of stability. Instability and volatility tend to inhibit capacities for future planning. Perhaps most important, adaptation depends on having access to resources financial, human, social, physical, natural in order to accommodate changing demands on land's productive capacity. Although many smallholders are capable of finding niches in today's more open markets, they do not have access to investments in technology and research that benefited agriculture in more industrial countries in recent decades (Cotter, 1994; Loker, 1996; Yapa, 1996). Small farmers in Latin America face challenges related to participating in commercial markets, such as lack of credit, rising input costs, lack of access to agricultural research, poor rural extension services, and water scarcity. All these factors will complicate their ability to employ the types of adaptations to climatic change often proposed for industrial agriculture.

Thus, as attention is given to the challenge of adaptation, it will become even more important to understand the role of climate in accentuating existing vulnerabilities, and, conversely, to understand the role of policy trends and economic stress in increasing farmers' sensitivity to climate impacts in the near future. Although these interactions may still defy modeling efforts, the recent past can serve as a useful reference for the near and perhaps medium term strategy formulation. The implications of policy change, economic uncertainty, and climatic change for future food security and food sovereignty are examined in the next section by examining Mexican maize production and policy.

12.5 MAIZE AND AGRICULTURAL VULNERABILITY IN MEXICO

Maize is the staple grain of Mexican cuisine and is the most essential contribution to the nutrition and sustenance of Mexico's population. By some estimates, maize contributes up to 50% of total calories consumed in Mexico, and up to 70% in rural areas (Fritscher Mundt, 1999).

Maize also has great cultural significance. From its initial domestication, it has had a central role in the evolution of Mesoamerican religion, mythology, social organization, and economy (León-Portilla, 1988). In the mid-1990s, it was estimated that anywhere from 2 to 3 million Mexican farmers were involved in white maize production, primarily on farms of less than 5 hectares (Nadal, 1999). In 1998, despite two decades of declining producer prices and increased maize imports, over half of Mexico's total cultivated area was planted in maize, primarily under rainfed conditions (SAGARPA, 2000). Today, Mexicans consume over 11 million metric tons of maize annually (SAGARPA, 2000).

Maize is thought to have been first domesticated in Mexico's central highlands, in the valley of Tehuacan, and subsequently adapted by smallholder farmers to a wide range of environmental conditions and social uses (Wellhausen et al., 1952). Although the subhumid highlands are known as the geographic center of Mexico's traditional maize farming systems, maize is as common in the semi-arid and arid northern states as in Mexico's southern tropical states. In the north, commercial maize is grown extensively under irrigation on large plots of over 100 ha. And in the central and southern states, maize often is part of complex agro-ecosystems in which it is primarily used for household consumption.

Any analysis of the attributes of the numerous local maize varieties illustrates the malleability of the crop as well as the extensive adaptive knowledge of the farmers who select their maize varieties according to their food preferences, economic needs, and variability of environmental conditions (Bellon, 1991, 1995; Trujillo, 1990). Differences in the attributes of local maize varieties have long been used by farmers to adjust to climatic variability. Typically, the varieties that farmers report to be the most resistant to drought impacts or frost risk are those that have been the least developed commercially and, in years of good climate conditions, those which tend to have lower yields than certified seeds (Eakin, 1998, 2002).

Household food security is the primary objective of most maize producers in Mexico, although the 1.5 to 2.0 metric tons required to feed the average household are often quite difficult to obtain from the small areas under household production (de Janvry et al., 1997; Eakin, 2002). Yields tend to be highly variable despite farmers' management of maize varieties according to climatic conditions (Figure 12.1). The summer rainy season of Mexico's highlands (May to September) is usually just sufficient for the growth cycle of maize. Much of eastern Mexico is affected by a mid-summer drought, whose intensity and timing can exhibit considerable variability. Frost is also a limiting factor in the highlands, preventing

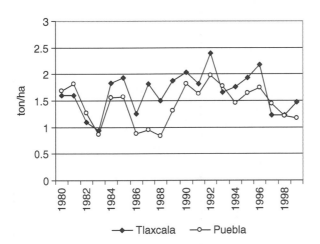

Figure 12.1 Average rainfed maize yields, Tlaxcala and Puebla, Mexico. (Based on data from Centro de Estadística Agropecuaria (CEA). 2003. Sistema de Información Agropecuaria de Consulta (SIACON). Versión 1.1. Secretaria de Agricultura, Gandería, Desarrollo Rural, Pescaría y Alimentación, Mexico City. http://www.siap. sagarpa.gob.mx/)

farmers from planting some slow-maturing varieties and limiting the flexibility of planting dates (Eakin, 1998).

Maize has always been an important element of national food security, and thus featured in both agricultural and social welfare policy. However, the 1980s and 1990s marked a dramatic shift in state policy. After the economic crisis of 1982, Mexico abandoned the goal of maize self-sufficiency, a policy objective that had been particularly important in the 1970s (Appendini, 1994; Escalante and Redón, 1987). By the mid to late 1980s, Mexico's elaborate and expensive system of urban food subsidies and guaranteed producer prices was largely dismantled (Appendini, 1998). Instead, the country turned toward models of comparative advantage, foreign investment, and free trade to address domestic food concerns. Mexico's adoption of a neoliberal policy involved a gradual but decisive withdrawal of state-supported agricultural services and inputs, crop price guarantees, and input subsidies, and a gradual opening of domestic agricultural markets to imports and foreign investment. As a result, in comparison with the high yields and large subsidies enjoyed by U.S. producers, Mexican maize is essentially noncompetitive in today's national and international markets.

The designers of the agricultural portion of the North American Free Trade Agreement (NAFTA) of 1994 expected that Mexico's farmers would make their livelihoods in alternative high-value export crops, such as specialty vegetables and tropical fruits, or move out of agriculture into gainful employment in alternative sectors (Dussel Peters, 2000). Under NAFTA, Mexico was permitted to protect its maize market for 15 years by imposing a tariff on all maize imported above an annually established import quota. However, in response to the failure of domestic production to keep up with demand, maize imports increased significantly in the 1990s. Although an important portion of these imports exceed the established annual tariff-free quotas, Mexico frequently opted not to enforce this protective policy and the tariffs were not collected (Dussel Peters, 2000; Fritscher Mundt, 1999). Thus, Mexican commercial maize farmers felt the impact of NAFTA almost immediately. Easy entry of imported maize to Mexican markets and open market pricing have reduced the ability of Mexican farmers to market their maize internally. This is true despite the fact that most maize produced in Mexico is white maize, and imports from the United States are typically lower-quality yellow maize. As mentioned above, 98% of annual maize imports are now from the United States (Dussel Peters, 2000).

Given that the price of yellow maize is quite sensitive to the global demand for livestock feed and to U.S. production and agricultural policy, prices are reportedly now more variable than they were in the past (Fritscher Mundt, 1999). Mexican maize prices in real terms have fallen by 13% since the implementation of NAFTA, and in 1997 the real price of maize was 54% of its 1990 value (Dussel Peters, 2000). Many farmers now agree that, "Maíz ya no es negocio" (Maize is no longer profitable) (Eakin, 2002).

Unfortunately for Mexican farmers, these inauspicious terms of trade and changes in sector policy coincided with an unusual frequency and intensity of ENSO events in the latter half of the 1990s. These events were partially the cause of significant crop losses (Figure 12.2). The mild El Niño of 1995 marked the start of a 3-year drought that lasted through the spring of 1998 (Magaña R., 1999). Forest fires, initiated by farmers who were clearing land for plowing, were so severe in the spring of 1998 that the skies of Arizona and Texas became clouded with ash and particulates. The landfall of hurricane Pauline in the fall of 1998 brought torrential rains and provoked landslides in Mexico's central highlands, causing the most damage in highly marginal agricultural regions in which subsistence maize production is critical for local food security. And unusual frost events in the middle of central Mexico's growing season reduced maize yields in parts of central Mexico in both 1998 and 1999 (Eakin, 2002). Even irrigated regions faced water shortages and crop losses. The impact of these events on domestic maize production have been used to explain Mexico's increasing maize imports in the 1990s (SAGARPA, 2000).

Given this combination of circumstances, it is perhaps not surprising that Mexico's rural population (approximately

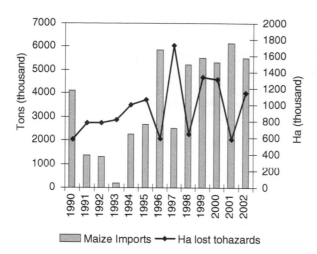

Figure 12.2 Maize imports and area lost to hazards, 1990–2002. (From Fox Quesada, V. 2003. Tercer Informe de Gobierno. Presidencia, Gobierno de Mexico, Mexico City.)

25% of the total) has not adapted to the open markets of the 1990s as smoothly as originally hoped. The collapse of domestic maize market has occurred in a context of declining incomes, increased livelihood insecurity, and persistent poverty in rural areas (Hernández Laos and Velásquez Roa, 2003; Kelly, 1999). Average gross income per hectare of rainfed maize since 1995 has declined by more than half (Hernández Laos and Velásquez Roa, 2003). By some estimates, 65% of rural households in Mexico are now unable to meet their basic needs (Hernández Laos and Velásquez Roa, 2003).

Surprisingly, despite the level of hardship found in agriculture, Mexico's rural population does not appear to have shifted into other formal economic sectors for the simple reason that there are few alternatives available. Instead, subsistence maize production has increased. Indeed, some fear that income-stressed households have expanded production onto marginal lands in order to meet their consumption requirements (de Janvry et al., 1997; Nadal, 1999). Recent case studies have illustrated that annual investments in subsistence production may in fact be necessary for some rural households to afford to diversify into alternative nonfarm activities, or even into more commercial crops (Eakin, 2002). Instead of leaving agriculture, households are diversifying into more informal economic activities or supporting migration to the United States by some of their members as a central means of rural survival (de Janvry and Sadoulet, 2001; Hernández Laos and Velásquez Roa, 2003).

Climate change in Mexico is occurring in this socioeconomic context. Increased climatic variability, including more frequent and perhaps more prolonged drought events, increased risk of torrential rainfall and flooding, and increasing maximum temperatures, can all be considered plausible for Mexico under different climate change scenarios (Conde et al., 1997). A trend of rising temperatures during April and May, the driest months of the year, is raising alarms about the country's capacity to provide sufficient water to meet current demands (Magaña, 2005). Global circulation models project that the geographic area considered unsuitable for maize production will increase in the future, by 8% to 37% (Conde et al., 1997). Others speculate that rainfed maize yields may decline by 20% to 61% under different climate change scenarios, largely because of reduced grain-filling periods caused by higher temperatures (Liverman et al., 1992). Irrigated production may become particularly stressed by increased competition for water, and increases in water evaporation caused by higher temperatures (Liverman and O'Brien, 1991).

The models that produced the above results did not take into consideration increases in the frequency of ENSO events, or possible future increases in climatic variability. Recent research shows that despite general warming trends, anomalous frost events continue to affect maize yields in the highland regions, and that they may be linked to ENSO (Eakin, 2002; Morales and Magaña, 1999). The impact of climatic change on the behavior of the mid-summer drought, a climatic feature of the highland regions, is also difficult to model. However, it has been shown to be quite sensitive to ENSO (Magaña et al., 1999). Variations in the timing and intensity of the mid-summer drought have also been shown to have important impacts on agricultural yields (Pereyra Diaz et al., 1994).

Adaptations such as increased fertilizer applications, improvements in irrigation infrastructure and efficiency, and new, more drought-tolerant seed varieties, will theoretically help Mexican farmers adapt to some of the worst impacts of climate change (Conde et al., 1997). However, expansion in irrigation is not a viable option for most regions, although improved efficiency in irrigation may provide substantial benefits for areas already with irrigation infrastructure. Chemical fertilizers have become more costly without public input subsidies and are too expensive for most smallholders.

Although farmers in Mexico's northern states are also likely to be affected by rising temperatures and decreasing rainfall, they are relatively better connected to agricultural markets, production supports, irrigation, and technical assistance than their counterparts in Mexico's central highlands and southern tropical states. In order to adapt simultaneously to new market challenges and climatic risks, smallholders in central and southern Mexico will need access to inexpensive technologies that will simultaneously improve their productivity and reduce their input costs. They will also need access to seed stocks that combine the resilience of the diverse varieties that they have traditionally used to manage risk, with new commercial advantages. They will require information about markets as well as assurances that taking new commercial risks will not result in greater hunger, debt, or increased poverty.

Some interesting and innovative projects in Mexico are designed to enhance smallholders' more successful traditional risk management strategies while helping them enter and hold their own in commercial markets. Cortés et al. (2004) describes one such project, involving agricultural intensification, intercropping, carbon sequestration, and commercial fruit production. In February 2003, the Secretary of Agriculture, Livestock, Rural Development, Fisheries, and Food (SAGARPA) also announced its support for a collaborative applied research project to improve local maize varieties. The lessons from these projects may be particularly instructive as

Mexico strives to improve its capacity for agricultural adaptation, but only if these projects are used to inform national policy.

Given that public support for agricultural research has been largely withdrawn, and that the budget for public investment in agriculture has been reduced (de Janvry et al., 1995; Myhre, 1994), it may be more difficult to use successful local experiments to improve rural livelihood stability and agricultural policy at the national level. Justification for investments in maize farmers will be particularly challenging when trade policy assumes that Mexico has no future in maize production. Yet without alternative sources of livelihood for rural residents, local maize production will remain a critical element of rural food security. A return to the old protectionist and distortionary polices of earlier decades, or to demand that Mexico attempt to become maize self-sufficient again, is unreasonable. However, the smallholder in Mexico cannot be assumed away.

In order for Mexican smallholders to be able to face both climatic risk and the challenges of open markets, rural policy needs to be reoriented and public agricultural research revitalized. If such efforts were to result in improved access by rural smallholders households to education, to domestic markets, to appropriate low-cost technologies and to environmental information (the essential resources of the global age), the resilience of rural populations to climatic risk would also improve, regardless of whether they remain in agriculture or find alternative sources of livelihood.

12.6 CONCLUSION

The impact of climate on present and future food security cannot be evaluated solely through use of crop yields estimates and models of future production and food deficits. Uncertainties in such estimates are large. There is also a clear need to understand how climatic factors might interact with future food and agricultural markets, sector and trade policy, and political instability, as well as the relevance of historical resource inequities and past development trends. The high sensitivity of tropical production to climatic extremes and variability, the importance of the sector to the livelihood security of significant proportions of the total population in the tropics, and the disadvantaged position of tropical food producers in globalized agricultural markets, present difficult challenges for the future food security of these regions. Even in countries such as Mexico, in which famine is only a very distant memory, malnutrition and rural poverty have become chronic problems.

Some populations that are currently highly vulnerable to food insecurity have, in the past, shown considerable resilience and innovation when confronted by risk and uncertainty. However, not enough is being done globally to strengthen that capacity and to build on existing knowledge. We do not yet know the types of farm systems or the types of agricultural technologies that will be most appropriate and most flexible in the future. Despite this uncertainty, the technology preferences and research priorities of the industrialized world are disturbingly becoming decisive for all.

The trend toward less public investment in agricultural research in developing nations and the concentration of basic grain production among a handful of nations may be undermining what remains of the capacity of rural populations in the tropics to feed themselves or to have the resources with which to purchase what they require.

Despite the fact that global food production may keep up with global population growth, as models tend to predict, current regional and local trends in income disparity, persistent poverty, and food insecurity are cause for considerable concern. While trade has increased significantly across the developing world, a recent CIMMYT (Centro Internacional de Mejoramiento de Maíz y Trigo) report argues that trade alone will be unable to accommodate the growing maize needs in developing nations (Pingali and Pandey, 2000). According to this assessment, countries in which maize is the staple grain face particularly difficult challenges. Over the period 1995 to 2020 maize demand in developing nations will increase by an average of 50% (Pingali and Pandey, 2000). Yet the political and economic trends observed in Mexico and mirrored in other Latin American nations (FAO, 2001) suggest that meeting these needs with improvements in domestic production will be difficult.

It is likely that rural Mexico's declining command over food and its persistent poverty would have occurred without the climatic events of the 1990s. However, droughts that affected the commercial agricultural districts of Mexico's northern states, and the frosts, floods, and water stress in Mexico's highlands apparently contributed to both the political decision to increase Mexico's reliance on foreign grain and to reduce public investments in a crop considered to have a limited future in Mexico (SAGARPA, 2000). Mexico's domestic production of maize, and the access of its citizenry to the quantity and type of maize that it requires, is thus being doubly threatened by both climatic uncertainties and economic realities. As a result although white maize continues to be the preferred crop for millions of small farmers, and is critical in the diet of millions of Mexicans the role of domestically produced maize in Mexico's future food security is uncertain.

As adaptation to climate change is introduced into national policy debates, the tendency may be to create adaptation policies, or in other words, distinct programs and policy initiatives designed to specifically address climatic threats to sector development or infrastructure. Separating adaptation to climatic risk from adaptation to economic challenges may be useful in climate research, but poorly reflects the reality experienced by many of the world's farmers. Changing the focus of analysis from the model to the field, and from the crop to the farmer may help to understand the complex equation of global change and local response to it. It is hoped that this understanding will hold the key to more sustainable and humane development, as well as to greater resilience and adaptation by rural smallholders in Mexico and elsewhere.

REFERENCES

Agnew, C.T. 1998. Climate, agriculture and vegetation in the tropics. In C.C. Webster and P.N. Wilson, Eds. *Agriculture in the Tropics*. Blackwell Science, Oxford.

Altieri, M. and Trujillo, J. 1987. The agroecology of corn production in Tlaxcala, Mexico. *Hum. Organ.,* 15:189–220.

Appendini, K. 1994. Transforming food policy over a decade: the balance for Mexican corn farmers in 1993. In C.H.D. Alcántara, Ed. *Economic Restructuring and Rural Subsistence in Mexico: Corn and the Crisis of the 1980s*. Center for U.S.–Mexican Studies, University of California, San Diego.

Appendini, K. 1998. Changing agrarian institutions: interpreting the contradictions. In W. Cornelius and D. Myhre, Eds. *The Transformation of Rural Mexico: Reforming the Ejido Sector*. Center for U.S.–Mexico Studies, University of California, San Diego.

Bellon, M.R. 1991. The ethnoecology of maize variety management: a case study from Mexico. *Hum. Ecol.,* 19:389–418.

Bellon, M.R. 1995. Farmer's knowledge and sustainable agroecosystem management: an operational definition and an example from Chiapas, Mexico. *Hum. Organ.,* 54:263–272.

Benson, C. and Clay, E. 1996. The impact of drought on sub-Saharan African economies: a preliminary assessment. Overseas Development Institute, London.

Blaikie, P., Cannon, T., Davis, I., and Wisner, B. 1994. *At Risk: Natural Hazards, People's Vulnerability and Disaster*. Routledge, London.

Brush, S., Bellon Corrales, M., and Schmidt, E. 1988. Agricultural development and maize diversity in Mexico. *Hum. Ecol.,* 16:307–328.

Burton, I. 1997. Vulnerability and adaptive response in the context of climate and climate change. *Climatic Change,* 36:185–196.

Buttel, F.H. 1997. Some observations on agro-food change and the future of agricultural sustainability movements. In D. Goodman and M. Watts, Eds. *Globalising Food*. New York: Routledge.

Cane, M. 2001. Understanding and Predicting the World's Climate System. Paper presented at Impacts of El Niño and Climate Variability on Agriculture Conference, Beltsville, MD.

Centro de Estadística Agropecuaria (CEA). 2003. Sistema de Información Agropecuaria de Consulta (SIACON). Versión 1.1. Secretaria de Agricultura, Ganadería, Desarrollo Rural, Pescaría y Alimentación. http://www.siap.sagarpa.gob.mx/.

Chen, R. and Katz, R. 1994. Climate change and world food security. *Global Environ. Change,* 4:3–6.

Comision Economica para America Latina y El Caribe (CEPAL). 2000. *Panorama Social de América Latina.* Santiago, Chile.

Conde, C., Liverman, D., Flores, M., Ferrer, R., Arajo, R., Betancourt, E., Villarreal, G., and Gay, C. 1997. Vulnerability of rainfed maize crops in Mexico to climate change. *Climate Res.,* 9:17–23.

Cortés, J.I., Turrent, A., Díaz, P., Jiménez, L., Hernández, E., and Mendoza, R. 2004. Chapter 23, this volume.

Cotter, J. 1994. Salinas de Gotari's agricultural policy and scientific exchange: some lessons from before and during the Green Revolution. In E.C. Ochoa and D.E. Lorey, Eds. *Estado y Agricultura en México: Antecedentes e Implicaciones de las Reformas Salinista.* Universidad Autonoma Metropolitana, Mexico City, pp. 39–55.

de Janvry, A., Chiriboga, M., Colmenares, H., Hintermeister, A., Howe, G., Irigoyen, R., Monares, A., Rello, F., Sadoulet, E., Secco, J., Pluijm, T.v.d., and Varese, S. 1995. Reformas del Sector Agricola y el Campesinado en Mexico. Fondo Internacional de Desarrollo Agrícola y Instituto Interamericano de Cooperacíon para la Agricultura, San José, Costa Rica.

de Janvry, A., Gordillo, G., and Sadoulet, E. 1997. Mexico's Second Agrarian Reform: Household and Community Responses. Center for U.S.–Mexico Studies, University of California, San Diego.

de Janvry, A. and Sadoulet, E. 2001. Income strategies among rural households in Mexico: the role of off-farm activities. *World Dev.,* 29:467–480.

Denevan, W.M. 1980. Latin America. In G.A. Klee, Ed. *World Systems of Traditional Resource Management.* Halsted Press, New York, pp. 217–256.

Dilley, M. and Heyman, B. 1995. ENSO and disaster: droughts, floods, and El Niño/Southern Oscillation warm events. *Disasters*, 19:181–193.

Dixon, J., Gulliver, A., and Gibbon, D. 2001. *Farming Systems and Poverty: Summary*. Food and Agriculture Organization, Rome; World Bank, Washington, DC.

Doolittle, W. 1989. Arroyos and the development of agriculture in northern Mexico. In J.O. Browder, Ed. *Fragile Lands in Latin America: Strategies for Sustainable Development*. Westview Press, Boulder, CO.

Downing, T.E., Ed. 1996. *Climate Change and World Food Security*. Springer-Verlag, Berlin.

Dreze, J. and Sen, A. 1989. *Hunger and Public Action*. Clarendon Press, Berlin.

Dussel Peters, E. 2000. El Tratado de Libre Comercio de Norteamerica y el Desempeo de la Economia en Mexico. LC/MEX/L.431. Comision Economica para America Latina y El Caribe, Mexico City.

Eakin, H. 1998. Adapting to Climate Variability in Tlaxcala, Mexico: Constraints and Opportunities for Small-scale Maize Producers. Master's thesis, University of Arizona, Tucson.

Eakin, H. 1999. Seasonal climate forecasting and the relevance of local knowledge. *Phys. Geogr.*, 20:447–460.

Eakin, H. 2000. Smallholder maize production and climatic risk: a case study from Mexico. *Climatic Change*, 45(1): 19–36.

Eakin, H. 2002. Rural Households' Vulnerability and Adaptation to Climatic Variability and Institutional Change. Ph.D. dissertation, University of Arizona, Tucson.

Escalante, R. and Redón, T. 1987. Neoliberalismo a la Mexicana: su impacto sobre el sector agropecuario. *Problemas del Desarrollo*, 75:115–151.

Food and Agriculture Organization. 2003. *World Agriculture: Towards 2015/2030*. Earthscan, London.

Food and Agriculture Organization. 2001. *Analysis of the Medium-term Effects of Hurricane Mitch on Food Security in Central America*. FAO, Rome.

Fox Quesada, V. 2003. Tercer Informe de Gobierno. Presidencia, Gobierno de México, Mexico City.

Friedmann, H. 1994. Distance and durability: shaky foundations of the world food economy. In P. McMichael, Ed. *The Global Restructuring of Agro Food Systems*. Cornell University Press, Ithaca, NY.

Fritscher Mundt, M. 1999. El maíz en México: auge y crisis en los noventa. *Cuadernos Agrarios*, 17–18:142–163.

Gledhill, J. 1995. *Neoliberalism, Transnationalization and Rural Poverty*. Westview Press, Boulder, CO.

Gommes, R. 1993. Current climate and population constraints on world agriculture. In H.M. Kaiser and T.E. Drennen, Eds. *Agricultural Dimensions of Global Climate Change*. St. Lucie Press, Delray Beach, FL.

Helms, S., Mendelsohn, R., and Neumann, J. 1996. The impact of climate change on agriculture. *Climatic Change*, 33:1–6.

Hernández Laos, E., and Velásquez Roa, J. 2003 *Globalizaciun, Desigualdad y Pobreza: Lecciones de la Experiencia Mexican*. Universidad Autónoma Metropolitana, Mexico City.

Intergovernmental Panel on Climate Change. 2001. *Climate Change 2001: Impacts, Adaptation and Vulnerability*. Cambridge University Press, London; New York.

Kelly, T. 1999. *The Effects of Economic Adjustment on Poverty in Mexico*. Ashgate Publishing, Brookfield, VT.

Kelly, T. 2001. Neoliberal reforms and rural poverty. *Latin American Perspectives*, 28:84–103.

Leichenko, R. and O'Brien, K. 2002. The dynamics of rural vulnerability to global change: the case of Southern Africa. *Mitigation and Adaptation Strategies for Global Change*, 7:1–18.

León-Portilla, M. 1988. El maíz: nuestro sustento, su realidad divina y humana en Mesoamerica. *America Indígena*, XLVIII:477–502.

Liverman, D., Dilley, M., O'Brien, K., and Menchaca, L. 1992. *The impacts of global warming on Mexican maize yields*. U.S. Environmental Protection Agency, Washington, DC.

Liverman, D. and O'Brien, K. 1991. Global warming and climate change in Mexico. *Global Environ. Change*, 1(5):351–364.

Loker, W.M. 1996. Campesinos and the crisis of modernization in Latin America. *J. Political Ecol.*, 3:69–88.

Magaña, V.O., Ed. 1999. Los Impactos de El Niño en México. – Secretaria de Educacíon Pública (SEP) and Consejo National de Ciencias y Tecnología (CONACYT), Mexico City.

Magaña, V. Universidad Nacional Autonoma de México, personal communication, Jan. 11, 2005.

Magaña, V., Amador, J., and Medina, S. 1999. The midsummer drought over Mexico and Central America.*J. Climate*, 12:1577–1588.

Marsden, T. 1997. Creating space for food: the distinctiveness of recent agrarian development. In D. Goodman and M. Watts, Eds. *Globalizing Food: Agrarian Questions and Global Restructuring*. Routledge, London.

McMichael, P., Ed. 1994. *The Global Restructuring of Agro-Food Systems*. Cornell University Press, Ithaca, NY, and London.

Morales, T. and Magaña, V. 1999. Unexpected frosts in central Mexico during summer. Paper presented at the 11th Conference on Applied Climatology, Dallas, TX.

Myhre, D. 1994. The politics of globalization in rural Mexico: campesino initiative to restructure the agricultural credit system. In P. McMichael, Ed. *The Global Restructuring of Agro-Food Systems*. Cornell University Press, Ithaca, NY, and London, pp. 145–169.

Nadal, A. 1999. *Maize in Mexico: Some Environmental Implications of the North American Free Trade Agreement (NAFTA)*. Environment and Trade Series 6, Commission on Environmental Cooperation, Montreal.

O'Brien, K. and Vogel, C., Eds. 2003. *Coping with Climate Variability: User Responses to Seasonal Climate Forecasts in Southern Africa*. Ashgate Publishing, Brookfield, VT.

O'Brien, K.L. and Leichenko, R.M. 2000. Double exposure: assessing the impacts of climate change within the context of economic globalization. *Global Environ. Change*, 10:221–232.

Parry, M. 1990. *Climate Change and World Agriculture*. Earthscan, London.

Parry, M., Arnell, N., McMichael, T., Nicholls, R., Martens, P., Kovats, S., Livermore, M., Rosenzweig, C., Iglesias, A., and Fisher, G. 2001. Millions at risk: defining critical climate change threats and targets. *Global Environ. Change*, 11:181–183.

Parry, M. and Carter, T. 1998. *Climate Impact and Adaptation Assessment*. Earthscan, London.

Pereyra Diaz, D., Angulo Cordova, Q., and Palma Grayeb, B.E. 1994. Effect of ENSO on the mid-summer drought in Veracruz State, Mexico. *Atmsfer*, 7:111–119.

Phillips, J.G., Makaudze, E., and Unganai, L. 1998. Current and potential use of climate forecasts for resource-poor farmers in Zimbabwe. Paper presented at the Impacts of Climate Variability on Agriculture: Regional Effects and Use of Climate Forecasts in Crop Management, Beltsville, MD.

Pingali, P.L. and Pandey, S. 2000. *Meeting World Maize Needs: Technological Opportunities and Priorities for the Public Sector*. Centro Internacional de Mejoramiento de Maíz y Trigo, Mexico City.

Raynolds, L. 1997. Restructuring national agriculture, agro-food trade, and agrarian livelihoods in the Caribbean. In D. Goodman and M. Watts, Eds. *Globalising Food*. Routledge, New York.

Reilly, J.M. 1995. Climate change and global agriculture: recent findings and issues. *Am. J. Agric. Econ.*, 77:727–733.

Reilly, J.M. and Schilmmelpfenning, D. 1999. Agricultural impact assessment, vulnerability and the scope for adaptation. *Climatic Change*, 43:745–788.

Ribot, J.C., Najam, A., and Watson, G. 1996. Climate variation, vulnerability and sustainable development in the semi-arid tropics. In J.C. Ribot, A.R. Magalhaes, and S.S. Panagides, Eds. *Climate Variability, Climate Change and Social Vulnerability in the Semi-arid Tropics*, Cambridge University Press, London; New York, pp. 13–51.

Rosenzweig, C. and Liverman, D. 1992. Predicted effects of climate change on agriculture: a comparison of temperate and tropical regions. In S. K. Majumdar, Ed. *Global Climate Change: Implications, Challenges and Mitigation Measures*. Pennsylvania Academy of Sciences, Pittsburgh, pp. 342–361.

Rosenzweig, C. and Parry, M. 1993. Potential impacts of climate change on world food supply: a summary of a recent international study. In H. Kaiser and T.E. Drennen, Eds. *Agricultural Dimensions of Global Climate Change*. St. Lucie Press, Delray Beach, FL.

Sen, A. 1981. *Poverty and Famines*. Clarendon Press, Oxford.

Sen, A. 1990. Food, economics and entitlements. In J. Dreze and A. Sen, Eds. *The Political Economy of Hunger*. Clarendon Press, Oxford.

Trujillo, J.A. 1990. Adaptación de sistemas tradicionales de producción de maíz a las condiciones "siniestrantes" de Tlaxca. *Historia y Sociedad en Tlaxcala*, October, pp. 67–70.

Wellhausen, E.J., Roberts, L.M., and Hernandez X.E. 1952. *Races of Maize in Mexico*. Harvard University, Cambridge, MA.

Whitmore, T.M. and Turner II, B.L. 1992. Landscapes of cultivation in Mesoamerica on the eve of the conquest. *Ann. Assoc. Am. Geogr.*, 82:402–425.

Wilken, G. 1982. *Agroclimatic Hazard Perception, Prediction and Risk-Avoidance Strategies in Lesotho*. Department of Geography, Natural Hazards Research Group, Colorado State University, Boulder.

Wilken, G. 1987. *Good Farmers: Traditional Agricultural Resource Management in Mexico and Central America*. University of California Press, Berkeley.

World Bank. 1996. Drought in Malawi: From Crisis Response to Strategic Management. Working Paper. Southern Africa Department, World Bank, Washington, DC.

Yapa, L. 1996. Improved seeds and constructed scarcity. In R. Peets and M. Watts, Eds. *Liberation Ecologies*. Routledge, New York.

13

Effects of Global Climate Change on Agricultural Pests: Possible Impacts and Dynamics at Population, Species Interaction, and Community Levels

ANTHONY JOERN, J. DAVID LOGAN, AND
WILLIAM WOLESENSKY

CONTENTS

Agricultural pests and diseases will not be taken into account unless the ecology of weeds, pests, and diseases under global climate change are explicitly addressed. It is an area of potentially major agricultural and forestry impact; but unfortunately one which has not progressed very much.

B. Walker
Global Change and Terrestrial Ecosystems (1996)

13.1 INTRODUCTION

On average, Earth's climate has warmed significantly (~0.6°C) over the last century, and additional rapid changes are anticipated (Houghton et al., 2001). General circulation models for assessing future climate, coupled to direct measurements, predict that atmospheric CO_2 levels will continue to increase (Schneider et al., 1992; Vitousek, 1994). More locally, significant regional climate changes in the primary agricultural regions of North America are anticipated (Rind et al., 1990; Schneider et al., 1992; Schneider, 1993; Reddy and Hodges, 2000), including increased summer temperatures (4°C to 8°C), decreased precipitation, and a significant drop in available soil water (ca. 40% to 50%), each associated with a significant change in seasonality. Some analyses indicate that both environmental changes and system responses can be sudden rather than gradual, further exacerbating the impact (Rietkerk and van de Koppel, 1997; van de Koppel et al., 1998; Alley et al., 2003; Gu et al., 2003). Human activity has greatly affected atmospheric concentrations of patterns of deposition of gaseous pollutants (e.g., SO_x, NO_x) that could affect insects in agro-ecosystems (Brown, 1995). In this context, important questions remain about responses of important natural and agricultural systems, including the issue of how agricultural pests and resulting food security will be affected. Diverse arthropod pests and plant pathogens significantly reduce agricultural production at present (Pimentel, 1991). How pests and their interactions with other organisms and the environment will respond to these changes, and whether such changes can be predicted is a big concern — one that requires consideration of regional and local climate changes that transcend global averages (Cammell and Knight, 1992; Lansberg and Smith, 1992; Harrington and Stork, 1995; Walther et al., 2002). The problem is complex and there are many possible ways that climate change from increased CO_2 can affect the outbreak potential of insect herbivores in agricultural systems. In general, if CO_2 continues to rise, average temperatures will also increase and precipitation will become more variable, suggesting that effects from temperature-dependent processes and plant responses to environmental stresses are the keys for understanding insect pests. One must also be cognizant of altered land use in the face of climate change as farmers and ranchers are likely to adapt and plant the most appropriate crops for new environmental and economic conditions.

Altered temperature and variable food quality (e.g., from changed CO_2 levels and other atmospheric pollutants) will likely be primary but not exclusive drivers affecting agricultural pests (Fajer, 1989; Fajer et al., 1991; Ayres, 1993; Lindroth, 1996a, 1996b). Because insect populations often respond to plant foliar quality, plant communities, and vegetation structure, which in turn are expected to change with CO_2 concentrations, effects of climate change on insect pests may initially be best understood as responses by plants that are then tracked by insects. However, we must also recognize that insect herbivores are attacked by a number of natural enemies, and these important interspecific interactions may also be vulnerable to environmental changes. Here, we examine a range of possible ecological responses that must be addressed to assess responses to agricultural insect pests to increased CO_2 and climate change. The task will be very challenging if predictive insights are required because so little is currently known. Even though insect pests are poorly studied in the context of global climate change, scientists are beginning to amass many examples. Because "pests" are really nothing more than species living where they are not wanted, capable of reaching high densities with economic impact (Nothnagle and Schultz, 1987), no loss of insight results from our approach of relying on other, better-studied taxa to develop understanding.

Ample evidence documents that organisms are responding to global climate change (Harrington and Stork, 1995; Dukes and Mooney, 1999; Harrington et al., 1999; Hughes, 2000; Walther et al., 2001, 2002; Warren et al., 2001; Parmesan and Yohe, 2003; Root et al., 2003). Responses to changing environments include more variable population dynamics, altered phenologies, shifts in biogeographic distributions, and disrupted species interactions because of changes affecting one or more participants. Much remains to be learned about multispecies responses, especially those that affect species interactions and food web dynamics and which have important implications for pest responses. To predict responses by insects, we believe that this challenge must include mechanistic approaches applied carefully to the problem (Lawton, 1991, 1995; Hassell et al., 1993; Kareiva et al., 1993; Gutierrez et al., 1994; Gutierrez, 1996, 2000; Bezener and Jones, 1998; Davis et al., 1998). This approach should incorporate the impact of direct and indirect effects of environmental changes (e.g., temperature, atmospheric gases, and resulting land use) to food quality, insect physiological processes, and interactions among species. We also examine briefly the impact of altered temperature and food quality on food web dynamics, often called tritrophic interactions (Rosenheim, 1998), with each link that may be impacted directly and indirectly by anticipated changes in climate. To provide opportunities for predicting responses, we focus when possible on impacts of climate change to the interactions between pests, their food plants, and their predators using temperature-dependent physiological and population-based models (Gutierrez et al., 1994; Gutierrez, 2000) coupled to field and laboratory observations and experiments.

A big question with regard to food security concerns the likelihood that insect herbivore populations will remove critical amounts of plant tissue and adversely affect food, fiber, and forage production in the face of increased atmospheric CO_2 levels. Focus on plant effects rather than just insect herbivore population responses adds one more level in the assessment process, but indicates that emphasis on tracking insect responses may be misleading. Other insect pests are more important as vectors of disease, and environmental changes that increase levels of transmission may be more important than direct tissue loss (Lines, 1995). In all cases, however, population and community dynamics of insects are driven by multiple abiotic and biotic ecological factors that act simultaneously and are often strongly impacted by food quality and temperature (Joern and Gaines, 1990; Belovsky and Joern, 1995; Harrington et al., 1999). Key life history responses of survivorship, growth, development, and reproduction are significantly affected by variable host plant nutritional quality (Joern and Behmer, 1997, 1998); habitat thermal characteristics and the capability for thermoregulation (Casey, 1993; Coxwell and Bock, 1995; Harrison and Fewell, 1995; Lactin et al., 1995); and interactions with other herbivores and predators (Belovsky and Joern, 1995; Cornell and Hawkins, 1995; Chase, 1996; Rosenheim, 1998; Harrington et al., 1999; Oedekoven and Joern, 2000). Each of these factors provides a useful temperature and/or CO_2-dependent link to the direct impacts of climate change on individual responses at different levels in the food chain, and provides a way to mechanistically link climate change to population and important host plant–herbivore and predator–prey interactions.

13.2 PHYSIOLOGICAL ECOLOGY AND NICHE-BASED RESPONSES

Physiological responses often determine our ability to predict species (population) responses to changes in environmental conditions in a reasonably analytical fashion (Dunham, 1993; Gutierrez, 1996). For example, abiotic (e.g., temperature or humidity) and biotic conditions (e.g., protein in food) may combine to affect developmental rate, growth, survival, and reproduction. In a niche-based framework (Maguire, 1973; Chase and Leibold, 2003), knowing an organism's response within an environmental state-space allows one to predict individual fitness and ultimately population responses (Figure 13.1); these types of responses have been worked out in some detail for some taxa for certain niche axes (Birch, 1953; Clancy and King, 1993; Busch and Phelan, 1999). Except under unusual situations when the entire structure of the ecosystem shifts (e.g., complete defoliation), insect pests will not have much impact on overall environmental conditions, so changes external to these agro-ecosystems are expected to drive population responses in a niche-based view. If critical environmental conditions change, then populations change as well by increasing or decreasing in abundance, depending on whether conditions are more or less favorable (Figure 13.1). Basic studies document the power of the relationship between developmental rate and temperature (Figure 13.2) or food quality (Figure 13.3), and the effects on population dynamics.

To predict likely responses of insect pest populations to global environmental changes from the anticipated increased CO_2 levels in the atmosphere, we examine the interactive effects of temperature and food quality on population processes as a model for developing predictions regarding insect pest responses to changing climates. In addition to effects from overall increases in crop productivity, the C:N content of food is predicted in response to increased atmospheric CO_2, resulting in leaf material that is generally of lower primary nutritional quality (relative amounts of protein and carbohydrates) to many insect herbivores (Fajer 1989; Fajer et al.,

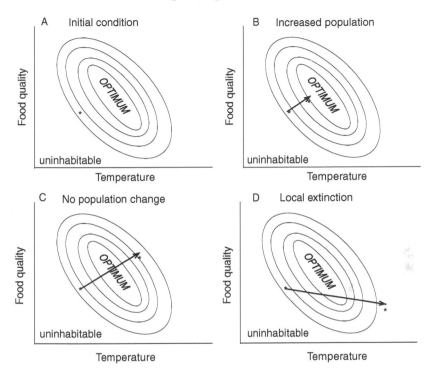

Figure 13.1 (a) Individual or population performance based on two niche axes, temperature and food quality. Optimum combinations support highest levels of performance, such as survival, development, or fecundity, resulting in maximal population growth rates. Some combinations are unsuitable for individuals to persist in the habitat. (b through d) Possible effects of changing environments from initial conditions on performance. (b) Population does better under new conditions. (c) Population does about the same. (d) New conditions lead to local extinction.

1991). For many insect herbivores, however, changes in the amount and types of plant defenses through altered secondary chemistry may be as important as changes in primary nutritional quality (Lincoln et al., 1984, 1986; Lincoln and Couvet, 1989; Stamp, 1990; Stamp and Bowers, 1990a, 1990b; Stamp, 1993; Yang and Stamp, 1995).

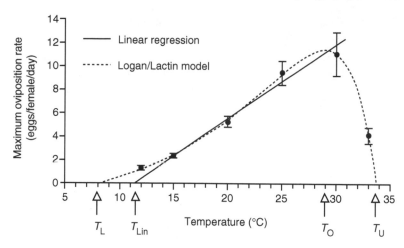

Figure 13.2   Effect of temperature on critical life history rates (see Wermelinger and Seifert, 1999). Data show oviposition rates of the spruce bark beetle (*Ips typographus*) over a range of temperatures. Note that the nonlinear response at higher temperatures corresponds to models proposed by Logan et al. (1976) and Lactin et al. (1995).

13.2.1   Critical Role of Body Temperature in Insect Biology and Development

Temperature directly affects virtually all biological rate processes in insects (Logan et al., 1976; Wagner, 1984a, 1984b; Chappell and Whitman, 1990; Stamp and Casey, 1993; Lactin et al., 1995; Lactin and Johnson, 1996b, 1998a, 1998b), and population responses may vary dramatically in response to climate change (Coxwell and Bock, 1995). Shifts may favor increased food consumption and digestion, more rapid development and faster growth rates, increased survival, or higher fecundity (Huey and Kingsolver, 1989; Kingsolver, 1989; Lawton, 1991; Yang and Joern, 1994b; Harrison and Fewell, 1995; Lactin et al., 1995; Woods and Kingsolver, 1999; Peterson et al., 2000). It is key to remember that the small size of insects makes it more important to recognize microclimate rather than macroclimate when assessing responses to global climate change (Casey, 1993; Lactin and Johnson, 1998b). Operative thermal environments relevant for insects may vary in ways that are not obvious from general descriptions of local weather conditions. For example, what is the effect of a 2°C increase at the global level to an insect living within the boundary layer (~5 mm) of a corn leaf?

Insects are ectotherms of generally small size, and must rely on external heat sources and sinks to control body temperature (T_b). If they cannot thermoregulate, T_b equals the temperature of the surrounding environment. Some insects can exert significant control through physiological and biochemical means, but with energetic costs (Heinrich, 1993). More often, however, insects control T_b using a variety of anatomical and behavioral means called thermoregulation (Casey, 1981, 1988; Chappell and Whitman, 1990; Casey, 1992; Lactin and Johnson, 1996a, 1996b, 1997, 1998a, 1998b) to keep T_b around 38°C for as much of the time as possible. These include manipulating body temperatures using incoming solar radiation and through microhabitat selection (Anderson et al., 1979; Chappell and Whitman, 1990; Aarssen, 1992; Casey, 1993). The quantity of solar radiation absorbed is relatively independent of body temperature, but strongly linked to absorptivity by the animal's surface, typically on the order of 70% to 75% for grasshoppers (Porter, 1969; Anderson et al., 1979; Wilmer, 1981). By manipulating the surface area and portion of the body that is exposed to the sun's rays coupled with judicious use of shade or microhabitats at varying heights above the ground (with and without wind), body temperatures can be significantly regulated within narrow limits (Casey, 1988, 1992). These relationships can now be predicted in the field (Lactin and Johnson, 1996). Because insects have so much control over body temperatures, and microclimates have so much impact, broad-scale predictions about the impact of temperature changes must be tempered and scaled to this much finer level.

13.2.2 Consequences of Altered Plant Nutrition

Environmental stresses have large effects on the nutritional quality of host plants to insect herbivores. The nutritional

A

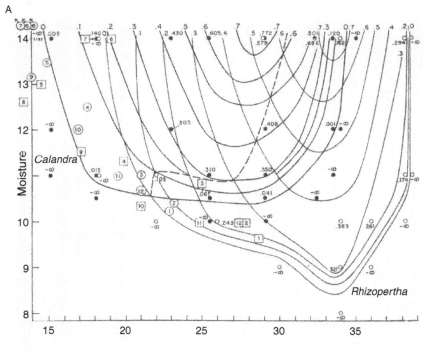

Temperature

B

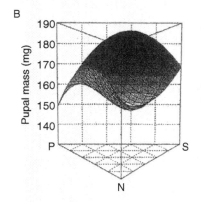

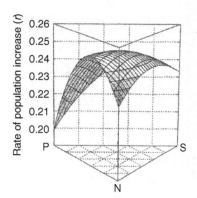

Figure 13.3 (opposite page)   Performance niche responses. (A) Population growth responses by the grain beetles *Calandra oryzae* and *Rhizopertha dominica* for temperature and moisture content of wheat. (Data from Birch, L.C. 1953. *Ecology,* 34:698–711; figure from Maguire, B. Jr. 1973. *Am. Naturalist,* 107:213–246. With permission.) The dashed line indicates conditions that determine competitive outcomes when beetles co-occur. (B) Herbivore performance by two arthropod herbivores (soybean looper, *Pseudoplusia includens* pupal mass, left; two-spotted spider mite, *Tetranychus urticae,* right) on soybeans with fertilizer containing different proportions of key minerals (S, P, and N). (From Busch, J.W., and L. Phelan. 1999. *Ecol. Entomol.,* 24:132–145. With permission.) Note that the two herbivores show very different responses to the treatments; good conditions for one may not be good for the other.

quality of host plants will likely vary in response to altered concentrations of atmospheric gases (e.g., ozone or CO_2 levels), particularly because they can lead to altered C:N:P ratios of leaf material which affects feeding (Lincoln et al., 1984, 1986; Bazzaz et al., 1987; Lincoln and Couvet, 1989; Bazzaz, 1990; Ayres, 1993; Roth and Lindroth, 1995; Watt et al., 1995; Lindroth, 1996b). Altered plant quality affects feeding behavior and, when insect herbivores are food-limited demographic responses (Joern and Gaines, 1990; Slansky, 1993; Joern and Behmer, 1997, 1998). This can be incredibly important as herbivores must maintain homeostatically balanced elemental ratios (biological stoichiometry, such as C:N:P ratios), a goal made more difficult when elemental ratios of food differ greatly from those of consumers (Sterner and Elser, 2002; Logan et al., 2003a, 2003b). Cold-adapted poikilothermic organisms contain 30% to 50% more N, P, protein, and RNA than warm exposed conspecifics (Woods et al., 2003). The implications of these results for global climate change are not settled, but could be important.

Naturally occurring plants are very responsive to environmental changes and are often stressed, with the chemical makeup foliar contents in remarkable flux (Buwai and Trlica, 1977a, 1977b; Bokhari, 1978; Bokhari and Trent, 1985a, 1985b; Mole and Joern, 1993; Mole et al., 1994). In addition to obvious and routine stress from nutrient-poor soils in many grasslands and agricultural settings, drought and herbivory often add opportunities for stress (Williams, 1979; Marrs, 1983; Bokhari and Trent, 1985a; Fitter, 1986; Mooney et al., 1991; White, 1993). Global warming and increased variability in precipitation will likely result in increased drought stress. Multiple environmental stresses affect the physiological state of the host plant, and significantly alter patterns of resource allocation among plant tissues (Chapin, 1980; Gershenzon, 1984; Bloom et al., 1985; Coley et al., 1985; Bazzaz et al., 1987; Chapin et al., 1987; Pearcy et al., 1987; Mooney et al., 1991).

13.2.2.1 Plant Stress and Foliar Quality for Herbivores

Elevated atmospheric CO_2 can alter the relevant nutrient content of plant tissues, usually increasing the C:N ratio (Cave et al., 1981; Lincoln et al., 1986; Osbrink et al., 1987; Fajer, 1989; Bazzaz, 1990; Johnson, 1990; Fajer et al., 1991; Johnson, 1991; Ayres, 1993) in response to reasonably well-understood resource allocation dynamics in plant metabolism (Ayres, 1993). Insect herbivores may alter feeding behavior and demographic attributes in response to these nutritional changes (Osbrink et al., 1987; Akey, 1989; Fajer, 1989; Johnson, 1990; Fajer et al., 1991; Johnson, 1991; Behmer and Joern, 1993; 1994; Yang and Joern, 1994a). Results to date indicate that altered plant quality from increased CO_2 can affect insect feeding, but responses are highly variable among insect herbivore taxa and can be small or equivocal (Watt et al., 1995).

13.2.2.2 Temperature, Food Quality, and Diet Processing

T_b in ectotherms influences food acquisition and processing capabilities (Huey and Stevenson, 1979; Crowder, 1983; Karasov, 1984; Zimmerman and Tracy, 1989; Yang and Joern, 1994a, 1994b). For example, more energy is obtained by a desert spider from selection of favorable environments than from sites with more prey because proper temperatures facilitate optimal digestion and use of nutrients (Riechert and Tracy, 1975). In herbivorous insects, temperature clearly affects foraging, digestion, and performance (Scriber and Lederhouse, 1983; Stamp, 1990; Stamp and Bowers, 1990a, 1990b; Casey, 1993; Stamp, 1993; Yang and Joern, 1994b; Harrison and Fewell, 1995; Lactin and Johnson, 1995). Tiger swallowtail caterpillars exhibit increased consumption as temperatures increase, resulting in increased growth until respiratory expenditures became too high; digestibility was unaffected (Scriber and Lederhouse, 1983). Lower food quality often results in increased consumption by insect herbivores as well (Yang and Joern, 1994a, 1994b), but even this is limited compared to the anticipated effects.

13.2.2.3 Climate-Based Insect Population Dynamics and Range Shifts

Conventional wisdom predicts that strong correlations between weather and insect densities will exist, and that these correlations explain population fluctuations (Capinera, 1987). For example, statistically significant correlations between population change and abiotic factors have been identified for grasshoppers (Gage, 1977; Hardman, 1982; Logan and Hilbert, 1983; Johnson and Worobec, 1988; Capinera and Horton, 1989; Fielding and Brusven, 1990), although responses vary according to geographic location. Variation in grasshopper population densities from arid grasslands is positively correlated with winter and spring precipitation (Nerney and Hamilton, 1969; Capinera and Horton, 1989; Fielding and Brusven, 1990), while temperature has a greater influence on northern populations (Johnson and Worobec, 1988; Capinera and Horton, 1989).

13.3 SPECIES INTERACTIONS AND THE COMPLEXITY OF COMMUNITY-LEVEL FOOD WEBS

It is reasonable to expect that climate change will affect the temporal and spatial associations between species interacting at different trophic levels (Porter, 1995; Sutherst et al., 1995; Harrington et al., 1999; Gutierrez, 2000; Walther et al., 2002). What happens to our predictions as we increase the number of species that must be included in our analyses?

Understanding the impact of climate changes on herbivore–predator interactions may be as or more important than plant–herbivore interactions (Cornell et al., 1998). Based on life table analyses, natural enemies acting in a top-down fashion were found to be the most important source of mortality overall in exophytic insects in agro-ecosystems. Both natural enemies and plant factors were important sources of mortality for endophytic insect herbivore populations. As described below, networks of interactions provide many opportunities for compensatory responses (Rosenheim, 1998; Oedekoven and Joern, 2000) and thresholds (Belovsky and Joern, 1995), context-dependent processes that might be especially sensitive to changing climates and seemingly act suddenly, resulting in large changes with only small changes in environmental conditions. For example, host plant quality may alter the impact of predation (Oedekoven and Joern, 2000), or the presence of predators may alter foraging by insect herbivores (Lima, 1998; Losey and Denno, 1998; Eubanks and Denno, 2000b; Danner and Joern, 2003). In turn, changes in local climate may alter the magnitude or even the qualitative expression of these already complex interactions.

Spiders exert a significant limiting influence on grasshopper populations through size-selective predation (Belovsky and Slade, 1993; Oedekoven and Joern, 1998, 2000). Responses by each species are affected directly by temperature (Kemp, 1986; Li and Jackson, 1996; Lactin and Johnson, 1998b), with important implications for understanding seasonal coincidence (phenologies), individual growth characteristics, daily activity schedules, microhabitat use, and spider feeding characteristics (functional responses and size relationships). Population densities of both participants will greatly affect the success of these predator–prey interactions. Each of these attributes can alter effective predator control of prey populations as each is affected by temperature.

What happens to insect herbivore populations in the face of climate change when more than one interacting species is affected? At present, no answer to this basic question really exists. The advantage of a mechanistic approach (Dunham et al., 1989; Huey, 1991; Lawton, 1991; Dunham, 1993; Lawton, 1995; Gutierrez, 1996) is that it permits one to extrapolate results to other sites and years, unlike approaches that merely correlate climatic attributes with population responses. A modeling hierarchy is needed that links the influence of temperature variability to physiological processes underlying feeding, digestion, and resulting demographic consequences (growth, development, survival, and reproduction) on different quality food. Temperature-based seasonal phenology and daily activity cycles and, in turn, individual-based life history responses over a range of food quality in both grasshoppers and spiders can then be examined.

13.3.1 Competition and Abiotic Conditions

Much remains to be learned about responses by pest insect populations to anticipated climate shifts (Lawton, 1995; Lindroth, 1996b; Davis et al., 1998; Harrington et al., 1999). Insect pest populations can rapidly and unexpectedly build to large numbers in response to environmental changes, including changes in weather (Cappuccino and Price, 1995; Dempster and McLean, 1998), resulting in significant plant damage (Hewitt and Onsager, 1983; Barbosa and Schultz, 1987). Potential impacts of climate change on these insect populations will reflect the combined, integrated response to the direct effects from altered physical environments that influence physiological processes, coupled with the indirect consequences of altered food plant quality and availability resulting from both the direct impact of CO_2 levels acting on plants and accompanying drought stress resulting from climatic shifts (Lansberg and Smith, 1992). The combined direct and indirect effects from climate change will alter key demographic attributes, thus resulting in unanticipated population-level consequences.

Competitive interactions among taxa represent primary interactions in natural communities, forming the basis of much ecological research. When key resources such as food are limiting, species abundances are lowered and some species may not be able to coexist with others (competitive exclusion). Interspecific competition can be common among insect herbivores (Denno et al., 1995). In most agricultural systems, however, it is probably less important than in natural communities (Cornell et al., 1998), although there are important examples. How will competitive interactions be affected by direct and indirect affects of altered CO_2 levels?

For poikilotherms such as arthropod pests, physical conditions of the environment are critical as they can readily reverse the outcome of competitive interactions. For example, the well-studied, stored-grain beetle *Tribolium confusum* outcompetes *Tribolium castaneum* when grown at a temperature of 34°C and low relative humidity (RH = 30%); when RH = 70% but temperature remains at 34°C, the outcome is reversed. *T. castaneum* usually wins when RH is high at high temperatures, but not when temperatures are more moderate (24°C). Similar results are observed for the grain beetles *Rhizopertha* and *Callandra* based on other niche axes (Figure 13.3b) (Birch, 1953; Maguire, 1973). Grain beetle performance varies according to internal niche structure in this case (Maguire, 1973), and reversals in the outcome of interspecific interactions such as competition due to physical conditions such as temperature and relative humidity greatly complicates the picture. Studies on a range of organisms indicate that the interaction between competitive interactions and physical conditions can be important (Tilman et al., 1981). Climate shifts will likely affect any number of plant–arthropod and arthropod–arthropod competitive interactions in agricultural settings.

13.3.2   Predator–Prey Interactions

Climatic conditions (especially microclimate) also have a very big impact on arthropod predator–prey interactions. This interaction is important because natural predators and parasitoids in agricultural settings often limit abundance levels of insect herbivore pests on crops (Cornell et al., 1998; Rosenheim, 1998; Gutierrez, 2000), including successful biocontrol efforts. The chain of events linking predators with prey species that affect predation success is large, and disruption at any level can significantly impact the likelihood that agricultural pests will increase.

13.3.2.1 Mediating Effects of Microclimate

An excellent example illustrating the importance of microclimatic conditions on spider mite predator–prey interactions is the interaction between the serious corn pest, Banks grass mite *Oligonychus pratensis*), and the predatory mite *Neoseiulus fallacis*, which can limit Banks grass mite in a standard predator–prey interaction (Perring et al., 1986; Berry et al., 1991a, 1991b). Temperature and humidity at the leaf surface defined the population dynamics of these two species, and greatly affect the ability of the predatory mite to hold the herbivorous Banks grass mite to low population levels. Outbreaks of Banks grass mite occurred during hotter years when corn was moderately drought stressed, in part because physical conditions determined whether the predatory mite population could increase sufficiently to exert control. Under cooler, wetter conditions, the predatory mite could control Banks grass mite and decrease economic losses. Banks grass mite populations are favored by warmer, dry conditions, whereas the predatory mite was favored by cooler, humid conditions. A short period (days) of hot, dry weather is sufficient for the Banks grass mite to escape the controlling influence of predation and cause significant damage to corn. Other studies further document the interaction between temperature and prey density in determining the predator's efficiency at capturing prey (functional response) (Hardman and Rogers, 1992). To transfer results to field situations, we emphasize that conclusions depend on conditions within the boundary layer at the leaf surface — the microclimate relevant to these very tiny mites — rather than the standard reference conditions measured by most weather stations and conditions predicted by GCC. While it is clear that microclimatic conditions drive this important predator–prey interaction, it is less clear how to scale these results in the context of predicted global climate change.

13.3.2.2 Context-Dependent Interactions between Predators and Prey

Food acquisition and activity is greatly affected by risk from predation (Lima, 1998), and such interactions can greatly alter performance outcomes of predator prey interactions. In choice tests, Plantago-feeding caterpillars of the butterfly *Junionia coenia* select plants with the lowest concentrations of iridoid glycocides, a plant defensive compound (Stamp and Bowers, 2000). Caterpillars fed equally on leaves with high and low iridoid glycoside levels in the presence of stinkbug predators (*Podisus maculiventris*), presumably spending less time feeding and more time in hiding or escape behaviors. Grasshoppers feed much less in the presence of wolf spiders than in predator-free environments, affecting survival, growth, developmental rate, and reproduction (Rothley et al., 1997; Danner and Joern, 2003). Eating higher-quality food mediates the predation risk effect such that grasshoppers with predation risk fed high-quality food have the same performance levels as individuals with lower-quality food but no predation risk (Danner and Joern, 2003). And, pea aphids in alfalfa exhibit a high propensity to drop from plants in the presence of the foliar foraging predaceous big-eyed bugs or ladybird beetles (*Coccinella septempunctata*) (Losey and Denno, 1998). The propensity by aphids to drop is also influenced by the quality of the resource to be abandoned and the risk of mortality in the new microhabitat. What these examples illustrate is the need to be aware of coexisting species in addition to physical features of an organism's environment to fully understand and predict responses. Of course, this makes it difficult to predict how changing environments will affect herbivores because different interactions may dominate in different settings, and the expression of these interactions is context dependent.

13.3.3 Food Web (Tritrophic) Interactions

What are the consequences of food web dynamics and trophic cascades for crop production and food security? For the most part, empirical research indicates that insect herbivore populations are regulated by both natural enemies (top-down effects) and food resources (bottom-up effects). Moreover, natural enemies of herbivores are not top predators in most agroecosystems, and they typically are attacked by another tier of pathogens, parasites and predators. While the above examples emphasize interactions between populations of two species, most naturally occurring communities, including agroecosystems, are much more complex. One observes more species participants, a greater variety of interactions among coexisting species, and the need to consider multiple trophic levels (Rosenheim, 1998). In a sense, combinations of species interactions often interact in tritrophic comparisons. Moreover, relationships among species become more difficult to assign as when species both compete and interact as predators and prey (intraguild predation). Effective biological control may be disrupted by natural enemies of the effective biological control agent itself (Rosenheim, 1998), or food plant quality may be altered by plant pathogens such that insect herbivores do better or worse depending on conditions. In food webs, top-down and bottom-up effects from species interactions "collide," introducing a variety of indirect effects that cannot be predicted let alone understood, unless all participants are both recognized and considered in analyses. Finally, there is an important spatial element that dictates the strength of species interactions and the persistence of species participants in a community, a point not covered in detail here but certainly important for understanding effects of insect pests on crop, fiber, and forage production. This complexity provides a broad array of possible outcomes. To the degree that webs of interactions are affected by climatic conditions and influencing the outcome of sets of often nonlinear interactions, global climate change may have remarkable effects on agricultural pests. Unfortunately, the outcomes may not be predictable. Above examples indicate how species interactions can be erased or even reversed as physical conditions of the environment change. What does this do to community-level interactions in which the end results are highly reticulated? A couple of examples indicate possible outcomes to illustrate the complexity of the problem.

A classic example that highlights the importance and complexity of predicting interactions between trophic levels is the trophic cascade, a process in which indirect interactions among species are often more important to the final outcome than direct interactions (Schmitz, 1997, 1998, 2003). In a trophic cascade, natural enemies affect plant production indirectly by controlling abundance of herbivore populations. Any changes in the interaction between herbivores and natural enemies, such as the addition of hyperparasites or predators that control natural enemies of herbivores, alter primary production. Trophic cascades are ubiquitous in both natural and managed ecosystems. In terrestrial ecosystems, food webs can be highly reticulated, with many species occurring and interacting within and between trophic levels. Consequently, it is essential that the ecological balance sheet be carefully drawn to make accurate predictions, although some recent work indicates that complex food webs often collapse into more tractable food chains from the standpoint of assessing trophic cascades.

Omnivory provides another example of complicating processes within arthropod communities in agro-ecosystems. Omnivores feed on both plants (herbivory) and other consumers (predation). Big-eyed bugs, *Geocoris punctipes* (Heteroptera: Geocoridae) prey on eggs of the corn earworm, *Helicoverpa zea* (Lepidoptera, Noctuidae) and pea aphids, *Acyrthrosiphum pisum* (Homoptera, Aphidae), in addition to plants. The big-eyed bugs need a mixture of both plants and insect prey for normal development. In controlled experiments in the presence of high-quality plants (lima bean pods), big-eyed bugs fed more on the bean pods, which reduced the number of prey taken by them, and pea aphids reached much larger size (Eubanks and Denno, 2000b), presumably having more potential to damage crops. However, the critical prediction that herbivore populations would be more abundant under field conditions when lima bean plants had pods was not upheld; pea aphid populations were lower. The presence of high-quality plant parts attracted more big-eyed bugs to the field, thus increasing predation pressure overall on aphid populations, even though per capita pressure from each predator was less. Increased prey densities did not have the same attractive effect to big-eyed bug populations (Eubanks and Denno, 1999). To further complicate the interaction, big-eyed bugs preferentially selected mobile pea aphids over corn earworm eggs, even though the corn earworm eggs were significantly more nutritious and were required to complete development (Eubanks and Denno, 2000a). In the field, however, the presence of lima bean pods was the greatest predictor of big-eyed bug density, and was responsible for maintaining bug populations during low-density periods.

13.4 CONCLUSIONS

Much remains to be learned about responses by pest insect populations in agro-ecosystems to anticipated climate shifts (Cammell and Knight, 1992; Lawton, 1995; Lindroth, 1996b; Cannon 1998; Davis et al., 1998; Harrington et al., 1999). Insect pest populations can build up to large numbers rapidly and unexpectedly in response to environmental changes, including weather (Lansberg and Smith, 1992; Cappuccino and Price, 1995; Dempster and McLean, 1998), often resulting in significant plant damage (Hewitt and Onsager, 1983). Potential impacts on insect pest populations will reflect the combined, integrated responses to direct effects from altered physical environments that influence physiological responses, coupled to indirect consequences of variable food quality resulting from elevated CO_2 and associated drought stress (Lansberg and Smith, 1992). Geographic range shifts of arthropod pests and their natural enemies in response to changing temperatures or plant communities are likely. Temperature and CO_2 levels are only part of the story (Lawton, 1995), however, and species interactions that are very difficult to predict in deterministic environments, let alone changing ones, may be most important. Combined direct and indirect effects from species interactions will alter key demographic responses by insect populations as well as interactions among populations within and between trophic levels.

The effects of elevated CO_2 on insect herbivores can be summarized according to a standard model described in Figure 13.4; this model ignores effects of species interactions within and between trophic levels. In a nutshell, elevated temperature resulting from increases in greenhouse gases will have direct effects on insect performance, which in turn affects population processes and species interactions. CO_2 will most likely have its greatest direct impact on food quality; food quality will be reduced for most insect herbivores. In many cases, insects can compensate for lowered N levels by eating more leaf or root material, but often not completely. Tissue loss in turn affects crop productivity, although plants also exhibit significant tolerance to losses from herbivores and

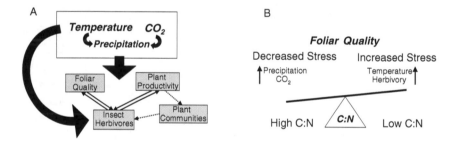

Figure 13.4 General model of effects of climate change on insect herbivores. Multiple interactions that are anticipated to operate within boxes, including a variety of indirect ones, are not represented, but are discussed in detail in the text. (A) Temperature has two main paths in its impact on insect herbivores, through direct effects on insects by affecting metabolic rates, microhabitat selection and time budgets, and indirect effects on food quality. (B) Food nutritional quality (e.g., C:N, or concentrations of secondary chemicals) responses to abiotic conditions including temperature in association with precipitation and CO_2. Such changes can greatly influence insect herbivore population dynamics.

pathogens, so the impact of consumption must be translated into actual effects on food-producing capacity.

Important but often unstated assumptions underlie scenarios proposed to deal with the impact of climate changes on agricultural pests:

1. The crops remain where they are and will suffer the effects of an altered physical environment.
2. Climate change is gradual.
3. Species assemblages will track conditions as a community unit.
4. Ecological interactions are deterministic, mostly linear and continuous (without thresholds) that could lead to alternate states with small changes in environmental conditions.

We expect that each assumption will not hold under many situations, and relaxing any of these assumptions makes it difficult to predict responses by agricultural pests as suggested by niche-based scenarios of Section 13.2.

Agronomists and agricultural economists expect that farmers and ranchers will adapt to new climatic conditions by growing the most appropriate crops for local conditions and cropping patterns will track changes in climate. Corn may be grown farther north than is now the case, such that the resulting environmental conditions in these new regions will be much the same as they are now. Agricultural pests will likely follow after some time lag. If so, the resulting impacts from pests on crops may not change, or they may be more affected by changes caused by final species in the assemblage in new regions rather than because of changes in prior species or community interactions due to climate.

Table 13.1 indicates possible changing conditions that may affect insect population abundance, and hence effects on crop productivity. In some cases, responses are reasonably deterministic and anticipate outcomes to changing climates. Many other situations are not so predictable with available information and understanding of interactions among species, and additional empirical research will be required to fill in the necessary gaps to support reasonable predictions.

Table 13.1 Factors Affecting the Likelihood that Insect Pest Populations Will Increase or Decrease in Response to Changing Environmental Conditions

Increased risk from agricultural pests following climate change is likely if the following conditions are met:

I. Species continue to inhabit a given location under changed global conditions and key participants (e.g., herbivores and host plants) continue to co-occur. Shifts in a species range with climate change may also cause new problems if a previously absent pest can now live in an area after climate changes that already has its host plant, or its previously successful natural enemy disappears.

II. Resources affecting population size (e.g., carrying capacity) *are not reduced* by:

    a. Climatic adversity disrupting insect populations or meta-population dynamics, causing local extinction (Hanski, 1998).

    b. Disruption of phenological timing between herbivores and critical stages of host plants (Cammell and Knight, 1992; Cannon, 1998; Harrington et al., 1999).

    c. New physical conditions slow rates of herbivore growth, development or reproduction because they have moved from more to less optimal states (Logan et al., 1976; Wermelinger and Seifert, 1999) (Figures 13.1 and 13.2).

III. Increased probability of outbreaks are expected if:

    a. Pest was previously enemy limited, and density-dependence feedback from enemies decreases in impact (Hassell et al., 1993; Gutierrez et al., 1994; Cornell and Hawkins, 1995).

    b. Pest was previously food limited, and density-dependent feedback from host plant decreases (Auerbach et al., 1995; van der Meijden et al., 1998).

    c. Prior limitation by physical conditions change to be more favorable for pest population growth, without compensating for increases in density dependence from food or enemies (den Boer, 1998; Logan et al., 2003).

    d. Pest outbreaks are exacerbated by positive feedbacks from the host that occurs because of feeding under new environmental conditions (Mattson, 1980; Mattson and Haack, 1987; Joern, 1992; White, 1993).

    e. Pests increase the number of generations in a normal growing season as the window of suitable conditions is expanded (univoltine to bivoltine, or bivoltine to trivoltine, etc.).

Note: References are representative, not exhaustive.
Source: Based on Lansberg, J., and M.S. Smith. 1992. *Aust. J. Botany*, 40:565–577.

Some important issues are not covered in this chapter. Pollination is critical for many agricultural situations, resulting from both diffuse and specialized interactions between plant floral phenology, presence of pollinators, and the need for genetic outcrossing to support seed production. Mutualistic interactions such as pollination differ somewhat from those described above, but the underlying approach to their study will be similar. Also, evolutionary responses to changing conditions have been ignored. Predictions basically assume that species will retain attributes characteristic of prior environmental conditions. Adaptive stasis is highly unlikely as natural selection will certainly lead to changes in the underlying adaptive ability to function in new environments. For example, phonological decoupling between previously interacting species such as insect herbivores and parasitoids may initially lead to pest outbreaks. After a relatively small number of generations, these species (or others) may reestablish the control of natural enemies on the pest as the natural enemy populations evolve to track prey. Anticipated evolutionary responses can become very complex (Malcom, 1993; Thomas et al., 2001), and are not considered further here, but the capability for evolutionary responses affecting species interactions must not be forgotten. Finally, the ability to predict responses by agricultural pests to climate or other types of environmental change will be very different depending on whether the changes are gradual, or whether they are sudden. Current species interactions are much more likely to survive intact under gradual changes.

The basic theme of this chapter concerns the difficulty in predicting responses by agricultural pest insects to climate change, especially those promoted by elevated CO_2. There is no longer any doubt that important responses by insect herbivores to climate change will occur. The biggest problem lies in our ability to predict the outcome of these changes with regard to food security. Clearly, more research on this important topic is needed.

ACKNOWLEDGMENTS

Our research is supported by the Office of Science, Biological and Environmental Research Program, U.S. Department of Energy, through the Great Plains Regional Center of the National Institute for Global Environmental Change, under cooperative agreement DE-FC03-90ER61010. We are grateful for the logistical support provided by Cedar Point Biological Station (University of Nebraska).

REFERENCES

Aarssen, L.W. and D.R. Taylor. 1992. Fecundity allocation in herbaceous plants. *Oikos*, 65:225–232.

Akey, D.H. and B.A. Kimball. 1989. Growth and development of the beet armyworm on cotton grown in an enriched carbon dioxide atmosphere. *Southwestern Entomologist*, 14:255–160.

Alley, R.B., J. Marotske, W.D. Nordhaus, et al. 2003. Abrupt climate change. *Science*, 299:2005–2010.

Anderson, R.V., C.R. Tracy, and Z. Abramsky. 1979. Habitat selection in two species of short-horned grasshoppers: the role of thermal and hydric stresses. *Oecologia*, 38:359–374.

Auerbach, M.J., E.F. Connor, and S. Mopper. 1995. Minor miners and major miners: population dynamics of leaf-mining insects. In N. Cappuccino and P.W. Price, Eds. *Population Dynamics: New Approaches and Syntheses*. Academic Press, San Diego, pp. 83–110.

Ayres, M.P. 1993. Plant defense, herbivory and climate change. In Kareiva, P.M., J.G. Kingsolver, and R.B. Huey, Eds. *Biotic Interactions and Global Change*. Sinauer Associates, Sunderland, MA, pp. 75–94.

Barbosa, P. and J.C. Schultz, Eds. 1987. *Insect Outbreaks*. Academic Press, New York.

Bazzaz, F.A. 1990. Response of natural ecosystems to the rising global CO_2 levels. *Annu. Rev. Ecol. Systematics*, 21:167–196.

Bazzaz, F.A., N.R. Chiariello, P.D. Coley, and L.F. Pitelka. 1987. Allocating resources to reproduction and defense. *BioScience*, 37:58–67.

Behmer, S.T. and A. Joern. 1993. Dietary selection by the generalist grasshopper, Phoetaliotes nebrascensis (Orthoptera: Acrididae) based on the need for phenylalanine. *Functional Ecol.*, 7:522–527.

Behmer, S.T. and A. Joern. 1994. The influence of proline on diet selection: sex-specific feeding preferences by the grasshoppers Ageneotettix deorum and Phoetaliotes nebrascensis (Orthoptera: Acrididae). *Oecologia,* 98:76–82.

Belovsky, G.E. and A. Joern. 1995. Regulation of grassland grasshoppers: differing dominant mechanisms in time and space. In P.W. Price, Ed. *Novel Approaches for the Study of Population Dynamics: Examples from Insect Herbivores.* Academic Press, New York, pp. 359–386.

Belovsky, G.E. and J.B. Slade. 1993. The role of vertebrate and invertebrate predators in a grasshopper community. *Oikos,* 68:193–201.

Berry, J.S., T.O. Holtzer, and J.M. Norman. 1991a. Experiments using a simulation model of the Banks grass mite (Acrai: Tetranuchidae) and the predatory mite Neoseiulus fallacis (Acari: Phytoseiidae) in a corn microenvironment. *Environ. Entomol.,* 20:1074–1078.

Berry, J.S., T.O. Holtzer, and J.M. Norman. 1991b. MiteSim: a simulation model of the Banks grass mite (Acari: Tetracnychidae) and the predatory mite, Neoseiulus fallacis (Acari: Phytoseiidae) on maize: model development and validation. *Ecol. Modelling,* 53:291–317.

Bezener, T.M. and T.H. Jones. 1998. Plant-insect herbivore interactions in elevated atmospheric CO_2; quantitative analyses and guild effects. *Oikos,* 82:212–222.

Birch, L.C. 1953. Experimental background to the study of the distribution and abundance of insects. I. The influence of temperature, moisture, and food on the innate capacity for increase of three grain beetles. *Ecology,* 34:698–711.

Bloom, A.J., F.S. Chapin III, and H.A. Mooney. 1985. Resource limitation in plants — an economic analogy. *Annu. Rev. Ecol. Systematics,* 16:363–392.

Bokhari, U.G. 1978. Nutrient characteristics of blue grama herbage under the influence of added water and nitrogen. *J. Range Manage.*, 31:18–22.

Bokhari, U.G. and J.D. Trent. 1985a. Proline concentration in water stressed grasses. *J. Range Manage.*, 38:37–38.

Bokhari, U.G. and M.J. Trent. 1985b. Proline concentrations in water-stressed grasses. *J. Range Manage.*, 38:37–38.

Brown, V.C. 1995. Insect herbivores and gaseous air pollutants — current knowledge and predictions. In R. Harrington and N.E. Stork, Eds. *Insects in a Changing Environment*. Academic Press, London, pp. 220–249.

Busch, J.W. and L. Phelan. 1999. Mixture models of soybean growth and herbivore performance in response to nitrogen-sulfur-phosphorous nutrient interactions. *Ecol. Entomol.*, 24:132–145.

Buwai, M. and M.J. Trlica. 1977a. Defoliation affects root weights and total nonstructural carbohydrates of blue grama and western wheatgrass. *Crop Sci.*, 17:15–17.

Buwai, M. and M.J. Trlica. 1977b. Multiple defoliation effects on herbage yield, vigor and total nonstructural carbohydrates of five range species. *J. Range Manage.*, 30:164–171.

Cammell, M.E. and J.D. Knight. 1992. Effects of climate change on the population dynamics of crop pests. *Adv. Ecol. Res.*, 22:117–162.

Cannon, R.J.C. 1998. The implictions of predicted climate change for insect pests in the UK, with emphasis on non-indigenous species. *Global Change Biol.*, 4:785–796.

Capinera, J.C. 1987. Population ecology of rangeland grasshoppers. In J.C. Capinera, Ed. *Integrated Pest Management on Rangeland: A Shortgrass Perspective*. Westview Press, Boulder, CO, pp. 162–182.

Capinera, J.C. and D.R. Horton. 1989. Geographic variation in effects of weather on grasshopper infestation. *Environ. Entomol.*, 18:8–14.

Cappuccino, N. and P.W. Price, Eds. 1995. *Population Dynamics: New Approaches and Syntheses*. Academic Press, San Diego, CA.

Casey, T.M. 1981. Behavioral mechanisms of thermoregulation. In B. Heinrich, Ed. *Insect Thermoregulation*. John Wiley & Sons, New York, pp. 79–113.

Casey, T.M. 1988. Thermoregulation and heat exchange. *Adv. Insect Physiol.,* 20:120–146.

Casey, T.M. 1992. Biophysical ecology and heat exchange in insects. *Am. Zoologist,* 32:225–237.

Casey, T.M. 1993. Effects of temperature on foraging of caterpillars. In N.E. Stamp and T. Casey, Eds. *Caterpillars: Ecological and Evolutionary Constraints on Foraging*. Chapman & Hall, New York.

Cave, G.L., C. Tolley, and B.R. Strain. 1981. Effect of carbon dioxide enrichment on chlorophyll content, starch content, and starch grain structure in Trifolium subterraneum leaves. *Physiol. Plant,* 51:171–174.

Chapin, F.S.I. 1980. The mineral nutrition of wild plants. *Annu. Rev. Ecol. Systematics,* 11:233–260.

Chapin, F.S.I., A.J. Bloom, C.B. Field, and R.H. Waring. 1987. Plant responses to variable environments. *Annu. Rev. Ecol. Systematics,* 16:363–392.

Chappell, M.A. and D.A. Whitman. 1990. Grasshopper thermoregulation. In R.F. Chapman and A. Joern, Eds. *Biology of Grasshoppers*. Wiley Interscience, New York.

Chase, J.M. 1996. Abiotic controls of trophic cascades in a simple grassland food chain. *Oikos,* 77:495–506.

Chase, J.M., and M.A. Leibold. 2003. *Ecological Niches: Linking Classical and Contemporary Approaches*. University of Chicago Press, Chicago.

Clancy, K.M. and R.M. King. 1993. Defining the western spruce budworm's nutritional niche with response surface methodology. *Ecology,* 74:442–454.

Coley, P.D., J.P. Bryant, and F.S. Chapin III. 1985. Resource availability and plant antiherbivore defense. *Science,* 230:895–899.

Cornell, H.V. and B.A. Hawkins. 1995. Survival patterns and mortality sources of herbivorous insects: some demographic trends. *Am. Naturalist,* 145:563–593.

Cornell, H.V., B.A. Hawkins, and M.E. Hochberg. 1998. *Ecol. Ento-mol.,* 23:340–349.

Coxwell, C.C. and C.E. Bock. 1995. Spatial variation in diurnal surface temperatures and the distribution and abundance of an alpine grasshopper. *Oecologia,* 104:433–439.

Crowder, L.B. and I.J. Magnuson. 1983. Cost-benefit analysis of temperature and food resource use: a synthesis with examples from fish. In W.P. Aspey and S.I. Lustick, Eds. *Behavioral Energetics: The Cost of Survival in Vertebrates.* Ohio State University Press, Columbus, pp. 189–221.

Danner, B.J. and A. Joern. 2003. Stage-specific behavioral responses of Ageneotettix deorum (Orthoptera: Acrididae) in the presence of Lycosid spider predators. *J. Insect Behav.,* .

Davis, A.J., J.H. Lawton, B. Shorrocks, and L.S. Jenkinson. 1998. Individualistic species responses invalidate simple physiological models of community dynamics under global environmental change. *J. Anim. Ecol.,* 67:600–612.

Dempster, J.P. and I.F.G. McLean, Eds. 1998. *Insect Populations in Theory and in Practice.* Kluwer, Dordrecht.

den Boer, P.J. 1998. The role of density-independent processes in the stabilization of insect populations. In J.P. Dempster and I.F.G. McLean, Eds. *Insect Populations in Theory and in Practice.* Kluwer, Dordrecht, pp. 53–80.

Denno, R.F., M.S. McClure, and J.R. Ott. 1995. Interspecific interactions in phytophagous insects: competition reexamined and resurrected. *Annu. Rev. Entomol.,* 40:297–331.

Dukes, J.S. and H.A. Mooney. 1999. Does global change increase the success of biological invaders? *Trends Ecol. Evol.,* 14:135–139.

Dunham, A.E. 1993. Population responses to environmental change: operative environments, physiologically structured models and population dynamics. In Kareiva, P.M., J.G. Kingsolver, and R.B. Huey, Eds. *Biotic Interactions and Global Change.* Sinauer Associates, Sunderland, MS, pp. 95–119.

Dunham, A.E., B.W. Grant, and K.L. Overall. 1989. Interface between biophysical and physiological ecology and the population ecology of terrestrial vertebrate ectotherms. *Physiological Zool.,* 62:335–355.

Eubanks, M.D. and R.F. Denno. 1999. The ecological consequences of variation in plants and prey for an omnivorous insect. *Ecology*, 80:1253–1266.

Eubanks, M.D. and R.F. Denno. 2000a. Health food versus fast food: the effects of prey quality and mobility on prey selection by a generalist predator and indirect interactions among prey species. *Ecol. Entomol.*, 25:140–146.

Eubanks, M.D. and R.F. Denno. 2000b. Host plants mediate omnivore-herbivore interactions and influence prey suppression. *Ecology*, 81:936–947.

Fajer, E.D. 1989. The effects of enriched carbon dioxide atmospheres on plant-insect herbivore interactions: growth responses of larvae of the specialist butterfly Junonia coenia (Lepidoptera: Nymphalidae). *Oecologia*, 81:514–520.

Fajer, E.D., M.D. Bowers and F.A. Bazzaz. 1991. The effects on enriched CO_2 atmospheres on the buckeye butterfly, Junonia coenia. *Ecology*, 72:751–754.

Fielding, D.J. and M.A. Brusven. 1990. Historical analysis of grasshopper (Orthoptera: Acrididae) population responses to climate in southern Idaho, 1950–1980. *Environ. Entomol.*, 19:1786–1791.

Fitter, A.H. and R.K.M. Hay. 1986. *Environmental Physiology of Plants*, 2nd ed. Academic Press, New York.

Gage, S.H. and M.K. Mukerji. 1977. A perspective of grasshopper population distribution in Saskatchewan and interrelationship with weather. *Environ. Entomol.*, 6:469–479.

Gershenzon, J. 1984. Changes in the levels of plant secondary metabolites under water and nutrient stress. *Recent Adv. Phytochem.*, 18:273–320.

Gu, L., D.D. Baldocchi, S.C. Wofsky, et al. 2003. Response of a deciduous forest to the Mount Pinatubo eruption: enhanced photosynthesis. *Science*, 299:2035–2038.

Gutierrez, A.P. 1996. *Applied Population Ecology*. John Wiley & Sons, New York.

Gutierrez, A.P. 2000. Crop ecosystem responses to climate change: pests and population dynamics. In K.R. Reddy and H.F. Hodges, Eds. *Climate Change and Global Crop Productivity*. CABI Publishing, New York.

Gutierrez, A.P., N.J. Milles, S.J. Schreiber, and C.K. Ellis. 1994. A physiologically based tritrophic perspective on bottom up–top down regulation of populations. *Ecology*, 75:2227–2242.

Hanski, I. 1998. Spatial structure and dynamics of insect populations. In J.P. Dempster and I.F.G. McLean, Eds. *Insect Populations in Theory and in Practice*. Kluwer, Dordrecht, pp. 3–27.

Hardman, J.M. and M.L. Rogers. 1992. Effects of temperature and prey density on survival, development, and feeding rates of immature Typhlodromus pyri (Acari: Phytoseiidae). *Environ. Entomol.*, 20:1089–1096.

Hardman, J.M. and M.K. Mukerji. 1982. A model simulating the population dynamics of the grasshoppers (Acrididae) Melanoplus sanguinipes (Fabr.), M. packardii Scudder, and Camnula pellucida (Scudder). *Res. Population Ecol. (Kyoto)*, 24:276–301.

Harrington, R. and N.E. Stork, Eds. 1995. *Insects in a Changing Environment*. Academic Press, London.

Harrington, R., I. Woiwood, and T. Sparks. 1999. Climate change and trophic interactions. *Trends Ecol. Evol.*, 14:146–150.

Harrison, J.F. and J.H. Fewell. 1995. Thermal effects on feeding behavior and net energy intake in a grasshopper experiencing large diurnal fluctuations in body temperature. *Physiological Zool.*, 68:453–473.

Hassell, M.P., H.C.J. Godfray, and H.N. Comins. 1993. Effects of global change on the dynamics of insect host-parasitoid interactions. In P. Kareiva, J.P. Kingsolver, and R.B. Huey, Eds. *Biotic Interactions and Global Climate Change*. Sinauer Associates, Sunderland, MA, pp. 402–423.

Heinrich, B. 1993. *The Hot-Blooded Insects*. Harvard University Press, Cambridge, MA.

Hewitt, G.B. and J.A. Onsager. 1983. Control of grasshoppers on rangeland in the United States — A perspective. *J. Range Manage.*, 36:202–207.

Houghton, J.T., Y. Ding, D.J. Griggs, et al. 2001. *Climate Change 2001: The Scientific Basis.* Contribution of Working Group I to the Third Assessment Report of the Intergovernmental Panel on Climate Change. Cambridge University Press, London; New York.

Huey, R.B. and R.D. Stevenson. 1979. Integrating thermal physiology and ecology of ectotherms: a discussion of approaches. *Am. Zoologist,* 19:357–366.

Huey, R.B. 1991. Physiological consequences of habitat selection. *Am. Naturalist,* 137S:S91–S115.

Huey, R.B. and J.G. Kingsolver. 1989. Evolution of thermal sensitivity of physiological performance. *Trends Ecol. Evol.,* 4:131–135.

Hughes, L. 2000. Biological consequences of global warming: is the signal already apparent? *Trends Ecol. Evol.,* 15:56–61.

Joern, A. 1992. Host plant quality: demographic responses of range grasshoppers to stressed host plants. In U.S. Department of Agriculture, Animal, and Plant Health Inspection Service, PPQ Grasshopper IPM Project FY 1992 Annual Report. U.S Department of Agriculture, Washington, DC, pp. 65–72.

Joern, A. and S.T. Behmer. 1997. Importance of dietary nitrogen and carbohydrates to survival, growth, and reproduction in adult Ageneotettix deorum (Orthoptera: Acrididae). *Oecologia,* 112:201–208.

Joern, A. and S.T. Behmer. 1998. Impact of diet quality on demographic attributes in adult grasshoppers and the nitrogen limitation hypothesis. *Ecol. Entomol.,* 23:174–184.

Joern, A. and S.B. Gaines. 1990. Population dynamics and regulation in grasshoppers. In R.F. Chapman and A. Joern, Eds. *Biology of Grasshoppers.* John Wiley & Sons, New York, pp. 415–482.

Johnson, D.L. and A. Worobec. 1988. Spatial and temporal computer analysis of insects and weather: grasshoppers and rainfall in Alberta. *Memoirs Entomol. Soc. Can.,* 146:33–46.

Johnson, R.H. and D.E. Lincoln. 1990. Sagebrush and grasshopper responses to atmospheric carbon dioxide concentration. *Oecologia,* 84:103–110.

Johnson, R.H. and D.E. Lincoln. 1991. Sagebrush carbon allocation patterns and grasshopper nutrition: the influence of CO_2 enrichment and soil mineral nutrition. *Oecologia*, 87:127–134.

Karasov, W.H. 1984. Interhabitat differences in energy acquisition and expenditure in a lizard. *Ecology*, 65:235–247.

Kareiva, P., J.G. Kingsolver, and R.B. Huey. 1993. *Biotic Interactions and Global Climate Change*. Sinauer Associates, Sunderland, MA.

Kemp, W.P. 1986. Thermoregulation in three rangeland grasshopper species. *Can. Entomol.*, 118:335–343.

Kingsolver, J.G. 1989. Weather and population dynamics in insects: integrating physiological and population ecology. *Physiological Zool.*, 62:314–334.

Lactin, D.J., N.J. Holliday, D.L. Johnson, and R. Craigen. 1995. Improved rate model of temperature-dependent development by arthropods. *Environ. Entomol.*, 24:68–75.

Lactin, D.J. and D.L. Johnson. 1995. Temperature-dependent feeding rates of Melanoplus sanguinipes nymphs (Orthoptera: Acrididae) in laboratory trials. *Environ. Entomol.*, 24:1291–1296.

Lactin, D.J. and D.L. Johnson. 1996a. Behavioural optimization of body temperature by nymphal grasshoppers (Melanoplus sanguinipes, Orthoptera: Acrididae) in temperature gradients using incandescent bulbs. *J. Thermal Biol.*, 21:231–238.

Lactin, D.J. and D.L. Johnson. 1996b. Effects of insolation and body orientation on internal thoracic temperature of nymphal Melanoplus packardii (Orthoptera: Acrididae). *Environ. Entomol.*, 25:423–429.

Lactin, D.J. and D.L. Johnson. 1998a. Convective heat loss and change in body temperature of grasshopper and locust nymphs: relative importance of wind speed, insect size, and insect orientation. *J. Thermal Biol.*, 23:5–13.

Lactin, D.J. and D.L. Johnson. 1998b. Environmental, physical, and behavioural determinants of body temperature in grasshopper nymphs (Orthoptera: Acrididae). *Can. Entomol.*, 130:551–577.

Lansberg, J. and M.S. Smith. 1992. A functional scheme for predicting the outbreak potential of herbivorous insects under global atmospheric change. *Aust. J. Botany*, 40:565–577.

Lawton, J.H. 1991. From physiology to population dynamics and communities. *Functional Ecology,* 5:155–161.

Lawton, J.H. 1995. The response of insects to environmental change. In R. Harrington and N.E. Stork, Eds. *Insects in a Changing Environment.* Academic Press, San Diego, pp. 3–26.

Li, D. and R.R. Jackson. 1996. How temperature affects development and reproduction in spiders: a review. *J. Thermal Biol.,* 21:245–274.

Lima, S.L. 1998. Nonlethal effects in the ecology of predator–prey interactions. *BioScience,* 48:25–34.

Lincoln, D.E. and D. Couvet. 1989. The effect of carbon supply on allocation to alleleochemicals and caterpillar consumption of peppermint. *Oecologia,* 78:112–114.

Lincoln, D.E., D. Couvet, and N. Sionit. 1986. Response of an insect herbivore to host plants grown in carbon dioxide enriched atmospheres. *Oecologia,* 69:556–560.

Lincoln, D.E., N. Sionit, and B.R. Strain. 1984. Growth and feeding responses of Pseudoplusia includans (Lepidoptera: Noctuidae) to host plants grown in controlled carbon dioxide atmospheres. *Environ. Entomol.,* 13:1527–1530.

Lindroth, R.L. 1996a. Consequence of elevated atmospheric CO_2 for forest insects. In C. Kerner and F.A. Bazzaz, Eds. *Carbon Dioxide, Populations and Communities.* Academic Press, San Diego, CA, pp. 347–361.

Lindroth, R.L. 1996b. CO_2-mediated changes in tree chemistry and tree-Lepidoptera interactions. In G.W. Koch and H.A. Mooney, Eds. *Carbon Dioxide and Terrestrial Ecosystems,* Academic Press, San Diego, CA, pp. 105–120.

Lines, J. 1995. The effects of climate and land-use changes on the insect vectors of human disease. In R. Harrington and N.E. Stork, Eds. *Insects in a Changing Environment.* Academic Press, London.

Logan, J.A., D.H. Wolkind, S.C. Hoyt, and L.K. Tanigoshi. 1976. An analytic model for description of temperature dependent rate phenomena in arthropods. *Environ. Entomol.,* 5:1133–1140.

Logan, J. A. and D.W. Hilbert. 1983. Modeling the effects of temperature on arthropod population systems. In W.K. Lauenroth, G.V. Skogerbee, and M. Flug, Eds. *Analysis of Ecological Systems: State of the Art Ecological Modeling.* Elsevier, Amsterdam, pp. 113–122.

Logan, J.D., A. Joern, and W. Wolesensky. 2003a. Chemical reactor models of optimal digestion efficiency with constant foraging costs. *Ecol. Modelling,* 168:25–38.

Logan, J.D., A. Joern, and W. Wolesensky. 2003b. Mathematical model of consumer homeostasis control in plant-herbivore dynamics. *Math. Computer Modelling,* .

Losey, J.E. and R.F. Denno. 1998. Interspecific variation in the escape responses of aphids: effect on risk of predation from foliar-foraging and ground-foraging predators. *Oecologia,* 115:245–252.

Maguire, B. Jr. 1973. Niche response structure and the analytical potentials of its relationship to the habitat. *Am. Naturalist,* 107:213–246.

Malcom, S.B. 1993. Prey defense and predator foraging. In M.J. Crawley, Ed. *Natural Enemies.* Blackwell Scientific Publications, Oxford, pp. 458–475.

Marrs, R.H., R.D. Roberts, R.A. Skefington, and A.D. Bradshaw. 1983. Nitrogen and the development of ecosystems. In J.A. Lee, Ed. *Nitrogen as an Ecological Factor.* Blackwell Scientific, Oxford, pp. 113–136.

Mattson, W.J. 1980. Herbivory in relation to plant nitrogen content. *Annu. Rev. Ecol. Systematics,* 11:119–161.

Mattson, W.J. and R.A. Haack. 1987. The role of drought stress in provoking outbreaks of phytophagous insects. In P. Barbosa and J.C. Schultz, Eds. *Insect Outbreaks.* Academic Press, San Diego, CA, pp. 365–407.

Mole, S. and A. Joern. 1993. Foliar phenolics of Nebraska Sandhills prairie graminoids: between-years, seasonal, and interspecific variation. *J. Chem. Ecol.,* 19:1861–1874.

Mole, S., A. Joern, M.H. O'Leary, and S. Madhavan. 1994. Spatial and temporal variation in carbon isotope discrimination in prairie graminoids. *Oecologia,* 97:316–321.

Mooney, H.A., W.E. Winner, and E.J. Pell, Eds. 1991. *Response of Plants to Multiple Stresses*. Academic Press, San Diego, CA.

Nerney, N.J. and A.G. Hamilton. 1969. Effects of rainfall on range forage and populations of grasshoppers, San Carlos Apache Reservation, Arizona. *J. Econ. Entomol.*, 62:329–333.

Nothnagle, P.J. and J.C. Schultz. 1987. What is a forest pest? In J.C. Schultz, Ed. *Insect Outbreaks*. Academic Press, San Diego, CA, pp. 59–80.

Oedekoven, M.A. and A. Joern. 1998. Stage-based mortality of grassland grasshoppers (Acrididae) from wandering spider predation. *Acta Oecologia,* 19:507–515.

Oedekoven, M.A. and A. Joern. 2000. Plant quality and spider predation affects grasshoppers (Acrididae): food-quality-dependent compensatory mortality. *Ecology,* 81:66–77.

Osbrink, W.E., J.T. Trumble, and R.E. Wagner. 1987. Host suitability of Phaseolus Iunata for Trichoplusia ni (Lepidoptera: Noctuidae) in a controlled carbon dioxide atmosphere. *Environ. Entomol.,* 16:639–644.

Parmesan, C. and G. Yohe. 2003. A globally coherent fingerprint of climate change impacts across natural systems. *Nature,* 421:37–42.

Pearcy, R.W., O. Bjorkman, M.M. Caldwell, J.C. Keeley, R.K. Monson, and B.R. Strain. 1987. Carbon gain by plants in natural environments. *BioScience,* 37:21–29.

Perring, T.M., T.O. Holtzer, J.L. Toole, and J.M. Norman. 1986. Relationships between corn-canopy microenvironments and Banks grass mite (Acari: Tetranachidae) abundance. *Environ. Entomol.,* 15:79–83.

Peterson, C., H.A. Woods, and J.G. Kingsolver. 2000. Stage-specific effects of temperature and dietary protein on growth and survival of Manduca sexta caterpillars. *Physiological Entomol.,* 25:35–40.

Pimentel, D. Ed. 1991. *Handbook of Pest Management in Agriculture*. 3 vols., 2nd ed. CRC Press, Boca Raton, FL.

Porter, J. 1995. The effect of climate change on the agricultural environment for crop insect pests with particular reference to the European corn borer and grain maize. In R. Harrington and N.E. Stork, Eds. *Insects in a Changing Environment.* Academic Press, San Diego, pp. 93–123.

Porter, W.P. and D.M. Gates. 1969. Thermodynamic equilibria of animals with the environment. *Ecol. Monogr.,* 39:245–270.

Reddy, K.R. and H.F. Hodges, Eds. 2000. *Climate Change and Global Crop Productivity.* CABI Publishing, New York.

Riechert, S.E. and C.R. Tracy. 1975. Thermal balance and prey availability: bases for a model relating web-site characteristics to spider reproductive success. *Ecology,* 56:265–284.

Rietkerk, M. and J. van de Koppel. 1997. Alternate stable states and threshold effects in semi-arid grazing systems. *Oikos,* 79:69–79.

Rind, D., R. Goldberg, J. Hansen, C. Rosenzweig, and R. Ruedy. 1990. Potential evapotranspiration and the likelihood of future drought. *J. Geol. Resour.,* 95:9983.

Root, T., J.T. Price, K.R. Hall, S.H. Schneider, C. Rosenzweig, and J.A. Pounds. 2003. Fingerprints of global warming on wild animals and plants. *Nature,* 421:57–60.

Rosenheim, J.A. 1998. Higher-order predators and the regulation of insect populations. *Annu. Rev. Entomol.,* 43:421–447.

Roth, S.K. and R.L. Lindroth. 1995. Elevated atmospheric CO_2: effects on phytochemistry, insect performance, and insect-parasitoid interactions. *Global Change Biol.,* 1:173–182.

Rothley, K.D., O.J. Schmitz, and J.L. Cohon. 1997. Foraging to balance conflicting demands: novel insights from grasshoppers under predation risk. *Behav. Ecol.,* 8:551–559.

Schmitz, O.J. 1997. Press perturbations and the predictability of ecological interaction in a food web. *Ecology,* 78:55–69.

Schmitz, O.J. 1998. Direct and indirect effects of predation and predation risk in old-field interaction webs. *Am. Naturalist,* 151:327–342.

Schmitz, O.J. 2003. Top predator control of plant biodiversity and productivity in an old-field ecosystem. *Ecology,* Letters 6:156–163.

Schneider, S.H. 1993. Scenarios of global warming. In P.M. Kareiva, J.G. Kingsolver, and R.B. Huey, Eds. *Biotic Interactions and Global Change.* Sinauer Associates, Sunderland, MS, pp. 9–23.

Schneider, S.H., L. Mearns, and P.H. Gleick. 1992. Climate-change scenarios for impact assessment. In T.E. Lovejoy and L. Hannah, Eds. *Global Warming and Biological Diversity.* Yale University Press, New Haven, CT, pp. 38–55.

Scriber, J.M. and R.C. Lederhouse. 1983. Temperatures as a factor in the development and feeding ecology of tiger swallowtail caterpillars, Papilio glaucus (Lepitodptera). *Oikos,* 40:95–102.

Slansky, F. Jr. 1993. Nutritional ecology: the fundamental quest for nutrients. In N.E. Stamp and T.M. Casey, Eds. *Caterpillars: Ecological and Evolutionary Constraints on Foraging.* Chapman and Hall, New York, pp. 29–91.

Stamp, N.E. 1990. Growth vs. molting time of caterpillars as a function of temperature, nutrient concentration, and the phenolic rutin. *Oecologia,* 82:107–113.

Stamp, N.E. 1993. A temperate region view of the interaction of temperature, food quality, and predators in caterpillar foraging. In N.E. Stamp and T.M. Casey, Eds. *Caterpillars: Ecological and Evolutionary Constraints on Foraging.* Chapman & Hall, New York.

Stamp, N.E. and M.D. Bowers. 1990a. Body temperature, behavior, and growth of early spring caterpillars (Hemileuca lucina: Saturniidae). *J. Lepidopterists Soc.,* 44:143–155.

Stamp, N.E. and M.D. Bowers. 1990b. Variation in food quality and temperature constrain foraging of gregarious caterpillars. *Ecology,* 71:1031–1039.

Stamp, N.E. and T.M. Casey, Eds. 1993. *Caterpillars: Ecological and Evolutionary Constraints on Foraging.* Chapman and Hall, New York.

Stamp, N.E. and M.D.Bowers. 2000. Foraging behaviour of caterpillars given a choice of plant genotypes in the presence of insect predators. *Ecol. Entomol.,* 25:486–492.

Sterner, R.W. and J.J. Elser. 2002. *Ecological Stoichiometry: The biology of Elements from Molecules to the Biosphere.* Princeton University Press, Princeton, NJ.

Sutherst, R.W., G.F. Mawwald, and D.B. Skarratt. 1995. Predicting insect distributions in a changed climate. In R. Harrington and N.E. Stork, Eds. *Insects in a Changing Environment.* Academic Press, San Diego, pp. 59–91.

Thomas, C.D., E.J. Bodsworth, R.J. Wilson, A.D. Simmons, Z.G. Davies, M. Musche, and L. Conrad. 2001. Ecological and evolutionary processes at expanding range margins. *Nature,* 411:577–581.

Tilman, D., M. Mattson, and S. Langer. 1981. Competition and nutrient kinetics along a temperature gradient: an experimental test of a mechanistic approach to niche theory. *Limnol. Oceanography,* 26:1020–1033.

van de Koppel, J., M. Rietkerk, and F.J. Weissing. 1998. Catastrophic vegetation shifts and soil degradation in terrestrial grazing systems. *Trends Ecol. Evol.,* 12:352–356.

van der Meijden, E., R.M. Nisbet, and M.J. Crawley. 1998. The dynamics of an herbivore-plant interaction, the cinnabar moth and ragwort. In J.P. Dempster and I.F.G. McLean, Eds. *Insect Populations in Theory and in Practice.* Kluwer, Dordrecht, pp. 291–308.

Vitousek, P. M. 1994. Beyond global warming: ecology and global change. *Ecology,* 75:1861–1876.

Wagner, T.L., H. Wu, P.J.H. Sharpe, and R.N. Coulson. 1984a. Modeling distributions of insect development time: a literature review and application of the Weibull function. *Ann. Entomol. Soc. Am.,* 77:475–483.

Wagner, T.L., H. Wu, P.J.H. Sharpe, R.M. Schoolfield, and R.N. Coulson. 1984b. Modeling insect developmental rates: a literature review and application of a biosphysical model. *Ann. Entomol. Soc. Am.,* 77:208–225.

Walker, B. and W. Steffen, Eds. 1996. *Global Change and Terrestrial Ecoystems.* Cambridge University Press.

Walther, G.-R., C.A. Burga, and P.J. Edwards, Eds. 2001. *"Fingerprints" of Climate Change: Adapted Behaviour and Shifting Species Range.* Kluwer/Plenum Press, New York.

Walther, G.-R., E. Post, A. Menzel, C. Parmesan, T.J.C. Beebee, J.-M. Fromentin, et al. 2002. Ecological responses to recent climate change. *Nature,* 416:389–395.

Warren, M.S., J.K. Hill, J.A. Thomas, et al. 2001. Rapid responses of British butterflies to opposing forces of climate and habitat change. *Nature,* 414:65–69.

Watt, A.D., J.B. Whittaker, M. Docherty, G. Brooks, E. Lindsay, and D.T. Salt. 1995. The impact of elevated CO_2 on Insect Herbivores. In R. Harrington and N.E. Stork, Eds. *Insects in a Changing Environment.* Academic Press, London.

Wermelinger, B. and M. Seifert. 1999. Temperature-dependent reproduction of the spruce bark beetle Ips typographus, and the analysis of the potential population growth. *Ecol. Entomol.,* 24:103–110.

White, T.C.R. 1993. *The Inadequate Environment: Nitrogen and the Abundance of Animals.* Springer-Verlag, Berlin.

Williams, R.J., K. Boersma, and A.L. van Ryswyk. 1979. The effect of nitrogen fertilization on water use by crested wheatgrass. *J. Range Manage.,* 29:180–185.

Wilmer, P.G. and D.M. Unwin. 1981. Field analysis of insect heat budgets: reflectance, size, and heating rates. *Oecologia,* 50:250–255.

Woods, H.A. and J.G. Kingsolver. 1999. Feeding rate and the structure of protein digestion and absorption in Lepidopteran midguts. *Arch. Insect Biochem. Physiol.,* 42:74–87.

Woods, H.A., W. Makino, J.B. Cotner, S.E. Hobbie, J.F. Harrison, K. Acharya, and J.J. Elser. 2003. Temperature and the chemical composition of poikilothermic organisms. *Functional Ecol.,* 17:237–245.

Yang, Y. and A. Joern. 1994a. Compensatory feeding in response to variable food quality by Melanoplus differentialis. *Physiological Entomol.,* 19:75–82.

Yang, Y. and A. Joern. 1994b. Influence of diet, developmental stage, and temperature on food residence time. *Physiological Zool.,* 67:598–616.

Yang, Y. and N.E. Stamp. 1995. Simultaneous effects of night-time temperature and an alleleochemical on performance of an insect herbivore. *Oecologia,* 104:225–233.

Zimmerman, L.C. and C.R. Tracy. 1989. Interactions between the environment and ectothermy and herbivory in reptiles. *Physiological Zool.,* 62:374–409.

14

Food Security and Production in Dryland Regions

B.A. STEWART

CONTENTS

14.1 INTRODUCTION

Approximately 45% of the world's total land area is considered drylands. The extent of drylands in various regions ranges from a low of 20% in North Africa and the Near East to 95% in North Asia, east of the Urals (Table 14.1). An estimated 38% of the world's 6 billion people live in dryland areas. Perhaps more important is that many of the people living in dryland areas are actively involved in agriculture. For example, of the 3.36 billion people living in the Asia and Pacific region, almost 60% are engaged in agriculture, many of whom are in dryland areas. This compares to only 3% of the people in North America who are engaged in agriculture and 13.5% in Europe (Table 14.1). Consequently, dryland areas are of great socioeconomic importance, and are likely to become more so in the future. The areas included in Table 14.1 cover a wide range of water conditions, although water is limited in all these areas for portions of the year. As population pressure increases, the net result is that more people move into less-favored areas. In essence, people move to the droughts instead of the droughts moving to the people. Hazell (1998) stated that despite some out-migration, population size continues to grow in many less-favored areas, but crop yields grow little or not at all. Hazell (1998) reported that about 500 million people live in less-favored areas, mostly in Asia and Sub-Saharan Africa; and if current trends persist, by 2020 more than 800 million people will live in less-favored lands. Hazell stressed that it is becoming increasingly clear that, on poverty and environmental grounds alone, more attention must be given to less-favored lands in setting priorities for policy and public investment. The rapid population growth in arid and semi-arid regions has placed tremendous pressure on the natural resource base. Often, the inevitable result of increasing population in resource-poor areas is land degradation. Droughts, which are common to these areas, exacerbate degradation processes. Land use practices in drylands are often opportunistic. Dryland farmers tend to maximize take-off during good periods and minimize loss during dry periods.

Cereals directly supplied around 57% of calories in the global human diet in 2000 (Food and Agriculture Organization [FAO], 2000b). The meat and dairy foods produced from animals that consumed 33% of world cereals provide additional calories. Together, wheat (*Triticum aestivum*), rice (*Oryza*

Table 14.1 Land Area by Dryland Type, Total Population, Population in Drylands, and Agricultural Population, in World Regions

Regions	World Land Area (% of 134 million km²)	Arid (%)	Semi-arid (%)	Dry Subhumid (%)	Total World Population (% of 6 billion)	Population in Drylands (%)	Agricultural Population (%)
Asia and Pacific	21.5	6	15	17	56	44	59.5
Europe	5.4	<0.5	13	16	12	18	13.5
North Africa and Near East	9.5	4	11	5	5	44	44.3
North America	14.8	12	28	23	5	19	3.0
North Asia, east of Urals	15.6	11	51	33	4	89	17.4
South and Central America	15.4	11	6	10	8	24	23.0
Sub-Saharan Africa	17.7	6	13	19	10	36	65.0
World (Total)	100	7	20	18	100	38	45.8

Note: Lands classified as hyperarid make up an additional 19% of world land area, but are not included as dryland areas because the Food and Agriculture Organization assumes that these lands are too dry to sustain agriculture.
Source: Adapted from Food and Agriculture Organization. 2000a. *Land Resource Potential and Constraints at Regional and Country Levels.* World Soil Resources Report 90. Food and Agriculture Organization, Rome.

sativa), and maize (*Zea mays*) make up approximately 85% of the world's cereal production. Wheat and rice are by far the most widely consumed cereals in the world, while maize is important for both direct and indirect human consumption. Approximately 67% of maize production is used as animal fodder. The FAO (1996) reported that 48% of cereal production in developing countries (excluding China) came from irrigated lands. This is in contrast to many developed countries where cereals, particularly wheat and maize, are largely grown without irrigation. It is estimated that 60% of wheat produced in developing countries is irrigated, while only 7% of wheat and 15% of maize are irrigated in the United States (U.S. Department of Agriculture, 1997). Other major cereal-producing areas such as Canada, Australia, and Europe are mainly nonirrigated.

World food production over the past four decades has more than kept pace with the rapid growth in population. Between 1961 and 1997, world population increased by 89% and food production per person increased by 24%, while food prices fell by 40% in real terms (Wood et al., 2000). This is a remarkable achievement on a worldwide basis, but global statistics do not reflect the wide range of differences between and within individual countries. The agronomic technologies that allowed the increases in world food production were largely based on high-yielding varieties, fertilizers, pest control, and irrigation. Irrigation has been particularly important in developing countries where the total irrigated area increased from 102 million hectares in 1961 to 207 million in 1999. This compares to 37 million hectares in the developed countries in 1961, and 67 million in 1999 (FAO, 2000b). Worldwide, about 17% of cropland is irrigated, and accounts for 40% of food and fiber production (Wood et al., 2000). On average, irrigation claims nearly 70% of world water abstraction and over 90% in agricultural economies in the arid and semi-arid tropics, but less than 40% in industrial economies in the humid temperate regions (FAO, 1996). With the cost of developing additional irrigated lands ranging from US$2,000 to US$10,000 per hectare (FAO, 1995, 1997a, 1997b, 1999), it is imperative that alternative water management and production system be considered for at least a part of the anticipated food demand.

14.2 CROP PRODUCTION IN DRYLAND REGIONS

Cereal yields in dryland regions, particularly when produced without irrigation, are low because of the lack of water. As already stated, high-yielding varieties, fertilizers, pest control, and irrigation have been mainly responsible for large increases in worldwide food and fiber production. When irrigation is not available in dryland regions, the lack of water limits production, and the benefits of the other technologies are largely muted. FAO (1996) reported that the 1988–1990 average yield of wheat in developing countries in semi-arid regions was 1100 kg ha^{-1} for wheat, 1130 kg ha^{-1} for maize, and 650 kg ha^{-1} for sorghum (*Sorghum bicolor*). In contrast, the worldwide average yields were 2561 kg ha^{-1} for wheat, 4313 kg ha^{-1} for maize, and 1439 kg ha^{-1} for sorghum. Only about 10% of wheat, 8% of maize, and 35% of sorghum grown in developing countries were produced in semi-arid regions (FAO, 1996).

The worldwide average wheat yield has increased 2.5 times during the past four decades, and maize yield has increased 2.2 times. These crops are mostly grown in the more favored areas or under irrigation. While statistics are not available for worldwide yields when these crops are grown in dryland areas, it is well known that average yields have not increased dramatically because of low and erratic precipitation. Worldwide yields of sorghum, which is grown more widely in dryland regions, have increased only 1.6 times during the past four decades. The yield of millet (*Panicum millaceum*), grown almost entirely in dryland regions, has increased only 1.26 times during the same time period. It is well understood and documented that the large benefits from technologies such as fertilizers, high-yielding varieties, and pest control have largely occurred under irrigation or in the more-favored precipitation regions. Figure 14.1 illustrates the small gains from technology inputs under limited water

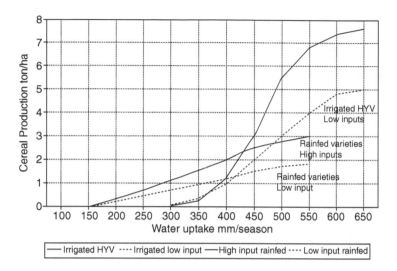

Figure 14.1 Effects of added inputs on the water use efficiency of cereal production. (From Koohafkan, P. 2000. *Food Security, and Sustainable Development in the Middle East Region*. Food and Agriculture Organization, Rome.)

conditions in comparison to the large gains obtained when water is not limiting.

14.3 SOIL ORGANIC MATTER

Although the benefits from technologies under dryland conditions have been relatively small, average crop yields would almost certainly be lower without them. This is because the capture, storage, and use of water in low and erratic precipitation regions have decreased because of the loss of soil organic matter (SOM). It is well documented that SOM levels decline when land is converted from grassland or forest ecosystems to cropland. The decline is most rapid in the first few years following conversion, and then continues at slower rates until a new steady state is reached. After 50 to 100 years, SOM levels are often 50% to 60% lower than the initial levels (Cole et al., 1993). These large losses of SOM have significant negative impacts on the soil structure and water-holding capacity of the soil, which makes limited precipitation even more limiting.

Soil management practices have significant effects on both the rate and extent of SOM decline and restoration. Stewart et al. (1991) pointed out that the maintenance of SOM becomes more difficult as temperatures increase and the amounts of precipitation decrease. The reasons are many, but are dominated by the fact that SOM decomposition is accelerated with rising temperatures, and the production of biomass to replenish SOM reserves declines as water becomes more limiting.

A long-term study at Bushland, TX — where annual average precipitation is 475 mm and annual average potential evapotranspiration is about 1900 mm — illustrates how quickly the SOM content can decline under semi-arid conditions (Table 14.2). Twenty-nine years of continuous cropping with wheat resulted in a 34% reduction in SOM content. An even greater loss, 39%, occurred in the wheat-fallow system where only one wheat crop was grown every 2 years. The apparent reason for more loss in the wheat-fallow system was because there was not as much crop biomass produced in the wheat-fallow system as in the continuous cropping system. Tillage intensity was also a factor. One-way disk tillage is more intensive than subtillage with sweeps. Subtillage using sweeps, often called stubble-mulch tillage, was initiated as a wind erosion practice during the infamous "Dustbowl" era of the 1930s in the U.S. Great Plains. With subtillage, sweeps are pulled about 10 cm beneath the surface, which cuts the roots of the weeds but does not invert the soil, so that some of the crop residues remain on the soil surface. Approximately 20% of the residues are incorporated with each subtillage operation, and there are usually about four tillage operations during the 15-month fallow period. Thus, even after four tillage operations, enough crop residues generally remain on the surface to significantly reduce wind erosion. Although stubble-mulch tillage was developed for controlling wind erosion, it soon became evident that it had a positive influence on stored soil water. The data in Table 14.2 clearly indicate that more soil water was present in the profile at time of seeding

Table 14.2 Effect of Various Cropping Systems on SOM, PAW, and Yield of Winter Wheat at Bushland, TX

Cropping System	SOM 1941 (%)	SOM 1970 (%)	Nitrate-N[a] (kg ha⁻¹)	PAW[b] (mm)	Average Yield (kg ha⁻¹)	29-Year Yield Range (kg ha⁻¹)
Continuous wheat						
One-way disk tillage[c]	2.44	1.61	417	91	593	0 to 1915
Subtilled with sweeps	2.44	2.02	179	103	694	0 to 2312
Wheat fallow[d]						
One-way disk tillage	2.44	1.49	519	128	944	0 to 2427
Subtilled with sweeps	2.44	1.81	325	154	1058	0 to 2589
Delayed subtilled with sweeps[e]	2.44	2.24	88	144	1038	0 to 2440

Notes: PAW = plant available soil water content at seeding; SOM = soil organic matter.
[a] Nitrate-N in 180-cm soil profile at end of experiment.
[b] Average plant-available water in 180-cm soil profile at seeding time.
[c] Winter wheat seeded annually at approximately October 1 and harvested at approximately July 1 the following year.
[d] Winter wheat seeded at approximately October 1 every second year; approximately 15 months fallow between crops and yields shown must be divided by 2 to indicate annual land production.
[e] Tillage was delayed for approximately 10 months following wheat harvest; weeds and volunteer wheat were allowed to grow during the 10-month period.
Source: Adapted from Johnson, W.C., C.E. Van Doren, and E. Burnett. 1974. Summer fallow in the southern Great Plains. In *Summer Fallow in the Western United States*. Conservation Research Report 17. U.S. Department of Agriculture, Agricultural Research Service, Washington, DC, pp. 86–109.

when stubble-mulch tillage was used in comparison to one-way disk tillage, and this increased soil water storage resulted in higher yields. The experiment also included a treatment that did not have any tillage until about 10 months following the wheat harvest. Weeds and volunteer wheat plants were allowed to grow during the summer and fall, and then stubble-mulch tillage was used in the spring. This treatment was by far the most effective in maintaining SOM, because there was a lot of biomass produced that was later incorporated into the soil surface. The average yield was also comparable to the other treatments because there were still about 5 months during the wettest part of the year to capture and store precipitation in the soil profile before seeding wheat. This system is not aesthetically acceptable, however, because enormous amounts of weed seed are produced that spread to other fields by wind, birds, insects, and other forms of transport. It is useful, however, to show the importance of returning crop residues to the soil and reducing the number and intensity of tillage operations for maintaining SOM in the soil. The study also demonstrated the relationship between SOM loss and mineralization of soil nitrogen (N). It is well established that SOM contains about one part of N for every ten parts of carbon (w/w). As the SOM decomposes, N is mineralized and made readily available for use by growing plants. When SOM decomposition is rapid as is often the case in semi-arid regions, N can be released in excess of plant needs because the first limiting factor for crop production in these regions is water. Therefore, N can accumulate in the soil profile in large amounts as shown in Table 14.2. This excess nitrogen is subject to losses by leaching and denitrification. In time, N becomes limiting in these regions because SOM decomposition slows to the point that insufficient N is released to meet the needs of crop production, particularly during years of above-average precipitation. Phosphorus (P), sulfur (S), and other nutrients are also contained in SOM and are released during its decomposition. In the short term, the decomposition of SOM is very positive for crop production, but the long-term effects are devastating because of reduced soil structure, plant-available soil water storage, and fertility. Perhaps even more important, these losses make it extremely difficult to restore SOM because C can only be sequestered as SOM when there is sufficient N, P, and other essential elements available.

14.4 SEQUESTERING SOIL CARBON

There is worldwide interest in the sequestration of soil C to reduce the amount of C going into the atmosphere as carbon dioxide. An estimated 40 to 50 Pg of carbon (Pg C = 1 petagram carbon = 1 billion metric tons of carbon) have been released from the soil worldwide (Lal et al., 1997). Most of this loss has resulted from the conversion of forest and grassland eco-systems into agro-ecosystems for the production of food and fiber.

Crop production requires relatively large amounts of N, P, S, and other essential plant nutrients. Decomposition of SOM is the primary source of plant nutrients for the first few years after crop production is initiated. As SOM decomposition begins to slow significantly, manures or fertilizers are often required to supplement nutrient requirements. Legumes are sometimes used to supplement N needs, but the use of legumes is limited in semi-arid regions. Nitrogen is usually the first nutrient that becomes limiting as SOM decomposition slows, and P is often the second essential nutrient that limits crop production.

The many calls for sequestering soil C seldom discuss the amounts of other nutrients that must be available because C is only one of many SOM constituents. Himes (1997) stated that on a weight basis, the C/N ratio in soil organic matter was 12/1, C/P was 50/1, and C/S was 70/1. Therefore, to sequester 10,000 kg of C in humus, 833 kg of N, 200 kg of P, and 143 kg of S are required. The amounts of nutrients necessary to produce high crop yields, and at the same time sequester large amounts of C, are difficult to estimate but they are huge. As already stated, it is believed that 40 to 50 Pg of C have been released from the soil. Therefore, approximately 4 Pg of N and 1 Pg of P were likely mineralized, and most of the mineralized N and P have either been removed from the land with harvested crops or lost by runoff, erosion, leaching, and other processes.

Some rather optimistic estimates have been made about how much of the C lost from soils can be restored. Paustian et al. (1995) estimated that the 13.8 million ha in the U.S. Conservation Reserve Program (CRP) could sequester about 25 Tg of C over a 10-year period. This would require approximately 2.5 Tg of N, or 18 kg per year for each hectare. Nitrogen fertilizers are not generally used on these lands, and there are also few, if any, legumes present because most of the CRP lands are located in semi-arid regions. There is some

N deposited with precipitation, but far below the amounts needed. Lal et al. (1997) indicated that widespread adoption of conservation tillage could lead to global C sequestration of 1.48 to 4.90 Pg, and restoration of degraded soils could lead to an additional 3 Pg year^{-1}. Again, the authors did not discuss the N requirements or the source. To gain some perspective, however, the restoration of 5 Pg of C in 1 year would require approximately 0.5 Pg of N, which is roughly six times the total worldwide usage of fertilizer N in 2000. About 0.1 Pg of P would also be required, and this is roughly seven times the total amount of fertilizer P used worldwide in 2000 (Fertilizer Institute, 2003). Consequently, sequestration of large amounts of C into SOM will require enormous amounts of N, P, and other nutrients, and these may be limiting in many cases, particularly in semi-arid regions where fertilizers are used sparingly and legumes are limited.

The potential for carbon sequestration in SOM will likely be higher in the first few years than in succeeding years. Decomposition of SOM is rapid the first few years after cultivation is initiated, and then it slows as it reaches a new equilibrium. The reverse is likely to occur with sequestration. If tillage is reduced or eliminated, there will be some C sequestered, but the amounts sequestered in succeeding years will likely slow as N, P, and possibly other nutrients become less available. For instance, C.A. Robinson (West Texas A&M University, unpublished data, 1996) sampled four sites on Pullman silty clay loam, a fine, mixed thermic Torrertic Paleustoll, located in a semi-arid region near Claude, TX (Figure 14.2). One site had been cropped to dryland wheat for more than 50 years with occasional fallow years. Tillage was primarily performed with a tandem disk plow. During the last few years before sampling in 1993, anhydrous ammonia at the rate of 22 kg ha^{-1} was applied in the fall prior to seeding wheat. The other three sites were grassland: one native grassland field, one previous cropland field returned to grass 37 years before sampling as part of the U.S. Department of Agriculture Soil Bank program of the 1950s, and a field that had been returned to grass 7 years before sampling as part of the CRP initiated in 1985 by the 1985 U.S.D.A. Food Security Act. Significant

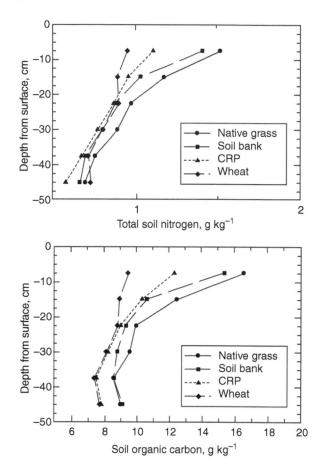

Figure 14.2 Soil organic carbon (top) and total nitrogen (bottom) by depth as affected by management systems. (Drawn from C.A. Robinson, West Texas A&M University, unpublished data, 1996.)

amounts of C and N were sequestered in the SOM (Figure 14.2). The data clearly show that substantial amounts of C can be retained even under semi-arid conditions. Precipitation at the sites averages approximately 500 mm annually. The native grassland site had significantly more soil organic carbon (SOC) and N at each depth than the CRP site and the wheat site. The Soil Bank site was intermediate. The increase in SOC was mostly in the surface 15 cm, although some differences at lower depths were largely associated with differences in bulk density. The results were similar to those of Ibori et al. (1995) on abandoned fields in northeastern Colorado. Others, however, found that SOC differences due to tillage were limited to the top 7 cm (Havlin et al., 1990; Potter et al., 1997). Nitrogen accumulations showed similar trends, but there was less recovery of N than C, particularly at the lower depths. The CRP site that had been in grass for 7 years recovered 48% of the C, but only 25% of the N relative to the Soil Bank site that had been in grass for 37 years. This illustrates that the initial rate of C accumulation in cropland soils returned to grass is considerably greater than in future years, and this rate may be partially constrained by N, or possibly P, as discussed earlier. No legumes were found at any of the sites.

14.5 FOOD SECURITY IN DRYLAND AREAS

The number of people in dryland regions, particularly in developing countries, will continue to increase in the next few years according to population estimates. The effect of climate change on these regions is not known, but there is little evidence that these areas will become more favorable. Because water is the first limiting factor for crop production in dryland regions, any additional warming will have a negative effect unless it is accompanied by increased precipitation. There is a growing belief that global climate change is making weather patterns more extreme. Climate extremes are already a major problem in dryland regions because annual precipitation can be from two to three times the average in wet years and substantially less than one-half in dry years. These extremes often result in devastating droughts and severe wind erosion in the dry years, and rampant water erosion in the wet years. Crop yields also vary greatly in these regions, and neither farmers nor governments can predict with confidence the production from a farm or region. Grain yields in dryland regions for a given year can range from zero to three or more times the average yield. Therefore, it is extremely difficult and will become increasingly more so with

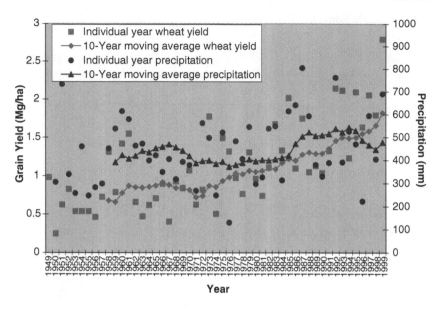

Figure 14.3 Relationship of annual precipitation and wheat yields showing improved water use efficiency. (From Z. Wu, A.W. Colette, and B.A. Stewart, West Texas A&M University, unpublished data, 2002.)

increasing populations for dryland regions to be self-sufficient in terms of food and fiber production.

However, crop yields in dryland regions can be increased by enhancing water use efficiency. An example of benefits that can accrue from improved water management practices is shown in Figure 14.3. These data are farmer yields of wheat grown in Deaf Smith County, TX, where average annual precipitation is about 450 mm, and annual potential evapotranspiration is about 1800 mm. Therefore, drought is a common occurrence and severe water stress occurs every year. The yearly precipitation amounts show a range from less than 200 mm to more than 800 mm. A 10-year moving average (each yearly point is the average precipitation amount for the year shown plus the 9 previous years) line of annual precipitation is also shown, and although there is some variation, the average annual amount has remained relatively stable.

County average wheat yields for each year are also shown along with a line showing the 10-year moving averages. It is noteworthy that the moving yield average and moving precipitation average closely paralleled each other until the early 1970s. Since that time, the moving average grain yield increased essentially every year, and the average yield has more than doubled. No single factor is responsible, but it clearly shows that use efficiency of the precipitation has dramatically increased. Water management is the first factor that must be addressed in dryland regions because other technologies such as improved cultivars and fertilizers are usually not beneficial without improved water management. In the early 1970s, the cost of oil and other energy sources increased rapidly, and there was a concerted effort by researchers and extension personnel to promote less tillage and more herbicide usage. This change in priorities increased the amounts of crop residues remaining on the soil surface, which resulted in more soil water storage during fallow periods that was used by the subsequent crop. A small amount of additional water used during the growing season can increase yields significantly because the threshold amount of water required for grain production has already been met, as shown in Figure 14.1.

14.6 DISCUSSION

The extent to which global climate change is occurring is not clear, but it is well documented that carbon dioxide (CO_2) levels in the atmosphere are steadily increasing. The increased levels of CO_2 will likely have a positive effect on food and fiber production in areas where water does not become more limited. However, in areas that become hotter and drier as a result of global climate change, yields will likely decline even though CO_2 levels for photosynthesis are more favorable. Rosenzweig (2000) surmises that if atmospheric buildup of greenhouse gases continues without limit, sooner or later it is bound to warm the Earth's surface. Such a warming trend cannot but affect the regional panoply of temperature and precipitation governing natural and agricultural systems. Some areas — possibly the northern United States,

Canada, and Russia — may become more productive, but other areas such as the Middle East and parts of Asia may become less productive. Dryland regions are already severely challenged because of high temperatures, and additional warming will indeed be a challenge to these regions unless average precipitation is significantly increased. There is also the possibility that the already highly variable precipitation amounts in these regions will become even more variable. This would result not only in more severe droughts but also more water erosion and flooding during extreme precipitation events.

Jones and Thornton (2003) stated that climate change could potentially lead to a 10% drop in developing-country maize production over the next 50 years. They projected that the lost production would not be across the board or evenly spread, but that losses would vary widely from one agro-ecosystem to another. For example, they predicted that large maize-producing countries such as Brazil and Mexico will be hit hard, while Chile and Ecuador will likely be relatively unaffected, and other countries may even experience increased yields. Latin America will likely see a reduction approaching 25% due to higher temperatures and decreasing rainfall. These authors forecast that Nigeria, South Africa, and Tanzania will lose upward of 20% of total production. Maize is particularly important in many parts of the world, and less maize means less grain for poor people, less feed for farm animals, and less milk and meat for hungry households. The researchers, however, emphasized that the projected decreases in maize production could be lessened by better crop varieties and land management systems.

Global food and fiber production may not be severely threatened because gradual adjustments can be made to pro-duce more in favored areas. There will, however, be areas where food and fiber production will be limited, as is already the situation today in many dryland regions. Policies should be identified and implemented that can mitigate the threat of global climate change. Conservation agriculture (FAO, 2002), based on the principles of avoiding mechanical soil disturbance, maintaining a permanent soil cover, and crop rotation is a sound strategy that is being promoted and applied in most parts of the world. Conservation agriculture is practiced on 45 million ha, mostly in South and North America (FAO, 2002). This accounts, however, for only about 3% of the 15 billion ha of arable land worldwide. Conservation agriculture is practiced from the humid tropics to almost the Arctic Circle and on all kinds of soils (FAO, 2002). According to the FAO (2002), "So far the only area where the concept has not been successfully adapted is the arid areas with extreme water shortage and low production of organic matter. In these areas both humans and animals compete with the soil for crop residues."

Although applying all conservation agriculture principles to dryland areas is a challenge, substantial benefits can occur by reducing tillage and managing crop residues as already illustrated in Figure 14.3. Some C can be sequestered in these regions to slowly begin restoring some of the lost SOM. Increases will likely be slow and in many cases insignificant because of high temperatures that hasten decomposition and the shortage of N, and perhaps P, that must be available to form SOM. Even small increases in dryland regions must be pursued because continued soil degradation will make these already water-deficient areas even more deficient.

14.7 CONCLUSIONS

Crop production in dryland regions has always been and always will be high risk because of low and highly variable precipitation. It is becoming clear, however, that risk can be reduced and yields can be increased by the adaptation and adoption of practices that maintain at least some plant residues on the soil surface. Maintaining crop residues on the soil surface will reduce erosion, increase water storage and use, enhance SOM, and improve soil physical properties. The adoption of such practices in dryland areas will not be fast or easy. Tillage is an ingrained part of agriculture culture, and it will take many years to change. It also requires more knowledge and better management, and generally a higher level of inputs that increases risk in dryland cropping systems. It is becoming increasingly clear, however, that the sustainability of dryland cropping in many regions will require major changes to prevent further soil degradation. The challenge is even greater in many developing countries in semi-arid regions because of the need for using crop residues for animal feed and household fuel. These short-term gains usually take priority, and that is easily understood. However, the long-term sustainability of crop production in some of these regions may very well depend on policymakers taking difficult and painful actions about how crop residues are managed. At some point, it is essential that some organic matter be returned to the soil.

REFERENCES

Cole, C.V., K. Flach, J. Lee, D. Sauerbeck, and B.A. Stewart. 1993. Agricultural sources and sinks of carbon. In J. Wisniewski and R.N. Sampson, Eds. *Terrestrial Biospheric Carbon Fluxes: Quantification of Sinks and Sources of CO_2*. Kluwer, Dordrecht, pp. 111–119.

Fertilizer Institute. 2003. World Fertilizer Use. Available at: *www.tfi.org/Statistics/worldfertuse.asp*.

Food and Agriculture Organization. 1995. Irrigation in Africa in Figures. Water Reports 7. FAO, Rome.

Food and Agriculture Organization. 1996. Prospects to 2010: agricultural resources and yields in developing countries. In Vol. 1, Technical Background Documents 1–5. World Food Summit, FAO, Rome, pp. 26–36.

Food and Agriculture Organization. 1997a. Irrigation in the Near East Region in Figures. Water Reports 9. FAO, Rome.

Food and Agriculture Organization. 1997b. Irrigation in the Countries of the Former Soviet Union in Figures. Water Reports 15. FAO, Rome.

Food and Agriculture Organization. 1999. Irrigation in Asia in Figures. Water Reports 18. FAO, Rome.

Food and Agriculture Organization. 2000a. *Land Resource Potential and Constraints at Regional and Country Levels.* World Soil Resources Report 90. FAO, Rome.

Food and Agriculture Organization. 2000b. Agriculture: Towards 2015/30. Technical Interim Report. Global Perspective Studies Unit. FAO, Rome.

Food and Agriculture Organization. 2002. Intensifying Crop Production with Conservation Agriculture. Available at: *www.fao.org/ag/ags/AGSE/main.htm.*

Havlin, J.L., D.E. Kissel, L.D. Maddux, M.M. Classen, and J.H. Long. 1990. Crop rotation and tillage effects on soil organic carbon and nitrogen. *Soil Sci. Soc. Am. J.,* 54:448–452.

Hazell, P. 1998. *Why Invest More in the Sustainable Development of Less-Favored Lands?* IFPRI Report, vol. 20, no. 2. International Food Policy Research Institute, Washington, DC.

Himes, F.L. 1997. Nitrogen, sulfur, and phosphorus and the sequestering of carbon. In R. Lal, J.M. Kimble, R.F. Follett, and B.A. Stewart, Eds. *Soil Processes and the Carbon Cycle.* Advances in Soil Science. CRC Press, Boca Raton, FL, pp. 315–319.

Ibori, T., I.C. Burke, W.K. Lauenroth, and D.P. Coffin. 1995. Effects of cultivation and abandonment on soil organic matter in northeastern Colorado. *Soil Sci. Soc. Am. J.,* 49:352–356.

Johnson, W.C., C.E. Van Doren, and E. Burnett. 1974. Summer fallow in the southern Great Plains. In *Summer Fallow in the Western United States.* Conservation Research Report 17. U.S. Department of Agriculture, Agricultural Research Service, Washington, DC, pp. 86–109.

Jones, P. and P. Thornton. 2003. Scenario Projects 10 Percent Fall in Developing Country Maize Production. International Center for Tropical Agriculture, Cali, Colombia, and International Livestock Research Institute, Nairobi, Kenya. Available at: *www.futureharvest.org/news/maize_model.shtml.*

Koohafkan, P. 2000. *Food Security, and Sustainable Development in the Middle East Region.* Food and Agriculture Organization, Rome.

Lal, R., J. Kimble, and R. Follett. 1997. Land use and soil C pools in terrestrial ecosystems. In R. Lal, J.M. Kimble, R.F. Follett, and B.A. Stewart, Eds. *Management of Carbon Sequestration in Soil*. CRC Press, Boca Raton, FL, pp. 1–10.

Paustian, K., C.V. Cole, E.T. Elliott, E.F. Kelly, C.M. Yonker, J. Cipra, and K. Killian. 1995. Assessment of the contributions of CRP lands to C sequestration. *Agron. Abstr.*, 87:136.

Potter, K.N., O.R. Jones, H.A. Torbert, and P.W. Unger. 1997. Crop rotation and tillage effects on organic carbon sequestration in the semiarid southern Great Plains. *Soil Sci.*, 162:140–147.

Robinson, C.A. 1996. West Texas A&M University, unpublished data.

Rosenzweig, C. 2000. Climate Change Will Affect World Food Supply. National Aeronautics and Space Administration/Goddard Institute for Space Studies. Available at: *www.globalchange.org/editall/.2000winter1.htm*.

Stewart, B.A., R. Lal, and S.A. El-Swaify. 1991. Sustaining the resource base of an expanding world agriculture. In R. Lal and F.J. Pierce, Eds. *Soil Management for Sustainability*. Soil and Water Conservation Society, Ankeny, IA, pp. 125–144.

U.S. Department of Agriculture. 1997. 1997 Census of Agriculture. National Agricultural Statistics Service, U.S. Department of Agriculture, Washington, DC.

Wood, S., K. Sebastian, and S.J. Scherr. 2000. Agroecosystems: Pilot Analysis of Global Ecosystems. International Food Policy Research Institute and World Resources Institute, Washington, DC.

Wu, Z., Colette, A.W., and Stewart, B.A. 2002. West Texas A&M University, unpublished data.

15

Climate Change and Crop Production: Challenges to Modeling Future Scenarios

EUGENE S. TAKLE AND ZAITAO PAN

CONTENTS

15.1 INTRODUCTION

Assessing the impact of climate change on crops requires an overview of the various aspects of climate that are projected to change and an evaluation of the individual and combined effects of these changes. Uncertainty in our understanding of the impacts of climate change on crop production arises from our uncertainty about change in the climate itself and also from uncertainty in translating those changes into changes in crop production. In this chapter, we provide an overview of our ability to simulate future scenario climates at regional scales with a particular eye toward the climate needs for crop modeling.

Food production is a global enterprise that links production and crop failures in one region with global markets and farmer choices in far distant regions. However, for this chapter we will focus on the United States, where climate data of relatively good spatial and temporal resolution are available over a large region and where regional climate modeling has been done with applications to crop modeling and yields.

Current efforts to study impacts of changing climate on sustainability and productivity of intensive agriculture in the U.S. Midwest are hampered by scale of resolution of future climate information, by use of a single routine to represent all crops, and by limited representation of hydrological processes to account for water and nutrient transport from individual fields, slopes, and watersheds (Mearns et al., 2001; Tsvetsinskaya et al., 2001). In particular, Mearns et al. (2001) evaluated the impact of spatial variability of both climate and soil properties on crop yield, and concluded that, compared to climate, the spatial scale of soil properties had a larger impact on variance and autocorrelation of yields, but a smaller impact on mean yields. From these results on yields, we can infer that the spatial variability of soil parameters also will have large impact on sustainability factors such as nitrogen loss, sediment loss, and change in soil organic matter content.

The impact on crops of changes in temperature and precipitation cannot be separated from impacts of other global changes. Shaw et al. (2003) evaluated the combined effects of warming, increased precipitation, increased nitrogen deposition, and increased CO_2 on California annual grassland, and found that the first three alone and together increased net primary production. Increased CO_2 alone also increased net primary productivity (NPP). However, in combination with other simulated changes, increased CO_2 generally decreased the enhancing impact of other factors, suggesting that single factor changes on ecosystems are not simply additive. This may be a consequence of the reduced root allocation by grasses grown in enhanced CO_2 environments. Shaw et al. (2003) concluded that multifactor responses must be evaluated. Analogous studies need to be conducted on agricultural crops to evaluate whether enhanced yields projected under increases in atmospheric CO_2 are realistic when taken in combination with other global change factors. Mearns et al. (1997) examined the impact of changes in both mean and variance of climate on output of a crop model, and demonstrated the importance of including variability.

Changnon and Hollinger (2003) analyzed data from an experiment in Illinois that artificially enhanced rainfall for each precipitation event during the growing season. Rains enhanced by 10% to 40% had little effect except in dry years. Furthermore, all increases had positive effects in a dry year and negative effects in a wet year. They also pointed out the important role of timing of precipitation events and temporal sequences of temperature throughout the growing season. Extrapolating their results to explore impacts of possible future increases in precipitation, they conclude that rainfall increases of 10% will have little impact on corn yields except in wet years when such increases may actually lower yields. They project rainfall increases of 25% and 40% to increase yields by 3% and 9%, respectively, except in unusually dry years. These increases are much less than the 15% yield increase for corn projected for 2030 by the National Assessment Synthesis Team (2000).

From these and other recent studies, we conclude that significant challenges lie ahead as we attempt to quantify food-production security and sustainability more realistically in future scenarios of climate. In this chapter, we present a brief overview of the current status and limitations of our ability to construct accurate simulations of future scenario climates at the scale of resolution needed to assess regional impacts. We also examine how inaccuracies in some critical climate inputs to crop models are reflected in the yields they produce.

15.2   MODELING COMPONENTS

Global climate models represent the only tool available to assess future impacts of changes in global radiative forcing (e.g., increasing greenhouse gas concentrations, presence of sulfate aerosols, and changing land-use characteristics) on climate. Global climate models all project the global mean temperature of the planet to increase over the next 100 years, but at different rates depending on the model. Approximations used in climate models and lack of fine-scale resolution are sources of these differences in global mean quantities. Uncertainty in our understanding of future scenario climates used for studying impacts on agriculture arises first from disagreement among global climate models, and second from lack of a clear and preferred path to translate changes in the global climate down to the regional and local scales.

15.2.1   Global Climate Models

Current global models project the global mean temperature to rise 2.5°C to 6°C, over the next 100 years with comparable or larger variation on regional scales. Changes in precipitation have large variation among models, with some indicating substantially lower mid-continental precipitation and others providing increases. Narrowing the range of uncertainty in determining impacts of climate change under such disagreement requires use of ensembles of global models to capture the uncertainty and to bracket the range of possible outcomes.

Downscaling changes in the global climate to regional and local scales is needed for assessing impact on crop production. Computing resources available now and in the next decade do not permit global models to run at resolutions sufficiently high to provide detailed climate information for assessing regional impacts of climate change on agriculture. While global climate models will approach 100-km resolution in the foreseeable future, the U.S. Climate Change Science Program (CCSP) noted that it is unlikely that in the near future they routinely will reach the 10-km scale that regional climate models (RCMs) can simulate (U.S. CCSP, 2003).

Alternative methods are needed to downscale global climate model results to spatial scales of importance to decision making in agriculture. Two candidate methods for achieving needed resolution are regional climate modeling (dynamical downscaling) and statistical downscaling. For defining future scenario climates, both methods rely on global climate model information to provide large-scale states of the climate system that are consistent with global physical conservation principles under conditions that include changes of external forcing (e.g., increases of atmospheric greenhouse gases and sulfate aerosols and changes in land use). The U.S. CCSP (2003) emphasizes RCMs as a means of achieving climate information at resolutions higher than are available from global models.

15.2.2 Regional Climate Models

Global models, typically having horizontal resolution of about 250 km, do not allow sufficient detail to determine climatic information for decision making in regions of complex orography such as California (Figure 15.1). Increasing resolution by a factor of 5 (Figure 15.1) produces a substantial improvement in representing orographic features of the California landscape, but even this resolution is insufficient for some purposes. Determining the appropriate resolution for this application will require careful evaluation of the impact of resolution changes on representing specific climate elements in specific circumstances.

RCMs represent the prevalent current approach to dynamical downscaling (U.S. CCSP, 2003). Following the pioneering work of F. Giorgi and colleagues (see Giorgi and Mearns, 1991, 1999, for reviews), regional climate modeling

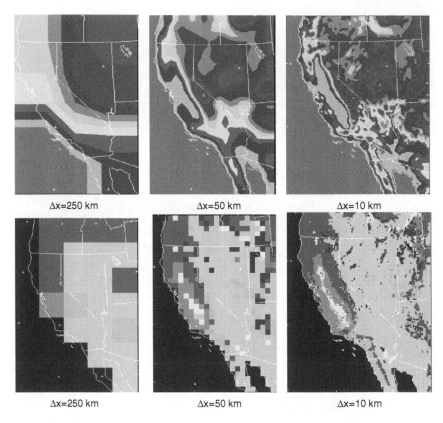

Δx=250 km   Δx=50 km   Δx=10 km

Δx=250 km   Δx=50 km   Δx=10 km

Figure 15.1   Illustration of terrain heights (top) and land use (bottom) at three different horizontal grid resolutions (Δx).

has emerged as a valuable approach in creating climate information at resolutions higher than are available from global models (Intergovernmental Panel on Climate Change [IPCC], 2001, chapter 10, "Regional Climate Information — Evaluation and Projections"). RCMs driven by lateral boundary conditions supplied by reanalysis data show considerable skill compared to global models in representing the spatial structure of temperature and precipitation in areas of complex orography (Leung et al., 2003). The higher resolution of RCMs also allows simulation of fine-scale dynamics (atmospheric jets, drainage flows, frontal structure, and regional convergence and divergence patterns) and temporal changes associated with smaller horizontal scales (temporal variability in wind speed, precipitation frequency, precipitation intensity, and cloudiness).

Regional models use results of global models as lateral boundary conditions and simulate (by use of essentially the same procedures as global models) scenario climates that are dynamically consistent with surface and external radiative forcing and lateral boundary conditions provided by the global model. Geographic features are represented with more detail in regional models as can be seen in Figure 15.1. Intercomparison studies of results of several regional models driven at lateral boundaries by the National Centers for Environmental Prediction/National Center for Atmospheric Research (NCEP/NCAR) reanalysis (NNR) (Kalnay et al., 1996) have helped reveal the strengths and weaknesses of this approach to downscaling global climate information (Takle et al., 1999; Anderson et al., 2003).

The agriculturally intensive central United States is unique in the country in that summertime mesoscale convective precipitation (Wallace and Hobbs, 1977) is dependent on nocturnal water vapor flux convergence (Anderson et al., 2003). Neither the NNR (Higgins et al., 1997) nor global climate models (Ghan et al., 1995) capture this essential mechanism. Finer grid spacing is needed to resolve the fine-scale dynamical processes that lead to timing, location, intensity, and amounts of precipitation (Anderson et al., 2003). Most, but not all, regional models are able to capture the nocturnal maximum in hourly precipitation in this region (Anderson et al., 2003), which is an indication that nocturnal moisture convergence at the outflow of the low-level jet is being simulated. For this reason, we expect that regional climate models can offer better representation of climate factors of critical importance to agriculture compared to low-resolution climate models.

The recently issued (24 July 2003) Strategic Plan for the CCSP (2003) calls for more regional climate modeling through Objective 16: "Accelerate the development of scientifically based predictive models to provide regional and fine-scale climate and climate impact information relevant for scientific research and decision support applications."

15.2.3 Statistical Methods

The computational demands of regional climate modeling have stimulated the search for more efficient methods of obtaining site-specific scenarios of future climates. Statistical downscaling (IPCC, 2001, chapter 10) assumes that regional climate is determined by two factors: the large-scale climatic state and local factors. Features of the large-scale climate and known local factors such as elevation are input to a statistical procedure, usually linear regression, that has been adjusted to yield the local climate. Many forms of statistical downscaling have been applied, such as simple regression and sophisticated artificial neural networks. The procedure requires good input data, both for predictors and the predictands, in order to arrive at an acceptably accurate statistical model.

Wilby and Wigley (1997) provide a review of statistical downscaling, and give a variety of predictors and predictands for applications to climate. The success of statistical downscaling is highly location dependent, but Wilby and Wigley (2000) confirmed results of previous research that showed strong correlation for winter precipitation and circulation predictors and precipitation for regions near oceanic sources of moisture. Kidson and Thompson (1998) assert that empirical downscaling methods provide skill comparable to that of dynamical modeling for application to the current climate. In a direct comparison of statistical downscaling to regional climate modeling (Wilby et al., 1999), we found advantages for both methods with neither method clearly superior in our hydrological application.

Statistical procedures tend to have low computational demand, which makes statistical downscaling attractive. However, some cautions should be recognized. When downscaling future climate, it must be assumed that statistical relations based on observed climate of the past will hold in the future. The statistical relations likely are strongly limited to those regions with sufficient data for calibration and validation. The adherence of statistical relations to physical laws may be obscure, limiting insight into climatic processes. These features complicate attempts to conduct systematic appraisals of statistical downscaling.

15.2.4 Crop Models

Most crop models, such as CROPGRO (Hoogenboom et al., 1994), which has evolved from the earlier soybean model SOYGRO, are process-oriented models that consider crop development, crop–soil carbon and nitrogen balances and soil–water balance. The hourly leaf-level photosynthesis formulation of CROPGRO allows for responses to changes in weather conditions that can be supplied by a regional climate model on subdaily time scales. Within the ensemble of CERES crop models, CERES-Maize is a model for corn developed originally by J.T. Ritchie and colleagues (see Jones and Kiniry, 1986), and is based on radiation-use efficiency as modified by nitrogen stress, water stress, plant population, and temperature. CROPGRO and CERES-Maize can be used in sequence to simulate multiyear crop production because they use the same balances for soil water and soil nitrogen, and recently have been coupled with the CENTURY soil organic matter model (Parton et al., 1987). The two crop models have been combined into a single model called the Cropping System Model (CSM) (Jones et al., 2003), representing the newest model development in crop modeling. We employed the CSM model for the study reported herein.

The climate model (whether dynamical or statistical) provides the crop model (CERES-Maize for our analysis) with daily totals of shortwave radiation and rainfall, as well as daily maximum and minimum temperatures. On the basis of these daily inputs, the crop model calculates daily updated values of plant biomass, evaporation, and transpiration, as well as the remaining water, carbon, and nitrogen in the soil profile. At the end of the growing season, the model supplies a total grain yield.

15.3 MODELING FUTURE CLIMATES AND CROP GROWTH

15.3.1 Temperature

Determining climate information for assessing future crop growth, by whatever methods, requires more than monthly mean values of temperature and precipitation. For a changing or variable climate, higher-order statistics, such as variance, extremes, and persistence, may have more significant influences than changes in the mean (Takle and Mearns, 1995; Mearns et al., 1997). For example, a month having a run of several consecutive days of extreme high temperatures (higher autocorrelation of the time series) has much more impact on crops than a month with a day or two of extreme high temperatures scattered throughout the month but with the same monthly mean. Takle and Mearns (1995) reported that a 3.5-year simulation with a regional climate model produced a reduction in the standard deviation and increase in the autocorrelation, in addition to an increase in the mean, for a future scenario climate for July daily maximum temperatures in the Midwest. Although the simulated climate record was short, and the quality of the climate models was lower than currently available, these results point out the need for examining more than mean values for assessing agricultural impacts.

Timing of changes also is critical. Heat stress during the vulnerable corn-pollination period (July in the U.S. Midwest) can have a particularly significant negative impact on yields (Shaw, 1983; Carlson, 1990). Mearns et al. (1984) calculated that a 1.7°C rise in mean maximum July temperature for Des Moines, IA, increases the probability of a heat wave (5 or more consecutive days with temperature above 35°C) from 6% to 21%.

15.3.2 Precipitation

Interannual variability in precipitation is a key factor that accounts for a large measure of interannual variability in crop yields. As with temperature, monthly mean values and timing are important factors controlling impact of precipitation on crop yields. Precipitation is a key input variable to the crop model that accounts for a large measure of interannual variability in crop yields. We evaluated the capability of a regional climate model (RegCM2) (Giorgi et al., 1993) to simulate precipitation for the growing season at Ames, IA. RegCM2 was driven with lateral boundary conditions over a 10-year period (1979–1988) supplied by the NNR (see Pan et al., 2001b, for more details). This 10-year period starts with the first availability of weather satellite imagery and matches the Atmospheric Model Intercomparison Project experiment (Gates, 1992), which allows comparisons of our regional model results with those of global models. Results shown in the upper plot of Figure 15.2, indicate that in some years the model is able to simulate the seasonal total quite well, but in some years (notably 1983) the model fails to capture the large seasonal total. There is a general tendency for the model to predict lower values than observed in all years. Table 15.1 gives a summary of totals for the 10-year observed and simulated periods.

The temporal distribution of rainfall intensities also influences crop development. Numerous light rainfall events that fail to provide moisture to deep roots may be far less beneficial than a single event of the same total amount. Our ability to simulate future crop development and yield by use of crop models will depend strongly on our ability to correctly simulate timing and amount of future precipitation. To examine the characteristics of model-simulated precipitation events, we plotted the distribution of daily total rainfall amounts during May to August as produced by the regional climate model RegCM2. Results (Figure 15.3) revealed that the model simulates too many low-precipitation events and not enough events in the range most usable by a crop such as corn, which develops a deep root system by the middle of the growing season. By use of the same regional model simulations, Kunkel et al. (2002) found annual and interannual extremes of precipitation to be reasonably well represented by the regional model.

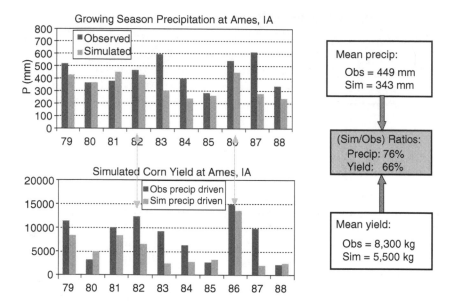

Figure 15.2 Growing season precipitation as simulated by RegCM2 for 10 years as compared to observed amounts for Ames, IA.

Table 15.1 Growing Season Precipitation and Yields Observed and Simulated by RegCM2 and CERES-Maize When Supplied with 10-Year Climate Data Sets for 1979–1988 Period and 10-Year Climate Data Sets of Contemporary (contmp) and Future Scenario (futscen) Climates as Simulated by HadCM2

Variables	Simulation Runs	Mean	Standard Deviation	Normalized Standard Deviation
Precipitation	Observed	449	114	0.25
(mm)	RegCM2/NNR	343	89	0.26
Yields (kg ha^{-1})	Observed	8,381	1,214	0.14
	CERES/Observed weather	8,259	4,494	0.54
	CERES/RegCM2/NNR	5,487	3,796	0.69
	CERES/RegCM2/Had CM2 contmp	5,002	1,777	0.36
	CERES/RegCM2/Had CM2 futscen	10,610	2,721	0.26

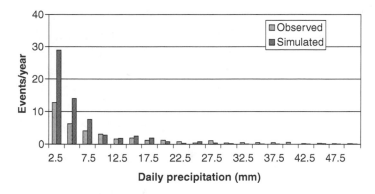

Figure 15.3 Histogram of daily rainfall amounts as simulated by RegCM2 for May to August in 1979–1988 period.

In Figure 15.4, we plotted just the range of daily totals considered to be most effective in promoting crop development. Although the model simulates a large number of events in this range, the distribution is skewed toward lower daily totals with too few at the higher end of the range.

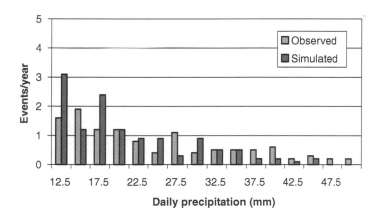

Figure 15.4 Histogram of precipitation events per season as simulated by RegCM2 in the range most usable by corn for May to August in 1979–1988 period.

15.3.3   Soil Moisture

Soil moisture values for the 1981–1988 period simulated by RegCM2 were compared with observations of soil moisture for the state of Illinois as reported by Hollinger and Isard (1994) and Robock et al. (2000). Composite monthly mean values for the 8-year period (Figure 15.5) for the top 10-cm layer show a winter–spring recharge of soil moisture and a growing-season drawdown. Simulation of the deep soil (figure not shown) reveals a persistent dry bias in the model, whereas the simulation of the top layer follows the observations much more closely, except for an early fall recharge rate that is too low. Interannual variability of moisture in the top layer (Figure 15.6) shows minima in three summers (1983, 1984, and 1988). Simulated moisture tends to capture dry and wet years but with lower interannual variability. There is no apparent tendency for long-term drift in simulated soil moisture. See Pan (2001a) for further details.

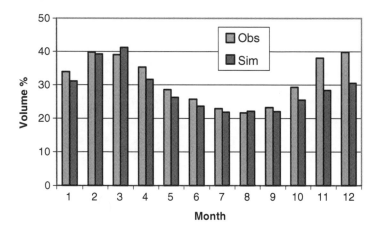

Figure 15.5   Monthly composites of observed and simulated volumetric soil moisture in top 10 cm for 1981–1988 period averaged over 19 observer sites in Illinois.

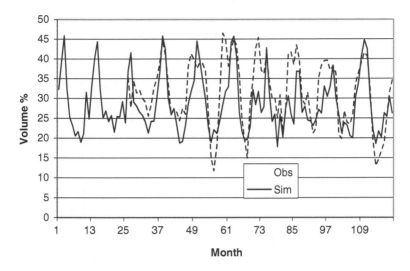

Figure 15.6 Time series of observed and simulated monthly volumetric soil moisture in top 10 cm averaged over 19 sites in Illinois.

15.3.4 Representing Crop Growth

We conducted a pilot study for central Iowa where CERES-Maize was used to simulate crop yields when supplied with 10-year data sets representing (1) observed weather conditions, (2) a climate produced by a regional climate model driven by the NNR, (3) a contemporary climate produced by a global model downscaled by the regional model, and (4) a future scenario climate (2040–2049) produced by a global model downscaled by the regional model. Results of each of these were compared to observed yields in this region. The validation data on corn yields for Ames, IA were taken from annual yields for the north-central reporting district of Iowa, and therefore represent a regional average rather than results from a single locale.

Results in Table 15.1 and shown in the lower panel of Figure 15.2 suggest that the crop model has higher interannual variability (standard deviation) than the observed values, but produces a mean yield close to observed yield. The climates produced by the regional model produced mean yields well below observed yields and standard deviations well above observed levels. When results of a global model for the contemporary climate are used to drive the regional model, the mean yield is modestly reduced, but the standard deviation is substantially reduced. The combination CERES/RegCM2 produces a doubling of maize yields for Ames under the future scenario climate and reduced variance. The reason for the large increase in yield is not completely clear, but may be a combination of more soil moisture and higher mean temperatures, but with little or no increase in the daily maximum temperature during critical parts of the growing season in the scenario climate.

From this pilot study, we conclude that the crop model is very sensitive to biases introduced by a regional climate model. The crop model seems to react more strongly than actual crops to interannual variability in precipitation. This sensitivity is exacerbated by inability of the climate models to simulate the proper distribution of rainfall intensities. The consequences of higher daytime maximum temperatures are amplified by crops (Mearns et al., 1984), so any termperature biases in the climate models will substantially impact yields. By these means, the crop model has exposed weaknesses in the climate model for simulating agricultural impacts of climate change.

Crops are not passive acceptors of externally imposed climate, but rather are being recognized for their role as a coupling mechanism between the soil and atmosphere. Sparks et al. (2002) suggest that extensive plantings of corn and soybeans in the Midwest are contributing to higher dew-point temperatures, which exacerbate heat waves (Kunkel et al., 1996). Kalnay and Cai (2003) compare a reanalysis of upper air data with surface observations to conclude that about half of the observed decrease in diurnal temperature range over the last 50 years is due to changes in land use, primarily agriculture. These and other reports suggest a need for more direct coupling of interactive ecosystem models, such as crop models for the central United States, that incorporate biophysical processes into weather and climate models for more accurate simulation of crop impacts on local weather. This, in turn, should lead to more accurate simulation of crop physiology and yield.

15.4 SUMMARY

A strategy is needed for translating changes and variability of global climate into impacts on agriculture. Global models, which are needed to evaluate global consequences of anthropogenic influences such as changes in greenhouse gas concentrations, do not supply climate information of sufficient resolution to meet this need. Regional climate modeling is conceptually more appealing than statistical downscaling, and captures some fine-scale dynamical processes that are unresolved by global models. However, inaccurate information on the timing and magnitude of precipitation events, and biases in temperature produced by the regional models are amplified by a crop model to produce large discrepancies between simulated and observed yields. If, as data seem to show, crops are more efficient than native perennial vegetation at recycling moisture to the atmosphere, then the physiological processes represented within crop models are needed within the climate model to more accurately represent the linkage between soil and atmosphere for simulating climates in intensively cultivated areas. For areas of intensive agriculture, such as the U.S. Midwest, these advances in climate modeling will be particularly beneficial.

ACKNOWLEDGMENTS

We are grateful to our colleagues R.W. Arritt, W.J. Gutowski, Jr., and C.J. Anderson of the regional climate modeling team at Iowa State University, and J.H. Christensen of the Danish Meteorological Institute, for their contributions to this research. We also acknowledge the research support of the Iowa State University Agronomy Department Endowment Program and the U.S. Department of Energy/Biological and Environmental Research (BER) through the Great Plains Regional Center of the National Institute for Global Environmental Change (DE-FC02-03ER63613).

REFERENCES

Anderson, C.J., R.W. Arritt, E.S. Takle, et al. 2003. Hydrological processes in regional climate model simulations of the central United States flood of June–July 1993. *J. Hydrometeorol.,* 4:584–598.

Carlson, R.E. 1990. Heat stress, plant-available soil moisture, and corn yields in Iowa: a short- and a long-term view. *J. Prod. Agric.,* 3:293–297.

Changnon, S.A. and S.E. Hollinger. 2003. Problems in estimating impacts of future climate change on Midwestern corn yields. *Climatic Change,* 58:109–118.

Gates, L.W. 1992. The Atmospheric Model Intercomparison Project. *Bull. Am. Meteorol. Soc.,* 73:1962–1970.

Ghan, S.J., X. Bian, and L. Corsetti. 1995. Simulation of the Great Plains low-level jet and associated clouds by general circulation models. *Monthly Weather Rev.,* 124:1388–1408.

Giorgi, F., M.R. Marinucci, G.T. Bates, and G. De Canio. 1993. Development of a second-generation regional climate model (RegCM2), I. Boundary-layer and radiative transfer. *Monthly Weather Rev.,* 121:2794–2813.

Giorgi, F. and L.O. Mearns. 1991. Approaches to the simulation of regional climate change: a review. *Rev. Geophys.,* 292:191–216.

Giorgi, F. and L.O. Mearns. 1999. Introduction to special section: Regional climate modeling revisited. *J. Geophys. Res.,* 104:6335–6552.

Higgins, R.W., Y. Yao, E.S. Yarosh, J.E. Janowiak, and K.C. Mo. 1997. Influence of the Great Plains low-level jet on summer time precipitation and moisture transport over the central United States. *J. Climate,* 10:481–507.

Hollinger, S.E. and S.A. Isard. 1994. A soil climatology of Illinois. *J. Climate,* 7:822–833.

Hoogenboom, G., J.W. Jones, P.W. Wilkens, et al. 1994. Crop models. In G.Y. Tsuji, G. Uehara, and S. Balas, Eds. DSSAT Version 3, Vol. 2. University of Hawaii, Honolulu, HI, pp. 95–244.

Intergovernmental Panel on Climate Change. 2001. Climate Change. 2001: *The Scientific Basis*. Contribution of Working Group I to the Third Assessment Report of the Intergovernmental Panel on Climate Change. Houghton, J.T., Y. Ding, D.J. Griggs, M. Noguer, P.J. van der Linden, X. Dai, K. Maskell, and C.A. Johnson, Eds. Cambridge University Press, London; New York.

Jones, J.W., G. Hoogenboom, C. Porter, et al. 2003. DSSAT: cropping system model. *Eur. J. Agron.*, 18:235–265.

Jones, C.A. and J.A. Kiniry. 1986. *CERES-Maize: A Simulation Model of Maize Growth and Development*. Texas A&M University Press, College Station, TX.

Kalnay, E., M. Kanamitsu, R. Kistler, et al. 1996. The NCEP/NCAR 40-year reanalysis project. *Bull. Am. Meteorol. Soc.*, 77:437–471.

Kalnay, E. and M. Cai. 2003. Impact of urbanization and land-use change on climate. *Nature*, 423:528–531.

Kidson, J.W. and C.S. Thompson. 1998. A comparison of statistical and model-based downscaling techniques for estimating local climate variations. *J. Climate*, 11:735–753.

Kunkel, K.E., K. Andsager, X.-Z. Liang, R.W. Arritt, E.S. Takle, W.J. Gutowski Jr., and Z. Pan. 2002. Observations and regional climate model simulations of extreme precipitation events and seasonal anomalies: a comparison. *J. Hydrometeorol.*, 3:322–334.

Kunkel, K.E., S.A. Changnon, B.C. Reinke, and R.W. Arritt. 1996. The July 1995 heat wave in the Midwest: a climatic perspective and critical weather factors. *Bull. Am. Meteorol. Soc.*, 77:1507–1518.

Leung, R., L.O. Mearns, F. Giorgi, and R.L. Wilby. 2003. Regional climate research needs and opportunities. *Bull. Am. Meteorol. Soc.*, 84:89–95.

Mearns, L.O., R.W. Katz, and S.H. Schneider. 1984. Extreme high-temperature events: changes in their probabilities with changes in mean temperature. *J. Climate Appl. Meteorol.*, 23:1601–1613.

Mearns, L.O., W. Easterling, C. Hays, and D. Marx. 2001. Comparison of agricultural impacts of climate change calculated from high and low resolution climate change scenarios. Part I. The uncertainty due to spatial scale. *Climatic Change*, 51:131–172.

Mearns, L.O., C. Rosenzweig, and R. Goldberg. 1997. Mean and variance change in climate scenarios: methods, agricultural applications, and measures of uncertainty. *Climatic Change,* 35:367–396.

National Assessment Synthesis Team. 2000. *Climate Change Impacts in the U.S.: Overview.* Cambridge University Press, London; New York.

Pan, Z., R.W. Arritt, W.J. Gutowski Jr., and E.S. Takle. 2001a. Soil moisture in a regional climate model: simulation and projection. *Geophys. Res. Lett.,* 28:2947–2950.

Pan, Z., J.H. Christensen, R.A. Arritt, W.J. Gutowski Jr., E.S. Takle, and F. Otenio. 2001b. Evaluation of uncertainties in regional climate change simulations. *J. Geophys. Res.,* 106:17,735–752.

Parton, W.J., D.S. Schimel, C.V. Cole, and D.S. Ojima. 1987. Analysis of factors controlling soil organic matter levels in Great Plains grasslands. *Soil Sci. Soc. Am. J.,* 51 :1173–1179.

Robock, A., K.Y. Vinnikov, G. Srinivasan, J.K. Entin, S.E. Hollinger, N.A. Speranskaya, S. Liu, and A. Namkhai. 2000. The global soil moisture data bank. *Bull. Am. Meteorol. Soc.,* 81:1281–1299.

Shaw, M.R., E.S. Zavaleta, N.R. Chiariello, E.E. Cleland, H.A. Mooney, and C.B. Field. 2003. Grassland responses to global environmental changes suppressed by elevated CO_2. *Science,* 298:1987–1990.

Shaw, R.H. 1983. Estimates in yield reduction in corn caused by water and temperature stress. In C.D. Raper and P.J. Kramer, Eds. *Crop Reactions to Water and Temperature Stresses in Humid, Temperate Climates.* Westview, Boulder, CO, pp. 49–66.

Sparks, J., D. Changnon, and J. Starke. 2002. Changes in the frequency of extreme warm-season surface dewpoints in Northeastern Illinois: implications for cooling system design and operation. *J. Appl. Meteorol.,* 41:890–898.

Takle, E.S., W.J. Gutowski Jr., R.W. Arritt, et al. 1999. Project to intercompare regional climate simulations (PIRCS): description and initial results. *J. Geophys. Res.*, 104:19,443–462.

Takle, E.S. and L.O. Mearns. 1995. Midwest temperature means, extremes, and variability: analysis of results from climate models. In G.R. Carmichael, G.E. Folk, and J.L. Schnoor, Eds., *Preparing for Global Change: A Midwestern Perspective*. Academic Publishing, Amsterdam, pp. 135–142.

Tsvetsinskaya, E.A., L.O. Mearns, and W.E. Easterling. 2001. Investigating the effect of seasonal plant growth and development in three-dimensional atmospheric simulations. Part I. Simulation of surface fluxes over the growing season. *J. Climate*, 14:692–709.

U.S. Climate Change Science Program. 2003. Strategic Plan for the Climate Change Science Program. Final Report, July 2003. Available at: *www.climatescience.gov / Library / Stratplan2003 / final / default.htm*.

Wallace, J.M. and P.V. Hobbs. 1977. *Atmospheric Science: An Introduction*. Academic Press, New York.

Wilby, R.L., L.E. Hay, W.J. Gutowki Jr., R.W. Arritt, E.S. Takle, Z. Pan, G.H. Leavesley, and M. Clark. 1999. Hydrological responses to dynamically and statistically downscaled general circulation model output. *Geophys. Res. Lett.*, 27:1199–1202.

Wilby, R.L. and T.M.L. Wigley. 1997. Downscaling general circulation model output: a review of methods and limitations. *Prog. Phys. Geogr.*, 21:530–548.

Wilby, R. and T.M.L. Wigley. 2000. Precipitation predictors for downscaling: observed and general circulation model relationships. *Int. J. Climatol.*, 20:641–661.

Section IV

Soil Carbon Dynamics and Farming/Cropping Systems

16

Soil Carbon Sequestration: Understanding and Predicting Responses to Soil, Climate, and Management

JAMES W. JONES, VALERIE WALEN,
MAMADOU DOUMBIA, AND
ARJAN J. GIJSMAN

CONTENTS

16.1 INTRODUCTION

Managing agricultural lands to increase soil organic carbon (SOC) could help counter the rising atmospheric CO_2 concentration as well as reduce soil degradation and improve crop productivity. However, soils, climate, and management practices vary over space and time, creating an almost infinite combination of factors that interact and influence how much carbon is stored in soils. Thus, quantifying soil carbon sequestration under widely varying conditions is complicated. Furthermore, SOC changes slowly over time; experiments for quantifying carbon gain under different practices must be conducted over a number of years. Due to the human and financial resources and time needed to conduct such experiments, it may not be practical to rely on this approach alone to provide needed information. Further complicating the picture is climate change. As temperature and atmospheric CO_2 increase and rainfall changes, new combinations of factors will occur that have not been studied. For these reasons, models are needed to complement information gained from experiments to help understand and predict SOC and food production responses to soil, climate, and management combinations.

Biophysical models integrate crop, soil, weather, and management practice information and predict the consequent biomass and yield components as well as changes in soil nutrients and carbon (Cole et al., 1987; Moulin and Beckie,

1993; Singh et al., 1993; Probert et al., 1995; Gijsman et al., 2002a; Jones et al., 2003). By simulating responses for a number of years, it is possible to estimate potential changes in productivity and SOC. Through a series of computer experiments, using models along with local soil and climate information, one could identify cropping systems that would meet productivity and SOC sequestration goals. However, there are a number of uncertainties associated with models and their use, and if one does not adequately address these uncertainties, simulated results will be meaningless. These uncertainties are due to the fact that models are simplifications of reality, there are uncertainties in model parameters, and there are uncertainties in inputs used in computer experiments. Thus, work is needed to ensure that models can reproduce responses measured in real experiments. This may require one to estimate crop and soil parameters (e.g., Mavromatis et al., 2001; Gijsman et al., 2002b), to adjust other relationships to adapt the model for the region in which it is to be used (e.g., du Toit et al., 1998), and possibly conduct new research to evaluate predictions. If the model accurately describes yield and SOC responses measured in real experiments in the region, one will have more confidence in its ability to predict responses under other combinations of soil, weather, and management practices.

Biophysical models may also be useful in monitoring SOC changes over time and space to fulfill carbon contract requirements. Although this is not a common use of agricultural models, methods developed in other fields of science and engineering can be applied to help quantify and verify soil carbon sequestration. Once models have been adapted for a region, they can be used to predict changes in soil C under weather conditions that occur each year and for management practices actually used at lower cost than empirical research (Bationo et al., 2003). However, model predictions are uncertain, even if inputs are accurate. Spatial variability of inputs adds to the uncertainties outlined above, which results in propagation of prediction errors over space and time. Measurements of carbon also are uncertain and costly; errors may be much larger than annual changes in SOC. Thus, by combining

measurements with model predictions, more accurate estimates of SOC can be obtained (Jones et al., 2004; Koo et al.,
2003; Bostick et al., 2003).

Existing models are useful tools for understanding and
predicting SOC changes if they are combined with measurements and used carefully. Our objective is to demonstrate the
use of biophysical models in combination with data for two
different types of uses. In the first demonstration, we explore
options for increasing yield and SOC in a maize farming
system in Mali. West and Post (2002) found a global average
C sequestration rate of 570 kg ha^{-1} year^{-1} for no-till vs. conventional tillage when they analyzed data from 67 long-term
experiments from around the world. In a 10-year study in
Burkina Faso, soil C increase averaged 116 and 377 kg ha^{-1}
year^{-1} for treatments with low and high levels of both inorganic fertilizer and manure, respectively (Pichot et al., 1981).
Lal (2000) observed annual rates of soil C increase under no-
till management ranging from 363 kg ha^{-1} year^{-1} to more than
1000 (for one severely depleted soil) over a 3-year experiment
aimed at restoring soil carbon in western Nigeria. Because
soil C in western African soils is known to be depleted
(Bationo et al., 2003), the hypothesis used to guide our study
was that ridge tillage (RT) combined with manure, nitrogen
fertilizer applications, and residue management will increase
soil carbon by 0.20% in 10 years (about 500 kg ha^{-1} year^{-1})
relative to levels under conventional tillage (CT) management. The DSSAT-CENTURY model is used to simulate
annual maize growth and yield as well as changes in SOC for
10 years. But first, care is taken to adapt the model to maize
cultivars, soil, management, and climate conditions of Mali
using available, although limited, data. In the second demonstration, the hypothesis is that model predictions of soil carbon can be combined with *in situ* measurements to improve
estimates of soil carbon sequestration. An ensemble Kalman
filter approach is used to assimilate observations over time
into a simple model to increase accuracy of SOC estimates
and to improve future predictions for specific fields.

16.2 COMBINING MODELS AND DATA TO ASSESS OPTIONS FOR SOIL C SEQUESTRATION

Two of the biggest constraints for improving household food security in West Africa are retention of rainwater in the field and improvement of soil quality (Kaya, 2000; Lal, 1997a, 1997b; Ringius, 2002; Bationo et al., 2003). The practice known as ridge tillage or *aménagement en courbes de niveau* (Gigou et al., 2000) was designed to address these issues concurrently, and is thought to have potential for sequestering SOC. This is logical since ridge tillage increases crop biomass production and grain yield in Mali (Gigou et al., 2000). Unfortunately, data for evaluating SOC sequestration potential in West Africa are scarce (Lal, 1997a, 1997b; Pieri, 1992; Ringius, 2002). Estimates of SOC sequestration potential are needed to help guide research and to give donors confidence that their investments will succeed. Soil C measurements taken by Yost and colleagues (Yost et al., 2002; Neely and Uehara, 2002) show that SOC levels in Mali are very low (ranging from 0.13% to 0.88% of soil mass) in the top 20 cm, and that fields that have been under ridge tillage for several years tend to have higher SOC levels than fields under conventional tillage. However, few measurements have been made to date, and thus no conclusions can be made regarding how much SOC will increase under RT, nor how long it might take to achieve that increase. Our hypothesis was that RT, coupled with other soil management practices, could increase soil C in the top 20 cm of soil by 5 metric tons ha^{-1} over 10 years. Objectives were (1) to adapt the DSSAT-CENTURY maize model for simulating RT vs. CT management systems in Mali using available data, and (2) to conduct a 10-year computer experiment to make preliminary estimates of potential SOC sequestration amounts under CT vs. RT management systems.

This study demonstrates the adaptation of a cropping system model for studying management options for increasing soil carbon in Oumarbougou, Mali (Lat 12.18 N, Long 5.14

W). Rainfall in the region is 900 to 1000 mm per year, falling unimodally from June to October (Roncoli et al., 2002). The cultivated soils in the area are characterized as red sandy soils (*bogo bile*), generally alfisols with high sand content and low organic C and N. The area is highly prone to runoff and erosion, as is the case in much of West Africa (Bielders et al., 1996; Daba, 1999; Rockstrom et al., 1998; Zhang and Miller, 1996).

16.2.1   Model Adaptation to Local Conditions

The Decision Support System for Agrotechnology Transfer (DSSAT), with its suite of CERES- and CROPGRO-based crop models, was developed to help researchers understand crop responses to various management options, soils, and weather conditions (Tsuji et al., 1998; Jones et al., 2003). The CEN-TURY soil organic matter model, originally developed to simulate soil C dynamics in temperate grasslands (Parton et al., 1987), has since been used in a wide range of conditions including tropical systems (Paustian et al., 1992; Parton et al., 1988, 1994; Woomer, 1993; Anderson and Ingram, 1993; International Centre for Research in Agroforestry, 1994). Recently, the CENTURY model was linked with the DSSAT cropping system model to improve capability for simulating cropping systems with low inputs (Gijsman et al., 2002a; Jones et al., 2003). This linked DSSAT–CENTURY model was used in this study.

Answers are sought to the following questions: (1) Does the model adequately simulate growth and yield of the crops under the soils, climate, and management conditions being considered? (2) Does the model adequately simulate changes in soil processes (including SOC) under those same conditions? One can be relatively sure that existing models, even robust, widely used models like DSSAT, will not perform well in a new location unless an effort is made to adapt them to local conditions. Adapting the model requires: (1) assembly of local data on soil, weather, and crop performance under field conditions, (2) estimation of crop model parameters for local cultivars, (3) estimation of critical soil parameters not normally measured (particularly soil hydraulic properties), and (4) evaluation of model ability to simulate crop (e.g., phenological development, yield, and biomass) and soil (e.g., water, SOC, and N) responses under local conditions.

Although appropriate data for this area were limited (i.e., no long-term experiments with observed SOC changes), available data were used to simulate these preliminary estimates of SOC sequestration. In this study, continuous use of maize was assumed to demonstrate the approach. The first step was to adapt the model for simulating maize cultivars normally grown in Mali agronomic experiments. The second step was to adjust runoff characteristics for CT vs. RT so that published differences in runoff and crop yield between these two systems were correctly simulated. The final step was to adjust initial C fractions so that SOC under CT was at steady state. These procedures allowed us to confirm that the model correctly simulates absolute yield levels as well as differences between the two systems that are being compared.

16.2.1.1 Soil Data

Soil samples collected by Mamadou Doumbia and Russ Yost in March 2002 were used to develop necessary soil profile inputs to the model. A composite of soils sampled from the fields of Zan Diarra, (Lat 12.55 N, Long 6.47 W) and of Yaya Diassa (Lat 11.14 N, Long 5.35 W) was used to create a soil input file with parameters listed in Table 16.1. Soil water-

Table 16.1 Selected Soil Inputs in Zan Diarra Samples and Yaya Diassa Soils

Soil Depth (cm)	SOC (%)	Sand (%)	Silt (%)	Clay (%)	pH	Wilting Point[a] (cm^3 cm^{-3})	Field Capacity[b] (cm^3 cm^{-3})	Bulk Density (g cm^{-3})
0–20	0.24[a]	72.4	21.4	6.2	5.34	0.069	0.176	1.44
20–40	0.22	52.9	25.3	21.8	4.93	0.213	0.297	1.49

[a] Initial soil C was assumed to be 7016 kg ha^{-1} in the top 20 cm.
[b] Calculated using the method described by Jagtap et al. (2004).
Source: From M. Doumbia, personal communication, 2003.

holding characteristics were estimated from soil texture using the nearest neighbor method of Jagtap et al. (2004). SOC composition in the DSSAT–CENTURY model is initialized by partitioning total C into three pools based on rates of decomposition: microbial, slow, and stable, with default fractions for grassland and previously-cultivated soils of 02:64:34 and 02:54:44, respectively. Since we had no measurements that would allow us to estimate these fractions directly, we assumed that soil C under CT was at a steady state. Thus, we varied these fractions for CT simulations until we achieved a steady-state level of SOC. When fractions of 02:41:57 were used for Mali soil, climate, and CT management, SOC remained at 0.24% for the 10-year period of simulations (see results for CT in Figure 16.1).

16.2.1.2   Weather Data

Historical daily weather data are needed for simulating experiments conducted in the past and evaluating model predictions vs. observations. Observed daily weather data were obtained in order to compare simulated maize results with

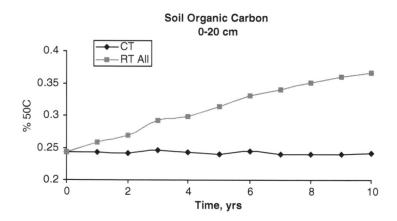

Figure 16.1   Plot of annual change in soil organic carbon (SOC) over 10 years under conventional tillage and fully implemented ridge tillage using initial SOC composition of calibrated stability (02:41:57).

Table 16.2 Calibration of Local Maize Variety Sotubaka:
Three Years of Observed and Simulated Grain Yield and
Days to Anthesis

	Time to Anthesis (days)			Maize Grain Yield (kg ha^{-1})		
	1999	2000	2001	1999	2000	2001
Simulated	62	61	57	5486	4138	5514
Observed	63	61	58	5100	3900	6070

Source: From Coulibaly, Ntji, personal communication, 2002.

those obtained by Coulibaly (i.e., Table 16.2). We also generated 10 years of daily weather data by interpolation between nearest existing weather stations using MarkSim, version 1 (P. Jones et al., 2002). The generated daily data include rainfall, maximum temperature, minimum temperature, and solar radiation. Small amounts of N (13 kg ha^{-1} 100 cm^{-1} infiltrated rainfall) (Campbell, 1978; Pieri, 1992; Vitousek et al., 1997) were applied to all simulated crops according to infiltration of rainfall.

16.2.1.3 Agronomic Experiment Data

Agronomic yield trial data for a 3-year maize study were obtained from Njti Coulibaly in Mali, including soil, weather, and management of the crops in each year. That experiment was simulated using the DSSAT CERES-Maize model, and genetic coefficients were estimated using measured anthesis dates and yields for the 3 years (Jones et al., 2002a). Data in Table 16.2 demonstrate that the model describes anthesis dates and yields across the 3 years with errors less than 10%. Although additional tests are desirable, this exercise demonstrated that the model can simulate growth and yield responses to typical growing conditions in Mali under conventional management.

Detailed measurements from experiments comparing RT vs. CT were not available. Thus, a computer experiment was conducted over a 10-year period: (1) to adjust field runoff parameters for RT vs. CT, and (2) to compare predicted grain and biomass yield values for RT and CT with those responses reported by Gigou et al. (2000). Maize was planted each year in the computer experiment as soon as the soil in the top 30 cm of soil reached 60% of plant available water, but no earlier than June 18 to ensure adequate moisture for germination. Harvest was assumed to occur at maturity, which was simulated for each season. Plant density was set at three plants m^{-2} in all simulations, and row width was 75 cm. Crops simulated under CT had no manure or N fertilizer applications, and 90% of the crop residue was removed after harvest. Management of RT included application of inorganic N (40 kg ha^{-1} applied in two doses), return of 90% of crop residue to the soil, and addition of 3 metric tons ha^{-1} of (dry) manure.

Runoff parameters for the RT field were set by assuming that the ridges were sufficiently constructed to reduce runoff by 45% relative to CT management (M. Doumbia, personal communication, 2003; Gigou et al., 2000). Runoff curve numbers of 90 and 96 were selected for RT and CT, respectively. Simulated runoff for the 10 years was 45% less for RT vs. CT management, and differences in yield and residue biomass between CT and RT closely matched differences reported by Gigou et al. (2000) (Table 16.3).

16.2.2 Simulation of Soil C Sequestration Potential, RT vs. CT

After soil and crop parameters were adjusted to mimic CT vs. RT as described above, simulated changes in SOC were compared over the 10 simulated years (Figure 16.1). SOC for CT

Table 16.3 Simulated Maize Grain and Biomass Production Under CT and RT Plus Amendments and Percent Increase, Compared with Similar Results in 2000 Study

	Maize Grain (kg ha^{-1})			Maize Biomass (kg ha^{-1})		
	CT	RT	% Increase	CT	RT	% Increase
Simulated 10-year average	2651	3565	34	6007	7750	29
Gigou et al. (2000)	2603	3599	38	6339	8704	37

Notes: CT = conventional tillage; RT = ridge tillage.

remained nearly constant for the initial C fractions used in the simulations but the RT-all treatment increased by 54%, from 0.24% to 0.37% in the top 20 cm of soil over the 10 years (10,527 to 7016 kg C ha^{-1}; bulk density was 1.44 g cm^3). This increase amounted to 3511 kg ha^{-1}, or about 351 kg ha^{-1} year^{-1}, and was about 10% of the total carbon added to the soil from residues and manure during the 10 years. This result indicates that the potential for SOC sequestration for the conditions studied may be less than the 0.20% increase that was hypothesized, and less than the average rates for no till vs. CT reported by West and Post (2002), but similar to values obtained by Pichot et al. (1981) in Burkina Faso and by Lal (2000) in the no-till treatment in Nigeria.

Two other treatments were simulated in the computer experiment to estimate how much SOC sequestration would occur under RT only (no amendments added), and under RT with 40 kg ha^{-1} N and return of 90% of crop residue (RT, Fl, R, no manure addition). Data in Table 16.4 indicate that yield increased under the RT only treatment, but that SOC did not increase. Although roots added C to the soil in all treatments, simulated results show that roots of maize alone would not be enough to increase SOC in this environment. For RT plus N fertilizer and residue return to the field, yield increased about the same as for RT plus all amendments, but the SOC increase was only 2058 kg ha^{-1}, or about 60% of the amount sequestered when manure was included.

Table 16.4 Grain Yield, Crop Residue, and Change in SOC After 10 Years Under Each Management System[a]

Treatment	CT	RT Only	RT, F, R	RT All
Mean grain yield (kg ha^{-1})	2651	3006	3592	3565
Mean harvest residue (kg ha^{-1})	6007	6875	7750	7751
Change in SOC (kg ha^{-1} [10 year]$^{-1}$])	−46	−355	2058	3511
Ending SOC (%)	0.24	0.23	0.32	0.37

[a] For top 20 cm of soil, bulk density = 1.44 g cm^{-3}.
Notes: CT = conventional tillage; F = 40 kg ha^{-1} of N fertilizer added; R = 90% of crop residue left on the field each year; RT = ridge tillage; SOC = soil organic carbon.

These preliminary results agree with the trend observed by Pieri (1992) in long-term studies in West Africa that, in treatments without manure, soil C remained constant or declined, and that fertilizer alone aggravated the condition. This also agrees with the study in Western Nigeria (Lal, 1997a, 1997b), which illustrated the importance of residue and tillage operations on SOC and maize grain yield. Under those conditions, the no-till plus mulch treatment had the greatest effect with a doubling of SOC during the first 4 years of the study. However, it is notable that all treatments showed initial increases followed by subsequent declines in SOC within 8 years. Several important differences existed in his study, however, namely, that average rainfall was higher (1200 mm), two crops were planted each year and more fertilizers were applied.

16.3 COMBINING MEASUREMENTS AND MODELS FOR ESTIMATING SOC SEQUESTRATION

If soil C sequestration is to become an accepted mechanism for reducing atmospheric CO_2 levels, a soil carbon accounting system needs to be developed (Antle and Uehara, 2002). Mass of carbon accumulation in soils is of interest, so measurements will include field sampling, laboratory determination of carbon, and its conversion to mass basis using soil bulk density. Thus, errors in such measurements would include errors associated with each step. As a result, yearly changes in soil C are small relative to errors associated with the measurement process, and such measurements are expensive. Biophysical models can be used to estimate SOC and its changes under different weather, soil, and management practices (Parton et al., 1988, 1994; Jones et al., 2002b). However, although these models may produce precise estimates, they are imperfect and parameters for specific field situations are uncertain. Thus, errors exist in estimates of SOC from field measurements and from model predictions.

Techniques exist to combine models and measurements to obtain better estimates of system states and model parameters. The Kalman filter (Maybeck, 1979; Welch and Bishop, 2002) approach first uses a model to predict the state of a system, and then uses measurements to update the estimates in an optimal way, taking into account errors in measurements and predictions. Variations of the Kalman filter, originally developed for linear models, have been developed for non linear models (e.g., Albiol et al., 1993; Graham, 2002). One variation, the ensemble Kalman filter (Burgers et al., 1998; Eknes and Evensen, 2002; Margulis et al., 2002), was used by Jones et al. (2004) to evaluate its use for estimating SOC and a decomposition rate parameter over time for a single field using a nonlinear model.

Here the use of ensemble Kalman filter (EnKF) methodology to combine measurements with models to estimate SOC is demonstrated. Analyses in this chapter focus on estimation of SOC over time (years) for a single field following the work of Jones et al. (2004); they do not address spatial variability or aggregation of estimates over space.

16.3.1 Soil Carbon Model

A simple discrete-time model is used to simulate SOC (X_t, kg ha^{-1}) as it changes over time, using a time step of 1 year. It is assumed that only one pool of C exists in the soil, and that biomass organic carbon (U_t, kg ha^{-1}) may increase this pool, while during the same annual time step, microbial activity decomposes both X_t and U_t. The model also has one parameter (SOC decomposition rate constant, R, year^{-1}) that is constant over time, but is not known with certainty. The resulting model has one state variable (X_t) and one parameter (R), which are estimated using the EnKF. Uncertainties in model predictions of SOC and R are assumed. State equations for the nonlinear, stochastic model follow:

$$X_t = X_{t-1} - R \cdot X_{t-1} + b \cdot U_{t-1} + \varepsilon_t$$
$$R = R_0 + \eta \tag{16.1}$$

where

X_t = soil organic carbon in year t (kg[C] ha⁻¹)

R = rate of decomposition of existing SOC (year⁻¹)

R_0 = initial estimate of SOC decomposition rate (year⁻¹)

b = fraction of fresh organic C that is added to the soil in year t that remains after 1 year

U_t = amount of C in crop biomass added to the soil in year t (kg[C] ha⁻¹ year⁻¹)

ε_t = model error for SOC (kg[C] ha⁻¹)

0 = error in estimate of decomposition rate R (year⁻¹)

Model error (ε_t) includes uncertainties in U and b, as well as uncertainties due to the fact that the model is a simplification of reality. It is also assumed that model and parameter errors are normally distributed and uncorrelated. Thus,

$$\varepsilon_t \sim N(0, \sigma_\varepsilon^2) \qquad\qquad (16.2)$$

$$\eta \sim N(0, \sigma_\eta^2)$$

where

σ_ε^2 = variance of model error for X_t

σ_η^2 = variance of model error for R

The model error (ε_t) is a random process that changes over time but is uncorrelated with time (i.e., white noise), whereas decomposition rate parameter error (η) is a random variable that does not change with time.

16.3.2 Soil Carbon Measurements

Soil C measurements (Z_t) may be made each year or less frequently, but measurements of R are not possible. Thus, the model has two variables that are to be estimated, but only one is observable. Furthermore, it is assumed that the SOC measurement error is normally distributed, independent in time and independent from X and R. A time series of measurements was generated using two steps to demonstrate the approach. First, a time series of true values of SOC (X_t) was computed using Equation 16.1 with the true value of the parameter R for the hypothetical field. Then, a time series of measurements was generated by randomly sampling from the distribution of Z (ε_z) and adding this random error to "true" SOC values at each discrete time step. Thus, Z_t was generated by

$$Z_t = X_t + \varepsilon_{z,t} \tag{16.3}$$

where

Z_t = measurement of SOC in year t, kg[C]/ha

$\varepsilon_{z,t}$ = error in measurement, $\varepsilon_{z,t} \sim N(0, \sigma_z^2)$

where σ_z^2 is variance of SOC measurement error. In real applications of the EnKF, actual measurements would be used.

16.3.3 The Ensemble Kalman Filter: Combining Model and Measurements

The Kalman filter is a set of mathematical equations that are used to obtain optimal estimates of the state of a system. There are two types of equations in a Kalman filter: (1) time update equations, and (2) measurement update equations (Welch and Bishop, 2002). The time update equations project forward in time the current predictions of the system state and covariance. The measurement update equations provide feedback by incorporating a new measurement to obtain an improved estimate of system state and covariance. In a discrete-time Kalman filter, a linear stochastic model is used to project the state and covariance estimates forward to the next time step. At measurement times, the model-projected state and covariance values are updated by using the measurement and its covariance characteristics. A Kalman gain matrix is computed to update estimates of system state and covariance. This process is repeated over time in a recursive fashion, projecting values for each discrete time step and updating those estimates for time steps when measurements are available.

The ensemble Kalman filter (EnKF) follows this same general approach for nonlinear models, but relies on Monte Carlo methods to project state and covariance values between measurement times (Burgers et al., 1998; Marguilis et al., 2002). The SOC model (Equation 16.1) is nonlinear due to multiplication of R and X, and both "states" of the system are estimated. The equations to update X_t and R_t at each time step in the EnKF are:

$$\text{Updated } X_t = \text{Predicted } X_t + K_X (Z_t - \text{Predicted } X_t)$$

$$(16.4)$$

$$\text{Updated } R_t = \text{Predicted } R_t + K_R (Z_t - \text{Predicted } X_t)$$

where K_X and K_R are Kalman gains for X_t and R_t, computed at each time that measurements are used to update the variables.

For this particular problem (i.e., the specific model, the variables to be estimated, and the measurements that are made), these gain factors can be computed as follows (Jones et al., 2004):

$$K_X = \frac{\sigma_{X,t}^2}{\sigma_{X,t}^2 + \sigma_Z^2} \qquad (16.5)$$

$$K_R = \frac{\sigma_{XR,t}}{\sigma_{X,t}^2 + \sigma_Z^2}$$

where $\sigma_{X,t}^2$ is the variance of soil C predictions at time t, and $\sigma_{XR,t}$ is the covariance between X and R estimates at time t. These variance and covariance values are estimated before state estimates are updated.

Note that although R is not measured, the measurement of SOC provides information for refining the estimate of R via the covariance term. Also note that these gains vary with time; they are recalculated each time a measurement is made. If measurements are not made in a particular year, model predictions provide estimates of SOC and the update step is omitted.

The Kalman gain variables are used to weight the updated estimate on the basis of error variances. Note, for example, that if measurement error variance (σ_Z^2) is very small relative to model prediction variance ($\sigma_{X,t}^2$), then K_X approaches 1.0, and the updated X_t (Equation 16.4) will be approximately the value (Z_t) that was measured. In contrast, if measurement error is large relative to prediction error, K_X

will be closer to 0.0, and the updated estimate will be near the predicted value. Furthermore, if the covariance term used to compute K_R is small, the updated R_t will remain near its estimate from the previous step. However, if the covariance term is large, differences between measured and predicted SOC will result in adjustments to R in the update step.

16.3.4 Example Results

Jones et al. (2004) presented a sensitivity analysis to assess the benefits of using this EnKF to estimate SOC over time, relative to measurements alone, for different combinations of model parameters, errors, and initial conditions. Here, base case results from this study are summarized, and it is shown how errors of SOC estimation vary over time under different frequencies of measurements (and thus, frequencies of updating estimates of X and R).

To demonstrate numerical values, realistic parameters and error terms were selected for a case study. The variance of SOC measurements (σ_Z^2) was set at 500,000 (a standard deviation of 707 kg[C] ha^{-1} or a standard error of measurement of 0.0253% C on a mass basis). The initial value of SOC was assumed to be 16,000 kg[C] ha^{-1} in the top 20 cm of soil, which is about 0.6% carbon on a mass basis (Yost et al., 2002). Variance of this initial SOC estimate was assumed to be 20,000 (kg[C] ha^{-1})^2, a standard deviation of 141 kg[C] ha^{-1}. It was also assumed that the model error variance (σ_ε^2) was 20,000 (kg[C] ha^{-1})^2. The value of R_0 was assumed to be 0.020 (based on a range of values reported by Pieri, 1992, and Bationo et al., 2003); the variance in this parameter (σ_η^2) was assumed to be 0.0001. The value of U_t was set at a relatively high value of 2000 kg[C] ha^{-1}, constant across all years, and the value of b at 0.20.

Equation 16.3 was used to generate measurements (Z_t) for t = 1 through 30 years for a hypothetical field for which SOC is to be estimated, using an initial value of SOC of 16,000 kg[C]/ha and an R value of 0.010. The difference between R_0 and R for a particular field conceptually represents the variability among fields that belong to the population of fields

Table 16.5 Values of Parameters, Initial Conditions, and Inputs for Example Ensemble Kalman Filter

Variable	Definition	Units	Value
X_0	True value of soil organic carbon at time 0	kg[C]/ha	16,000
R	True value of mineralization parameter	1/year	0.010
$\sigma_{\hat{z}}^2$	Variance of measurement, constant over time	$(kg[C]/ha)^2$	500,000
σ_{ε}^2	Variance in model estimates of soil organic carbon, each year time step	$(kg[C]/ha)^2$	20,000
R_0	Initial estimate of soil C decomposition parameter	1/year	0.020
σ_{η}^2	Variance of decomposition rate parameter	$(1/year)^2$	0.0001
U_t	Input of C to the soil each year (assumed constant)	kg[C]/ha	2,000
b	Proportion of annual soil C input that remains after 1 year	—	0.20

Source: Adapted from Jones, J.W., W.D. Graham, D. Wallach, W.M. Bostick, and J. Koo. 2004. *Trans. ASAE,* 47(1):331–339.

that has a mean value of R_0. The updated estimates of R_t should converge to the value for the specific field, starting from the initial value of 0.020. The values of parameters and initial conditions used to implement the EnKF are summarized in Table 16.5. The EnKF was used to estimate X and R, and their variances for each year of the 30 years for which measurements were generated. Annual changes in SOC estimated from measurements ($Z_t - Z_{t-1}$) and from EnKF estimates (Updated X_{t+1} – Updated X_t), were compared with true values that were generated.

Figure 16.2 shows EnKF estimates of SOC for the inputs used (Table 16.5), as well as annual measurements (generated as discussed above), and "true" values of SOC for the 30-year case study. Estimates made by the EnKF are smooth, and in most years are closer to the "true" values than measured values. Estimates of R evolved from an initial estimate of 0.020 year^{-1} to values near the "true" value of 0.010 after about 6 years (not shown), and remained near that value for

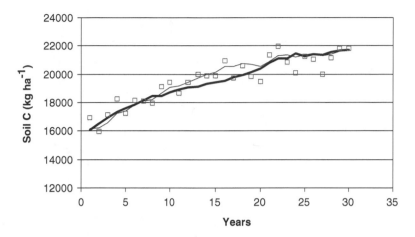

Figure 16.2 Changes in soil organic carbon (SOC) over time based on measurements (open symbols) and EnKF (heavy line) compared with "true" values of SOC (light line). (Modified from Jones, J.W., W.D. Graham, D. Wallach, W.M. Bostick, and J. Koo. 2004. *Trans. ASAE,* 47(1):331–339.)

the remainder of time in the case study. The effect of the EnKF is clear when one compares annual changes in SOC (Figure 16.3). Over the 30-year study, the EnKF estimates of annual changes were closer to "true" values in all years except one (year 10). EnKF estimates of annual changes in SOC were improved more than estimates of SOC vs. time when compared with measurements because of the smoothing process that occurs when model estimates are combined with measurements.

Three additional runs were made to demonstrate the effect of measurement frequency on standard error of SOC estimates. Errors for measurements made every year, every 2 years, every 3 years, and every 5 years are shown in Figure 16.4. Also shown is the standard error of SOC estimate obtained from using measurements alone (707 kg ha^{-1}). These results indicate that SOC estimation errors using the EnKF and measurements every 3 years would be less than errors based on measurements alone. They also show that errors in SOC estimates decrease over time, after an initial increase,

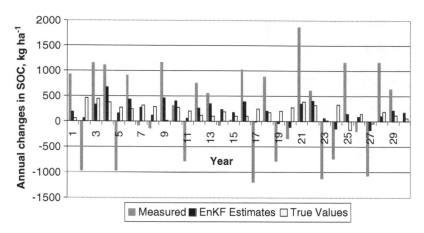

Figure 16.3 Annual changes in soil organic carbon comparing EnKF estimates with measured and true values. (Modified from Jones, J.W., W.D. Graham, D. Wallach, W.M. Bostick, and J. Koo. 2004. *Trans. ASAE,* 47(1):331–339.)

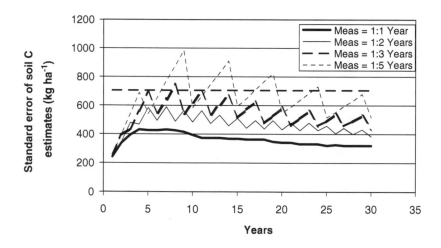

Figure 16.4 Effect of measurement frequency on errors of soil organic carbon (SOC) estimates. The heavy dashed line is the standard error of SOC estimates based on measurements alone. (Modified from Jones, J.W., W.D. Graham, D. Wallach, W.M. Bostick, and J. Koo. 2004. *Trans. ASAE,* 47(1):331–339.)

which is the result of more accurate estimates of R and lower model prediction error. Jones et al. (2004) reported on a more comprehensive analysis of the EnKF under different combinations of parameters, initial conditions, and errors in model and measurements. They found that estimates of SOC were better than measurements alone in all combinations when this simple model was used in the EnKF, although estimation errors decreased more under some combinations than others.

16.4 DISCUSSION

In this chapter, two types of model uses related to SOC sequestration were demonstrated. In the first one, a model was used to perform computer experiments to evaluate SOC sequestration potential as affected by different management practices in a particular location in Mali. Although data were limited for this region, the example demonstrated the value of using available data to adapt a model before using it to evaluate alternative management practices in a region. When the computer experiment was conducted using the same DSSAT maize model, but without using local data to adapt the model, results were clearly inconsistent with known yield and runoff responses of RT vs. CT and to realistic changes in SOC. Obviously more data and more work are needed to improve predictions and confidence in simulated results. Nevertheless, the preliminary estimates of SOC sequestration obtained in this study are certainly reasonable. They fall within the range of values reported elsewhere (see West and Post, 2002; Pichot et al., 1981; Lal, 2000), and simulations of runoff and yield of maize in RT vs. CT are consistent with local data and knowledge.

The second example represents a new use of biophysical models. The motivation for this use was that reliable estimates of SOC sequestration will be needed if landowners enter into contracts in which they are paid to sequester an agreed upon amount of carbon. In this case, a model was used to integrate measurements over time to improve estimates of SOC sequestration using an ensemble Kalman filter. This procedure also improved model predictions over time (errors were reduced) as new measurements were used, and a model parameter was more accurately estimated for the particular field. Through the recursive combination of model predictions and new observations, the model was better adapted to predict SOC levels at the specific site.

The second example showed the need for knowing errors associated with model predictions, and it demonstrated how to include uncertainties (stochastic features) in models for this purpose. The demonstration of this data assimilation technique was limited to a single field and to the use of a simple model. However, this approach is amenable to the use of more complex models, including the DSSAT-CENTURY model used in the first example. Koo et al. (2003) and Bostick et al. (2003) showed that one can use the EnKF to assimilate both crop biomass and SOC measurements to improve estimates of SOC using more complex models. This approach lends itself to the use of remote sensing data to improve SOC estimates by assimilating biomass data over space and time. Although the case study presented was for a single field, the EnKF can be expanded to estimate SOC over large areas that would be required in a carbon contract. This capability is currently being developed, following similar developments in the field of hydrology (Graham, 2002). This approach appears to have potential for other practical applications as well.

Crop and soil models can be very useful tools in SOC sequestration studies. But, they can also be misused. The phrase "all models are wrong; some are useful" is important to keep in mind. The main theme of this paper was the importance of using local data, however scarce, with such models to help better understand and predict SOC sequestration responses to soil, climate, and management. The procedure used to integrate local data with models was referred to as model adaptation. One aspect of that adaptation process is the estimation of soil, crop, and management parameters that allow the model to predict important variables for application to the problem being addressed. This process is sometimes referred to as model calibration, which may invoke criticisms from those whose aim is to have models that do not require modifications in order to predict system performance. Certainly, more work is needed to improve model capabilities; this will always be true. However, the diversity and spatial variability of land management, soils, climate, and genetic composition of crops will always create challenges to those who need to tailor agricultural management systems to achieve goals of individual farmers and of society. In this chapter, it has been shown that existing models can be used effectively to better understand and predict SOC sequestration. The importance of both data and models needs to be recognized as efforts are made to improve knowledge and tools for use in science and policymaking.

ACKNOWLEDGMENTS

This research is supported by the Soil Management Collaborative Research Program (SM CRSP) through a grant (LAG-G-00-97-00002-00) from the U.S. Agency for International Development and by a grant from the National Aeronautics and Space Administration titled "Carbon from Communities: A Satellite View" (Florida Agricultural Experiment Station Journal Series N-02376).

REFERENCES

Albiol, J., J. Robuste, C. Casas, and M. Poch. 1993. Biomass estimation in plant cell cultures using an extended Kalman filter. *Biotechnol. Prog.,* 9:174–178.

Anderson, J.M. and J.S.I. Ingram. 1993. *Tropical Soil Biology and Fertility: A Handbook of Methods.* CAB International, Wallingford, UK.

Antle, J.M. and G. Uehara. 2002. Creating incentives for sustainable agriculture: defining, estimating potential, and verifying compliance with carbon contracts for soil carbon projects in developing countries. In *A Soil Carbon Accounting System for Emissions Trading.* Special Publication SM CRSP 2002-4. University of Hawaii, Honolulu, HI, pp. 1–12.

Bationo, A., U. Mokwunye, P.L.G. Vlek, S. Koala, and B.I. Shapiro. 2003. Soil fertility management for sustainable land use in the West African Sudano-Sahelian zone. In Gichuru, M.P., A. Bationo, M.A. Bekunda, H.C. Goma, P.L. Mafongonya, D.N. Mugendi, et al., Eds. *Soil Fertility Management in Africa: A Regional Perspective.* Academy Science Publishers, Nairobi, pp. 253–292.

Bielders, C.L., P. Baveye, L.P. Wilding, L.R. Drees, and C. Valentin. 1996. Tillage-induced spatial distribution of surface crusts on a sandy paleustult from Togo. *Soil Sci. Soc. Am. J.,* 60:843–855.

Bostick, W.M., J. Koo, J.W. Jones, A.J. Gijsman, P.S. Traore, and B.V. Bado. 2003. *Combining Model Estimates and Measurements Through an Ensemble Kalman Filter to Estimate Carbon Sequestration.* ASAE Paper 033042. American Society of Agricultural Engineers, St. Joseph, MI.

Burgers, G., P.J. van Leeuwen, and G. Evensen. 1998. Analysis scheme in the Ensemble Kalman Filter. *Monthly Weather Rev.,* 126:1719–1724.

Campbell, K.L. 1978. Pollution in Runoff from Nonpoint Sources. Water Resources Research Center. University of Florida, Gainesville.

Cole, C.V., J. Williams, M. Shaffer, and J. Hanson. 1987. Nutrient and organic matter dynamics as components of agricultural production systems models. In *Soil Fertility and Organic Matter as Critical Components of Production Systems.* Special Publication 19, Soil Science Society of America/American Society of Agronomy, Madison, WI, pp. 147–166.

Coulibaly, Ntji. 2002. Personal communication, September 5, IER, Sotuba, Bamako, Mali.

Daba, S. 1999. Note on effects of soil surface crust on the grain yield of sorghum (Sorghum bicolor) in the Sahel. *Field Crops Res.,* 61:193–199.

Doumbia, M. 2003. Personal communication. SM CRSP Annual Review, February 17, Bambey, Senegal.

du Toit, A.S., J. Booysen, and J.J. Human. 1998. Calibration of CERES3 (maize) to improve silking date prediction values. *S. Afr. J. Plant Soil,* 15:61–65.

Eknes, M. and G. Evensen. 2002. An Ensemble Kalman Filter with a 1-D marine ecosystem model. *J. Marine Sys.*, 36:75–100.

Gigou, J., K.B. Traore, H. Coulibaly, M. Vaksmann, and M. Kouressy. 2000. Thème 2. Conservation de L'eau et du Sol. Aménagement en courbes de niveau et rendements des cultures en region Mali-Sud. Proceedings from Bamako, Mali.

Gijsman, A.J., G. Hoogenboom, W.J. Parton, and P.C. Kerridge. 2002a. Modifying DSSAT Crop Models for low-input agricultural systems using a soil organic matter-residue module from CENTURY. *Agron. J.*, 94:462–474.

Gijsman, A.J., S.S. Jagtap, and J.W. Jones. 2002b. Wading through a swamp of complete confusion: how to choose a method for estimating soil water retention parameters for crop models. *Eur. J. Agron.*, 18:75–105.

Graham, W.D. 2002. Estimation and prediction of hydrogeochemical parameters using Extended Kalman Filtering. In R.S. Govindaraju, Ed. *Stochastic Methods in Subsurface Contaminant Hydrology*. ASCE Publications, Reston, VA, pp. 327–363.

International Centre for Research in Agroforestry. 1994. *Slash-and-Burn: Update on Alternatives*. Vol. 1, no. 2. International Centre for Research in Agroforestry, Nairobi, Kenya.

Jagtap, S.S., U. Lall, J.W. Jones, A.J. Gijsman, and J.T. Ritchie. 2004. A dynamic nearest neighbor method for estimating soil water parameters. *Trans. ASAE*, 47(5):1437–1444.

Jones, J.W., K. Boote, and G. Hoogenboom. 2002a. Crop Modeling Team Trip Report: Mali and Ghana. NASA Carbon from Communities and the SM CRSP Soil Carbon Projects.

Jones, J.W., A.J. Gijsman, W.J. Parton, K.J. Boote, and P. Doraiswamy. 2002b. Predicting soil carbon accretion: the role of biophysical models in monitoring and verifying soil carbon. In A Soil Carbon Accounting System for Emissions Trading. Special Publication SM CRSP 2002-4. University of Hawaii, Honolulu, HI, pp. 41–68.

Jones, J.W., G. Hoogenboom, C.H. Porter, et al. 2003. The DSSAT cropping system model. *Eur. J. Agron.*, 18:235–265.

Jones, J.W., W.D. Graham, D. Wallach, W.M. Bostick, and J. Koo. 2004. Estimating soil carbon levels using an ensemble Kalman filter. *Trans. ASAE*, 47(1):331–339.

Jones, P.G., P.K. Thornton, W. Diaz, and P.W. Wilkens. 2002. Mark-Sim. A Computer Tool That Generates Simulated Weather Data for Crop Modelling and Risk Assessment. Ver 1. Centro Internacional de Agricultura Tropical, Cali, Colombia.

Kaya, B. 2000. Soil Fertility Regeneration Through Improved Fallow Systems in Southern Mali. Ph.D. diss. University of Florida, Gainesville.

Koo, J., W.M. Bostick, J.W. Jones, A.J. Gijsman, and J.B. Naab. 2003. Estimating soil carbon in agricultural using ensemble Kalman filter and DSSAT-CENTURY. ASAE Paper 033041. American Society of Agricultural Engineers, St. Joseph, MO.

Lal, R. 1997a. Long-term tillage and maize monoculture effects on a tropical alfisol in Western Nigeria. I. Crop yield and soil physical properties. *Soil Tillage Res.,* 42:145–160.

Lal, R. 1997b. Long-term tillage and maize monoculture effects on a tropical alfisol in Western Nigeria. II. Soil chemical properties. *Soil Tillage Res.,* 42:161–174.

Lal, R. 2000. Land use and cropping system effects on restoring soil carbon pool of degraded alfisols in Western Nigeria. In Lal, R., J.M. Kimble, and B.A. Stewart, Eds. *Global Change and Tropical Ecosystems.* Lewis Publishers, Boca Raton, FL, pp. 157–165.

Margulis, S.A., D. McLaughlin, D. Entekhabi, and S. Dunne. 2002. Land data assimilation and soil moisture estimation using measurements from the Southern Great Plains: 1997 field experiment. *Water Resour. Res.,* 38:1299.

Mavromatis, T., K.J. Boote, J.W. Jones, A. Irmak, D. Shinde, and G. Hoogenboom. 2001. Developing genetic coefficients for crop simulation models with data from crop performance trials. *Crop Sci.,* 41:40–51.

Maybeck, P.S. 1979. *Stochastic Models, Estimation, and Control.* Vol. 1. Academic Press, New York.

Moulin, A.P. and H.J. Beckie. 1993. Evaluation of the CERES and EPIC models for predicting spring wheat grain yield over time. *Can. J. Plant Sci.,* 73:713–719.

Neely, C. and G. Uehara. 2002. Carbon from Communities Progress Report. For Carbon from Communities, A Satellite View. NASA Project NAG13-2003.

Parton, W.J., D.S. Schimel, C.V. Cole, and D.S. Ojima. 1987. Analysis of factors controlling soil organic matter levels in Great Plain grasslands. *Soil Sci. Soc. Am. J.,* 51:1173–1179.

Parton, W.J., J.W.B. Stewart, and C.V. Cole. 1988. Dynamics of C, N, P and S in grassland soils: a model. *Biogeochemistry,* 5:109–131.

Parton, W.J., D.S. Ojima, C.V. Cole, and D.S. Schimel. 1994. A general model for soil organic matter dynamics: sensitivity to litter chemistry, texture and management. In R.B. Bryant and R.W. Arnold, Eds. Quantitative Modeling of Soil Forming Processes. Special Publication 39. Soil Science Society of America, Madison, WI, pp. 147–167.

Paustian, K., W.J. Parton, and J. Persons. 1992. Modeling soil organic matter in organic-amended and nitrogen-fertilized long-term plots. *Soil Sci. Soc. Am. J.,* 56:476–488.

Pichot, J., M.P. Sedogo, J.F. Poulain, and J. Arrivets. 1981. Fertility evolution in a tropical ferruginous soil under the effect of organic manure and inorganic fertilizer applications. *Agric. Trop.,* 37:122–133.

Pieri, C.J.M.G. 1992. *Fertility of Soils: A Future for Farming in the West African Savannah.* Springer-Verlag, Berlin.

Probert, M.E., B.A. Keating, J.P. Thompson, and W.J. Parton. 1995. Modelling water, nitrogen, and crop yield for a long-term fallow management experiment. *Aust. J. Exp. Agric.,* 35:941–950.

Ringius, L. 2002. Soil carbon sequestration and the CDM: Opportunities and challenges for Africa. *Climatic Change,* 54:471–495.

Rockstrom, J., P.-E. Jansson, and J. Barron. 1998. Seasonal rainfall partitioning under runon and runoff conditions on sandy soil in Niger. On-farm measurements and water balance modelling. *J. Hydrol.,* 210:68–92.

Roncoli, C., A. Berthé, C. Neely, K. Moore, H. Coulibaly, B. Traoré, and S. Kanté. 2002. A characterization of conditions and constraints relative to adoption and implementation of soil fertility management technologies in three sites in Southern Mali. Report prepared for Carbon from Communities Project.

Singh, U., P.K. Thornton, A.R Saka, and J.B. Dent. 1993. Maize modelling in Malawi: a tool for soil fertility research and development. In F.W.T. Penning de Vries, P.S. Teng, and K. Metselaar, Eds. *Systems Approaches for Agricultural Development*. Kluwer, Dordrecht, pp. 253–273.

Tsuji, G.Y., G. Hoogenboom, and P.K. Thornton, Eds. 1998. *Understanding Options for Agricultural Production System Approaches for Sustainable Agricultural Development*. Kluwer, Dordrecht.

Vitousek, M.P., J.D. Aber, R.W. Howarth, G.E. Likens, P.A. Matson, D.W. Schindler, W.J. Schlesinger, and D.G. Tilman. 1997. Human alterations of the global nitrogen cycle: sources and consequences. *Technical Rep. Ecol. Appl.*, 7:737–750.

Welch, G. and G. Bishop. 2002. An introduction to the Kalman Filter. Department of Computer Science, University of North Carolina, Chapel Hill.

West, T.O. and W.M. Post. 2002. Soil organic carbon sequestration rates by tillage and crop rotation: a global data analysis. *Soil Sci. Soc. Am. J.*, 66:1930–1946.

Woomer, P.L. 1993. Modelling soil organic matter dynamics in tropical ecosystems: model adoption, uses and limitations. In K. Mulongoy and R. Mercks, Eds. Phosphorus Cycles in Terrestrial and Aquatic Ecosystems. Regional workshop 4: Africa. SCOPE/UNEP. Proceedings of Symposium, Nairobi, Kenya. 1991. Saskatoon Institute of Pedology, University of Saskatchewan, Saskatoon.

Yost, R.S., P. Doraiswamy, and M. Doumbia. 2002. Defining the contract area: using spatial variation in land, cropping systems and soil organic carbon. In A Soil Carbon Accounting System for Emissions Trading. Special Publication SM CRSP 2002-4. University of Hawaii, Honolulu, HI, pp. 13–40.

Zhang, X.C. and W.P. Miller. 1996. Physical and chemical crusting processes affecting runoff and erosion in furrows. *Soil Sci. Soc. Am. J.*, 60:860–865.

17

Reducing Greenhouse Warming Potential by Carbon Sequestration in Soils: Opportunities, Limits, and Tradeoffs

JOHN M. DUXBURY

CONTENTS

Carbon sequestration in soils is a land-based option to reduce the greenhouse warming potential (GWP) of the atmosphere. It has the additional benefits of improving soil quality and the sustainability of agriculture. This chapter discusses the potential for carbon sequestration in soils from a conceptual framework based on aggregation of soils as the primary process that protects organic matter from biological oxidation to form carbon dioxide. Tillage of soils breaks aggregates, and therefore leads to loss of organic carbon (OC); this process can only be prevented or reversed by stopping tillage. No-tillage crop production is already happening on a large scale in several countries, most notably the United States and Brazil. Conversion from conventional tillage (CT) to no tillage (NT) has several other GWP effects, however, in addition to increasing carbon sequestration, and these effects have been not been adequately analyzed. Consequently, it is unclear what the GWP outcome of promoting carbon sequestration through a switch to NT agriculture actually is. Given the current high level of interest in promoting carbon-trading schemes based on carbon sequestration in soil, it is surprising that neither scientists nor policymakers have recognized the need for a comprehensive evaluation of GWP effects of the change from CT to NT. This chapter extends the carbon dioxide–based analysis of West and Marland (2002) for grain production in the United States to include other greenhouse gases, showing how the timeframe of analysis can affect GWP outcome.

17.1 CONCEPTUAL BASIS FOR CARBON SEQUESTRATION IN SOILS

The nonliving organic matter in soils is generally considered to be in three pools: (1) a labile pool that is free within the soil matrix, (2) a physically protected pool that is within larger aggregates, and (3) a chemically protected or mineral-associated pool that is within the smallest aggregates (e.g., Paustian et al., 1997; Jenkinson and Rayner, 1977). The specific aggregate sizes associated with these pools vary with soil type, but chemically protected organic matter can be considered to be within aggregates that are not disrupted by the normal forces that a soil is exposed to, namely, drying and wetting (and freeze-thaw) cycles and the mechanical forces of tillage. Consequently, the OC in these aggregates is very stable and is often referred to as the passive pool. Tillage leads to disruption of larger aggregates, partly by the mechanical forces of tillage, and partly because tillage is a mixing process that, over time, exposes all soil in the tilled layer to the harsh environment of the soil surface where wet-dry and/or freeze-thaw cycles are most frequent. The loss of OC associated with tillage-induced destruction of aggregates in agricultural soils can be considered as a conversion of OC from a protected pool to a labile pool that is then mineralized to CO_2. The factors that regulate the size of the various soil OC pools are shown in Table 17.1.

Most of the OC in the labile pool is derived from recent inputs such as crop residues, and the size of this pool varies with the level of organic inputs into the soil and climatic factors (temperature and moisture) that affect the rate of biological decomposition processes. The size of the labile OC pool is little affected by soil properties. It is the most important contributor to total soil OC in cold climates where both plant growth and biological decomposition processes are slow, but where the decomposition is slower than production, allowing an accumulation of substantial amounts of OC over long

Table 17.1 Organic Carbon Pools and Factors Controlling Pool Size

Organic Carbon Pool	Location in Soil	Factors Controlling Pool Size
Labile (free)[a]	Free in soil matrix	1. Amount and frequency of organic inputs 2. Temperature and moisture regimes
Physically protected (particulate, slow)	Within large aggregates	1. Soil texture 2. Tillage
Chemically protected (passive, mineral associated)	Within small aggregates	1. Soil texture 2. Soil mineralogy

[a] Parentheses give alternative names for pools.

periods of time. Thus, the average per ha carbon stock in soils of boreal forests (343 metric tons/ha) is almost three times that in tropical forests (123 metric tons/ha), whereas the vegetation stock in tropical forests (120 metric tons/ha) is about twice that in boreal forests (64 metric tons/ha) (Intergovernmental Panel on Climate Change [IPCC], 2001).

The bulk of the organic matter in noncultivated temperate and tropical region soils is associated with aggregates that provide varying degrees of protection against biological decomposition processes. The potential for formation of aggregates in soils depends on soil texture and mineralogy, and increases with the fineness (surface area) of soil particles. The stability of aggregates, and hence their associated OC, varies greatly with soil texture and mineralogy. Except for oxisols, which have very stable large aggregates, the dividing point between large and small aggregates, and between the physically protected and passive organic matter pools, is most likely in the 50- to 250-μm size range.

The total OC content of soils increases with the ability to form aggregates, and hence with increasing clay content. Figure 17.1 shows how tillage and soil texture relate to soil aggregation and carbon sequestration. Sandy soils with little capacity to form aggregates have low OC contents that vary little with cultivation. Finer-textured soils have higher OC contents that are reduced by tillage-induced destruction of aggregates. The upper and lower lines represent the range in the maximum and minimum OC contents, respectively, as a function of soil texture and tillage. The difference between the two lines represents the OC protected within aggregates. The minimum OC content is associated with the passive or stable OC pool. For a clay soil with an initial OC content at point A, the sequestration potential is the difference between the maximum value and the initial value, that is, B − A. In most situations, it is unlikely that the OC content of soils under NT agriculture can quite reach that of their original natural ecosystems, due to lower levels of residue return, and hence a smaller labile OC pool.

The equilibrium OC content that can be achieved by a given carbon sequestration practice will vary with the extent

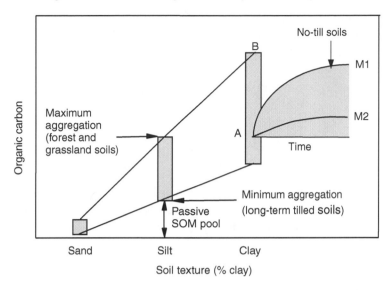

Figure 17.1 Influence of texture and aggregation on carbon sequestration in soils

of tillage. Carbon sequestration management regime M1, which represents the maximum OC level that can be reached under NT agriculture, gives a different pattern of carbon accumulation and a higher final or equilibrium OC level than regime M2, which represents a practice that has some tillage and hence is less effective.

Estimates of OC accumulation rates following a switch from CT to NT are fairly consistent. The IPCC (2000) suggests that average OC accumulation rates will range between 0.1 to 0.8 metric tons C/ha/year, with mean values of 0.4 to 0.5 metric tons C/ha/year for moist temperate and tropical environments. West and Marland (2002) estimate an average rate of 0.34 ±0.1 metric tons C/ha/year for cropland in the United States, and West and Post (2002) provide a global average of 0.57 ±0.14 metric tons C/ha/year by considering 67 long-term experiments around the world. However, the rate of OC accumulation depends on the rate at which macroaggregates are generated, which will vary as a function of soil texture and fertility, cropping system, and residue return levels, all affecting soil biological activity as well as physical–chemical processes that together affect soil structure. The rate at which C accumulates in soils when tillage is stopped or reduced will vary with management. However, the equilibrium OC level associated with a particular reduced tillage practice is independent of the rate of OC accumulation.

17.2 OPPORTUNITIES FOR CARBON SEQUESTRATION IN SOILS

The conceptual model presented in Figure 17.1 illustrates that carbon sequestration potential is related to soil texture and the intensity of tillage. The greatest potential for carbon sequestration is with fine-textured soils that have been carbon degraded due to intensive tillage. Many such soils exist in developed countries, where large tillage equipment is the norm. In tropical environments of the developing world, soils that are in a paddy rice–upland crop rotation are the most carbon degraded due to puddling of soils to reduce water infiltration, which destroys all but the most stable aggregates. Decomposition of released OC is then rapid in the following aerobic phase of the crop rotation. The lowest potential for carbon sequestration is with light textured soils, even though they may have very low OC levels.

   Erosion of topsoil where low OC subsoils become surface soils also creates a situation where carbon sequestration potential is high. Since carbon in eroded soils may well be conserved, the combination of erosion and building up of OC in former subsoils may well lead to greater overall carbon storage than originally existed before erosion occurred (Duxbury, 1995). This is not to say that erosion should be promoted, of course, since it has a range of negative side effects.

17.3 IMPACT OF TILLAGE MANAGEMENT ON GREENHOUSE GAS FLUXES

A change from CT to NT involves changes in management practices, such as chemical inputs and tractor use, that also generate CO_2, and it potentially alters fluxes of other greenhouse gases, especially nitrous oxide (N_2O) and methane (CH_4). The impact of all management changes on all greenhouse gases contributing to GWP must be evaluated in order to determine the net effect of changing tillage practice on GWP. These other factors may provide additional GWP benefits or they may offset carbon sequestration in soils, creating a situation where the net GWP benefits from a change in tillage practice are affected by tradeoffs. Furthermore, if the changes made in the additional sources of GWP are permanent, that is, continue indefinitely, they will affect the long-term GWP outcome of a change to NT as soil carbon sequestration continues only until a new OC equilibrium is reached in a soil. These two issues are analyzed in the following sections.

17.4 GWP ANALYSIS FOR CONVENTIONAL AND NO-TILLAGE MAIZE PRODUCTION IN THE UNITED STATES

A comprehensive analysis of the effect of a change from CT to NT on the carbon dioxide flux for grain production in the United States has showed that emissions associated with machinery were reduced, but those associated with agricultural inputs increased (West and Marland, 2002). These changes produced a net savings of 31 kg C/ha/yr in addition to that achieved by carbon sequestration in soil (337 kg C/ha/year) for a total carbon benefit of 368 kg C/ha/year. However, the impact of the change from CT to NT on contributions to greenhouse gas build-up from N_2O and CH_4 were not considered. A more complete analysis is undertaken here for rainfed corn. Additional sources of GWP (Table 17.2) were associated with increased emissions of N_2O from soil under NT and during manufacture of the additional 42.6 kg N/ha that is used in NT maize (West and Marland, 2002). Changes in GWP associated with emissions of N_2O and CH_4 from use of farm machinery and soil fluxes of CH_4 were small and are not presented.

The N_2O and CH_4 emissions associated with increased fertilizer N input under NT were estimated for production,

Table 17.2   Effect of Tillage Practice on Relative GWP
Sources from Rainfed Maize Production in United States

	kg C_{equiv}/ha/year	
GWP Component	Conventional Till	No-Till
Soil C sequestration	0	-337[a]
Carbon dioxide emissions		
* Agricultural inputs	+156[a]	+202[a]
* Machinery	+72[a]	+23[a]
Net C flux	+228	−112
Relative C flux	0	−340
Relative N_2O emission		
Additional N input (42.6 kg/ha)[a]		
* Manufacture	0	+11[b]
* Use	0	+64[c]
Switch to no-till	0	+219[d]
Total additional GWP flux	0	+294
Revised relative GWP flux	0	−46

[a] From West, T.O., and G. Marland. 2002. *Agric. Ecosyst. Environ.*, 91:217–232; positive and negative signs indicate carbon emission and sink, respectively.
[b] Based on 22% of fertilizer as NH_4NO_3, and N_2O release from Kramer, K.J., H.C. Moll, and S. Nonhebel. 1999. *Agric. Ecosyst. Environ.*, 72:9–16.
[c] Using Intergovernmental Panel on Climate Change formula of 1.25% fertilizer N released as N_2O from 90% of applied N.
[d] Average value from Smith, K.A., F. Conen, B.C. Ball, A. Leip, and S. Russo. 2002. Emissions of non-CO_2 greenhouse gases from agricultural land, and the implications for carbon trading. In J. van-Ham, A.P.M. Baede, R. Guicherit, and J.G.F.M. Williams-Jacobese, Eds. *Non-CO_2 Greenhouse Gases: Scientific Understanding, Control Options and Policy Aspects.* Proceedings of 3rd International Symposium, Maastricht, Netherlands, 21–23 January 2002. Millpress Science, Rotterdam, Netherlands.
Note: GWP = greenhouse warming potential.

transport, and use. Only N_2O emission during N fertilizer production and use added significant amounts of GWP (Table 17.2). Nitrous oxide emission during fertilizer manufacture is associated with production of nitric acid that, in turn, is used to manufacture ammonium nitrate. Ammonium nitrate accounts for 22% of the fertilizer N consumption in

U.S. agriculture (Brady and Weil, 1999), and it was assumed that this proportion was used on maize. The value for N_2O release during production of ammonium nitrate (4.68 g N_2O/kg NH_4NO_3) was taken from Kramer et al. (1999). Use emission of N_2O was based on the IPCC formula of 1.25% N_2O release from 90% of added fertilizer N, which is derived from the work of Bouman (1996).

An increase in emissions of N_2O from soils associated with the change from CT to NT was estimated by using the mean value of 1.65 kg N_2O-N/ha/year from Smith et al. (2002), who summarized published information. This does not duplicate the N_2O input from increased N use, as CT/NT comparisons were made at the same N input levels. All N_2O emission values were converted to CO_2 equivalents using a factor of 310 (IPCC, 1996) and then to CO_2–C.

When all three greenhouse gases (CO_2, N_2O, and CH_4) were included in the analysis, the net annual GWP benefit associated with the change from CT to NT was 46 kg CO_2–C/ha, compared to 340 kg CO_2–C/ha if only soil carbon sequestration is considered.

The revised estimate of the GWP benefit created by a change from CT to NT is largely driven by the estimate of increased N_2O emission from soil when NT is adopted. However, this value is quite uncertain due to the limited number of comparisons and the inadequate sampling methodologies used in most studies. Fluxes of N_2O from soils are highly variable in space and time (Mosier and Hutchinson, 1981; Duxbury et al., 1982), which presents a challenge to anyone generating annual flux data using chamber measurements. Only one of the reported comparisons of N_2O emissions from CT and NT practices used an automatic chamber system (Ball et al., 1999), which addresses temporal variability very well, and there are no reports using micrometeorological methods, which can address both temporal and spatial variability.

Nitrous oxide emissions from soil are episodic, and both short-term comparisons between CT and NT practices and estimates of annual N_2O fluxes would be improved by a focus on events that lead to bursts of N_2O emission. Such events are rainfall, freeze-thaw cycles, plowing, fertilizer application, and residue, and manure additions (Skiba et al., 2002; Yamulki and Jarvis, 2002; Davies et al., 2001; Van Bochove, 2000; Lemke et al., 1999; Mackenzie et al., 1997; Chen et al., 1995; Duxbury et al., 1982).

Most investigators recognize the variability issue but do not adjust sampling strategies, especially in terms of frequency, to provide confidence that annual flux values are accurate. Nevertheless, the majority of studies (Six et al., 2002; Skiba et al., 2002; Mackenzie et al., 1997; Hilton et al., 1994; Hutsch, 1991; Aulakh et al., 1984, Burford et al., 1981), but not all (Choudhary et al., 2002; Lemke et al., 1999; Jacinthe and Dick, 1997) find higher emissions of N_2O under NT. There is also evidence to support the contention that denitrification, the primary source of N_2O, is greater under NT (Linn and Doran, 1984; Rice and Smith, 1982; Staley et al., 1990).

It is also important to recognize that soil physical and biological properties are dynamic for many years following a shift from CT to NT. Soil physical conditions improve considerably with time, and the reestablishment of macro fauna, especially earthworms, leads to the development of macropore channels and much improved drainage. Organic matter and biological activity are concentrated close to the soil surface where gas exchange with the atmosphere is most rapid, promoting aeration of soils but also emission of N_2O when it is being generated. It is probable that N_2O emissions change over time after adoption of NT. Initially they are greater than for CT, as soils have poor structure and are poorly drained, but later they may become similar to or even less than CT as soil structure and drainage improve. Unfortunately, there has not been any systematic study of temporal effects of changing from CT to NT on N_2O emissions from soils.

Soil compaction, which increases N_2O emission from soils (Ball et al., 1999; Yamulki and Jarvis, 2002), is another important parameter to consider, and inadequate attention is given to this in experiments and also in current NT agriculture in the United States. No tillage, and hence carbon sequestration, is unlikely to be agronomically sustainable on many soils without the use of controlled traffic patterns to minimize compaction, and a mulch to prevent surface sealing by raindrops.

These practices are aimed at maintaining a desirable physical condition at the soil's surface, which would also be expected to reduce N_2O generation and emission. Researchers also need to be cognizant of potential differences in soil physical condition between researchers' plots and farmers' fields, and hence in their N_2O emissions. Overall, it is clear that much additional work is needed to determine the impact of any change from CT to NT on N_2O emission from soils. Such information is critical for determining whether a change from CT to NT is, on balance, an effective means of reducing GWP.

17.5 LONG-TERM GWP EFFECTS OF CHANGING TILLAGE PRACTICE

A change from CT to NT will lead to increasing carbon sequestration in soil for a limited time, depending on management and environmental factors. This will likely be for tens of years, after which a higher soil equilibrium organic carbon content will be reached. The rate of OC accumulation will follow a generally declining rate as OC accumulates, but actual patterns of OC accumulation may be complex, and can be expected to vary as soil faunal populations are reestablished and soil physical condition improves.

The long-term effect of the change from CT to NT on cumulative GWP will also be influenced by any permanent changes in greenhouse gas fluxes associated with the change in tillage practice. This concept is illustrated conceptually in Figure 17.2. The cumulative GWP benefit can be greater than that described by soil carbon sequestration alone, and will continue to rise indefinitely if other components contribute a net permanent reduction in GWP. In cases where other GWP components create a permanent net source of GWP that is initially less than the C sequestration benefit, the cumulative GWP will initially show a benefit before it reaches a maximum and then declines until the result is a net increase in GWP. Cumulative GWP outcomes will vary depending on the magnitude of the permanent GWP component and the timeframe considered. The analysis will be more complicated where a permanent GWP component changes over time from a source

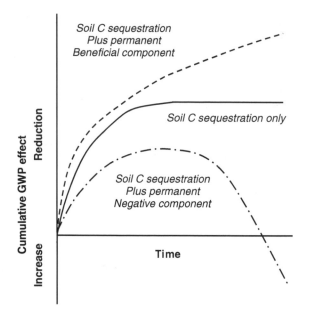

Figure 17.2 Long-term cumulative GWP effects of soil carbon sequestration plus an additional permanent GWP component associated with a change from conventional tillage to no-tillage.

to a sink, or vice versa, as is hypothesized for N_2O flux following conversion from CT to NT.

17.6 CONCLUSIONS

Carbon sequestration in soils, which is linked to the adoption of NT practice, is being promoted as a viable strategy for reducing GWP. The analyses presented here show that:

1. Carbon sequestration potential will largely be associated with heavier textured soils that have been carbon degraded due to intensive tillage.
2. The carbon sequestration benefit expected with a switch from CT to NT for rainfed corn production in the United States may be largely offset by increased emissions of N_2O from soils associated with the tillage change and from increased use of fertilizer N.

3. The cumulative GWP outcome associated with a switch from CT to NT will depend on whether changes in other GWP components permanently supplement or offest carbon sequestration, and the timeframe of analysis. The long-term outcome of conversion to NT could possibly be to increase GWP, even though there may be short-term GWP benefits.

These results highlight the need for more comprehensive analysis of the GWP effects of carbon sequestration in soils. There is no scientific basis for promoting carbon sequestration in the policy arena at the present time because its true potential to mitigate GWP has not been determined. Policymakers have the right to expect objective, rigorous scientific advice, and they, in turn, need to be better informed and more questioning about the various components that contribute to GWP.

It is also well established that fossil fuel combustion is, and will continue to be, the most important source of GWP. Anthropogenic emissions of N_2O and CH_4 are estimated to contribute 6% and 20%, respectively, to global radiative forcing (IPCC, 2001), and agriculture is the major contributor of these gases. Given these facts, in my view, agriculture can best contribute to reducing GWP by focusing on the production of plant biomass for use as an alternative to fossil fuel and on the reduction of agricultural sources of N_2O and CH_4.

ACKNOWLEDGMENTS

This chapter is a joint contribution by Cornell University's Agricultural Ecosystems Program, funded by a special grant from U.S. Department of Agriculture, Cooperative Research Extension and Education Service, and the Soil Management Collaborative Research Support Program funded by the U.S. Agency for International Development.

REFERENCES

Aulakh, M.S., D.A. Rennie, and E.A. Paul. 1984. Gaseous N losses from soils under zero-till as compared to conventional-till management systems. *J. Environ. Qual.,* 13:130–136.

Ball, B.C., A. Scott and J.P. Parker. 1999. Field N_2O, CO_2, and CH_4 in relation to tillage, compaction, and soil quality in Scotland. *Soil Tillage Res.*, 53:29–39.

Bouman, A.F. 1996. Direct emission of nitrous oxide from agricultural soils. *Nutrient Cycling Agroecosyst.*, 46:53–70.

Brady, N.C. and R. Weil. 1999. *The Nature and Properties of Soils*, 12th ed. Macmillan, New York.

Burford, J.R., R.J. Dowdell, and R. Crees. 1981. Emission of nitrous oxide to the atmosphere from direct-drilled and ploughed clay soils. *J. Sci. Food Agric.*, 32:219–223.

Chen, Y., S. Tessier, A.F. Mackenzie, and M.R. Laverdiere. 1995. Nitrous oxide emission from an agricultural soil subjected to different freeze-thaw cycles. *Agric. Ecosyst. Environ.*, 55:123–128.

Choudhary, M.A., A. Akramkhanov, and S. Saggar. 2002. Nitrous oxide emissions from a New Zealand cropped soil: tillage effects, spatial and seasonal variability. *Agric. Ecosyst. Environ.*, 93:33–43.

Davies, M.G., K.A. Smith, and A.J.A. Vinten. 2001. The mineralisation and fate of nitrogen following ploughing of grass and grass-clover sward. *Biol. and Fertil. of Soils*, 33:423–434.

Duxbury, J.M., D.R. Bouldin, R.E. Terry, and R.L. Tate. 1982. Emissions of nitrous oxide from soils. *Nature*, 298:462–464.

Duxbury, J.M. 1995. The significance of greenhouse gas emissions from soils of tropial agroecosystems. In R. Lal, J. Kimble, E. Levine, and B.A. Stewart, Eds. *Soil Management and the Greenhouse Effect*. Lewis Publishers, Boca Raton, FL.

Groffman, P.M. 1984. Nitrification and denitrification in conventional and no-tillage soils. *Soil Sci. Soc. Am. J.*, 49:329–334.

Hilton, B.R., P.E. Fixen, and H.J. Woodard. 1994. Effects of tillage, nitrogen placement and wheel compaction on denitrification rates in the corn cycle of a corn-oats rotation. *J. Plant Nutr.*, 17:1341–1357.

Hutsch, B. 1991. Influence of long term differences in soil management on denitrification losses. Proceedings VDFLUFA Congress, 16–21 September, Ulm, Germany. VDFLUA-Verlag, Darmstadt, Germany.

Intergovernmental Panel on Climate Change. 1996. *Climate Change 1995: The Science of Climate Change.* Second Assessment Report of the Intergovernmental Panel on Climate Change. Cambridge University Press, London; New York.

Intergovernmental Panel on Climate Change. 2000. *Special Report on Land Use, Land-Use Change and Forestry.* Cambridge University Press, London; New York.

Intergovernmental Panel on Climate Change. 2001. *Climate Change 2001: Synthesis Report.* Cambridge University Press, London; New York.

Jacinthe, P-A. and W.A. Dick. 1997. Soil management and nitrous oxide emissions from cultivated fields in southern Ohio. *Soil Tillage Res.,* 41:221–235.

Jenkinson, D.S. and J.H. Rayner. 1977. The turnover of soil organic matter in some of the Rothamsted classical experiments. *Soil Sci.,* 123:298–305.

Kramer, K.J., H.C. Moll, and S. Nonhebel. 1999. Total greenhouse gas emissions related to the Dutch cropping system. *Agric. Ecosyst. Environ.,* 72:9–16.

Lemke, R.L., R.C. Izaurralde, M. Nyborg, and E.D. Solberg. 1999. Tillage and N source influence soil-emitted nitrous oxide in the Alberta parkland region. *Can. J. Soil Sci.,* 79:15–24.

Linn, D.M. and J.W. Doran. 1984. Effect of water-filled pore space on carbon dioxide and nitrous oxide production in tilled and nontilled soils. *Soil Sci. Soc. Am. J.,* 48:1267–1272.

MacKenzie, A.F., M.X. Fan, and F. Cadrin. 1997. Nitrous oxide emission as affected by tillage, corn-soybean-alfalfa rotations and nitrogen fertilization. *Can. J. Soil Sci.,* 77:145–152.

Mosier, A.R. and G.L. Hutchinson. 1981. Nitrous oxide emissions from cropped fields. *J. Environ. Qual.,* 10:169–173.

Paustian, K., H.P. Collins, and E.A. Paul. 1997. Management controls on soil carbon. In E.A. Paul, K. Paustian, E.T. Elliot, and C.V. Cole, Eds. *Soil Organic Matter in Temperate Agroecosystems: Long-Term Experiments in North America,* CRC Press, Boca Raton, FL, pp. 51–72.

Rice, C.W. and M.S. Smith. 1982. Denitrification in no-till and plowed soils. *Soil Sci. Soc. Am. J.,* 46:1168–1173.

Six, J., C. Feller, K. Denef, S. Ogle, J.C.M. Sa, and A. Albrecht. 2002. Soil organic matter, biota, and aggregation in temperate and tropical soils: effect of no-tillage. *Agronomie,* 22:755–775.

Skiba, U., S. van Dijk, and B.C. Ball. 2002. The influence of tillage on NO and N_2O fluxes under spring and winter barley. *Soil Use Manage.,* 18:340–345.

Smith, K.A., F. Conen, B.C. Ball, A. Leip, and S. Russo. 2002. Emissions of non-CO_2 greenhouse gases from agricultural land, and the implications for carbon trading. In J. van-Ham, A.P.M. Baede, R. Guicherit, and J.G.F.M. Williams-Jacobese, Eds. *Non-CO_2 Greenhouse Gases: Scientific Understanding, Control Options and Policy Aspects.* Proceedings of 3rd International Symposium, Maastricht, Netherlands, 21–23 January 2002. Millpress Science, Rotterdam, Netherlands.

Staley, T.E., W.H. Caskey, and D.G. Boyer. 1990. Soil denitrification and nitrification potentials during the growing season relative to tillage. *Soil Sci. Soc. Am. J.*, 54:1602–1608.

Van Bochove, E., D. Prevost, and F. Pelletier. 2000. Effects of freeze-thaw and soil structure on nitrous oxide produced in a clay soil. *Soil Sci. Soc. Am. J.,* 64:1638–1643.

West, T.O. and G. Marland. 2002. A synthesis of carbon sequestration, carbon emissions, and net carbon flux in agriculture: comparing tillage practices in the United States. *Agric. Ecosyst. Environ.*, 91:217–232.

West, T.O. and W.M. Post. 2002. Soil organic carbon sequestration by tillage and crop rotation: a global data analysis. *Soil. Sci. Soc. Am. J.,* 66:1930–1946.

Yamulki, S. and S.C. Jarvis. 2002. Short-term effects of tillage and compaction on nitrous oxide nitric oxide, nitrogen dioxide, methane and carbon dioxide fluxes from grassland. *Biol. Fertil. Soils*, 36:224–231.

18

Management Practices and Carbon Losses via Sediment and Subsurface Flow

LLOYD B. OWENS AND MARTIN J. SHIPITALO

CONTENTS

18.1 INTRODUCTION

Management practices can have major impacts on soil organic carbon (SOC) levels, gains, and losses. There is wide acceptance that cultivating native land causes loss of soil organic matter. Davidson and Ackerman (1993) reported 20% to 40% loss of soil organic matter following the conversion of previously untilled soils to agricultural production. Changes in agricultural production are reversing this trend (Buyanovsky and Wagner, 1998). Conversion of land from plow tillage to long-term no-tillage management often increases soil organic C and N content (Doran, 1980, 1987; McCarty et al., 1995; McCarty and Meisinger, 1997).

The impacts of tillage and cropping rotations on SOC have been studied at several sites over various time periods. Increased SOC was measured in long-term (8 to 24 years) studies with no-tillage practices in the southeastern Coastal Plain of the United States (Hunt et al., 1996), Kentucky (Ismail et al., 1994), Illinois (Hussain et al., 1999), Iowa (Karlen et al., 1994), Nebraska (McCallister and Chen, 2000), North Dakota (Halvorson et al., 2002), Alberta, Canada (Nyborg et al., 1995), Saskatchewan, Canada (Campbell and Zentner, 1993; Campbell et al., 1995, 1996), and Argentina (Alvarez et al., 1995).

The SOC is lost from soil through several pathways, including mineralization following tillage, translocation with sediment, and leaching from the soil profile. Because tillage aerates soil and allows greater C mineralization (Eghball et al., 1994), a reduction in tillage intensity decreases SOC loss. Reicosky et al. (1995) identified large gaseous losses of soil C following moldboard plowing compared with relatively small losses with no till.

Among the multiple pathways of C loss from agricultural fields is C lost with sediment. Many studies of C loss from cropped fields were runoff plot (Massey and Jackson, 1952; Wan and El-Swaify, 1997; Zobisch et al., 1995) or erosion plot studies (Ambassa-Kiki and Nill, 1999; Kaihura et al., 1999) that had C as part of a group of nutrients instead of the central focus. Sediments enriched with organic C compared with surface soil have been observed in soils ranging from Wisconsin silt loams (Massey and Jackson, 1952) to clay soils in Hawaii (Wan and El-Swaify, 1997). In reviewing literature about C redistribution and loss by erosion, Gregorich et al. (1998) concluded that erosion usually resulted in decreased primary production and that SOC decreased because of a reduction in primary production.

Although numerous studies of SOC loss have been conducted, the concentrations and losses of total organic carbon (TOC) moving through soil profiles have received little attention. There are studies of dissolved organic carbon (DOC) in streams draining forested areas in New England. Values for DOC ranged between 0.3 and 2.0 mg/L in the stream tributaries of Hubbard Brook Experimental Forest in New Hampshire (Likens et al., 1977). In two paired streams draining forested watersheds in Maine, DOC concentrations of 2.0 and 2.2 mg/L were found (David et al., 1992). Owens et al. (1991) compared TOC concentrations in stormflow and baseflow from wooded, nonwooded, and mixed-management watersheds in Ohio. During the 10 years of study, average TOC concentrations in stormflow ranged from 12.8 (wooded watershed) to 29.7 mg/L (unimproved pasture watershed). Baseflow concentrations ranged from 5.5 (unimproved pasture watershed) to 8.5 mg/L (watershed with more than 50% pasture and meadow). Concentrations of TOC in developed springs under rotational pastures in east-central Ohio ranged from 2 to 8 mg/L (Owens et al., 1998). Jardine et al. (1990) investigated DOC transport through a forested Tennessee hillslope on the basis of single rainfall events. In shallow subsurface flow (up to 1-m depth), DOC ranged up to nearly 11 mg/L, and the DOC concentrations reported for lower subsurface flow were less than 2 mg/L.

This chapter provides an overview of the research conducted at the North Appalachian Experimental Watershed that focuses on C that is moved from agricultural fields with sediment, and C that moves in water through the soil profile.

18.2 CARBON LOSS WITH SEDIMENT

18.2.1 Carbon Concentration in Sediment

Surface runoff from the watersheds was automatically measured with H-flumes and sampled with Coshocton wheels (Brakensiek et al., 1979) that were modified to continuously deliver a proportional sample of runoff water and suspended sediment during each runoff event. Sediment was occasionally deposited in the flume floor and flume approach. This sediment was also collected.

Sediment losses occurred almost every year from each of six small watersheds (Table 18.1) during the 15-year study period, but there was sufficient sediment for C analyses

Table 18.1 Tillage Treatments and Selected Landscape and Soil Characteristics of Six Watersheds

Watershed #	Tillage	Area (ha)	Average Slope (%)	Dominant Soil[a]
WS 113	No till[b]	0.59	11	Coshocton SiL
WS 118	No till[b]	0.79	10	Coshocton SiL
WS 109	Chisel plow[b]	0.68	13	Rayne SiL
WS 123	Chisel plow[b]	0.55	7	Keene SiL
WS 115	Paraplow/disk[c]	0.65	7	Coshocton SiL
WS 127	Paraplow/disk[c]	0.68	9	Coshocton SiL

[a] Rayne: fine-loamy, mixed, mesic Typic Hapludult. Keene: fine-silty, mixed, mesic Aquic Hapludalf. Coshocton: fine-loamy, mixed, mesic Aquultic Hapludalf. SiL: sil loam. Soils were residual and formed from sandstone and shale bedrock. They are moderately well-drained and well-drained silt loams. For more detail on soils, see Edwards et al. (1993) and Kelley et al. (1975).

[b] Two-year corn/soybean-rye (*Zea mays* L./*Glycine max* (L.) Merr. – *Secale cereale* L.) rotation for 15 years beginning in 1984 (Edwards et al., 1993). Rye was sown in the soybeans as a winter cover crop.

[c] Two-year corn/soybean-rye rotation for 6 years beginning in 1984. In 1990, they were placed in a 3-year, reduced chemical input rotation (corn/soybean/wheat [*Triticum aestivum* L.]-clover [*Trifolium pretense* L.]).

Source: From Shipitalo, M.J., and W.M. Edwards. 1998. *Soil Tillage Res.*, 46:1–12. With permission.

Table 18.2 Weighted Average C Concentration on
Sediments Collected with the Coshocton Wheel (g/kg)[a,b]

	No Tillage	Chisel Plow	Paraplow (1984–1989)	Disk (1990–1998)
Corn years	29.6[a]	21.3[a]	18.9[a]	21.9[a]
Soybean years	24.1[b]	20.2[a]	20.5[a]	21.4[a]
Wheat/clover years	NA[c]	NA	NA	29.0[b]
Overall[d]	26.1x	20.7y	20.3y	22.0xy

[a] Annual cycle was May through April.
[b] Means within a column followed by the same letter (a, b) are not significantly different at the 0.05 level. Statistically significant differences were determined by general linear model procedures (SAS Institute, 1985).
[c] Not available: wheat/clover only grown in the disk treatment.
[d] Means in this row followed by the same letter (x, y) are not significantly different at the 0.05 level.
Source: From Owens, L.B., R.W. Malone, D.L. Hothem, G.C. Starr, and R. Lal. 2002a. *Soil Tillage Res.*, 67:65–73. With permission.

(determined by a dry-combustion method) in less than half of the watershed years. Average C concentration in sediments passing through the H-flume (wheel sediment) usually did not differ significantly among crops within the rotations (Table 18.2). Although sediment C concentration from years in soybeans was usually lower than from years in corn, this difference was significant only in the no-till watersheds. Moreover, the weighted average of C concentration in sediment under no-till was significantly greater than the C concentration in sediments from the other tillage practices (Table 18.2). The higher C concentration in the no-till sediments reflects the higher C concentration in the topsoil of the no-till watersheds than in the tilled watersheds.

Wheel sediments were enriched with C compared with flume floor sediments. Flume floor and flume approach sediments, however, only accounted for approximately 18% of the total sediment leaving the watersheds. The weighted-average C concentration in these sediments was 22.5 g/kg for no-till watersheds, and 17 g/kg for the chisel-plow watershed. Wheel sediments were also enriched compared with the surface soil. The enrichment ratio (ER) for sediments from the no-till and chisel-plow watersheds to the 0- to 2.5-cm soil layer was 1.5. The lowest ER (1.2) was for sediments from the paraplow/disk watersheds. In the disked watersheds, rills developed and allowed some erosion to occur from depths greater than 2.5 cm, where the C concentrations were lower. Thus, the overall C concentration in the sediments was probably lower than if the sediments had come only from the 0- to 2.5-cm layer (Owens et al., 2002a).

18.2.2   Carbon Transported via Sediments

Although differences and trends in sediment transport can be noted based on tillage practice or crop year (Table 18.3), the year-to-year variation was so great that there were not significant differences among the values reported. The standard deviation was almost always greater than the mean for each practice and year. This was also true for the transport of C in the sediments. The lowest annual sediment transport occurred with no till and the highest with disk. With the exception of the chisel-plow practice, average annual total sediment losses (combined wheel and flume floor sediment) were 2.8 to 8.5 times greater with soybeans than corn. Total sediment C transport had similar ratios (2.7 to 7.7 times) (Owens et al., 2002a).

Large events in June 1990 resulted in a monthly average sediment transport and carbon transport much greater than any other month. In spite of the large monthly differences, there were no significant differences among monthly averages because of the great variability in event sizes and occurrences. A few large events usually transport most of the sediment (Edwards and Owens, 1991). The monthly distribution of C transport closely followed the pattern of sediment transport because the monthly C concentrations did not vary greatly.

18.3   CARBON LOSS IN SUBSURFACE FLOW

Seven monolith lysimeters, each having a surface area of 8.1 m^2 and a depth of 2.4 m, were used to study TOC moving through the soil profile under a corn/soybean–rye rotation

Table 18.3 Average Annual Wheel and Total (Wheel Plus Flume Floor) Sediment and C Transport by Crop and Tillage Practice

Management	Wheel Sediment (kg/ha)		Total Sediment (kg/ha)	
	Sediment	C	Sediment	C
No Till (15 years)				
Corn	236 ±245	6.9 ±7.2	262 ±271	7.5 ±7.8
Soybean	682 ±789	17.4 ±19.0	800 ±850	20.0 ±20.2
Overall	459 ±617	12.1 ±15.1	531 ±686	13.8 ±16.6
Chisel-Plow (15 years)				
Corn	974 ±2141	17.4 ±34.0	1258 ±2954	22.0 ±46.6
Soybean	356 ±584	7.2 ±10.1	399 ±640	8.0 ±11.2
Overall	665 ±1573	12.3 ±25.2	828 ±2180	15.0 ±34.6
Paraplow (6 years)				
Corn	273 ±416	5.5 ±8.4	317 ±463	6.4 ±9.2
Soybean	737 ±830	16.3 ±21.0	883 ±937	19.1 ±23.0
Overall	505 ±672	10.9 ±16.2	600 ±791	12.7 ±18.7
Disk (9 years)				
Corn	302 ±444	6.8 ±9.7	348 ±539	7.8 ±11.8
Soybean	2193 ±2043	43.9 ±36.2	2956 ±2778	60.1 ±49.6
Wheat/clover	127 ±128	3.5 ±3.7	151 ±151	4.1 ±4.3
Overall	874 ±1529	18.1 ±28.4	1152 ±2076	24.0 ±39.0

Source: From Owens, L.B., R.W. Malone, D.L. Hothem, G.C. Starr, and R. Lal. 2002a. *Soil Tillage Res.*, 67:65–73. With permission.

with a chisel-plow tillage treatment. Developed springs were used to study TOC movement through well-drained and moderately well-drained, residual silt loam soils under pasture. Low TOC concentrations were found in both studies, with annual averages ranging between 0.5 and 3.2 mg/L (Table 18.4). These values are similar to the levels of DOC reported in New England streams and tributaries (David et al., 1992; Likens et al., 1977). With the exception of stormflow situations, much of the stream flow would be from subsurface return flow. Therefore, the water would be "filtered" much

Table 18.4   Flow-Weighted Average[a] Annual Total Organic Carbon Concentrations and Flux in Subsurface Flow

	Minimum	Maximum	Mean	Standard Deviation	Coefficient of Variation (%)
Flow-Weighted Concentrations (mg/L)					
Corn/soybean–rye rotations					
Y102 ABC[b]	0.5	3.2	1.3	0.6	44
Y103 ABCD[b]	0.5	3.0	1.4	0.5	37
All lysimeters[b]	0.5	3.2	1.3	0.5	40
Rotationally grazed pastures					
B Area[c]	1.0	3.0	1.6	0.5	30
D–E Area[d]	1.0	3.1	2.1	0.6	26
All watersheds	1.0	3.1	1.9	0.6	31
Total Organic Carbon Flux (kg/ha)					
Corn/soybean–rye rotations					
Y102 ABC[b]	1.6	12.4	4.6	2.4	52
Y103 ABCD[b]	1.5	11.9	4.5	2.4	52
All lysimeters[b]	1.5	12.4	4.5	2.4	52
Rotationally grazed pastures					
B Area[c]	1.5	11.9	4.4	2.1	47
D–E Area[d]	1.2	14.9	5.0	3.4	67
All watersheds	1.2	4.9	4.7	2.8	60

[a] Average for 10 years.
[b] Constructed without disturbing the soil profile and underlying fractured shale bedrock (Harrold and Dreibelbis, 1958).
[c] Forage was fertilized orchardgrass (*Dactylis glomerata* L.).
[d] Forage was legumes, principally alfalfa (*Medicago sativa* L.) mixed with orchardgrass.
Source: From Owens, L.B., R.W. Malone, D.L. Hothem, G.C. Starr, and R. Lal. 2002a. *Soil Tillage Res.*, 67:65–73. With permission.

like the lysimeter percolate and water from springflow developments.

    There were no TOC concentration trends during the 10-year study period with either the lysimeter percolate or springflow. This indicates that there were no differences in TOC concentrations based on the management practices of corn/soybean–rye rotations and rotationally grazed pastures. The TOC flux closely followed the quantities of percolate and springflow, indicating that the amount of subsurface water movement was much more important in determining TOC flux than TOC concentration (Owens et al., 2002b).

Average annual TOC concentrations in lysimeter percolate had greater coefficients of variation (Table 18.4) than TOC concentrations in springflow, even though the mean concentrations by management practice were within a single standard deviation. The greater variation with the lysimeters probably was influenced by the shorter flow paths, which would respond more quickly to climate and treatment variations. Coefficients of variation for the average annual TOC flux were greater than for concentrations.

The seasonal relationships between management practices and TOC concentrations and TOC transport were consistent with the observations for annual comparisons. There were some monthly TOC concentration variations in the subsurface flow from the corn/soybean-rye rotations and from the rotationally grazed pastures, but there were no seasonal variations. Coefficients of variations of the average monthly TOC concentrations for the lysimeters and watersheds (Table 18.5) were similar to the respective coefficients of variations for the annual averages. Plots of the average monthly flow and TOC transport show great similarity. This further indicates that the amount of flow is much more dominant in determining TOC transport than is TOC concentration.

18.4 IMPLICATIONS OF THESE C LOSS PATHWAYS FOR MANAGEMENT

An important aspect of relating management practices to carbon sequestration is quantifying the pathways of C loss from soil as well as measuring the increase in C storage and the net amount of C stored. The above review showed great variability in the amounts of sediment and C transported within a management treatment. Nevertheless, there was no

Table 18.5 Flow-Weighted Average[a] Monthly Total Organic Carbon Concentrations in Subsurface Flow

Minimum (mg/L)	Maximum (mg/L)	Mean (mg/L)	Standard Deviation (mg/L)	Coefficient of Variation (%)
Corn/Soybean——Rye Rotations				
Y102 ABC	0.5	4.5	1.6	0.8	51
Y103 ABCD	0.7	3.2	1.4	0.4	31
All lysimeters	0.5	4.5	1.5	0.6	42
Rotationally Grazed Pastures				
B Area	1.0	3.2	1.7	0.5	30
D–E Area	1.0	3.5	2.3	0.5	24
All watersheds	1.0	3.5	2.0	0.6	31

[a] Average for 10 years.
Notes: SD = Standard Deviation; CV = Coefficient of Variation.
Source: From Owens, L.B., R.W. Malone, D.L. Hothem, G.C. Starr, and R. Lal. 2002a. *Soil Tillage Res.*, 67:65–73. With permission.

significance with C concentration on sediments, except with no till, among the different tillage management practices or between corn and soybeans (Table 18.2). Thus, greater C losses occurred with greater sediment losses, i.e., sediment and C losses were greater with disk than with no-till (Table 18.3). This means that land management should be selected for soil loss reduction and not for reduction of C concentration on sediment. Reducing soil loss reduces C loss.

Carbon concentrations are much lower in subsurface flow than on sediments (Tables 18.2 and 18.4), and they did not differ greatly among the management practices studied. Likewise, C losses via subsurface flow are considerably less than via sediment (Tables 18.3 and 18.4). Average annual C losses via sediment were 1.4 to 4.8 and 1.7 to 13 times the average annual C losses via subsurface flow for corn and soybeans, respectively. These small losses of C in subsurface flow indicate that there are more important pathways of C loss to be addressed by management practices. Thus, reduction of C loss via subsurface flow should not be a major consideration when selecting land use practices for C sequestration.

18.5 CONCLUSIONS

Carbon concentrations on sediment lost from watersheds with different tillage practices (e.g., no till, chisel-plow, paraplow, and disk) varied little with time. Tillage practices and weather had major impacts on soil loss from field scale watersheds; however, they had much less impact on sediment C concentration. Management systems that control sediment loss have greater impact on reducing C loss via erosion than those that might change sediment C concentration.

Annual concentrations of TOC in lysimeter percolate from a corn/soybean–rye rotation and in springflow from pastured watersheds ranged between 0.5 and 3.2 mg/L. These concentrations are low in the range of published TOC concentrations in stream flow. No major trends in TOC concentrations were observed during the 10 years for either management treatment. Annual average losses of TOC were similar for both management practices, ranging from 1.2 to 14.9 kg/ha. TOC leaching losses are small compared with other pathways.

REFERENCES

Alvarez, R., R.A. Diaz, N. Barbero, O.J. Santanatoglia, and L. Blotta. 1995. Soil organic carbon, microbial biomass and CO_2-C production from three tillage systems. *Soil Tillage Res.*, 33:17–28.

Ambassa-Kiki, R. and D. Nill. 1999. Effects of different land management techniques on selected topsoil properties of a forest Ferralsol. *Soil Tillage Res.*, 52:259–264.

Brakensiek, D.L., H.B. Osborn, and W.J. Rawls, Coords. 1979. *Field Manual for Research in Agricultural Hydrology*. Agriculture Handbook, vol. 224. U.S. Department of Agriculture, Washington, DC.

Buyanovsky, G.A. and G.H. Wagner. 1998. Changing role of cultivated land in the global carbon cycle. *Biol. Fertil. Soils*, 27:242–245.

Campbell, C.A., B.G. McConkey, R.P. Zentner, R.P. Dyck, F. Selles, and D. Curtin. 1995. Carbon sequestration in a brown Chernozem as affected by tillage and rotation. *Can. J. Soil Sci.,* 75:449–458.

Campbell, C.A., B.G. McConkey, R.P. Zentner, F. Selles, D. Curtin. 1996. Long-term effects of tillage and crop rotations on soil organic C and total N in a clay soil in southwestern Saskatchewan. *Can. J. Soil Sci.,* 76:395–401.

Campbell, C.A. and B.G. Zentner. 1993. Soil organic matter as influenced by crop rotations and fertilization. *Soil Sci. Soc. Am. J.,* 56:1034–1040.

David, M.B., G.F. Vance, and J.S. Kahl. 1992. Chemistry of dissolved organic carbon and organic acids in two streams draining forested watersheds. *Water Resources Res.,* 28:389–396.

Davidson, E.A. and I.L. Ackerman. 1993. Changes in soil carbon inventories following cultivation of previously untilled soils. *Biogeochemistry,* 20:161–193.

Doran, J.W. 1980. Soil microbial and biochemical changes associated with reduced tillage. *Soil Sci. Soc. Am. J.,* 44:765–771.

Doran, J.W. 1987. Microbial biomass and mineralizable nitrogen distributions in no tillage and plowed fields. *Biol. Fertil. Soils,* 5:68–75.

Edwards, W.M. and L.B. Owens. 1991. Large storm effects on total soil erosion. *J. Soil Water Conserv.,* 46:75–78.

Edwards, W.M., G.B. Triplett, D.M. Van Doren, L.B. Owens, C.E. Redmond, and W.A. Dick. 1993. Tillage studies with a corn–soybean rotation: hydrology and sediment loss. *Soil Sci. Soc. Am. J.,* 57:1051–1055.

Eghball, B., L.N. Mielke, D.L. McCallister, and J.W. Doran. 1994. Distribution of organic and inorganic nitrogen in a soil under various tillage and crop sequences. *J. Soil Water Conserv.,* 49:201–205.

Gregorich, E.G., K.J. Greer, D.W. Anderson, and B.C. Lang. 1998. Carbon distribution and losses: erosion and deposition effects. *Soil Tillage Res.,* 47:291–302.

Halvorson, A.D., B.J. Wienhhold, and A.L. Black. 2002. Tillage, nitrogen, and cropping system effects on soil carbon sequestration. *Soil Sci. Soc. Am. J.,* 66:906–912.

Harrold, L.L., and F.R. Dreibelbis. 1958. *Evaluation of Agricultural Hydrology by Monolith Lysimeters 1944–55.* U.S. Department of Agriculture Technical Bulletin 1179. U.S. Government Printing Office, Washington, DC.

Hunt, P.G., D.L. Karlen, T.A. Matheny, and V.L. Quisenberry. 1996. Changes in carbon content of Norfolk loamy sand after 14 years of conservation or conventional tillage. *J. Soil Water Conserv.,* 51:255–258.

Hussain, I., K.R. Olson, and S.A. Ebelhar. 1999. Long-term tillage effects on soil chemical properties and organic matter fractions. *Soil Sci. Soc. Am. J.,* 63:1335–1341.

Ismail, I., R.L. Blevins, and W.W. Frye. 1994. Long-term no-tillage effects on soil properties and continuous corn yields. *Soil Sci. Soc. Am. J.,* 58:193–198.

Jardine, P.M., G.V. Wilson, J.F. McCarthy, R.J. Luxmoore, D.L. Taylor, and L.W. Zelazny. 1990. Hydrogeochemical processes controlling the transport of dissolved organic carbon through a forested hillslope. *J. Contaminant Hydrol.,* 6:3–19.

Kaihura, F.B.S., I.K. Kullaya, M. Kilasara, J.B. Aune, B.R. Singh, and R. Lal. 1999. Soil quality effects of accelerated erosion and management systems in three eco-regions of Tanzania. *Soil Tillage Res.,* 53:59–70.

Karlen, D.L., N.C. Wollenhaupt, D.C. Erbach, E.C. Berry, J.B. Swan, N.S. Eash, and J.L. Jordahl. 1994. Long-term tillage effects on soil quality. *Soil Tillage Res.,* 32:313–327.

Kelley, G.E., W.M. Edwards, L.L. Harrold, and J.L. McGuinness. 1975. *Soils of the North Appalachian Experimental Watershed.* U.S. Department of Agriculture Miscellaneous Publication 1296. U.S. Government Printing Office, Washington, DC.

Likens, G.E., F.H. Bormann, R.S. Pierce, J.S. Eaton, and N.M. Johnson. 1977. *Biogeochemistry of a Forested Ecosystem.* Springer-Verlag. New York.

Massey, H.F. and M.L. Jackson. 1952. Selective erosion of soil fertility constituents. *Soil Sci. Soc. Proc.,* 16:353–356.

McCallister, D.L. and W.L. Chen. 2000. Organic carbon quantity and forms as influenced by tillage and cropping sequence. *Commn. Soil Sci. Plant Anal.*, 31:465–479.

McCarty, G.W. and J.J. Meisinger. 1997. Effects of N fertilizer treatments on biologically active N pools in soils under plow and no tillage. *Biol. Fertil. Soils,* 24:406–412.

McCarty, G.W., J.J. Meisinger, and J.M.M. Jennsikens. 1995. Relationships between total-N, biomass-N, and active-N in soil under different tillage and N fertilizer treatments. *Soil Biol. Biochem.*, 27:1245–1250.

Nyborg, M., E.D. Solberg, S.S. Malhi, and R.C. Izaurralde. 1995. Fertilizer N, crop residue, and tillage alter soil C and N content in a decade. In R. Lal, et al., Eds. *Advances in Soil Science: Soil Management and Greenhouse Effect*, Lewis Publishers/CRC Press, Boca Raton, FL, pp.93–100.

Owens, L.B., W.M. Edwards, and R.W. Van Keuren. 1991. Baseflow and stormflow transport of nutrients from mixed agricultural watersheds. *J. Environ. Qual.*, 20:407–414.

Owens, L.B., R.W. Malone, D.L. Hothem, G.C. Starr, and R. Lal. 2002a. Sediment carbon concentration and transport from small watersheds under various conservation tillage practices. *Soil Tillage Res.*, 67:65–73.

Owens, L.B., G.C. Starr, and D.R. Lightell. 2002b. Total organic carbon losses in subsurface flow under two management practices. *J. Soil Water Conserv.*, 57:74–81.

Owens, L.B., R.W. Van Keuren, and W.M. Edwards. 1998. Budgets of non-nitrogen nutrients in a high fertility pasture system. *Agric. Ecosyst. Environ.*, 70:7–18.

Reicosky, D.C., W.D. Kemper, G.W. Langdale, C.L. Douglas Jr., and P.E. Rasmussen. 1995. Soil organic matter changes resulting from tillage and biomass production. *J. Soil Water Conserv.*, 50:253–261.

SAS Institute. 1985. *SAS/STAT Guide for Personal Computers*, 6th ed. SAS Institute, Cary, NC.

Shipitalo, M.J. and W.M. Edwards. 1998. Runoff and erosion control with conservation tillage and reduced-input practices on cropped watersheds. *Soil Tillage Res.*, 46:1–12.

Wan, Y. and S.A. El-Swaify. 1997. Flow-induced transport and enrichment of erosional sediment from a well-aggregated and uniformly textured Oxisol. *Geoderma,* 75:251–265.

Zobisch, M.A., C. Richter, B. Heiligtag, and R. Schlott. 1995. Nutrient losses from cropland in the Central Highlands of Kenya due to surface runoff and soil erosion. *Soil Tillage Res.,* 33:109–116.

19

Measuring and Monitoring Soil Carbon Sequestration at the Project Level

R. CÉSAR IZAURRALDE

CONTENTS

19.1 INTRODUCTION

The possibility of using improved farming practices to mitigate the increase in atmospheric CO_2 through soil carbon sequestration (SCS) reached international consensus during meetings leading to the Intergovernmental Panel on Climage Change (IPCC) First Assessment Report (1990). The rationale for this consensus was that by fostering the adoption of improved farming practices, it would be possible not only to enhance agricultural productivity but also to make soils act as sinks for atmospheric CO_2. A significant amount of scientific and practical evidence has accumulated to date in support of this consensus (Kern and Johnson, 1993; Lal et al., 1995a, 1995b, 1998a, 1998b; Paul et al., 1997; Powlson et al., 1996). In chapter 23 of the IPCC Second Assessment Report, Cole et al. (1996) reported estimates of SCS potential that could be achieved during a 50- to 100-year span. Under a program of global adoption, the international team of authors estimated that approximately 40 Pg C could be removed from the atmosphere via SCS at rates of 0.4 to 0.8 Pg C y^{-1} depending on the length of sequestration considered. This global estimate was made based on the assumption that it was possible to recover about two-thirds of the 55 Pg C historically lost from the soil organic carbon (SOC) pool by implementing land use conversions (e.g., conversion of marginal agricultural land to permanent vegetation) and improved management practices (e.g., no-tillage agriculture).

When evaluated within a global environmental and energy framework that considered a portfolio of technologies (e.g., use of biofuels, improved energy efficiency, hydrogen production), SCS was found to be competitive particularly as an early starter of climate change mitigation technologies (Rosenberg and Izaurralde, 2001). By the mid-1990s, energy industries in western Canada became interested in analyzing the extent to which SCS could be used to offset industrial CO_2 emissions. Thus, a group of Canadian energy industries sponsored, together with the federal and the provincial government of Saskatchewan, the Prairie Soil Carbon Balance Project (PSCBP), which successfully documented — province wide and at field scale — changes in SOC after 3 years of adoption of direct seeding (no tillage) practices (McConkey et al., 2000). The program was successful in demonstrating that it was possible to detect, with statistical power, changes in SOC as small as 1.2 Mg ha^{-1} occurring after the first 3 years of implementing a SCS practice.

Since SOC is a direct indicator of soil quality and fertility, soil scientists and agrologists have long been studying, monitoring, and mapping SOC under plot and field conditions (e.g., Jenkinson, 1988; Stevenson and Cole, 1999). However, if large-scale soil C offset projects were to be implemented, these would require methodologies to measure and monitor SOC changes that are not only applicable at a relatively low cost but also accurate enough to satisfy the requirements of an emerging carbon trading market. Thus, pilot and feasibility projects on carbon sequestration are being developed worldwide to learn about scientific and operational aspects of their implementation. The Food and Agriculture Organization (*www.fao.org / ag / agl / agll / carbonsequestration*) and the World Resources Institute (*http:/ / climate.wri.org / sequestration.cfm*) maintain online databases with descriptions of ongoing agricultural and forestry carbon sequestration projects. These databases reveal the existence of relatively few field pilot projects that include SCS as a main objective.

There are economic and technical facets that need to be resolved in order to advance the field of SCS from a feasible to a mature mitigation technology. Several economic aspects of SCS practices have to be evaluated, especially those dealing with their cost-effectiveness relative to other C sequestration practices (e.g., afforestation, reforestation). Marland et al. (2001) analyzed in detail the various policy and economic issues that should be evaluated in order to determine the success of SCS programs. McCarl and Schneider (2001) compared the economic competitiveness of various forms of greenhouse gas mitigation strategies including SCS, afforestation, and biofuels offset. These authors found SCS to be competitive with other mitigation options at low carbon prices.

As with any other natural variable, the detection of changes in SOC is often associated with uncertainty due to the interactive effects of many factors controlling its dynamic such as amount and quality of C input to soil, climatic, and edaphic conditions, as well as land use and soil management practices. Another concern arises from the fact that soils have a finite capacity to store SOC and, eventually, the C stored via SCS practices could be re-released to the atmosphere when policy and market incentives or personal decisions happen to favor the application of non–C-sequestering practices (e.g., return to an intensive tillage regime, conversion of marginal agricultural land from perennial vegetation to crop production).

These and other concerns will have to be addressed during the development of methods to measure and monitor SCS at the project level. It is likely, however, that if SCS practices were to be applied at a global scale, there would be a variety of methods adapted to local conditions. Thus, international efforts will be required to ensure the comparability of results across environments and methodologies. Regardless of the methods adopted, however, these are likely to have three components, namely, direct measurements, computer modeling, and remotely sensed monitoring. The objective of this chapter is to review advances in methodologies to measure SCS at the project level. We begin with a short discussion of the conceptual design proposed by Post et al. (1999) of a plan for monitoring and verifying SCS at regional scales and then follow this conceptual design with a discussion on methodological advances for measuring, modeling, and monitoring SCS at the project level.

19.2 CONCEPTUAL DESIGN FOR MONITORING AND VERIFYING SOIL CARBON SEQUESTRATION AT THE PROJECT LEVEL

Post et al. (1999) outlined the elements of a plan for monitoring and verifying SCS at regional scales. The plan had four basic elements: (1) selection of landscape units suitable for measuring and monitoring SCS, (2) development of measurement protocols, (3) utilization of remote sensing information and simulation models, and (4) development of a methodology to scale up results to represent the entire SCS project. The selection of landscape units for measuring and monitoring SCS will depend on the general responsiveness of a region to SCS practices such as climate and soil properties, management history, and availability of research data. Further, an initial estimation is required on how many farms could be involved in a project. For example, Izaurralde et al. (1998) estimated that a project promising to deliver 500 Tg C in the form of SCS would require the participation of about 1000 farms. This assumed a conservative SCS rate of 1 Mg C ha^{-1} and an average farm size of 500 ha. This element will require the participation and interaction of regional agronomists, soil consultants, and farmer organizations who as a group can decide on the selection of the best pilot areas and the extent to which the results can be extrapolated.

Another element of the plan deals with the development of methods for detecting changes in SOC that would take place during several years after the implementation of an SCS practice. Changes in SOC can be detected as changes in stocks or fluxes. Changes in stocks can be calculated from direct measurements of SOC stocks present when the SCS practice is implemented and a number of years after its implementation. Eddy covariance methods can also be used to determine net ecosystem exchange, which is the difference between net primary productivity and soil respiration, and thus provide an indirect but accurate estimate of SOC change (Baldocchi and Wilson, 2001). Eddy covariance methods are attractive because they would provide estimates of SCS over fields, but their use will remain experimental until they prove their applicability for these types of projects. A methodology for direct measurement of SOC changes should include the following: (1) selection of sampling sites; (2) soil sampling protocols (depth and volume of sampling); (3) ancillary soil measurements (bulk density); (4) sample treatment and analyses; (5) ancillary field measurements (plant biomass and yield, plant residues, and soil erosion); and (6) calculation of results including adjustments for marginal costs of SCS. Details and examples for some of these components will be described in the next section.

Satellite remote sensing and simulation models have key roles in monitoring and predicting SCS at field scale and for extrapolating SCS from field to regional scales (Post et al., 1999). Remote sensing methods can be useful for monitoring net primary productivity, estimating crop yields and leaf area index (LAI), and — in combination with ground-based data — deriving maps of land cover and use history. Currently, however, it is not possible to directly estimate SOC stocks unless ground-based information on SOC values is available (Chen et al., 2000).

Simulation models have a key role to play in understanding and predicting SCS at project and regional scales. Currently, there are numerous soil and ecosystem models capable of describing SOC trajectories in response to varying climate, soil, and management conditions. McGill (1996) reviewed ten soil organic matter (SOM) models, and classified them in terms of their treatment of environmental (weather) drivers; temporal scale (day, month); vertical distribution (i.e., soil horizons); soil properties (clay content); and litter description (kinetic vs. biochemical). Subsequently, Smith et al. (1997) compared the performance of nine of them against 12 data sets from seven long-term experiments representing a range in land uses (grassland, cropland, woodland), climatic conditions within the temperate region, and treatments (nutrient sources and rotations). While no one model performed consistently better than the others across all data sets, a group of them (RothC, CANDY, DNDC, CENTURY, DAISY, and NCSOIL) had lower model errors than another group (SOMM, ITE, and Verberne). Well-tested models will be particularly useful not only for understanding interactions between biophysical and management variables but also for projecting SCS over large areas. The issue of upscaling is very important because it requires the integration of information from a variety of sources such as field measurements, geographic information systems and associated databases, computer models, and remote sensing.

Table 19.1 characterizes a series of feasibility and pilot projects designed to evaluate SCS under a variety of climate, soil, and management conditions. The table also provides

Table 19.1 Examples of Feasibility and Pilot Projects on Soil Carbon Sequestration

Region	Reference	Land Use	Land Ownership	Scale	Land Management Change	Soil Sampling	Soil Carbon Analysis	Remote Sensing	Modeling	Scale-Up Area
Saskatchewan, Canada	McConkey (2000)	Cropland	Private	Field/region	Direct seeding/cropping intensification	Stratified random sampling	Dry combustion	No	CENTURY	4 to 5 million ha
Pacific Northwest, United States	Scholz (2004)	Cropland	Private	Field/region	Direct seeding/cropping intensification				CQUESTER	2600 ha
Oaxaca, Mexico	Vergara Sánchez et al. (2004)	Crop/natural fallow secondary forest	Private/communal	Watershed	Fruit tree intercrops with annual crops/conservation tillage	Gridded within fields	Dry combustion	SPOT	CENTURY, Roth-C	1 million ha
Pampas, Argentina	Casas (2003)	Cropland	Private	Field/region	Direct seeding	Paired sampling	Wet combustion	Landsat		3.3 million ha
Senegal	Tieszen (2004)	Agroforestry	Communal	Field	Nutrient management, N fixation agroforestry	Random, re-sampled	Wet and dry combustion	Yes	CENTURY	
Mali	Venteris (2004)	Agroforestry	Communal	Field	Tree conservation/ridge tillage	Random within fields	Wet and dry combustion, loss of ignition, infrared spectroscopy	Quick-Bird, Landsat	DSSAT-CENTURY, EPIC	
Kazakhstan	Tieszen (2004)	Cropland	State/private	Region	Agriculture to grassland		Carbon flux towers	AVHRR, Landsat	CENTURY	Northern Kazakhstan

Notes: SPOT = Système Probatoire pour l'Observation de la Terre; AVHRR = Advanced Very High Resolution Radiometer.

information about management practices, sampling techniques, analytical procedures, and the use of models and remote sensing tools. Details on soil sampling, soil analysis, and the use of simulation models and remote sensing for assessing SCS are discussed in the next section.

19.3 DETECTING AND PREDICTING SOIL ORGANIC CARBON CHANGES: SAMPLING, ANALYSIS, MODELS, AND REMOTE SENSING

19.3.1 Soil Sampling

Soil sampling for determination of SOC content represents the most direct way to determine whether a particular soil management or land use practice has caused a net change in SOC storage. Soils are three-dimensional natural bodies that vary in time and space in response to biophysical properties (i.e., climate, topography, parent material, and vegetation) and management (fertilization, tillage, land use change, etc.). Detailed soil surveys can greatly assist in the design of a sampling scheme. Once the general properties about the field to be sampled are obtained (soil series, soil phases, landscape characteristics, etc.), the next step consists of deciding the sampling plan. According to Petersen and Calvin (1996), the sampling plan should help select which units of the soil population are to be included in the sample. The best sampling plan would be the one that provides the maximum precision at a given cost or, conversely, the lowest cost under a specified precision and error. There are three sampling plans recommended: (1) simple random, (2) stratified random sample, and (3) systematic. In the simple random plan, each unit of the total n units being sampled from the population has an equal chance of being selected. A stratified random scheme would be more desirable when there are variations that can be predicted from knowledge about soil or landscape properties. This stratification would help reduce a source of variation of the sampling error. A systematic sampling plan occurs when soil samples are taken at regular distances from each other, either in one or two dimensions. In principle, systematic sampling would yield more precise results than the first two sampling schemes (Petersen and Calvin, 1996).

The number of samples (n) to be taken from a field depends on the variability in SOC content as well as on minimum difference that needs to be detected. For example, Izaurralde et al. (1998) used a one-tailed t test to calculate the number of soil samples needed to detect, with a 90% confidence, a 0.1% increase in SOC with a known variance of 3.3 (g C kg^{-1})^2. They calculated that for each representative parcel of land the baseline sampling would require 54 samples, while the final sampling would require another 54 samples, a large number indeed. Similar calculations were carried out by Garten and Wullschleger (1999), who evaluated the statistical power to detect significant SOC differences under switchgrass (*Panicum virgatum* L.). They calculated the smallest difference in SOC that could be detected between two means for a given variance, significance level, statistical power, and numbers of samples. They concluded that while differences of about 5 Mg SOC ha^{-1} were detectable with reasonable numbers of samples (n = 16) and good statistical power (1 − β = 0.90), the smallest difference in SOC inventories (1 Mg SOC ha^{-1}) would be detectable only with large numbers of samples (n > 100).

In order to reduce the number of samples required and to minimize soil variability, Ellert et al. (2001) proposed a high-resolution method to detect temporal changes in SOC storage by comparing the quantities from a sampling microsite (4 × 7 m) at two sampling times separated by periods of 4 to 8 years. In this method, one to six microsites are selected in such as way so as to represent the dominant soils found in fields ranging from 30 to 65 ha. Guidance for the location of the microsites is obtained from experienced pedologists. The location of each microsite is recorded by survey methods, including geographic positioning systems. The authors also made useful recommendations regarding core size and number, time of sampling, depth of sampling, and ancillary measurements. This methodology was successfully applied to the PSCBP in 1997 to 2000, and allowed for the statistically significant detection of SOC storage gains as small as 1.2 Mg ha^{-1}, only 3 years after the implementation of a SCS practice.

Upscaling point measurements of SOC storage to the field level requires confidence in the assumption that the properties of the point measurements, including their measurement errors, will hold across the area of prediction. This confidence has been growing by an increased understanding of the relationships of soils in the landscape. The spatial dependence of soil attributes, including SOC content, has been studied with a variety of techniques or tools, including soil and topographical surveys, geostatistical techniques, remotely sensed data interpretation, as well as ground and monitoring devices. Like other disciplines, soil science has greatly benefited from advances in computation and information technology (e.g., McBratney et al., 2003). A few examples of these approaches follow.

Pennock et al. (1987) proposed a segmentation procedure to describe landscapes into functional units (i.e., landform segments such as shoulder, backslope, footslope, and depression). Pennock and Corré (2001) used it to study the comparative effects of cultivation on soil distribution and SOC storage, and to understand the main landscape features controlling soil emissions of N$_2$O. This approach was used in the PSCBP to help delineate the sampling areas for monitoring SOC changes. MacMillan et al. (2000) expanded on Pennock's approach and developed a model, which based on digital elevation models (DEMs) and fuzzy rules, could identify up to 15 morphologically defined landform facets. A consolidation in the number of landforms can be obtained to provide units at a farm field scale that are relevant for benchmark soil testing, application of simulation models, and precision farming.

Geostatistical methods are being increasingly used to predict soil attributes. Odeh et al. (1994) compared various interpolation methods (e.g., multilinear regression, kriging, co-kriging, and regression kriging) in their ability to predict soil properties from landform attributes derived from a DEM. The two regression-kriging procedures tested performed best, and thus showed promise for predicting sparsely located soil properties from dense observations of landform attributes derived from DEM data. Triantafilis et al. (2001) had success in using regression kriging to predict soil salinity in cotton (*Gossipium hirsutum* L.) fields with electromagnetic induction data. They attributed the success of the method to the incorporation of regression residuals within the kriging system. Hengl et al. (2004) tested a framework based on regression-kriging to predict SOM, soil pH, and topsoil depth from 135 soil profile observations from the Croatian national survey. These research results are promising, as they anticipate the possibility of implementing these algorithms in a GIS, thus enabling the interpolation of soil profile data from existing data sets (Hengl et al., 2004). The challenge remains, however, of developing rapid methods to accurately estimate SOC stocks in space and time (including uncertainties) at a relatively low cost.

19.3.2 Bulk Density

Soil bulk density (ρ_b, Mg m^{-3}) is the ratio of the mass of dry solids to a bulk volume of soil (Blake and Hartge, 1986). Its determination is essential to calculate the mass of soil organic carbon (SOC_m, Mg C m^{-3}) from SOC concentration (SOC_c, Mg C Mg^{-1}):

$$SOC_m = SOC_c \times \rho_b \qquad (19.1)$$

Although ρ_b is a relatively straightforward measurement, its evaluation can be subject to errors. Blake and Hartge (1986) and Culley (1993) offer excellent descriptions of the various methods that can be used to determine ρ_b. In the extractive methods, a soil sample of known (core method) or unknown volume (clod and excavation methods) is extracted, dried, and weighed (Blake and Hartge, 1986). Bulk density can also be determined *in situ* with the use of gamma radiation methods (Blake and Hartge, 1986). Instrument cost and radiation hazard may limit the utilization of gamma radiation methods in carbon sequestration projects.

For determination of ρ_b at various depths, which will be the case for carbon sequestration projects, Blake and Hartge (1986) recommend the use of hydraulically driven probes mounted on pickup trucks, tractors, or other vehicles, but certainly, hand-driven samplers are appropriate as well. The obvious goal with any sampling method for determining ρ_b is to avoid compressing the soil in the confined space of the sampler. Challenges are encountered when trying to determine ρ_b in soils containing coarse fragments, soils with large swell-shrink capacity, or high organic matter content (Lal and Kimble, 2001). Each of these challenges must be answered with specific solutions. Lal and Kimble (2001) briefly review these and other cases and recommend solutions. For example, the excavation method might work best for determining ρ_b in soils containing significant amounts of coarse fragments or soils with high organic matter content. The clod method might be the best method for soils that develop large cracks upon drying.

Can ρ_b be estimated by *in situ* measurements other than the gamma radiation probe? Time domain reflectometry (TDR), a technique originally designed for detecting failures in coaxial transmission lines, was first applied in soil science to measure soil water content (Topp et al., 1980). Theory and applications for TDR technology have expanded quickly since then as a way to measure mass and energy in soil (Topp and Reynolds, 1998). Ren et al. (2003) used a thermo-TDR probe to make simultaneous field determinations of soil water content, temperature, electrical conductivity, thermal conductivity, thermal diffusivity, and volumetric heat capacity. Knowledge of volumetric heat capacity (ρc) and soil water content (η) further allowed them to calculate other soil physical parameters such as ρ_b, air-filled porosity, and degree of saturation. They calculated ρ_b, as in Ochsner et al. (2001):

$$\rho_b = \frac{\rho c - \rho_w c_w \theta}{c_s} \qquad (19.2)$$

where c_s is the specific heat capacity of soil solids (kJ kg^{-1} K^{-1}), ρ_w is the density of water (kg m^{-3}), and c_w is the specific heat of water (kJ kg^{-1} K^{-1}). They tested their procedure in the laboratory with six column-packed soils ranging in texture from sand to silty clay loam with ρ_b ranging from 0.85 to 1.52

Mg m^{-3}. The ρ_b predicted with the thermo-TDR was able to explain slightly more than half of the variation in measured ρ_b, which suggests a method that, when improved, could deliver rapid measurements of ρ_b in the field.

Soil bulk density is a dynamic property; its value changes in response to applied pressure, soil water content, and SOM content. Up to a 20% change in ρ_b can occur with changes in soil water potential from 0.03 to 1.5 MPa (Lal and Kimble, 2001). Reporting ρ_b at standardized soil water content of 0.03 MPa is recommended. SOM content has a strong effect on ρ_b. Adams (1973) developed an equation to estimate ρ_b:

$$\rho_b = \frac{100}{\dfrac{\%OM}{0.244} + \dfrac{100 - \%OM}{\rho_m}} \tag{19.3}$$

where $\%OM$ is percent SOM, ρ_m is mineral bulk density (Mg m^{-3}), and the value 0.244 is the bulk density of organic matter (Mg m^{-3}). The bulk density of organic matter is fairly constant. However, the formula is difficult to apply because ρ_m is not usually known. Mann (1986) rearranged Adams's equation to calculate ρ_m from 121 pairs of soil samples with known values of SOM and ρ_b.

The Adams equation is difficult to solve directly because it has two unknowns (ρ_b, ρ_m). In principle, ρ_b could be estimated from knowledge of soil texture, soil particle density (ρ_s), and the packing arrangement of mineral particles. Here, a simple method is proposed for estimation of ρ_b based on soil texture, ρ_s, and packing arrangement information. Soil particle density is usually assumed to be 2.65 Mg m^{-3}, but there are slight variations depending on the textural composition. While the sand fraction has a ρ_s of 2.65 Mg m^{-3}, the clay and silt fractions have ρ_s of about 2.78 Mg m^{-3}.

If the fractional values of sand ($sand_f$), silt ($silt_f$), and clay ($clay_f$) are known, then ρ_s can be calculated as:

$$\rho_s = \rho_{sa} \times sand_f + \rho_{sc} \times (silt_f + clay_f) \tag{19.4}$$

where ρ_{sa} is the soil particle density of sand, while ρ_{sc} is the soil particle density of silt and clay. The next problem is to estimate a possible arrangement of these particles in the soil matrix.

Assuming a spherical shape for soil particles, there are various geometrical arrangements in which these particles can accommodate when packed. Sphere packing can be done in two and three dimensions, but only three-dimensional packing applies to soils. The densest packing is provided by the cubic close and the hexagonal close geometries (*http://mathworld.wolfram.com/SpherePacking.html*). These and other types of packing are defined by the packing density (η), which is the fraction of a volume filled by a given collection of solids. The packing density can be solved analytically for some types of arrangements; for others it cannot. For example, η for a cubic lattice arrangement is 0.524; it is 0.64 for a random arrangement, and 0.74 for a hexagonal close packing arrangement. (See *http://mathworld.wolfram.com/SpherePacking.html* for additional information on this topic.)

After selecting a value for η, mineral bulk density can be estimated as:

$$\rho_m = \rho_s \times \eta \qquad (19.5)$$

A modified Equation 19.1 is then used to calculate ρ_b at a given SOC concentration:

$$\rho_b = \frac{100}{\dfrac{SOC \times 1.724}{0.244} + \dfrac{100 - SOC \times 1.724}{\rho_m}}; \qquad (19.6)$$

$$0 \le SOC \le 58 \ (\text{g C kg}^{-1} \times 10^{-1})$$

where 1.724 is the conversion factor generally used to convert SOC into SOM. A theoretical example is shown in Figure 19.1 for three types of sphere packing (cubic lattice, random, and hexagonal). A test of the model is shown in Figure 19.2 against soil taxonomy data (U.S. Department of Agriculture, 1999). Gupta and Larson (1979) described a model that uses the same principles of sphere packing described here. This model is theoretically very good because it accounts for various particle sizes, but it requires complete information on soil fractions (very coarse sand, coarse sand, medium sand, fine sand, very fine sand, coarse silt, fine silt, and clay).

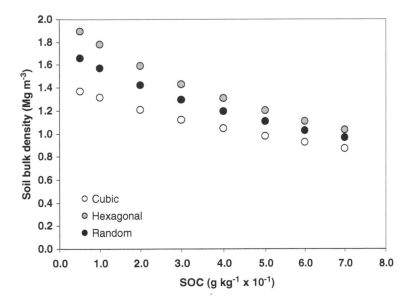

Figure 19.1 Soil bulk density estimated with three packing density models at different soil organic carbon concentrations and constant texture (0.33 clay, 0.33 silt, and 0.34 sand).

19.3.3 Analysis of Soil Organic Carbon

Soils contain carbon in two forms: organic and inorganic. Organic C is the main constituent of SOM. Inorganic C appears largely in carbonate minerals. Soil organic C is very dynamic, intensively reflects management influences, and exhibits turnover times that range from tens to hundreds of years (Six and Jastrow, 2002). Soil inorganic C, instead, is less responsive to management, and has greater turnover times than SOC. Thus, the emphasis in this section is to summarize methodologies to determine SOC concentration and to provide an update on emerging methodologies for quick and *in situ* determinations of soil C.

Detailed methodologies of organic C, inorganic C, and total C are provided by Nelson and Sommers (1996) and Tiessen and Moir (1993). Basically, SOC concentration can be determined by either wet or dry combustion. In the wet

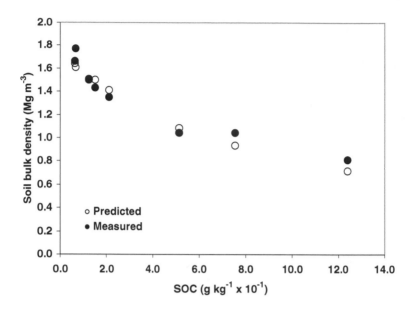

Figure 19.2 Predicted vs. measured soil bulk density using soil taxonomy data (Typic Haplustert, Typic Haplustalf, Typic Kandiudult, Pachic Argiustoll, Aeric Haplaquox, Typic Dystrudept, Typic Molliorthel, Eutric Fulvudand).

combustion procedure, a soil sample is treated with acid dichromate solution in a heated vessel, and then the CO_2 generated due to the oxidation of organic matter is evaluated either by titrimetric (indirect) or gravimetric (direct) methods. In titrimetric methods, the amount of organic C present in a soil sample is obtained by back titration of the unused dichromate with ferrous ammonium sulfate solution. This method is relatively easy to implement and has been used worldwide for many years. However, in this method, the digestion of organic matter is usually incomplete due to insufficient heating. Correction factors have been reported to correct for this incomplete oxidation, but these factors are soil dependent. Nelson and Sommers (1996) reported correction factors for 15 studies that averaged 1.24 ± 0.11, or a mean recovery of 81%. Improvements in recoveries are obtained with the wet oxidation method, with determination of CO_2 due to the higher digestion temperatures achieved.

In dry combustion, all forms of C (organic and inorganic, if present) are converted to CO_2 at high temperatures achieved in resistance (~1000°C) or induction furnaces (>1500°C) (Nelson and Sommers, 1996). Once generated, the CO_2 can be then assessed with a variety of spectrophotometric, volumetric, titrimetric, gravimetric, or conductimetric techniques. Dry combustion methods, instrumented in automated systems capable or performing multiple elemental analysis (C, H, N, or S), have become the standard in many laboratories worldwide. They are very accurate and exhibit minimal variability and low operational errors. Dry combustion instruments have a detection limit of about 10 mg C kg^{-1}, and their relative deviation (accuracy) decreases as soil carbon concentration increases. Figure 19.3 presents a comparison of total soil C (%), as measured by two dry combustion instruments. Because dry combustion determines total carbon, extra steps are required for reporting organic C concentration when carbonates are present in the soil sample. If this is the case, then the fraction of total C that is inorganic can be estimated with either an independent measurement of carbonate C or the total C analysis conducted on a carbonate-free soil sample previously treated with an acid solution. Dry combustion should be the preferred methodology for measuring SOC concentration in SCS projects. However, due to its relatively high initial cost (>\$20,000), the dry combustion methodology may be difficult to implement in developing countries participating in SCS projects and programs.

Assessment of SCS due to the implementation of alternative practices worldwide will require a technical effort to ensure that the results obtained are accurate and comparable, and include an estimation of the uncertainty associated with the measurements. Assessment of these changes will occur under many environmental conditions, and will have to be provided at a relatively low cost and may have to include numerous measurements within a field in order to detect more continuously the response of SOC to changes in management. With this in mind, various research groups in the United

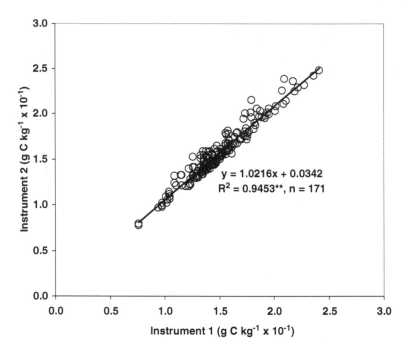

Figure 19.3 Total soil C as measured by two dry combustion instruments. (Data from Izaurralde et al. 2001b. *Soil Sci. Soc. Am. J.* 65:431–441.)

States have been advancing and developing instrumentation for fast, *in situ* measurements of soil C. Three methodologies have been advanced (adapted) so far to measure soil C: (1) laser-induced breakdown spectroscopy (LIBS) (Cremers et al., 2001); (2) mid-infrared (MIRS) and near-infrared (NIRS) spectroscopy (McCarty et al., 2002); and (3) inelastic neutron scattering (INS) (Wielopolski, 2002).

The LIBS method is based on atomic emission spectroscopy (Cremers et al., 2001). In this method, a laser is applied to a (soil) sample, converting it into plasma that emits light whose colors are spectrally resolved. Cremers et al. (2001) calibrated a LIBS instrument that measured total C in soils from east-central Colorado against measurements with a dry combustion apparatus, and used the calibration curve obtained to predict the total C of additional soil samples. Their results indicated that LIBS has a detection limit of 300 mg C kg^{-1} with a precision of 4% to 5%, and an accuracy ranging from 3% to 14%. The laboratory version of LIBS tested was capable of analyzing samples in less than a minute, with a daily throughput of more than 200 samples. The authors also reported the development of a field version of the LIBS instrument capable of analyzing soil C over large areas and also in depth. Martin et al. (2003) also used LIBS to measure total C and N in samples of soils that had or had not received acid washing to destroy carbonates. Like Cremers et al. (2001), Martin et al. (2003) obtained high correlations between total C measured by LIBS and dry combustion (r^2 = 0.962). The latter team, however, reported increased variability in C determinations in soils low in organic matter content due to spectral interference with iron whose peak (248.4 nm) appears very close to that of carbon (247.9 nm).

NIRS is a widely used technique used to characterize organic and inorganic compounds in the chemical, pharmaceutical, agricultural, semiconductor, and other industries. Dalal and Henry (1986) pioneered the use of near-infrared reflectance spectroscopy to determine water content, organic C, and total N in soils. Ben-Dor and Banin (1994) used NIRS to characterize the spectral reflectance of 91 Israeli soils for several soil properties, including carbonate and organic matter content. Because the NIRS approach is empirical, it requires the availability of calibration sets to match the spectral characteristics of the sample. They used 39 soils to calibrate the method, and 52 to validate it. Although predicted and measured SOM values were significantly correlated (r^2 = 0.51) within a range of 0% to 12%, NIRS underpredicted SOM concentration at the high end. More recently, McCarty and Reeves (2001) used NIRS and pyrolysis analysis to quantify SOC content from soils under conventional and no-tillage management in central Maryland. One objective of the study was to test whether these methods could be used to understand the spatial structure of SOC distribution in agricultural fields by sacrificing some accuracy in the point measurements. Their findings confirmed that NIRS offers a simple and rapid way for assessing SOC content, but at the expense of some loss in accuracy. McCarty et al. (2002) compared two infrared spectroscopic techniques (mid-infrared and near-infrared) for determining SOC content of 273 soils from the U.S. Great Plains. Overall, the MIRS method was a better predictor of total C (r^2 = 0.97) and carbonate C (r^2 = 0.99) concentration than the NIRS method (r^2 = 0.9 and r^2 = 0.96, respectively). The total C in the samples tested ranged from 0% to 10%, while that of carbonate C ranged from 0% to 7%. McCarty et al. (2002) concluded that the MIRS method yields better spectral information than the NIRS method. Because spectral analysis for soil C is nondestructive, requires no reagents, and is easily adaptable to automated and *in situ* measurements, it could become a key component of methodologies for assessing the spatial distribution of SOC at landscape and regional levels.

With the same objective of developing accurate, *in situ*, and practical methods to measure soil C, Wielopolski et al. (2002) adapted and tested a method based on neutron scattering principles. In the inelastic neutron-scattering (INS) method, 14 MeV neutrons are applied to the C nuclei present in soil, while recording on a NaI detector the 4.4 MeV gamma rays subsequently generated after the impacts. The C concentration in the soil sample is directly proportional to the intensity of the C peak detected in the gamma spectra. A calibration curve is obtained by running the INS on a sand vessel containing known amounts of C. Preliminary results indicate the INS instrument to be able to detect SOC changes of 100 g C m^{-2} with a 5% precision. Several field tests of this instrument were underway at the end of this writing.

19.3.4 Models

SOM models or agroecosystem models with specific treatment of SOC dynamics represent our conceptual and quantitative understanding of the mechanisms regulating C transformations in soil. Well-tested models have successfully been used to simulate SOC changes at various temporal (day to century) and spatial (site to national) scales. When well-initialized and calibrated (if necessary), simulation models have been capable of describing historical trends in SOM dynamics observed in long-term experiments worldwide (Paustian et al., 1992; Parton and Rasmussen, 1994; Grant et al., 2001). When driven with more generic soil and climate databases, these models have been able to provide insight on scaling issues (Izaurralde et al., 2001a) or regional trends in SOM evolution (Falloon et al., 1998). What level of model complexity is needed for project level applications? On the one hand, simple models, or simplified versions of complex models, can provide rapid assessment of the impact of management practices on SOC content under a given set of climate and soil conditions. On the other, models of increased complexity can yield an in-depth understanding of the processes regulating SOC dynamics in addition to providing ancillary information on the environmental impacts of carbon sequestration practices such as trace gases (Del Grosso et al., 2000) and erosion (Izaurralde et al., 2001c).

At the project level, however, there seems to be a scale missing in SOC prediction: the landscape or field scale. SOM models have been applied either at the site scale with specific information, or at the county or smaller spatial scales with generalized soil and climate databases. A key limitation in applying models at the landscape scale has been the input data needed to drive them. As discussed in previous sections, advances in methodologies to estimate soil parameters from ancillary, more easily acquired parameters (digital elevation maps), may facilitate the application of models in landscape, whole farm, mode. The APEX model (agricultural policy environmental extender) (Williams, 1999) is an example of one such model with this type of capability. APEX is a watershed-scale model capable of simulating plant growth, watershed hydrology, soil C dynamics (Izaurralde et al., 2001b), erosion, nutrient cycling and routing of water, nutrients, pesticides, and soil across connected fields under various land use and management practices. The APEX model contains all the functions of the EPIC model and can be applied to whole farms and small watersheds (<2500 km^2) including multiple fields, soil types, and landscapes. Further, APEX simulates routing of water, sediment, eroded C, nutrients, and pesticides across complex landscapes and channel systems to the watershed outlet.

Model evaluations will also be needed for rapid assessment of the possible impacts of management practices on soil C sequestration. CSTORE is an example of a simple model being developed with such an objective (Paustian, 2004). In this case, the model is intended for use in field-level prediction of SCS, and also as part of a decision-support system, with minimum data requirements. The CSTORE model is based on the widely used CENTURY model with a simpler representation of the SOM pools.

19.3.5 Remote Sensing

Remotely sensed data will be very useful for monitoring the evolution of SCS projects through spatiotemporal assessments of vegetation (type, cover, and productivity), water (soil moisture), and energy (soil temperature). In SCS projects, the soil surface is likely going to be covered most of the time by live or dead vegetation (crops, plant litter). Thus, remote sensing cannot be used for direct measurements of soil carbon stocks (Merry and Levine, 1995; vanDeventer et al., 1997) unless the soil is bare, and correlations among soil reflectance, soil color, and SOM concentration can be established (Bhatti et al., 1991; Chen et al., 2000; Henderson et al., 1989). Merry and Levine (1995) found a negative correlation ($r = -0.63$) between NDVI values derived from the AVIRIS (airborne visible infrared imaging spectroradiometer) sensor and soil carbon density determined from a detailed soil survey (scale 1:12,000) conducted in Orono, Maine. The negative sign of the correlation coefficient suggested that certain types of vegetation with low NDVI values were more adapted to grow in places with enhanced levels of soil carbon densities. (NDVI, the normalized difference vegetation index, is calculated as the ratio of the difference between the near-infrared and red bands over the sum of these bands. Use of the NDVI allows for meaningful comparisons of seasonal and interannual changes in vegetation growth and activity.)

A number of satellite and airborne sensors are capable of gathering data useful for estimating LAI, net primary productivity (NPP), crop yields, and litter cover (Table 19.2). For many years, the AVHRR (Advanced Very High Resolution Radiometer) and Landsat sensors have been the traditional sources of land cover data. More recently, the MODIS (MODerate-resolution Imaging Spectroradiometer) sensor onboard the Terra satellite has been obtaining data on land cover with a spatial and spectral resolution that was not possible with AVHRR. MODIS and AVHRR are able to provide data at rather frequent temporal resolution and a good spatial resolution for monitoring vegetation growth during the growing season. Conversely, the Landsat and SPOT (Système Probatoire pour l'Observation de la Terre) sensors allow for excellent spatial detail of land cover/crop identification, but might not the best sensors for monitoring seasonal vegetation dynamics. The commercial sensors IKONOS and Quickbird offer excellent spatial and temporal resolution, and could be very helpful for agricultural applications and SCS projects. Two airborne sensors might be particularly useful for SCS projects: AVIRIS and LIDAR (light detection and ranging) (Table 19.2).

Hyperspectral images provide spectral images that resemble those obtained by laboratory spectroscopic instruments (Shippert, 2004). Because they are overdetermined, hyperspectral images allow for the identification and distinction of spectrally similar, but unique, materials (Shippert, 2004). The remote sensing community has been advancing methodologies to interpret hyperspectral imagery, including the development of libraries for identifying minerals, substances, and vegetation (*http://rst.gsfc.nasa.gov/Intro/Part2_24.html*). The AVIRIS sensor, aboard the NASA-ER1 plane, has been providing hyperspectral data of immense value since 1987 (Table 19.2). Another hyperspectral sensor is Hyperion (*http://eo1.gsfc.nasa.gov/technology/hyperion.html*), part of EO-1, in operation since December 2000, which serves as testing unit for advancing the development of spaceborne hyperspectral instrumentation.

Remote sensing data have also been used with simulation models to derive spatially explicit estimates of cropland NPP

Table 19.2 Temporal and Spatial Resolutions of Various Satellite and Airborne Sensors Collecting Data Useful for Land Cover/Land Use Classifications, Vegetation Monitoring, and Topographic Mapping

Sensor in Operation Since	Use Category	Instruments, Number of Bands	Spectral Bands (nm)	Spatial Resolution/Swath (m/km)	Temporal Resolution (d)	Websites and Special Features
Satellite						
AVHRR, >20 years	Operational	Multispectral, 2 + 3 thermal	580–1100	1000/2485	1 bi-weekly composites	http://edcsns17.cr.usgs.gov/EarthExplorer/
MODIS, December 1999	Research	Multispectral, 2 / Multispectral, 5 / 29 bands for clouds, O_3, etc.	620–876 / 459–2155	250 / 500 / 1000/2330	2 / 2 / 10	http://modis.gsfc.nasa.gov/
ASTER December 1999	Research	Multispectral, 3 / Multispectral, 6	520–860 / 1600–2430	15/60 / 30/60	Pointing	http://www.science.aster.ersdac.or.jp/en/science_info/index.html
Landsat MSS / Landsat TM / Landsat ETM+, > 30 years	Operational	Multispectral, 4 / Multispectral, 6 / Multispectral, 6 / Panchromatic	500–1100 / 450–2350 / 450–2350 / 520–900	80 / 30 / 30 / 15	18 / 16	http://landsat7.usgs.gov/index.php / http://edcsns17.cr.usgs.gov/EarthExplorer
SPOT (2, 4), 13 years; SPOT 5, May 2002	Commercial	Multispectral, 4 / Panchromatic, 1 / Stereoscopic / Vegetation	500–1750 / 480–710 / 490–690 / 450–1750	20/60 / 2.5—10 / 10 / 1000	Pointing / 1	http://www.spotimage.fr/html/_167_.php / http://www.spot-vegetation.com/
Hyperion	Research	Hyperspectral, 220	400–2500	30/7.5	Pointing	http://eo1.gsfc.nasa.gov/technology/hyperion.html
IKONOS, September 1999	Commercial	Multispectral, 4 / Panchromatic, 1	450–880 / 450–900	4/7 / 1	3 pointing	http://www.spaceimaging.com/

QuickBird, October 2001	Commercial	Multispectral, 4 Panchromatic	450–900	2.44/16.5 0.61	Pointing	http://www.spaceimaging.com/
Airborne						
AVIRIS >16 years	Research	Hyperspectral, 224	380–2550	20/11 1	N/A	http://aviris.jpl.nasa.gov/
AISA	Commercial	Hyperspectral, varies	User selectable	Variable	NA	http://www.specim.fi/ (Available from several companies)
LIDAR, commercial since 1993	Research, Commercial	Active laser sensor, 1	1045–1065	> 0.75 horizontal 0.15 vertical	Dependent on flight schedule	http://ltpwww.gsfc.nasa.gov/eib/projects/airborne_lidar/lvis/ Several companies provide airborne LIDAR services
Multispectral	Operational	Similar to Landsat TM	400–900	~0.25 adjustable	Dependent on flight schedule	Several companies provide multispectral service using instruments of varying sophistication

(Prince et al., 2001; Lobell et al., 2002). Satellite estimates of residue cover (plant litter) would also be useful for assessing adoption of conservation practices (no till) and calibrating models. Daughtry et al. (2004) determined spectral reflectance characteristics of dry and wet crop residues and soils over the 400- to 2400-nm wavelength region. Since crop residue cover was linearly related to a cellulose absorption index, and this to NDVI, Daughtry et al. (2004) proposed a method to estimate soil tillage intensity based on these two indices. They concluded that it was possible to use advanced multispectral or hyperspectral imaging systems to conduct regional surveys of conservation practices.

19.4. SUMMARY AND CONCLUSIONS

This chapter has highlighted significant scientific and technological advances in the area of SCS achieved during the last 15 years. At the project level, the Saskatchewan project demonstrated that increases in SOC as small as 1.2 Mg C ha^{-1} can be accurately detected in periods as short as 3 years. A number of feasibility or pilot projects are underway worldwide under a variety of environmental and socioeconomic situations. To further advance the field of SCS, more projects like these will be necessary if we are to advance an internationally accepted and adaptable framework that can guide landowner, energy, and government groups in the development of SCS projects. The formation of a collaborative network for these types of SCS projects would be very helpful to compare methodologies in use across diverse environments and to exchange data for laboratory quality controls and verification of simulation models, among other purposes. These activities could go on for 5 to 8 years. These projects would also be useful to advance new methodologies that integrate many of the novel concepts discussed in this chapter as well as many yet to be discovered.

ACKNOWLEDGMENTS

I thank Rattan Lal for his encouragement in preparing this review. I am also grateful for the technical advice and information received from J.E. Amonette (United States), R. Casas (Argentina), J. Etchevers (Mexico), E.R. Hunt (United States), B.G. McConkey (Canada), S.D. Prince (United States), T. Scholz (United States), L. Tieszen (United States), and E. Venteris (United States). The preparation of this review was supported by the U.S. Department of Energy, Office of Science, Carbon Sequestration in Terrestrial Ecosystems program, the U.S. Department of Agriculture Consortium for Agricultural Soils Mitigation of Greenhouse Gases, and the National Aeronatics and Space Administration Office of Earth Science Applications Division.

REFERENCES

Adams, W.A. 1973. The effect of organic matter on the bulk and true densities of some uncultivated podzolic soils. *J. Soil Sci.,* 24:10–17.

Baldocchi, D.D. and K.B. Wilson. 2001. Modeling CO_2 and water vapor exchange of a temperate broadleaved forest across hourly to decadal time scales. *Ecol. Modeling,* 142:155–184.

Ben-Dor, E. and A. Banin. 1994. Visible and near-infrared (0.4–1.1 μm) analysis of arid and semiarid soils. *Remote Sensing Environ.,* 48:261–274.

Bhatti, A.U., D.J. Mulla, and B.E. Frazier. 1991. Estimation of soil properties and wheat yields on complex eroded hills using geostatistics and thematic mapper images. *Remote Sensing Environ.,* 37:181–191.

Blake, G.R. and K.H. Hartge. 1986. Bulk density. In *Methods of Soil Analysis, Part I. Physical and Mineralogical Methods.* Agronomy Monograph 9, 2nd ed. American Society of Agronomy, Soil Science Society of America, *Soc. Am.* Madison, WI, pp. 363–382.

Chen, F., D.E. Kissel, L.T. West, and W. Adkins. 2000. Field-scale mapping of surface soil organic carbon using remotely sensed imagery. *Soil Sci. Soc. Am. J.,* 64:746–753.

Cole, V., C. Cerri, K. Minami, A. Mosier, N. Rosenberg, and D. Sauerbeck. 1996. Agricultural options for mitigation of greenhouse gas emissions. In R.T. Watson, M.C. Zinyowera, and R.H. Moss, Eds. *Climate Change 1995: Impacts, Adaptations, and Mitigation of Climate Change: Scientific-Technical Analyses.* Contribution of Working Group II to the Second Assessment Report of the Intergovernmental Panel on Climate Change. Cambridge University Press, London; New York, pp. 744–771.

Cremers, D.A., M.H. Ebinger, D.D. Breshears, P.J. Unkefer, S.A. Kammerdiener, M.J. Ferris, K.M. Catlett, and J.R. Brown. 2001. Measuring total soil carbon with laser-induced breakdown spectroscopy (LIBS). *J. Environ. Qual.,* 30:2202–2206.

Culley, J.L.B. 1993. Density and compressibility. In M.R. Carter, Ed. Soil Sampling and Methods of Analysis. Canadian Society of Soil Science/Lewis Publishers, Boca Raton, FL, pp. 529–539.

Dalal, R.C. and R.J. Henry. 1986. Simultaneous determination of moisture, organic carbon and total nitrogen by near infrared reflectance spectrophotometry. *Soil Sci. Soc. Am. J.,* 50:120–123.

Daughtry, C.S.T., E.R. Hunt Jr., and J.E. McMurtrey III. 2004. Assessing crop residue cover using shortwave infrared reflectance. *Remote Sensing Environ.,* 90:126–134.

Del Grosso, S.J., W.J. Parton, A.R. Mosier, D.S. Ojima, A.E. Kulmala, and S. Phongpan. 2000. General model for N_2O and N_2 gas emissions from soils due to denitrification. *Global Biogeochem. Cycles,* 14:1045–1060.

Ellert, B.H., H.H. Janzen, and B. McConkey. 2001. Measuring and comparing soil carbon storage. In R. Lal, J.M. Kimble, R.F. Follett., and B.A. Stewart, Eds. *Assessment Methods for Soil Carbon.* Lewis Publishers, Boca Raton, FL, pp. 131–146.

Falloon, P.D., P. Smith, J.U. Smith, J. Szabó, K. Coleman, and S. Marshall. 1998. Regional estimates of carbon sequestration potential: linking the Rothamsted Carbon Model to GIS databases. *Biol. Fertil. Soils,* 27:236–241.

Garten, C.T. and S.D. Wullschleger. 1999. Soil carbon inventories under a bioenergy crop (switchgrass): measurement limitations. *J. Environ. Qual.,* 28:1359–1365.

Grant R.F., N.G. Juma, J.A. Robertson, R.C. Izaurralde, and W.B. McGill. 2001. Long-term changes in soil C under different fertilizer, manure and rotation: testing the mathematical model ecosystems with data from the Breton Plots. *Soil Sci. Soc. Am. J.*, 65:205–214.

Gupta, S.C. and W.E. Larson. 1979. A model for predicting packing density of soils using particle-size distribution. *Soil Sci. Soc. Am. J.*, 43:758–764.

Henderson, T.L., A. Szilagyi, M.F. Baumgardner, C.C.T. Chen, and D.A. Landgrebe. 1989. Spectral band selection for classification of soil organic-matter content. *Soil Sci. Soc. Am. J.*, 53:1778–1784.

Hengl, T., G.B.M. Heuvelink, and A. Stein. 2004. A generic framework for spatial prediction of soil variables based on regression-kriging. *Geoderma, 122:75–93.*

Intergovernmental Panel on Climate Change. 1990. *Climate Change: the IPCC Impacts Assessment.* IPCC Working Group II. Australian Government Publishing Service, Canberra, Australia.

Izaurralde, R.C., W.B. McGill, A. Bryden, S. Graham, M. Ward, and P. Dickey. 1998. Scientific challenges in developing a plan to predict and verify carbon storage in Canadian prairie soils. In R. Lal, J. Kimble, R. Follett, and B.A. Stewart, Eds. *Management of Carbon Sequestration in Soil.* CRC Press, Boca Raton, FL, pp. 433–446.

Izaurralde, R.C., K.H. Haugen-Kozyra, D.C. Jans, W.B. McGill, R.F. Grant, and J.C. Hiley. 2001a. Soil organic carbon dynamics: measurement, simulation and site to region scale-up. In R. Lal, J.M. Kimble, R.F. Follett, and B.A. Stewart, Eds. *Assessment Methods for Soil Carbon.* Lewis Publishers. Boca Raton, FL, pp. 553–575.

Izaurralde, R.C., W.B. McGill, J.A. Robertson, N.G. Juma, and J.J. Thurston. 2001b. Carbon balance of the Breton classical plots over half a century. *Soil Sci. Soc. Am. J.* 65:431–441.

Izaurralde, R.C., J.R. Williams, W.B. McGill, and N.J. Rosenberg. 2001c. Simulating soil carbon dynamics, erosion and tillage with EPIC. First National Conference on Carbon Sequestration, U.S. Department of Energy, National Energy Technology Laboratory, Washington, DC, 14–17 May. Available at: *www.netl.doe.gov/publications/proceedings/01/carbon_seq/5c2.pdf.*

Jenkinson, D.S., 1988. Soil organic matter and its dynamics. In A. Wild, Ed. Russell's Soil Conditions and Plant Growth, 11th ed. Longman, New York, pp. 564–607.

Kern, J.S. and M.G. Johnson. 1993. Conservation tillage impacts on national soil and atmospheric carbon levels. *Soil Sci. Soc. Am. J.,* 57:200–210.

Lal, R., J. Kimble, E. Levine, and B.A. Stewart, Eds. 1995a. *Soil Management and the Greenhouse Effect.* Lewis Publishers/CRC Press, Boca Raton, FL.

Lal, R., J. Kimble, E. Levine, and B.A. Stewart, Eds. 1995b. *Soils and Global Change.* Lewis Publishers/CRC Press, Boca Raton, FL.

Lal, R., J. Kimble, R. Follett, and B.A. Stewart, Eds. 1998a. *Management of Carbon Sequestration in Soil.* CRC Press, Boca Raton, FL.

Lal, R., J. Kimble, R. Follett, and B.A. Stewart, Eds. 1998b. *Soil Processes and the Carbon Cycle.* CRC Press, Boca Raton, FL.

Lal, R. and J.M. Kimble. 2001. Importance of soil bulk density and methods of its importance. In R. Lal, J.M. Kimble, R.F. Follett, and B.A. Stewart, Eds. *Assessment Methods for Soil Carbon.* Lewis Publishers/CRC Press, Boca Raton, FL, pp. 31–44.

Lobell D.B., J.A. Hicke, G.P. Asner, C.B. Field, and C.J. Tucker. 2002. Satellite estimates of productivity and light use efficiency in United States agriculture 1982–98. *Global Change Biol.,* 8:722–735.

Macmillan, R.A., W.W. Pettapiece, S.C. Nolan, and T.W. Goddard. 2000. A generic procedure for automatically segmenting landforms into landform elements using DEMs, heuristic rules and fuzzy logic. *Fuzzy Sets Syst.,* 113:81–109.

Mann, L.K. 1986. Changes in soil carbon storage after cultivation. *Soil Sci.,* 142:279–288.

Marland, G., B. McCarl, and U.A. Schneider. 2001. Soil carbon: policy and economics. *Climatic Change,* 51:101–117.

Martin, M.Z., S.D. Wullschleger, C.T. Garten, and A.V. Palumbo. 2003. Laser-induced breakdown spectroscopy for the environmental determination of total carbon and nitrogen in soils. *Appl. Optics,* 42:2072–2077.

McBratney, A.B., M.I. Mendonça Santos, and B. Minasny. 2003. On digital soil mapping. *Geoderma,* 117:3–52.

McCarl, B.A. and U.A. Schneider. 2001. The cost of greenhouse gas mitigation in U.S. agriculture and forestry. *Science,* 294:2481–2482.

McCarty, G.W. and J.B. Reeves III. 2001. Development of rapid instrumental methods for measuring soil organic carbon. In R. Lal, J.M. Kimble, R.F. Follett, and B.A. Stewart, Eds. *Assessment Methods for Soil Carbon*. Lewis Publishers, Boca Raton, FL, pp. 371–380.

McCarty, G.W., J.B. Reeves, V.B. Reeves, R.F. Follett, and J.M. Kimble. 2002. Mid-infrared and near-infrared diffuse reflectance spectroscopy for soil carbon measurement. *Soil Sci. Soc. Am. J.,* 66:640–646.

McConkey, B.G., B.C. Liang, G. Padbury, and R. Heck. 2000. Prairie Soil Carbon Balance Project: Carbon Sequestration from Adoption of Conservation Cropping Practices. Final Report to GEMCo. Agriculture and Agri-Food Canada, Swift Current, SK, Canada.

McGill, W.B. 1996. Review and classification of ten soil organic matter (SOM) models. In D.S. Powlson, P. Smith, and J.U. Smith, Eds. *Evaluation of Soil Organic Matter Models Using Existing Long-Term Datasets*. NATO ASI Series I, vol. 38. Springer-Verlag, Heidelberg.

Merry, C.J. and E.R. Levine. 1995. Methods to assess soil carbon using remote sensing techniques. In R. Lal, J. Kimble, E. Levine, and B.A. Stewart, Eds. *Advances in Soil Science: Soils and Global Change*. Lewis Publishers/CRC Press, Boca Raton, FL, pp. 265–274.

Nelson, D.W. and L.E. Sommers. 1996. Total carbon, organic carbon, and organic matter. In *Methods of Soil Analysis. Part 3. Chemical Methods*. Book Series no. 5. Soil Science Society of America/American Society of Agronomy, Madison, WI.

Ochsner, T.E., R. Horton, and T. Ren. 2001. Simultaneous water content, air-filled porosity, and bulk density measurements with thermo-time domain reflectometry. *Soil Sci. Soc. Am. J.*, 65:1618–1622.

Odeh, I.O.A., A.B. McBratney, and D.J. Chittleborough. 1994. Spatial prediction of soil properties from landform attributes derived from a digital elevation model. *Geoderma*, 63:197–214.

Parton, W.J. and P.E. Rasmussen. 1994. Long-term effects of crop management in wheat-fallow. 2. Century model simulations. *Soil Sci. Soc. Am. J.* 58:530–536.

Paul, E.A., K. Paustian, E.T. Elliott, and C.V. Cole, Eds. 1997. *Soil Organic Matter in Temperate Agroecosystems: Long-Term Experiments in North America*. Lewis Publishers/CRC Press, Boca Raton, FL.

Paustian, K., W.J. Parton, J. Persson. 1992. Modeling soil organic-matter in organic-amended and nitrogen-fertilized long-term plots. *Soil Sci. Soc. Am. J.*, 56:476–488.

Paustian, K. 2004. Personal communication, Colorado State University, Fort Collins, CO.

Pennock, D.J., B.J. Zebarth, and E. De Jong. 1987. Landform classification and soil distribution in hummocky terrain, Saskatoon, Canada. *Geoderma*, 40:297–315.

Pennock, D.J. and M.D. Corre. 2001. Development and application of landform segmentation procedures. *Soil Tillage Res.*, 58:151–162.

Petersen, R.G. and L.D. Calvin. 1996. Sampling. In *Methods of Soil Analysis. Part 3. Chemical Methods*. SSSA Book Series no. 5. Soil Science Society of America, Madison, WI.

Post, W.M., R.C. Izaurralde, L.K. Mann, and N. Bliss. 1999. Monitoring and verifying soil organic carbon sequestration. In N.J. Rosenberg, R.C. Izaurralde, and E.L. Malone, Eds. *Carbon Sequestration in Soils: Science, Monitoring and Beyond*. Battelle Press, Columbus, OH, pp. 41–66.

Powlson, D.S., P. Smith, and J.U. Smith, Eds. 1996. *Evaluation of Soil Organic Matter Models Using Existing Long-Term Datasets.* NATO ASI Series I, vol. 38, Springer-Verlag, Heidelberg.

Prince S.D., J. Haskett, M. Steininger, H. Strand, and R. Wright. 2001. Net primary production of US Midwest croplands from agricultural harvest yield data. *Ecol. Appl.,* 11:1194–1205.

Ren, T., T.E. Ochsner, and R. Horton. 2003. Development of thermo-time domain reflectometry for vadose zone measurements. *Vadose Zone J.,* 2:544–551.

Rosenberg, N.J. and R.C. Izaurralde. 2001. Introductory paper to the special issue on mitigation of climate change by soil carbon sequestration. *Climatic Change,* 51:1–10.

Scholz, T. 2004. Personal communication, Pacific Northwest Direct Seeding Association, Pasco, WA.

Shippert, P. 2004. Why use hyperspectral imagery? *Photogrammetry Eng. Remote Sensing,* 70:377–380.

Six, J. and J.D. Jastrow. 2002. Soil organic matter turnover. In R. Lal, Ed. *Encyclopedia of Soil Science.* Marcel Dekker, New York, pp. 936–942.

Smith P., J.U. Smith, D.S. Powlson, et al. 1997. A comparison of the performance of nine soil organic matter models using datasets from seven long-term experiments. *Geoderma,* 81:153–225.

Stevenson, F.J. and M.A. Cole. 1999. *Cycles of Soil,* 2nd ed. John Wiley & Sons, New York.

Tiessen, H. and J.O. Moir. 1993. Total organic carbon. In M.R. Carter, Ed. *Soil Sampling and Methods of Analysis.* Canadian Society of Soil Science/Lewis Publishers, Boca Raton, FL.

Tieszen, L. 2004. Personal communication, U.S. Geological Society, Sioux Falls, SD.

Topp, G.C., J.L. Davis, and A.P. Annan. 1980. Electromagnetic determination of soil water content: measurements in coaxial transmission lines. *Water Resour. Res.,* 16:574–582.

Topp, G.C. and W.D. Reynolds. 1998. Time domain reflectometry: a seminal technique for measuring mass and energy in soil. *Soil Tillage Res.,* 47:125–132.

Triantafilis, J., I.O.A. Odeh, and A.B. McBratney. 2001. Five geostatistical models to predict soil salinity from electromagnetic induction data across irrigated cotton. *Soil Sci. Soc. Am. J.,* 65:869–878.

U.S. Department of Agriculture, Soil Survey Staff. 1999. *Soil Taxonomy: A Basic System of Soil Classification for Making and Interpreting Soil Surveys,* 2nd ed. Agriculture Handbook 436. U.S. Department of Agriculture, Natural Resource Conservation Service, Washington, DC.

vanDeventer, A.P., A.D. Ward, P.H. Gowda, and J.G. Lyon. 1997. Using thematic mapper data to identify contrasting soil plains and tillage practices. *Photogrammetry Eng. Remote Sensing,* 63:87–93.

Venteris, E. 2004. Personal communication, U.S. Department of Agriculture, Beltsville, MD.

Vergara Sánchez, M.A., J.D. Etchevers, and M. Vargas. 2004. Variabilidad del carbono orgánico en suelos de ladera del sureste de México. *Terra,* 22:359–367.

Wielopolski, L., S. Mitra, G. Hendrey, H. Rogers, A. Torgert, and S. Prior. 2002. Non-destructive *in situ* soil carbon analysis: principle and results. Oak Ridge National Laboratory, Oak Ridge, TN. Available at: *http://cdiac2.esd.ornl.gov/wielopolski.pdf.*

Williams, J.R. 1995. The EPIC Model. In V.P. Singh, Ed. *Computer Models of Watershed Hydrology.* Water Resources Publications, Highlands Ranch, CO, pp. 909–1000.

Williams, J.R. 1999. *APEX: User's Guide and Technical Documentation.* Version 8190. Blackland Research Center, Texas Agricultural Experiment Station, Temple, TX.

20

Dynamics of Carbon Sequestration in Various Agroclimatic Zones of Uganda

MOSES M. TENYWA, MAJALIWA MWANJALOLO,
MATTHIAS K. MAGUNDA, RATTAN LAL, AND
GODFREY TAULYA

CONTENTS

20.1 INTRODUCTION

Soil organic carbon is one of the most important indicators of soil health and factors controlling soil productivity (Sanchez et al., 1997; Lal and Kimble, 2000). It stands a great chance of becoming a marketable commodity because of its influence on both soil productivity and atmospheric chemistry. The dynamics of soil organic carbon are controlled by many factors affecting the balance between its buildup and breakdown processes. These processes are influenced by climate, soil type, land use, and management regime (Tenywa et al., 1999). Although the impact of each of these factors is known, soil organic carbon dynamics driven by the above factors combined do not display an obvious trend. Therefore, a simple and robust tool for assessing soil organic carbon dynamics in various environments and guiding management and policy decisions across land uses in the course of time is needed.

Most of the conventional methods for quantification of soil organic carbon have limitations in capturing the dynamics of its sequestration processes resulting from spatial and temporal changes in one or more of the factor controls. Modeling techniques have been deemed useful in organizing and synthesizing multiple data sets and providing valuable insights for developing a scientifically defensible soil organic carbon sequestration accounting system. However, the available models can only be used for fine scale (Van Veen et al., 1984; Parton et al., 1994). The objective of this study was to integrate the knowledge of climate, soil, land use, and management conditions on variability of soil organic carbon in different agroclimatic zones of Uganda.

20.2 MATERIALS AND METHODS

Uganda ($30°E1°S$, $35°E4°N$) is located in East Africa (Figure 20.1). It is characterized by diverse agricultural systems that have evolved from the interaction of several factors governing land use (climate, soil, terrain, and socioeconomic factors) (Komutunga and Musiitwa, 2001). In Uganda, the climate varies with altitude, which ranges from 610 m in the rift valley to 4324 m on Mount Elgon. Mean annual rainfall ranges from 510 mm in the semi-arid northeast of the country to 2160 mm in the Ssese Islands of Lake Victoria.

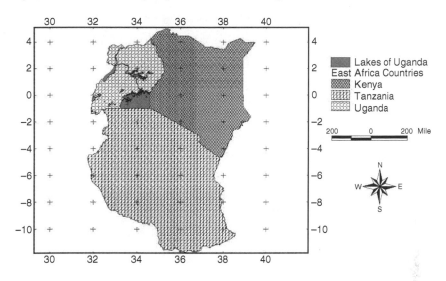

Figure 20.1 Map of East Africa

Based on climate, soils, population, and market orientation of production, 33 agroecological zones were identified in Uganda (Wortman and Eledu, 1999). For purposes of this study, broad divisions into agroclimatic zones were considered including eastern humid montane (Mt. Elgon), southwestern humid montane (Mt. Muhavura), subhumid (Lake Victoria crescent) and semi-arid (Lake Kyoga Basin). Long–term (1962 to 2002) averages of relative humidity, rainfall, and maximum and minimum temperature are presented in Table 20.1.

20.2.1 Agricultural Systems

The three broad agroclimatic zones (humid, subhumid, and semi-arid) in Uganda are characterized by different agricultural systems. The humid agroclimatic zone is found mainly in the highlands (Mt. Elgon in the east and Mt. Muhavura in the southwest). Mt. Elgon is a banana–coffee (arabica) agricultural system, while Mt. Muhavura is under the Irish (Idaho) potato–sorghum (annual crops) agricultural system. The subhumid Lake Victoria crescent is characterized by the banana–coffee (robusta) system. The semi-arid Lake Kyoga

Table 20.1 Characteristics of Uganda's Major Agroclimatic Zones

Agroclimatic Zones	Location	Land Use	RH (%)	R/fall (mm)	MaxT (°C)	MiniT (°C)	Textural Class
Eastern humid montane	Mt. Elgon	Banana (*Musa* Spp.)-coffee (*Coffea arabica*)	52	1523	26.3	16.0	C
Southwestern humid montane	Mt. Muhavura	Annual crops: potato (*Solanum tuberosum*), sorghum (*Sorghum bicolar*)	62	1035	24.0	11.0	C
Subhumid	Lake Victoria Crescent	Banana (*Musa* Spp.)-coffee (*Coffea canephora*) (Robusta)	NA	1211	28.0	16.0	SCL
Semi-arid	Lake Kyoga Basin	Annual crop cultivation-cattle raising	48	1415	30.6	17.9	SL

Notes: C = clay; NA = not available; R/fall = annual rainfall; RH = relative humidity; MaxT = maximum temperature; MinT = minimum temperature; SL = sandy clay; SCL = sandy clay loam.
Source: Department of Meteorology, Ministry of Lands, Water and Natural Resources, Kampala, Uganda.

basin is a mixed farming system with a good integration of cultivation of annual crops (e.g., finger millet, *Eleucine indica*; sorghum; cassava, *Manihot esculentum*; cotton, *Gossypium hirsutum*), and cattle raising.

20.2.2   Soils

Much of Uganda lies on a raised plateau between the western and eastern African rift valleys, characterized by basement complex rocks, granites, gneisses, and schists of the pre-Cambrian age. The rocks have been subjected to differential weathering. According to past surveys, a total of 138 soil units were recognized. For the purpose of this study focus is placed on soils of the agroclimatic zones. Humid montane agroclimatic zone (Mt. Elgon and Mt. Muhavura) soils are derived from volcanic ash and agglomerate; they can be classified as Andosols and vary along the landscape depending on altitude. Their textural class is predominantly clay. These soils are generally rich in organic matter. Soils in the subhumid agroclimatic zone are highly weathered and are variable but often have high clay content grouped under sandy clay loams (Lufafa et al., 2003). The soils are acidic with moderate levels of organic matter. In the semi-arid environment soils are of lacustrine origin and vary along the landscape. They are predominantly sandy with low levels of organic matter and largely classified as Acrisols and Ferralsols (Figure 20.2). In general, the soil organic carbon across the agroclimatic zones is closely related to land use (Table 20.2).

20.2.3   Deriving Model Parameters

A specific model incorporating a state-variable approach was used to predict soil organic carbon under different conditions. The model was based on the concept of buildup and breakdown processes that depend on the soil organic carbon content of the system. The soil organic carbon build-up rate (R) was described by the following mathematical function:

$$R = g\left(1 - \frac{C}{C_{\max}}\right) \tag{20.1}$$

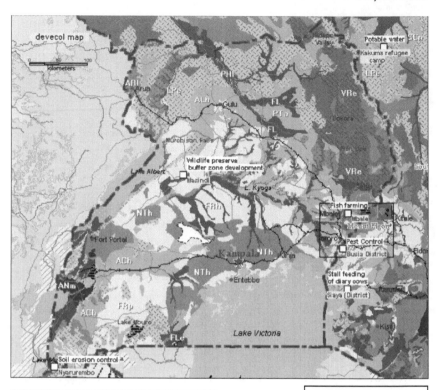

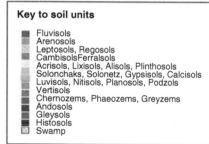

Key to soil units	Key to soil symbols
■ Fluvisols □ Arenosols □ Leptosols, Regosols ■ CambisolsFerralsols □ Acrisols, Lixisols, Alisols, Plinthosols □ Solonchaks, Solonetz, Gypsisols, Calcisols □ Luvisols, Nitisols, Planosols, Podzols □ Vertisols □ Chernozems, Phaeozems, Greyzems □ Andosols □ Gleysols ■ Histosols □ Swamp	ACh: Haplic Acrisols ANm: Mollic Andosols ARl: Luvic Arenosols FLe: Eutric Fluvisols FRh: Haplic Ferralsols FRp: PLinthic Ferralsols LPe: Eutric Leptosols NTh: Haplic Nitosols PHl: Luvic Phaeozems PTa: Albic Plinthosols VRe: Eutric Vertisols

Figure 20.2 Soils of Uganda. (Data from Food and Agriculture Organization. 1998. *The Soil and Terrain Database for Northeastern Africa: Crop Production System Zones of IGAD Sub Region.* FAO Land and Water Digital Media Series 2. Food and Agricultural Organization, Rome. Map courtesy of Development Ecology Information Service, Alexandria, VA.)

Table 20.2 Soil Organic Carbon (0- to 15-cm Depth) in Various
Land Uses Across Major Agroclimatic Zones in Uganda

Agroclimatic Zones	Location	Land Use	Soil Organic Carbon (%)	Range
Eastern humid montane	Mt. Elgon	Forest	2.7	0.7
		Woodland	2.4	0.2
		Annuals	2.7	0.8
		Banana	1.5	0.2
Southwestern humid montane	Mt. Muhavura	Fallow	1.3	0.6
		Grazing	5.3	1.7
Subhumid	Lake Victoria Crescent	Fallow	2.3	1.4
		Forest	3.3	3.3
		Woodland	4.0	6.8
		Grazing	3.2	4.7
		Annuals	2.4	1.4
		Banana	3.9	1.5
		Coffee	2.6	3.5
Semi-arid	Lake Kyoga Basin	Grazing	1.0	0.8
		Annuals	1.0	0.4

Source: Mulebeke, R. 2002. Validation of a GIS-USLE Model in a Banana-Based Microcatchment of the Lake Victoria Basin. Master's thesis, Makerere University, Kampala, Uganda; and Nantumbwe, C. 2002. Evaluation of Temporal Land Use and Soil Property Interrelationships on the Slopes of Mt. Elgon. Master's thesis, Makerere University, Kampala, Uganda. With permission.

where g is the relative soil organic carbon buildup rate (yr^{-1}), C is the soil organic carbon content (%) at time t, and C_{max} is the maximum attainable soil organic carbon content (%). Equation 20.1 implies that soil organic carbon buildup is exponential when soil organic carbon content is very small compared to C_{max}, and maximum R = g when soil organic carbon tends to zero. The rate of soil organic carbon buildup reduces and tends to zero as C tends to C_{max}. The rate of soil organic carbon breakdown (L) was described by the following function:

$$L = L_{max} (1 - \exp(-(C - C_r)/(C_s - C_r))) \qquad (20.2)$$

where L_{max} is the maximum attainable soil organic carbon loss rate (yr^{-1}), C_r is the minimum/passive soil organic carbon content (%), C_s is shape factor of the function, and C is the soil organic carbon content (%) at time t. C_r represents the minimum protected soil organic carbon beyond which further breakdown of soil organic carbon is not significant. The breakdown rate (L) is highest when soil organic carbon content is maximum. The soil organic carbon sequestration over time per unit soil organic carbon content is the difference between the rate of buildup (R) and breakdown (L), and is described by the function

$$\frac{dC}{dt} = (R - L)C \tag{20.3}$$

where C is the soil organic carbon content (%) at a time t.

The soil organic carbon content (area between the two functions) was then obtained by the analytical integration of Equation 20.3; thus,

$$\int \frac{dC}{C(R - L)} = \int dt \tag{20.4}$$

Numerical integration of Equation 20.4 was used to predict soil organic carbon at any time t, $C(t + \Delta t) = C(t) + \Delta C$, where $\Delta C = \Delta C_R - \Delta C_L$; ΔC is the change in soil organic carbon, $\Delta C_R = R*C*\Delta t$ the change in soil organic carbon due to the buildup process; and $\Delta C_L = L*C*\Delta t$ change in soil organic carbon due to the breakdown process.

20.2.4   Modeling Database

Data from the subhumid agroclimatic zone (Lake Victoria crescent) were used to test the predictive performance of the model. The data set was compiled from 1998 to 2002, using runoff plots (15×10 m) established in four major agricultural land use types (annual crops, banana, coffee, and range lands), under the Lake Victoria Environmental Management Project (LVEMP). Each land use type was replicated three times, except banana, which was replicated four times. Plots under banana were mulched and not tilled, while those under coffee were neither tilled nor mulched. Soil chemical properties including soil organic carbon were determined every season from 0– to 30-cm depths. Input of organic matter (litter

Table 20.3 Modeling Parameters for Specific
Sequestration Model

Agroclimatic Zones	Historical Maximum Soil C (0–30 cm) (C_{max}) (%)	Minimum Soil Organic Carbon (C_r) (%)
Humid	11.0	<2
Humid	5.0	<2
Subhumid	4.7	<1.5
Semi-arid	1.3	<0.5

Source: Compiled from Radwanski, S.A., Ed. 1960. *The Soils and Land-Use of Uganda.* Memoirs of the Research Division, Series 1, no. 4, Soils. Uganda Protectorate, Department of Agriculture, Kampala, Uganda.

and mulch) was also monitored seasonally. The amount of soil organic carbon lost through water erosion for each season was determined by multiplication of the seasonal soil organic carbon concentration by seasonal soil loss. We assumed that the minimum/passive soil organic carbon, C_r = 1.5% (Paul and Clark, 1996), and the decomposition rate of organic matter to be 50% (Lal and Kimble, 2000). The maximum attainable rate of soil organic carbon loss was estimated for each of the four land use types (annual crops, banana, coffee, and range lands) as the difference between the historical maximum soil organic carbon and the minimum/passive soil organic carbon.

The model parameter C_s was estimated using regression method. $C_s - C_r$ was the slope of $C(\ln(L_{max} - L)/L_{max})$. L_{max} was obtained as a product of $(C_{max} - C_r)$ and the mass of slice topsoil layer (0- to 30-cm depth). L was then taken to be the soil organic carbon content lost through erosion from the different land use types (annual crops, banana, coffee, and range lands). The relative growth rate of soil organic carbon (g) was the average of 4 years in g values computed from Equation 20.1. The parameters C_s and g are presented in Table 20.4. The change in soil organic carbon over a period of four years was then predicted and compared with observed data using regression analysis.

Table 20.4 Model Parameters Computed
for Sub-humid Lake Victoria Crescent

Land Use	Cs (Shape Parameter)	G (Soil Organic Carbon Relative Growth Rate)
Annuals	69.7	0.02
Banana	376.6	0.09
Coffee	338.0	0.05
Range land	135.8	0.46

20.3 RESULTS

Summary values from the regression analysis (Table 20.5)
indicate that the model performed well in predicting soil
organic carbon sequestration in annuals, banana, and coffee.
Over 50% of the variation in soil organic carbon was explained
by the model for annuals, banana, and coffee land use showing
that the assumption upon which the predictions were based
held true. Predictions were more accurate in banana and
coffee land use where there are better soil conservation practices (minimizing soil loss) and no tillage (minimizing oxidative soil organic carbon losses), unlike in annual crops. The
poor performance of the model on range lands may have been
due to errors in accounting for the soil organic carbon inputs
and losses. It was assumed that all the herbage in the range
lands constituted inputs, whereas in reality part of it is grazed
by livestock, which led to a net loss of input and/or diffuse
redistribution of the soil organic carbon inputs as dung. On

Table 20.5 Predictive
Performance of the Model

Land Use	Slope	Intercept	R^2
Annuals	−0.30	2.83	0.54
Banana	−0.19	4.29	0.98
Coffee	0.30	2.60	0.70
Range land	−0.07	4.41	0.07

the soil organic carbon losses, erosion was assumed to be the major loss pathway and yet annual sediment yields are low on this land use (Mulebeke, 2002).

In light of the Cs and g values, the model shows that there is a great potential for soil C sequestration in perennial crops followed by range lands compared to annuals.

20.4 CONCLUSION AND RECOMMENDATIONS

1. Soil organic carbon content decreased over the period (1962 to 2002) depending on the agroclimatic zone and land use. In the Lake Victoria Basin, the relative decline was of 50%, 27%, 21%, and 12% for annual crops, banana, coffee, and range lands, respectively.
2. Under average climatic conditions, perennial crops have higher potential for soil organic carbon sequestration than annual crops in Uganda's subhumid agroclimatic zone of the Lake Victoria crescent.
3. There is great potential for using the modeling approach to evaluate the influence of various factors on soil organic carbon sequestration.

Therefore, it is recommended that the model be tested across other agroclimatic zones, and that long-term experimental data be assembled to develop a defensive system for soil C accounting.

REFERENCES

Food and Agriculture Organization. 1998. *The Soil and Terrain Database for Northeastern Africa: Crop Production System Zones of IGAD Sub Region*. FAO Land and Water Digital Media Series 2. FAO, Rome.

Komutunga, E.T. and Musiitwa, F. 2001. Climate. In J.K. Mukiibi, Ed., *Agriculture in Uganda*. Fountain Publishers/CTA/NARO, Kampala, Uganda.

Lal, R. and Kimble, J.M. 2000. Tropical ecosystems and the global C cycle. In R. Lal, J.M. Kimble, and B.A. Stewart, Eds. *Global Climate Change and Tropical Ecosystems*. Advances in Soil Science. Lewis Publishers, Boca Raton, FL, pp. 3–32.

Lufafa, A., Tenywa, M.M., Isabirye, M., Majaliwa, M.J.G., and Woomer, P.L. 2003. Prediction of soil erosion in a Lake Victoria Basin catchment using GIS-based Universal soil loss model. *Agric. Syst.*, 76:883–894.

Mulebeke, R. 2002. Validation of a GIS-USLE Model in a Banana-Based Microcatchment of the Lake Victoria Basin. Master's thesis, Makerere University, Kampala, Uganda.

Nantumbwe, C. 2002. Evaluation of Temporal Land Use and Soil Property Interrelationships on the Slopes of Mt. Elgon. Master's thesis, Makerere University, Kampala, Uganda.

Parton, W.J., Woomer, P.L., and Martin, A. 1994. Modeling soil organic matter dynamics and plant productivity in tropical ecosystems. In Woomer, P.J., and Swift, M.J., Eds. *The Biological Management of Tropical Soil Fertility*. Wiley-Sayce, Chichester, United Kingdom, pp. 121–188.

Paul, E.A. and Clark, F.E., Eds. 1996. *Soil Microbiology and Biochemistry*, 2nd ed. Academic Press, New York.

Radwanski, S.A., Ed. 1960. *The Soils and Land-Use of Uganda*. Memoirs of the Research Division, Series 1, no. 4, Soils. Uganda Protectorate, Department of Agriculture, Kampala, Uganda.

Sanchez, P.A., Shepherd, K.D, Soule, M.J., Place, F.M., Mokwunye, A.U., Buresh, R.J., Kwesiga, F.R., Izac, A.M.N., Ndiritu, C.G., and Woorner, P.L. 1997. Soil fertility replenishment in Africa: an investment in natural capital. In Buresh, R.J., Sanchez, P.A., and Calhoun, F., Eds. *Replenishing Soil Fertility in Africa*. SSSA Special Publication 51. Soil Science Society of America, Madison, WI, pp., 1–46.

Tenywa, M.M., Isabirye, M.I., Lal, R., Lufafa, A., and Achan, P. 1999. Cultural practices and production constraints in smallholder banana-based cropping systems of Uganda's Lake Victoria Basin. *Afr. Crop Sci. J.*, 7:613–623.

Van Veen, J.A., Ladd, J.N., and Frissel, M.J. 1984. Modelling C and N turnover through microbial biomass in soil. *Plant Sci.*, 76:257–274.

Wortmann, C.S. and Eledu, C.A., Eds. 1999. *Uganda's Agroecological Zones: A Guide for Planners and Policy Makers*. Centro International de Agricultura Tropical, Cali, Colombia.

21

Soil Carbon Sequestration in Dryland Farming Systems

PARVIZ KOOHAFKAN, ANA REY, AND
JACQUES ANTOINE

CONTENTS

21.1 INTRODUCTION: FOOD AND AGRICULTURE ORGANIZATION AND CLIMATE CHANGE

Issues related to climate change are high on the agenda of the Food and Agriculture Organization of the United Nations (FAO). FAO is an active partner in the implementation of the different conventions related to climate change, notably UN Framework Convention on Climate Change (UN-FCCC), UN Convention to Combat Desertification (UN-CCD), and the Convention on Biological Diversity (UNCBD). FAO is particularly concerned with the effects of agricultural (including forestry and fisheries) practices on climate change, the role that agriculture can play to mitigate these effects, the impact that climate change may have on the food security situation in developing countries as well as adaptation strategies to face challeges of climate change.

Agricultural activities can be both a problem for and a solution to climate change concerns. Land use conversion and soil cultivation practices have contributed and continue to contribute to greenhouse gas (GHG) emissions in the atmosphere, and agriculture is thought to be a driving factor for about a third of the GHG emissions according to the Intergovernmental Panel on Climate Change (IPCC, 2000). Agricultural activities such as ploughing land and shifting cultivation practices ("slash and burn") release significant amounts of CO_2 in the air. A large part of the human-caused methane (CH4) emissions originates from the decomposition of organic matter in flooded rice paddies. About 25% of the world methane emission comes from livestock. In addition agriculture is responsible for about 80% of the nitrous oxide (NOx) emissions through the breakdown of fertilizer and manure and urine from livestock. On the other hand, the use of improved agricultural and forestry practices would significantly mitigate these emission levels, while at the same time store considerable amount of carbon in the soil and biomass.

In its current medium-term plan, FAO is implementing a series of activities within its so-called Priority Area for Interdisciplinary Action on Climate Change (PAIA-CLIM). An

Interdepartmental Working Group on Climate Change has been established to guide and coordinate the work of PAIA-CLIM in developing and disseminating normative and methodological approaches integrating climate change aspects into agricultural, forestry, and fisheries activities, and providing technical support to member countries on climate change questions and activities related to agriculture.

FAO plays a key role as a neutral mediator between the agricultural community and international agencies and institutions in the international negotiations process. Thereby, FAO contributes to the implementation of agreements that ensure the fair and balanced participation of all countries, particularly developing countries.

FAO collaborates on technical matters via the flow of information and the development of methodological tools. Furthermore, FAO has made available comprehensive data sets that can be very useful in the design and implementation of specific projects. Capacity building is another key activity in which FAO is heavily involved. FAO has vast experience assisting developing countries to consider climate change in the formulation of their national agricultural development policies. Specifically, this refers to the development of policies that create incentives for farmers and land users to limit their GHG emissions that promote carbon sinks in agriculture and forestry.

The work of FAO aims to identify, develop, and promote cultural practices that reduce agricultural emissions and sequester carbon, while contributing to improve the livelihoods of farmers, especially in the developing countries, through increased production and additional incomes from carbon credits under the mechanisms that have emerged since the Kyoto Protocol (KP). Specific activities include the promotion of practices that prevent desertification, land degradation, and loss of biodiversity, improve nutrient cycling, and increase carbon sequestration (CS), and consequently, food security. Examples are the intensive activities promoting conservation agriculture (FAO, 2002a), efficient plant nutrition management (FAO, 1998), soil biodiversity (Bennack et al., 2003), and assessment and prevention of land degradation (FAO, 2002b).

21.2 FOOD AND AGRICULTURE ORGANIZATION'S ACTIVITIES IN CARBON SEQUESTRATION IN DRYLAND FARMING SYSTEMS

In early 2002, the Land and Plant Nutrition Management Service (AGLL) of the Land and Water Development Division of FAO initiated a project cosponsored by the Global Mechanism (GM) on carbon sequestration incentive mechanisms to combat land degradation and desertification. The central objective of the program is to assist dryland countries in the formulation of policy and technical options for agricultural systems that can improve livelihood of poor farmers through CS. One output of the program is a knowledge base on soil CS in drylands. The knowledge base is to contain factual information on the potentiality of CS in drylands across different land use and management systems, as well as an overview of the policies and clarification of the different economic incentives regarding soil CS. AGLL is implementing the knowledge base in collaboration with the Universities of Essex and Lund.

There are many reviews and studies on the subject of CS, and some have highlighted the global and regional potential offered by drylands (e.g., Batjes, 1998; Rosenberg et al., 1999; Lal, 2001, 2003) and the scientific progress made on this issue (Izaurralde et al., 2001). However, studies on the potential of CS under local farming conditions in rural dryland communities in developing countries are few. The AGLL/GM project aims at contributing to filling the existing gap in knowledge at that level. This paper evaluates specific options for land management practices by analyzing some case studies carried out in several distinctive drylands areas of the world. The final aim of FAO is to facilitate the dissemination of such practices in soil CS programs in similar agroecological environments in other countries to improve food security and rural livelihood in those areas.

21.3 MAIN CHARACTERISTICS OF FARMING SYSTEMS IN DRYLANDS

Drylands have particular characteristics that will affect their capacity to sequester carbon. Drylands often experience high temperatures, low and erratic rainfall, minimal cloud cover, and small amounts of plant residues to act as surface cover to minimize radiation impact. As a result, soils in the drylands are, generally, both inherently low in organic matter and nutrients and rapidly lose large proportions of those small quantities as CO_2 when exposed by tillage and other conventional practices. Exposed and loosened soils are also highly prone to soil erosion, particularly rainfall patterns that include intense, storm precipitation after long dry periods. The key issue in drylands is therefore to maximize the capture, infiltration, and storage of rainfall water into soils by promoting conditions that accumulate organic matter and increase soil biodiversity. Drylands are particularly prone to soil degradation and desertification, with 70% of the agricultural land degraded. This means that soils have lost considerable amounts of carbon. As a consequence, the C stock of most dryland soils is less than 1%, and in many cases less than 0.5% (Lal, 2002). Increasing soil quality is therefore the main strategy for CS in drylands. Because drylands cover approximately 43% of the Earth's land surface (FAO, 2000), and dryland soils have lost carbon as a result of land degradation, they offer a great potential to sequester carbon (Scurlock and Hall, 1998; Rosenberg et al., 1999). Furthermore, soil carbon decomposition is also dependent on soil moisture, so dry soils are less likely to lose carbon (Glenn et al., 1993), and consequently the residence time of carbon in drylands is much longer than, for instance, in forest (Gifford et al., 1992). Whereas forest and intensive farming systems may be important carbon sinks, increasing carbon in degraded agricultural soils of dryland regions would also have direct environmental, economic, and social benefits to the local people and smallholders that depend on them.

Although most of the research on soil organic matter dynamics and processes has been conducted in temperate zones, several reviews have highlighted the potential offered by drylands and degraded lands to sequester carbon (Izaurralde et al., 2001; Lal, 2001). Agricultural productivity in drylands is not only limited by natural constraints, but also by low input management as a result of limited resources and technologies. The depletion in soil carbon in agricultural soils as a result of land misuse and soil mismanagement, can be reversed. Important strategies to improve productivity include (1) growing adapted species, (2) enhancing water-use efficiency and water retention in soils, (3) managing and enhancing soil fertility, and (4) adopting improved cropping systems (Lal, 2001). Improved cropping systems include crop rotations, planted fallows, residue mulch, conservation of trees, and growing leguminous species. Recommended practices involve soil water conservation and management, irrigation management, soil fertility management either by adding inorganic fertilizers or organic inputs, residues management, and reduced or zero tillage (Lal, 2003). Some of these practices are the main principle of conservation agriculture, which has been proven to be effective in increasing productivity and CS. Furthermore, the fact that conservation agriculture requires much less external inputs makes it more attractive to poor farmers. The case studies presented here analyse the effect of such practices on soil carbon stocks in various dryland systems.

21.4 CASE STUDIES IN DRYLANDS

Some global estimates have been made about the potential for CS in drylands. The total amount of C loss as a consequence of desertification may be 18 to 28 Pg (10^{15} g, or 1 gigaton) C (Lal, 2001). Assuming that two-thirds of the C lost (18 to 28 Pg) can be resequestered (IPCC, 1995) through soil and vegetation restoration, the potential of C sequestration through desertification control is 12 to 18 Pg C (Lal, 2002). The case studies presented here assess the effect of different management practices on soil carbon stocks in various dryland ecosystems.

The effect of climate and/or land use change can be predicted only through the use of accurate dynamic models. Given the difficulty of measuring changes in soil carbon stocks, modeling is a useful tool and has been used as an effective methodology for analysing and predicting the effect of land management practices on soil carbon stocks. A number of process-based models have been developed over the last two decades and are available (as reviewed by Smith et al., 1997). FAO has developed a model in collaboration with the University of Trent (Canada) as a methodological framework for the assessment of carbon stocks and prediction of CS scenarios that links SOC turnover simulation models (particularly CENTURY and RothC-26.3) to geographical information systems and field measurement procedures (FAO, 2004a).

For the case studies, the CENTURY 4.0 (Parton et al., 1987, 1988) model was used. It has been tested against a variety of long-term agricultural field trials and has also been used in a variety of climatic zones, including dryland regions. The ability of any model to predict accurately into the future depends on the accuracy and quality of the data used to parametize the model (climatic, soil, and land management data). Few studies contain sufficiently detailed information and the complete data set required for modeling purposes, particularly in dryland regions where such studies are scarce.

Data from four distinctly different dryland systems in Nigeria (Kano region), Kenya (Makueni district), India (Andhra Pradesh and Karnataka States), and Argentina (Monte Redondo and Santa Maria provinces) were used in investigations carried out by Essex University. Table 21.1 summarizes the main characteristics of these systems.

21.4.1 Case Study 1: Nigeria

Nigeria comprises some of the most densely inhabited areas of semi-arid West Africa (Harris, 2000). As a result, the soils of this region have been cultivated for long periods. Plant

Table 21.1 Main Characteristics of Study Sites

Case Study	1 Kano Region	2 Andrha Pradesh and Karnataka States	3 Makueni District	4 Tucuma, Catamarca, and Cordoba Provinces
Country	Nigeria	India	Kenya	Argentina
Soil type	Ferrugineous tropical soils, sandy, poor water holding capacity (WHC) and low-nutrient organic matter content	Alfisols and vertisols	Ferralsols naturally low in P	Haplic Phaeozems
Farming systems	Smallholder farming: Intensive: permanent annual or biannual cultivation (cropping intensity >60%) Less intensive: shrub/short-bush fallow regime (30% to 60%) Extensive: long-bush fallow and uncultivated areas (<30%)	Smallholder farming Integration among livestock, crops, and trees	Annual or multiple cropping	Grazed prairie Row cropping
Farming practice	Low-input systems Cattle manure	Cattle manure Tillage Soil fertility treatment	Short in livestock Very little fertilizer Crop residue management	Inorganic fertilizers No tillage
Study sites	**Futchimiram, Borno State**: Low-intensity agropastoral CP: 5-year cycle of grazing and millet cropping	**Lingampally** Large mixed dryland farming	**Darjani**	**Tucuman province**: graze prairie and row cropping

	Kaska, Yobe State:	Dagaceri, Jigawa State:	Tumbau, Kano close-settlement zone (CSZ):	Kaiani	Kymausoi	Athi Kamunyuni	Yedakulapaly	Metalkunta	Malligere	Tucuman province:	Catamarca province	Cordoba province
	Low-intensity agropastoral CP: 7-year cycle of grazing and millet-cowpea cropping	Intensive agropastoral (legumes and grains) CP: shrub with short-bush fallowing	Highly intensive agricultural CP: crop and livestock production system with intercropping of legumes, intensive manuring, and inorganic fertilizer				Large farmers using irrigation	Small mixed dryland farming Mixed crop and livestock	Small, mixed dryland farming	graze prairie and row cropping		
Main crops	Millet, sorghum, groundnut, sesame, cowpea			Maize and pulses Millet, cowpea, sorghum			Large agrodiversity, between 8 to 10 crops: paddy, sorghum, maize, millet, groundnut, coconut, cotton, etc.			Maize, sunflower, wheat, and soybean		
Natural vegetation	Open forest savannahs Grasslands						Grass, woodland system					

Note: CP = current practices.

production is limited by rainfall and nutrients (Breman and De Wit, 1983). The economy and infrastructure of northern Nigeria are not suited to high external inputs or fertilizers, and thus smallholder farming units operate as low-input systems. Legumes such as cowpeas are used to provide nitrogen inputs.

CENTURY was run for several practices at four sites: Futchimiram, Kaska, Dagaceri, and Tumbau for the last 50 to 60 years with alternate cycles of grazing and cropping (Table 21.1). Land degradation is a problem at all sites. Current practices (CP) were compared with continuous cultivation (CC), additions of inorganic fertilizer (IF), farmyard manure (FYM), plant residues (PR), and retained plant residues and grazing (NG). The predicted annual change in soil carbon for the various scenarios is presented in Figure 21.1.

The modeling exercise of the farming systems in the Nigerian case studies shows that soil carbon stocks can be increased from a low base with a variety of technologies and practices already available to farmers (Figure 21.1). The total amount of carbon that can be sequestered with the use of legumes, fallow periods, farmyard manure, and retention of plant residues varied between 0.1 to 0.3 metric tons C ha^{-1} $year^{-1}$. Figures for CS were slightly higher when trees were introduced. The use of inorganic fertilizers caused no change or loss of soil carbon. Continuous cultivation (reduced fallows) caused small carbon losses each year when no additional organic inputs were provided. However, when a cropping practice is accumulating significant amounts of carbon, fallowing will decrease the CS potential. Despite the intensification of the current systems, the levels of carbon were maintained. The main conclusion from these systems is that CS can only be achieved by increasing organic inputs into the soil.

21.4.2   Case Study 2: Kenya

Arid and semi-arid lands occupy two-thirds of Kenya (Nandwa et al., 1999). Erratic rainfall and poor fertility as a result of intensive cultivation are the main limiting factors of plant productivity. Droughts have also affected farming livelihoods.

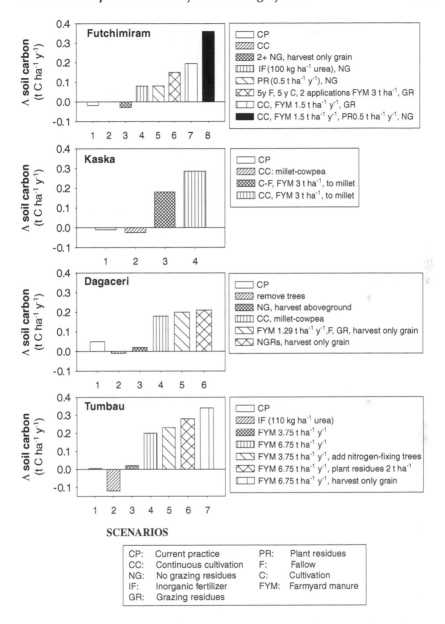

Figure 21.1 Average annual change in soil carbon stock under various practice scenarios (CENTURY) for the case study 1: Nigeria.

Annual or multiple cropping is practiced with occasional fallow periods. CENTURY was run to equilibrium using a grassland tree scenario with grass fires every 10 years and major fires every 30 years at four farmland sites: Darjani, Kaiani, Kymausoi, and Athi Kamunyuni. CPs were compared with CC, burn residues (BR), inorganic fertilizers (IF), and FYM applications (Figure 21.2).

The practices that led to increased CS were the addition of organic material in the form of farmyard manure and plant residues, particularly when systems are at steady state or near. Removal of fallows resulted in losses of 0.1 metric tons C $ha^{-1}y^{-1}$. In this case, the use of inorganic fertilizers was also inadequate as a sole source of plant nutrients. The combination of legumes in rotations, 2 to 4 metric tons ha^{-1} $year^{-1}$ of farmyard manure in addition to 0.6 metric tons ha^{-1} $year^{-1}$ of plant residues results in the highest rate of CS in all dryland cases: 0.7 metric tons C ha^{-1} $year^{-1}$.

21.4.3   Case Study 3: India

Over half of the farming population of India live in semi-arid regions. Crop yields have recently increased as a result of the green revolution. In this case, irrigation and inorganic fertilizers are expensive and inaccessible for the rural poor. Soil fertility treatments using legume cultivation and vermicompost production are increasingly used, leading generally to increased organic matter. Meanwhile, practices such as the use of inorganic fertilizers and the continuous cultivation of cereals lead to a substantial decline in soil carbon levels.

CENTURY was run for several practices at three sites: Lingampally, Metalkunta, and Malligere. The CP is 1 year of fallow and 4 years of cropping. The use of IF, FYM, green manure, vermicompost (V) and PR, trees, and legumes inclusion were compared with CP (Figure 21.3).

The use of farmyard manure, green manure, and vermicompost and plant residues produced increases in soil carbon of 0.2 to 0.4 ha^{-1} $year^{-1}$. Increases in soil carbon can also be obtained by leaving crop residues in the soil. Agroforestry practices substantially increased soil carbon by 0.9 metric

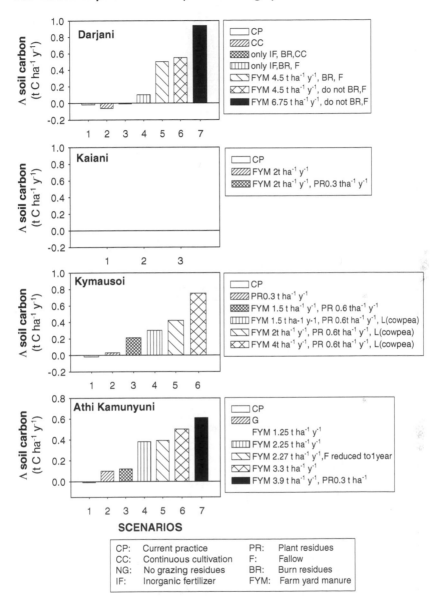

Figure 21.2 Average annual change in soil carbon stock under various practice scenarios (CENTURY) for the case study 2: Kenya.

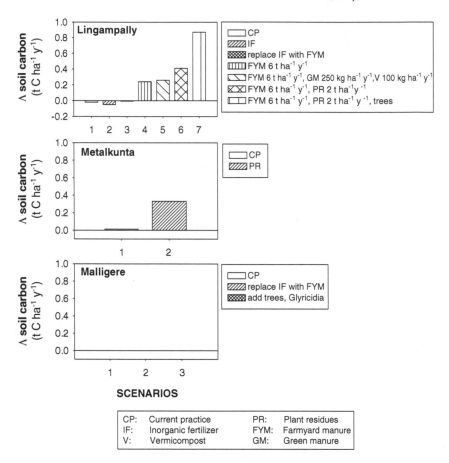

Figure 21.3 Average annual change in soil carbon stock under various practice scenarios (CENTURY) for the case study 3: India.

tons C ha^{-1} year^{-1}. Leguminous crops had also clear beneficial effects when included in rotations. The use of inorganic fertilizers resulted in carbon loss. Inorganic fertilizers and irrigation both have a carbon cost, reducing the amount of sequestered carbon. A full carbon accounting in these farming systems should consider the high-energy cost of nitrogen fertilizers (Pretty et al., 2002), the use of mechanized operations, and the transfer of carbon from one farm to another by livestock.

21.4.4   Case Study 4: Argentina

In recent years, Argentina has adopted reduced or zero-till-age practices especially in dryland regions, as a result of soil degradation. Two case studies under various conventional and zero-tillage systems were used: Monte Redondo and Santa Maria provinces (Figure 21.4). The use of CT, zero-tillage (ZT), IF, FYM, and green manure addition scenarios were compared.

Adopting zero tillage will halt the decline in soil carbon. However, to induce CS, organic inputs are needed (green and farmyard manures), and can be used to replace the inorganic fertilizer applications.

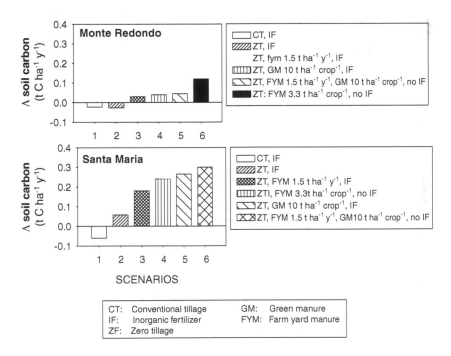

Figure 21.4 Average annual change in soil carbon stock under various practice scenarios (CENTURY) for the case study 4: Argentina.

The two cases studied in Argentina showed that carbon stocks were reduced considerably after cultivation. At all locations, sharp falls in carbon stocks occurred of about 15 metric tons ha^{-1}. The adoption of no-tillage practices has halted this decline and even led to a small increase in carbon levels on the order of 0.02 metric tons ha^{-1} year^{-1}. Rotations with significant periods for return to grasslands (4 years in 11) resulted in further carbon increases. The highest rates of CS occurred when zero tillage includes cultivation with green manures and addition of farmyard manures.

21.4.5 Overall Findings from Case Studies

The scenarios show that CS in tropical dryland soils can be achieved at the various sites (Table 21.2). The land management practices were chosen to be in accordance with the current farming systems and provide specific real cases for CS strategies in different dryland systems. For example, application rates of organic matter are commensurate with quantities that should be available to local farmers. However, at the field level, important tradeoffs may occur, preventing adoption of the best strategies for CS. Crop residues may be required for livestock feed or fuel, or during difficult times may be sold. Animal manures may be burned for fuel. Many socioeconomic factors will interact to determine which scenario or combination of scenarios is implemented in each growing season, all of which must be analyzed for each case.

Some of the results predict that soil carbon can be restored to precultivation levels or in certain circumstances to above these levels. The true "native soil carbon level" is often difficult to establish in those systems where agricultural activity has been present for at least several centuries or millennia, such as in the Nigeria and Kenya cases. To achieve quantities of soil carbon in excess of the "natural level" implies that the agricultural system has greater productivity than the native system, assuming that carbon is not being imported. The scenarios that predict the highest CS rates are often associated with the introduction of trees to the system. The inputs of carbon from trees are more resistant to

Table 21.2 Summary of Findings on C Stocks (metric tons ha^{-1}) and Rates (metric tons ha^{-1} year^{-1}) of Carbon Sequestration or Loss in Four Dryland Ecosystems Compared with Rates of CS in Drylands Reported by Lal (1999)

Management Practice	Nigeria	India	Kenya	Argentina	Lal
C stock before cultivation	8–23	15–20	33–41	50–70	—
Current C stocks	6–12	13–22	18–28	37–41	—
Conventional tillage	−0.05 to −0.01	−0.07 to 0.06	−0.3 to −0.1	−0.17 to 0.19	—
Green and farmyard manures, plant residues, and fallows in rotations	0.1 to 0.3	0.2 to 0.4	0.4 to 0.9	—	0.01 to 0.3
Effect of trees	Plus additional 0.05 to 0.15	Plus additional 0.05 to 0.15	—	—	—
Inorganic fertilizers as sole source of nutrients	−0.12 to 0.08	−0.01	−0.3	—	—
Zero tillage	—	—	—	0.02	0.1 to 0.2
Zero tillage + green or farmyard manure	—	—	—	0.1 to 0.25	0.15 to 0.3
Zero tillage + inorganic fertilizers	—	—	—	0.04	—

decomposition than those from herbaceous crops, and consequently can cause marked increases in the level of soil carbon
(Falloon and Smith, 2002). The highest rates of sequestration
(0.1 to 0.25 metric tons ha^{-1} year^{-1}) occur when zero-tillage
systems also include cultivation of green manures and additions of farmyard manure. Inorganic fertilizers were generally
inefficient in providing the necessary nutrients, and therefore
increase CS. CS is greatly enhanced by including cover crops
in the rotation cycle. Cover crops enhance soil biodiversity
which is known to increase CS. The modeling data from the
case studies are similar with rates of soil carbon sequestration
under various land management regimes in drylands reported
by Lal (1999), and reproduced in the last column of Table 21.2.

12.5 CONCLUSIONS

There are vast areas of dryland ecosystems in the world, many
in developing countries, where improvement of farming systems can add carbon to soils. Results from the various case
studies presented above show that several practices are available to increase carbon stocks in soils.

Whereas CS itself is not a priority in poor countries, land
management options that increase CS, enhance plant productivity, and prevent erosion and desertification are of major
interest in these regions. However, it is unlikely that current
mechanisms, such as the Clean Development Mechanism
(CDM), can provide the necessary funds for these regions
(FAO, 2002c). Although soils are the major terrestrial carbon
reservoir, and agriculture is recognized as one of the major
causes of GHG emissions to the atmosphere (IPCC, 2001),
neither soils nor land use practices are eligible under the first
commitment period of the Kyoto Protocol. This is partly
because of the difficulty of measuring and verifying carbon
sequestered and maintained in soils, particularly in drylands
where any changes in soil carbon will be small. However,
actions for soil fertility improvement through increased
organic matter in dryland soils are a real win–win situation
with higher agronomic productivity, reduced land degradation, and increased CS and biodiversity. Enhancing CS in

degraded agricultural lands, could have direct environmental, economic, and social benefits for local people. It will increase farmers' benefits as well as help mitigate global warming, at least during the next few decades until other alternative energy sources are developed. Therefore, initiatives that sequester carbon are welcomed for the improvement of degraded soils, plant productivity, and consequently food safety and alleviation of poverty in dryland regions, and are among the main priorities of FAO.

While a purely carbon market approach is not likely to be applicable to smallscale farming systems in developing countries, a multilateral approach for mobilization of resources under existing mechanisms is what is required. The Global Mechanism of the UN Convention to Combat Desertification (CCD) promotes such a multilateral path in implementing its mandate to increase the effectiveness and efficiency of existing financial resources and to explore new and additional funding mechanisms for the implementation of the convention. Specific emphasis is given to small scale farming systems in dryland areas of the developing countries. Multilateral approaches include sources to combat climate change with desertification funds, links with sustainable livelihoods and provision of visible benefits to local people, mobilizing resources from the private sector, and so on. Several UN conventions are closely linked and share common goals. The CCD is concerned with the degradation of extensive areas that show declining crop production and are insufficient for the needs of local people as a result of land degradation and reduction in carbon stocks. Therefore, the CCD, the Climate Change Convention (CCC), the Convention on Biodiversity (CBD), and the Kyoto Protocol all share a common goal, the proper management of soils to increase soil carbon. Therefore, an important goal of the FAO-GM program on CS is to promote synergies with the conventions and the private sector for the establishment of an environmental fund specifically targeted to CS projects in drylands. Opportunities exist for bilateral partnerships with industrial country institutions to initiate soil CS projects involving local communities also linked to global networks on CS. FAO believes that more effort should be put into exploring and exploiting those opportunities.

FAO will take part in the design and implementation of CS programs in tropical dryland countries based on the regional policies, and will bring the attention of governments to the possible benefits that CS measures could bring to dryland farming communities and society. FAO could also play an important role in providing secure institutional support for the implementation of CS programs that encourage the collaboration between local farmers and investors.

ACKNOWLEDGMENTS

The case studies presented here were carried out by P. Farage, J. Pretty, and A. Ball of the University of Essex (United Kingdom), L. Olsson, Lund University (Sweden), and P. Tschakert of the University of Arizona (United States) in collaboration with A. Warren of University College London (United Kingdom) and have been published in an FAO publication called *Carbon Sequestration in Dryland Soils* (FAO, 2004b). The information for the Nigerian and Kenyan case studies was supplied by M. Mortimore and M. Tiffen (Drylands Research, Crewkerne, United Kingdom), and the Indian case study used data collected by B. Adolph and J. Butterworth (Natural Resources Institute, Chatham, United Kingdom, in association with the Deccan Development Society, Hyderabad and Pastapur, the BAIF Institute of Rural Development, at Tiptur and Lakihalli, India). Details of the Argentine systems were provided by E.A. Rienzi (Faculty of Agronomy, University of Buenos Aires, Argentina). The Senegal case study is based on the work of P. Tschakert. The University of Edinburgh (United Kingdom) supported the FAO through an academic exchange programme for Dr. Ana Rey at the Land and Plant Nutrition Management Service (AGLL).

REFERENCES

Batjes, N.H. 1998. Mitigation of atmospheric CO_2 concentrations by increased carbon sequestration in the soil. *Biol. Fertil. Soils,* 27:230–235.

Bennack, D., Brown, G., Bunning, S., and Hungria da Cunha, M. 2003. Soil biodiversity management for sustainable and reductive agriculture. In Biodiversity and the Ecosystem Approach in Agriculture; Forestry and Fisheries. Proceedings, Ninth Regular Session of Commission on Genetic Resources for Food and Agriculture. Food and Agriculture Organization, Rome.

Breman, H., and De Wit, C.T. 1983. Rangeland productivity and exploitation in the Sahel. *Science,* 221:1311 –1347.

Falloon, P., and Smith, P. 2002. Simulating SOC changes in long-term experiments with RothC and CENTURY: a model evaluation for a regional scale application. *Soil Use Manage.,* 18:101–111.

Food and Agriculture Organization. 1998. *Guide to Efficient Plant Nutrition Management.* FAO, Rome.

Food and Agriculture Organization. 2000. *Carbon Sequestration Options Under the Clean Development Mechanism to Address Land Degradation.* World Soil Resources Report 92. FAO, Rome.

Food and Agriculture Organization. 2002a. *Conservation Agriculture. Case Studies in Latin America and Africa.* Soils Bulletin 78. FAO, Rome.

Food and Agriculture Organization. 2000b. *Land Degradation Assessment in Drylands — LADA Project 2002 (E).* World Soil Resources Report 97. FAO, Rome.

Food and Agriculture Organization. 2002c. Soil carbon sequestration for improved land management by Robert, M. World Soil Resources Report 96. FAO, Rome.

Food and Agriculture Organization. 2004a. *Assessing Carbon Stocks and Modelling Win-Win Scenarios through Land-Use Changes.* FAO, Rome.

Food and Agriculture Organization. 2004b. *Carbon Sequestraton in Dryland Soils.* World Soil Resources Report 102. FAO, Rome.

Gifford, R.M. Cheney, N.P., Nobel, J.C., Russell, J.S., Wellington, A.B., and Zammit, C. 1992. Australian land use, primary production of vegetation and carbon pools in relation to atmospheric carbon dioxide levels. *Bur. Rural Res. Proc.*, 14:151–187.

Glenn, E., Squires, V., Olsen, M., and Frye, R. 1993. Potential for carbon sequestration in drylands. *Water Air Soil Pollut.*, 0:341–355.

Harris, F. 2000. *Changes in Soils Fertility Under Indigenous Agricultural Intensification in the Kano Region.* Drylands Research Working Paper 36. Drylands Research, Crewkerne, United Kingdom.

Intergovernmental Panel on Climate Change. 1995. *Climate Change. 1995. The Science of Climate Change.* Working Group 1. Cambridge University Press, London; New York.

Intergovernmental Panel on Climate Change. 2000. *Land Use, Land-Use Change, and Forestry.* Special report of Intergovernmental panel on Climate Change. Cambridge University Press, London; New York.

Intergovernmental Panel on Climate Change. 2001. *Climate Change: The Scientific Basis.* Cambridge University Press, London; New York.

Izaurralde, R.C., Rosenberg, N.J., and Lal, R. 2001. Mitigation of climate change by soil carbon sequestration: issues of science, monitoring, and degraded lands. *Adv. Agron.*, 70:1–75.

Lal, R. 1999. Soil management and restoration for C sequestration to mitigate the greenhouse effect. *Prog. Environ. Sci.*, 1:307–326.

Lal, R. 2001. Potential of desertification control to sequester carbon and mitigate the greenhouse effect. *Climate Change,* 51:35–72.

Lal, R. 2002. The potential of soils of the tropics to sequester carbon and mitigate the greenhouse effect. *Adv. Agron.*, 76:1–30.

Lal, R. 2003. Global potential of soil carbon sequestration to mitigate the greenhouse effect. *Crit. Rev. Plant Sci.*, 22:151–184.

Nandwa, S.M., Gicheru, P.T., Qureshi, J.N., Kibunja, C., and Makokha, S. 1999. Kenya — country report. In H. Nabhan, A.M. Mashali, A.R. Mermut, Eds. Integrated Soil Management for Sustainable Agriculture and Food Security in Southern and East Africa, Proceedings of Expert Consultation, Harare, Zimbabwe, December 1997. Food and Agriculture Organization, Rome.

Parton, W.J., Schimel, D.S., Cole, C.V., and Ojima, D.S. 1987. Analysis of factors controlling soil organic levels of grasslands in the Great Plains. *Soil Sci. Soc. Am. J.,* 51:1173–1179.

Parton, W.J., Stewart, J.W.B., and Cole, C.V. 1988. Dynamics of C, N, P, and S in grassland soils: a model. *Biogeochemistry,* 5:109–131.

Pretty, J.N., Ball, A.S., Xiaoyun, L., and Ravindranath, N.H. 2002. The role of sustainable agriculture and renewable resource management in reducing greenhouse gas emissions and increasing sinks in China and India. *Trans. R. Soc. London A,* 360:1741–1761.

Rosenberg, N.J., Izaurralde, R.C., and Malone, E.L. 1999. *Carbon Sequestration in Soils: Science, Monitoring, and Beyond.* Battelle Press, Columbus, OH.

Scurlock, J.M.O., and Hall, D.O. 1998. The global carbon sink: a grassland perspective. *Global Change Biol.,* 4:229–233.

Smith, P., Powlson, D.S., Smith, J.U., and Elliot, E.T. 1997. Evaluation and comparison of soils organic matter models using datasets from seven long-term experiments. *Geoderma,* 81(special issue).

22

More Food, Less Poverty? The Potential Role of Carbon Sequestration in Smallholder Farming Systems in Senegal

PETRA TSCHAKERT

CONTENTS

22.1 INTRODUCTION

Amartya Sen's critique of the prevailing climate–food supply–famine framework (Sen, 1981) caused the common agrotechnical assumption that more food production will also provide more food for the rural poor and less famine to be increasingly challenged. Today, most agree that food insecurity is primarily a result of low household incomes, poverty, and lack of access rather than inadequate aggregate food supply (Watts, 1991; Maxwell and Frankenberger, 1992; Davies, 1996; Dilley and Boudreau, 2001; Gladwin et al., 2001). Food security is defined as "sufficient food consumption by all people at all times for a healthy and productive life" (Thomson and Metz, 1997). Chronic food insecurity is a long-term problem, caused by lack of income or assets at the household level to produce or buy sufficient and adequate food for the entire household (Gladwin et al., 2001). As a solution, complex approaches that link food security issues to livelihood systems have been proposed rather than encouraging smallholders to grow more food crops. These approaches focus on multiple livelihood strategies at the household level. They include alternative, nonfarm income-generating activities that diversify and increase poor people's income, enhance their food security, and make livelihoods more sustainable in the long run (Devereux, 1993; Sanchez, 2000; Gladwin et al., 2001).

For much of Sub-Saharan Africa, however, Sen's legacy remains less relevant. Annual population growth rates of 3%, which are among the highest in the world, and continuously low cereal yields have limited aggregate food production (Gladwin et al., 2001). Taking into account climate change, projections look rather bleak, particularly for drylands. In the Sahel, agricultural productivity is projected to decrease as climate conditions become more extreme (Watson et al., 1996). The Intergovernmental Panel on Climate Change (IPCC) predicts that the greatest risk of increased hunger exists for people who are dependent on small, subsistence-based agricultural systems in semi-arid and arid regions. Millet yields for Senegal, for instance, are expected to decrease by 63% to

79% under 2× CO_2-equivalent global circulation model scenarios (Watson et al., 1996).

Carbon sequestration by small-scale farmers in drylands could potentially be an important mitigation strategy to address global climate change, and related problems of declining soil fertility, land degradation, and food insecurity. Soil carbon sequestration could indeed be a win–win strategy, as argued by Lal (1999b, 2002a), Kimble et al. (2002), and the Food and Agriculture Organization (2002), among others. It would reflect an example of how to integrate one of the lessons learned from food security and livelihood research, namely, the need to diversity resources and income in order to more effectively spread risk. The emphasis on diversified production systems is consistent with adaptive strategies for drought and desertification (Watson et al., 1996).

Results from a case study of the potential of carbon sequestration for small-scale farming systems in Senegal are presented below. The case study addresses four major questions: (1) From a biophysical–technical perspective, what are the best soil management practices and land use options to sequester carbon and increase crop yields? (2) What practices are economically most feasible for small-scale farmers? (3) Would 'best' practices increase crop yields and reduce poverty? (4) How would best practices fit into the livelihood strategies of local smallholders?

22.2 RESEARCH SETTING

22.2.1 Description of Study Area

The research was conducted in the west-central part of Senegal that is known as the "Old Peanut Basin" (See Figure 22.1). The name goes back to the introduction of peanuts, or groundnuts, by the French colonial power at the end of the 19th century in an area initially overlapping with today's administrative regions of Diourbel, Louga, and Thiès (Pelissier and Laclavére, 1984). At present, the term "Peanut Basin" or "Bassin Arachidier" reflects a certain socioeconomic entity in Senegal (Stomal-Weigel, 1988). It is a gently undulating plain

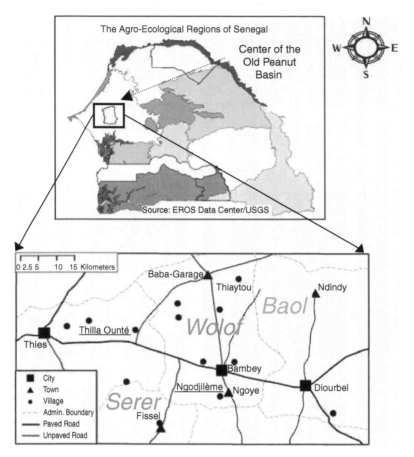

Figure 22.1 Diagram showing the research area and villages (underlined) in the Old Peanut Basin of Senegal.

with soils that contain large amounts of aeolian material. The main soil types are luvic Areneols, ferric Luvisols, and chromic Vertisols (Food and Agriculture Organization–UN Educational, Scientific, and Cultural Organization, 1974). Farmers distinguish typically between two types of soil: *dior* and *deck*. *Dior* soils contain more than 95% sand and less than 0.2% organic carbon and are usually found on former dune slopes. *Deck* soils are hydromorphic with 85% to 90% sand, and organic carbon contents of 0.5% to 0.8% (Badiane et al., 2000).

Annual precipitation ranges from 350 to 700 mm. It is barely suitable for rainfed agriculture although 90% of all arable lands are used for cultivation (Centre de Suivi Ecologique [CSE], 2000). The main crops are millet (*Pennisetum typhoideum*), groundnuts (*Arachis hypogaea*), sorghum (*Sorghum bicolor*), and cowpeas (*Vigna unguiculata*). The rainy season lasts from July to September or October. Spatial and temporal variation of rainfall are high, and episodic droughts and crop failures are common. The area under cultivation and crop yields also vary notably from year to year as indicated in Figure 22.2. The 1980–2001 mean millet yields were 527 kg ha^{-1}, and mean yields for groundnuts, the major cash crop, were 665 kg ha^{-1} (Direction de l'Analyse, de la Prévision et des Statistiques/Division de Statistiques, Documentation et Information Agricole [DAPS/DSDIA], 2001).

The scope of agricultural intensification within farming systems varies depending on population pressure, ethnic group affiliation, and availability of and access to land (Pelissier, 1966; Copans, 1988; Stomal-Weigel, 1988; Lericollais et al., 1998). Today, population density in the Old Peanut Basin

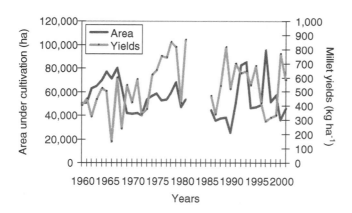

Figure 22.2 Area under production (ha) and yields (kg ha^{-1}) for millet, Département de Bambey, Région de Diourbel. (Data from DAPS/DSDIA.)

is higher than in any other agricultural zone in Senegal, ranging from 150 to 225 inhabitants km^{-2} (CSE, 2000). The two major ethnic groups are the Wolof and the Serer. Together they account for 88% of the population (CSE, 2000). Woody shrubs predominate and consist of *Balanites aegyptica (sump)* and *Guiera senegalensis (nger)*. The most common tree species are *Acacia raddiana (seing), Acacia nilotica (nebneb), Acacia seyal (surur), Adansonia digitata (gui)*, and *Faidherbia albida (kad)*.

22.2.2 Methods

Research was conducted at three sites, namely Thilla Ounté, Ngodjilème, and Thiaytou (Figure 22.1), from December 2000 to December 2001. These sites were part of an original group of 13. They were selected using a decision matrix derived from a stratified sampling procedure based on variations in soil types, agricultural activities, land use/land cover change as assessed through remotely sensed imagery, ethnic affiliation, village size, and proximity to major roads and markets.

One-day workshops on possible carbon sequestration projects were held with seven rural councils representing the '*communautés rurales*' in the research area. A market study was also conducted to assess prices of all major goods purchased and sold by farmers. Crop statistics, climate records, and reports on experimental farming sites were used to complement data gathered in the field.

The methods used in the villages were highly participatory and are equivalent to 'participatory rural appraisals' (Chambers, 1997). They included resource mapping at the village level to identify land use types and infrastructure (Lightfoot and Noble, 1993); individual household resource mapping; change matrices to assess dynamics in land use and management practices over time (Freudenberger and Schoonmaker, 1994); Venn diagrams to identify institutions influencing access to key resources (Guijt and Pretty, 1992); agricultural calendars with household labor hours; wealth ranking to stratify village households into resource endowment groups (Bellon, 2001); structured and semistructured interviews, focus groups, and discussions; and formal surveys at the field and household level (Bernard, 1994).

Soil and biomass carbon measurements were taken on seven fields in each village. They were selected using a stratified random sampling procedure and included both infields and outfields. Infields (*champs de case*) are directly adjacent to village compounds and typically receive more organic matter input. Outfields (*champs de brousse*) are fields farther away from the compounds. Herbaceous litter and root samples were collected from 42 randomly positioned replicate quadrates, and soil samples from 84 replicate quadrates from 0 to 20 cm and 20 to 40 cm profiles. Tree biomass was measured at diameter at breast height. Samples were analyzed at the Institut Sénégalais de Recherches Agricoles/Centre National de Recherche Agronomique, using the Walkley–Black method for soil samples. Past, present, and future carbon stocks and dynamics, the impact of land use and management practices, and various climate change conditions were simulated with CENTURY, a biogeochemical model (Parton et al., 1994). A detailed description of the model scenarios and input data can be found in Tschakert (2004a) and Tschakert et al. (2004).

A farmer-based cost–benefit analysis (CBA) was also performed, using data from the market and household surveys and group discussions (Tschakert, 2004b). Income-expenditure profiles for 15 management options and three resource endowment groups were created in Microsoft Excel. Net benefits and net present values at a 20% discount rate were calculated for 1 hectare of land over a 25-year period. Results from the CBA were integrated into a household cash-flow model designed in STELLA, an object-oriented graphical programming language (High Performance Systems, 2000). Model sectors and parameters are explained in Tschakert (2004b).

22.3 FOOD SECURITY AND LIVELIHOOD PROFILES

Smallholders in the Sahel rarely depend on agricultural production as their only source of household revenue (Gladwin et al., 2001; Mortimore and Adams, 1999). This is also true for those classified as subsistence farmers. In the Old Peanut Basin, farmers obtain on average less than 20% of their annual income from crops, although the actual amount varies considerably depending on the specific set of livelihood strategies pursued by individual households. As illustrated in Figure 22.3, noncrop sources of income include animal husbandry, off-farm activities, and remittances from seasonal and long-term migration. Relying on just one strategy alone is not sustainable for these households. The result is a mosaic of 'multienterprise rural household or production units' in which diversification of the resource endowment is a common practice in order to spread and reduce risk (Hunt, 1991).

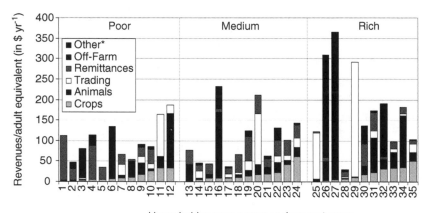

Figure 22.3 Annual revenues per adult equivalent and resource endowment group in 35 sample households in 2001. 'Other' includes income from local market sales, monetary gifts, traditional cultural services (*griots*), and revenues for *marabouts* (religious leaders). (Based on data from household surveys, 2001.)

The annual revenues per adult equivalent (AE) in the research sample varied from $30 to $365 as shown in Figure 22.3. (Using adult equivalents per household instead of entire households as the unit of analysis compensates for unequal household sizes. The computation factors used were adult = 1, child [aged less than 15 years] = 0.5, and temporary adult resident = 0.25.) The variation suggests that farmers are not a homogenous group. The importance of food and cash crop production depends on the availability and reliability of alternative sources of income, in addition to other components of a household's resource endowment, such as labor, land, and access to markets and extension services. Having alternative sources of income obviously reduces the pressure on food production. What cannot be produced can at least be purchased.

Small-scale farmers are also quite diverse when it comes to production and consumption. Table 22.1 shows such variation for three resource endowment groups in 2001. The poor household (household 5 in Figure 22.3) relies solely on its crops and remittances, and uses all of its land for cultivation. Because of the reduced labor force of these households (5.5 AE), labor hours spent on crops clearly exceed those of the medium and rich household. High labor inputs might also explain the fact that crop yields are highest for these households. However, millet production is not sufficient to feed the family, and additional food must be purchased for 4 months. The medium household (household 21 in Figure 22.3) has more land and a larger labor force. The number of labor hours spent on crops is fairly high. Yet, overall millet production in 2001 could not satisfy all household food needs. Finally, the rich household (household 31 in Figure 22.3) relies primarily on its cattle and trading to generate income. Food production is less of a priority for these households. Despite an abundance of land, natural regeneration through fallow practices, and a large labor force, labor hours spent on crops are low and millet yields are roughly half of those achieved by the poor household. From a livelihood perspective, this is not a problem because these households have sufficient income to purchase supplemental millet year round.

Table 22.1 Production and Consumption Profiles for Three Types of Households Based on Resource Endowment, 2001

	Poor Household	Medium Household	Rich Household
Total land available (ha)	1.42	7.69	17.32
Land in production (ha)	1.42	6.33	12.58
Fallow (ha)	0	0.4	4.74
Cropping pattern (ha)	0.74 millet	2.88 millet	6.20 millet
	0.49 sorghum	2.49 groundnuts	4.58 groundnuts
	0.19 groundnuts	0.,46 cowpeas	1.32 cowpeas
			0.49 watermelon
Inputs on millet	350 kg ha^{-1} household waste	1,280 kg ha^{-1} manure	27.7 kg ha^{-1} fertilizer
			27 cows stubble grazing (2000/2001)
Inputs on groundnuts	680 kg ha^{-1} household waste	0	27 cows stubble grazing (1999/2000)
Adult equivalents (AE)*	5.5	14.5	21
Labor hours millet/AE* (ha^{-1})	47	43	12
Labor hours groundnuts/AE* (ha^{-1})	120	98	32
Average millet yields (kg/ha^{-1})	893	579	451
Millet total production (kg)	660	1,300	2,900
Average groundnuts yields (kg ha^{-1})	773	322	652
Groundnuts total production (kg)	150	930	2,199
Expenditures millet purchase ($)	50	130	380
Months millet purchase	4	7	12
Income from crops ($)	5	19	22

Source: Based on fieldwork (household surveys) completed in 2001.

The above description of households suggests that addressing household food insecurity through soil carbon sequestration requires consideration of much more than simply increasing aggregate household crop yields. The types of carbon management activities that make most sense to farmers will depend on their individual assets and livelihood options and strategies. Potential increases in household food supply will be the main concern for some. For others it will be additional income from otherwise unprofitable types of land use, such as long-term set-aside lands. Given this large variation in resource management practices, more than one "best" practice may be needed. The most promising "best" practices in terms of carbon storage and increased crop yields for the Old Peanut Basin are discussed in the next section.

22.4 CARBON SCENARIOS AND SMALLHOLDERS' REALITY

22.4.1 Biophysical Potential

Past, current, and future soil and biomass carbon (C) dynamics under various management practices for the Old Peanut Basin are discussed in Tschakert (2004a) and Tschakert et al. (2004). Current soil C values were measured in the field and amounted to 11.3 metric tons ha^{-1} on average for the upper 0 to 20 cm soil layer. The average C stock in the 20 to 40 cm layer was 6.9 metric tons ha^{-1}, with a mean soil bulk density of 1.6 g cm^{-3}. Nowadays, such a low range is common for the area (Rabot, 1984; Badiane et al., 2000) and consistent with mean values for both the Sudano–Sahelian zone (Manu et al., 1991) and global averages for drylands, which range from 0.2% to 0.8% (Lal, 2002a, 2002b). However, measurements showed large variation between individual fields, with values ranging from 3.9 metric tons ha^{-1} to 29.6 metric tons ha^{-1} for upper soil C. This variation reflects different clay content and management practices. Observed tree C was 6.3 metric tons C ha^{-1}, while litter, root, and herbaceous C together accounted for 5 metric tons ha^{-1}.

The biophysical potential to increase current carbon stocks can be approximated by the amount lost over time. Results from C dynamics simulations with CENTURY suggest that total system C in the area, including soil to a 20-cm depth, has declined from 60 metric tons ha^{-1} under native savanna and before the introduction of agriculture to the present level of 17 metric tons ha^{-1}. This constitutes a total loss of 43 metric tons C ha^{-1} or 71% of the precultivation stocks. As shown in Figure 22.4, forest C decreased more rapidly than soil C. The results suggest a decline from 35 metric tons ha^{-1} in 1850 to 4.2 metric tons ha^{-1} in 2001 (0.2 metric tons C ha^{-1} year^{-1}) compared to 20 metric tons ha^{-1} and 11.9 metric tons ha^{-1} for soil in 1850 and 2001, respectively (0.05 metric tons C ha^{-1} year^{-1}).

How much C can in fact be sequestered through improved land use and management practices in order to boost today's low soil organic carbon values and approach historical values depends to a large extent on soil texture, precipitation, temperature, evapotranspiration, and the availability of nitrogen

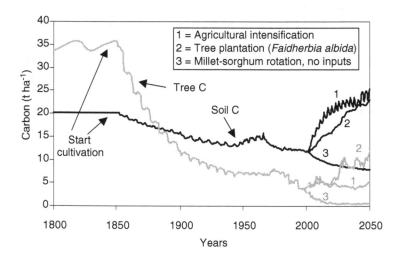

Figure 22.4 Changes in soil and tree carbon for the Old Peanut Basin, simulated with CENTURY. (From Tschakert, P. 2004a. Carbon for farmers: assessing the potential for soil carbon sequestration in the Old Peanut Basin of Senegal. Climatic Change, 67(2-3): 273–290. With permission.)

in the soil. Results from CENTURY simulations indicate that, over a 50-year period, total system C could be increased by 244% from a mean present value of 17.3 metric tons ha^{-1} to a maximum of 40.8 metric tons ha^{-1}. Simulated changes in soil and forest C for the 2001–2050 period are depicted in Figure 22.4. The largest gains in soil C can be expected under an optimal intensification scenario. (The optimum agricultural intensification scenario includes 3-year rotation cycle with groundnuts-millet-fallow, 150 kg of fertilizer [10-10-20] on groundnuts, 5 metric tons of sheep manure, and 2 metric tons of *Leucaena leucocephala* prunings on millet, 4 tons of manure on fallow, and improved millet seeds, as well as reducing tree pruning.) At the other extreme, soil C is likely to drop to 8.7 metric tons ha^{-1} under a millet–sorghum rotation with no external inputs. Significant increases in tree C can only be achieved under scenarios involving agroforestry (such as 250 to 300 nitrogen-fixing *Faidherbia albida* per hectare) or a conversion of croplands to tree plantations. Other management and land use practices are also likely to result in more soil C such as, for example, 3- to 10-year improved fallows in rotation with cropping cycles, the application of 4 to 10 metric tons of manure with and without mineral fertilizer, and the conversion of cropland to grassland with and without tree protection. Under these scenarios, soil C is expected to increase by 2 to 5.3 metric tons ha^{-1} over 25 years. Losses in soil C are projected for crop rotation without fallow and stubble grazing with no other inputs, reaching 1.7 to 3.9 metric tons ha^{-1} over 25 years.

Simulated changes in crop yields resulting from improved management practices are depicted in Figure 22.5. They varied from −62% to +200% for millet and −45% to +133% for groundnuts compared to 1980 and 2000 values (653 kg ha^{-1} and 707 kg ha^{-1}, respectively). This correlates well with the simulated changes in soil C. Under the worst-case scenario (millet–sorghum rotation with no external inputs), CENTURY suggested a drop from 653 kg ha^{-1} to less than 300 kg ha^{-1}. Clearly, this option would not be in the interest of farmers. Millet yields as high as 2 metric tons ha^{-1} and groundnut yields of 1.6 metric tons ha^{-1} could be reached under the

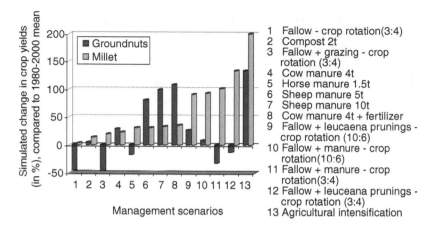

Figure 22.5 numbered legend:
1 Fallow - crop rotation(3:4)
2 Compost 2t
3 Fallow + grazing - crop rotation (3:4)
4 Cow manure 4t
5 Horse manure 1.5t
6 Sheep manure 5t
7 Sheep manure 10t
8 Cow manure 4t + fertilizer
9 Fallow + leucaena prunings - crop rotation (10:6)
10 Fallow + manure - crop rotation(10:6)
11 Fallow + manure - crop rotation(3:4)
12 Fallow + leuceana prunings - crop rotation (3:4)
13 Agricultural intensification

Figure 22.5 Simulated changes via CENTURY in crop yields (percent) as a result of improved management practices.

optimal agricultural intensification scenario. All gains in millet under fallow scenarios (scenarios 9 through 12) are based on the assumption that food crops are planted the first year after fallowing, a typical practice. Increases in groundnuts exceed those in millet (scenarios 6 to 8) when the first benefit from high manure and/or mineral fertilizer inputs made on the latter during the previous year. This reflects the typical 1-year lag effect of manure on millet (Badiane and Lesage, 1996).

When climate change is taken into account, however, such projected gains in both soil C and crop yields become more uncertain. Climate change simulations were also performed with CENTURY (Tschakert et al., 2004). They suggest that larger threats to C stocks and agricultural production loom in the future, and effective actions taken today may at best result in no additional losses in the future. Under extreme conditions, they might even be futile. For instance, optimal agricultural intensification under a 'high increase' climate change scenario would result in a decrease of soil C by 5% by the year 2025 and by 40% by the year 2100. (Climate change assumptions under the 'high increase' scenario [2000–2100] are: 2020 — temperature +2°C, precipitation –30%; 2050 — temperature

+3.8°C, precipitation −48%; and 2080 — temperature +5.8°C, precipitation −48%.) Crop yields are even more at risk. The model predicts a decline in both millet and groundnuts yields of 5.5% by the year 2025. By the year 2100, yields would drop by as much as 93%. In other words, the steep increase in soil C simulated for the first 20 years of intensification (+85%) would counteract the projected impact of climate change during the same period of time. However, after this period, climatic conditions would be too unfavorable and losses inevitable.

22.4.2 Economic Feasibility

Knowing how much carbon can be stored and the extent to which crop yields can be increased under various management scenarios is only the first step in a feasibility assessment. The second step is to evaluate whether farmers would in fact be able to afford to implement promising practices. Limited investment capital among smallholders has been recognized as major obstacle to adoption of soil fertility management strategies, regardless of their long-term financial and social profitability (Izac, 1997; Sanchez et al., 1997; Ayuk, 2001).

Results from a cost–benefit analysis of the most promising land use and management options (Tschakert, 2004b) suggest three important points. First, costs for poor farmers are expected to be considerably higher than those for medium and rich farmers. This is due to the fact that most poor farmers lack the key resources for improving current practices and, thus, would have to buy the necessary inputs. This discrepancy is illustrated in Table 22.2 which lists first-year costs for 15 'best' management practices. Second, first-year costs for inputs are likely to exceed costs for foregone labor and rent when converting cropland to alternative uses. Third, there is a large variation in costs between the various management practices. Some of the practices listed in Table 22.2 include income-generating activities to compensate farmers for foregone production and to make alternatives more attractive. Animal fattening (13) and live hedges with high-value *Acacaia leatea* seeds (4) require considerable upfront cash. (The second

Table 22.2 First-Year Costs for "Best" Management Practices
and Three Resource Endowment Groups ($ h^{-1})

Number	Management Practices	Poor Households	Medium Households	Rich Households
1	Compost 2 t	1,198	232	232
2	Conversion ag land to grassland	54	0	0
3	Grassland + protection kad* (baskets)	180	122	122
4	Grassland + protection kad* (live hedges)	1,124	974	974
5	Cattle manure 4 t	1,455	0	0
6	Cattle manure 4 t + fertilizer	1,511	56	56
7	Sheep manure 5 t	1,068	702	388
8	3 yr fallow + organic matter 2t	228	0	0
9	Sheep manure 10 t	2,060	1,696	1,381
10	3 yr fallow + laucaena prunings 2 t	323	106	106
11	Kad* plantation (250 trees/ha)	723	650	650
12	10 yr fallow + organic matter 2 t	225	0	0
13	10 yr fallow + 2 t org. matter + animal fattening	3,074	2,849	2,849
14	10 yr fallow + leucaena prunings 2 t	323	106	106
15	Optimum agricultural intensification	1,215	942	654

*Kad = *Faidherbia albida*
Source: Based on Tschakert, P. 2004b. The costs of soil carbon sequestration: an
economic analysis for smallholder farming systems in Senegal. *Agric. Syst.*,
81:227–253.

scenario is based on suggestions from Senegalese agronomists
summarized in Christopherson and Faye, 1999.) The same is
true for options involving the purchase of animals to generate
large amounts of manure.

A comparison of net present values (NPVs) also reveals
large differences between resource endowment groups on the

Table 22.3 Net Present Values for "Best" Management Practices After a 25-Year Period (in $ ha⁻¹), Based on Discount Rate of 20%

Number	Management Practices	Poor Households	Medium Households	Rich Households
1	Compost 2 t	−643	22	22
2	Conversion ag land to grassland	−603	−379	−260
3	Grassland + protection kad* (baskets)	−549	−319	−201
4	Grassland + protection kad* (live hedges)	2,155	2,476	2,595
5	Cattle manure 4 t	−2,042	474	474
6	Cattle manure 4 t + fertilizer	−1,128	720	720
7	Sheep manure 5 t	−424	12	131
8	3 yr fallow + organic matter 2 t	−440	99	99
9	Sheep manure 10 t	−119	395	433
10	3 yr fallow + laucaena prunings 2 t	−514	11	11
11	Kad* plantation (250 trees/ha)	−344	−116	2
12	10 yr fallow + organic matter 2 t	−707	−260	−136
13	10 yr fallow + 2 t org. matter + animal fattening	−1,116	−669	−546
14	10 yr fallow + leucaena prunings 2 t	−795	−344	−220
15	Optimum agricultural intensification	−1,037	11	408

*Kad = *Faidherbia albida*
Source: Based on Tschakert, P. 2004b. The costs of soil carbon sequestration: an economic analysis for smallholder farming systems in Senegal. *Agric. Syst.*, 81:227–253.

one hand, and the various management practices on the other hand (Table 22.3). (NPVs are discounted net costs and benefits. They indicate the profitability of an option. The discount rate used here was 20%.) Calculated for 1 hectare and for 25 years, NPVs ranged from −$2042 to +$2595. Only one practice would be lucrative for poor households (4) while medium and rich farmers would have a choice of profitable options. Some practices, such as the conversion of cropland to grassland or long-term fallow without a parallel income-generating component, are unlikely to yield positive returns, even for high-resource-endowment households.

Another way to appraise the value of "best" management strategies is to look at tradeoffs between options that increase soil C and those that enhance economic profitability (Gokowski et al., 2000, cited in Sanchez, 2000). The idea is to identify options that optimize these tradeoffs. Figure 22.6 depicts the tradeoffs for 15 "best" management options for the Peanut Basin.

As in Sanchez's example from Cameroon, there is no win–win situation (high C gains and high profitability) for small-scale farming systems in semi-arid Senegal. However, using more organic matter in the form of manure (4 to 5 metric tons ha^{-1}) and short-term improved fallows would optimize the tradeoffs between the profitability interest of farmers and the interests of global society to obtain higher C uptake and storage. The high C/very low profitability options (10 and 15) might only work for better-endowed farmers. The high C/negative profitability options (11 through 14), although highly attractive to global society, are economically not feasible for local farmers, unless substantial financial support is provided. The only practice that shows a positive NPV for all resource endowment groups (4), because of its income-generating component, yields only low C gains.

The final question is how estimated costs, profitability, and tradeoff scenarios compare to actual household revenues, expenditures, and overall budgets. Annual revenues and large variations were shown to exist between sample households. For the low-resource endowment group, 9 of the total 15 management practices exceed annual household income ($577). Medium and rich households, with an annual income of $1282 and $2102, respectively, would be able to afford more than half of the tested practices, but not those with very high upfront costs as indicated in Table 22.1. However, these calculations do not take into account household expenditures. In the research sample, expenditures ranged from $68 year^{-1} to

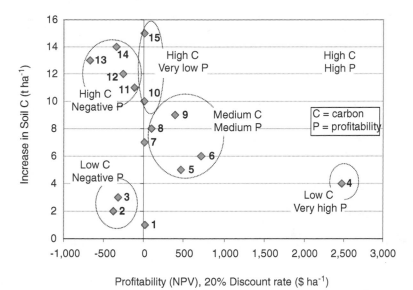

Figure 22.6 Gains in soil carbon vs. profitability (net present values) of several alternative land use types and management practices. Results are for a 25-year period and medium resource endowment farmers.

$425year^{-1} per AE. Nearly half of all household revenues were spent on food (mean 46%). Expenditures for clothing (12%) and family support and other social investments (6%) represented other important categories. It should be stressed that the large majority of household funds spent on food are spent on imported rice, which usually constitutes the basis for one meal per day, and not on millet.

Whether actual household budgets (revenues – expenditures) would be sufficient to cover initial investment costs of "best" management practices was assessed through a cash-flow analysis simulated with STELLA (Tschakert, 2004b). The results suggest that poor farmers would not have the necessary capital to invest in any of the proposed management practices after covering their basic household needs. Medium households are likely to be able to cover costs for two practices, and rich households costs for three options. These are practices with lowest rates of manure application. The shortage of first-year funds varies from $588 to $1217 on average. These figures fail to take access to credit and specific savings into account.

However, if external financial support were to be provided to complement local cash resources, especially at the beginning, household funds would probably experience an economic "takeoff" and build up crucial capital. Examples of such a takeoff are shown in Figure 22.7. Although such capital accumulation and time for takeoff vary from practice to practice, all three resource endowment groups can be expected to benefit. Simulated gains range from $1200 to $25000 at the end of a 25-year investment cycle, assuming that there are no "luxury investments," such as the construction of large brick houses. The simulation results also suggest that rich households would require only short-term external financial support (1 to 4 years), while poor households are likely to need financial commitment for longer periods of time, except for practices with immediate revenues and relatively low costs, such as the fattening of sheep.

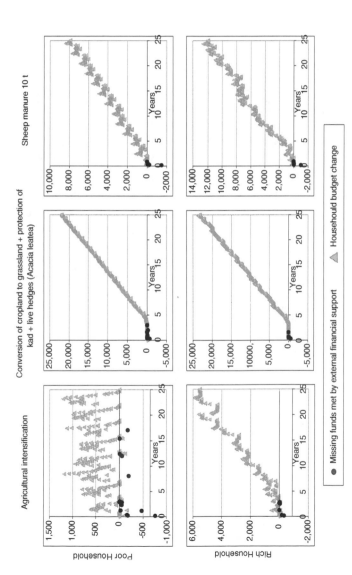

Figure 22.7 STELLA output showing monthly changes in household budgets ($) as a result of carbon sequestration practices, compared to a base scenario (household cash flows as they would occur without these practices). Also shown are "missing household funds" (the amount the household lacks to invest in a certain practice). Ideally, these missing amounts would be met by external funding sources. The examples are for poor and rich households. (Based on Tschkert, P. 2004b. The costs of soil carbon sequestration: an economic analysis for smallholder farming systems in Senegal. *Agric. Syst.*, 81:227–253.)

22.5 DISCUSSION AND CONCLUSION

The current notion that food security is determined by house-hold income, poverty, and access to resources rather than by aggregate increased food production puts the potential role of carbon sequestration in small-scale farming systems in a broad and complex context. One or two 'best' management practices that simultaneously boost soil carbon and crop yields, but fail to address any of the other multiple social goals of rural households, are most likely not the alternatives that are likely to be adopted by resource-poor farmers.

It has been shown that carbon stocks in the central part of Senegal are relatively low (on average, less than 20 metric tons C ha^{-1}), and that they are expected to remain at risk without adequate management practices. The same is true for crop yields. Results from CENTURY simulations indicated that improvements in C stocks in agricultural systems are possible but, that the overall rate of increase is likely to be modest. For soils, annual gains on croplands over 25 years amounted to 0.02 to 0.42 metric tons C ha^{-1} year^{-1}, which is slightly higher than other existing estimates of 0.1 to 0.3 metric tons C ha^{-1} year^{-1} (Lal, 1999a; Batjes, 2001). "Best" C management practices for soils included plantations with nitrogen-fixing trees (*Faidherbia albida*), the application of 4 to 10 metric tons of manure with and without fertilizer, agri-cultural intensification with manure, fertilizer, green manure from agroforestry, and short-term fallow, as well as 6 to 10 years of improved fallows in rotation with millet and ground-nuts. The same practices, with the exception of tree planta-tions, would also result in increased crop yields, ranging from +50% to +200% compared to current average yields. However, the impacts of climate change could seriously jeopardize gains in both soil carbon and crop yields, and losses over the next 100 years could be as high as 40% and 93% for both, respectively.

It may very well be in the interest of farmers to improve carbon stocks and, hence, soil fertility, especially if they are subsistence farmers. However, cropping is only one of multiple household livelihood strategies associated with flexible and adaptable risk management in drylands (Scoones et al., 1996;

Mortimore and Adams, 1999). For instance, only 6% of all sample households in the Old Peanut Basin obtained most of their income from farming. Most rely on remittances, which account for one-third of total average household revenue. Thus, in a complex livelihood system, a 50% to 200% increase in food production through 'best' carbon sequestration practices does not automatically mean enhanced food security. If applied on a larger scale, these practices might in fact compete with or undermine other livelihood practices that also contribute to food and livelihood security. In a risk-prone environment, such as the Sahel, secure livelihoods are those that are flexible and resilient.

In terms of economic feasibility, the results from this case study suggest that there may not be any significant low-cost options with ancillary benefits as indicated in preliminary studies (Ringius, 2002). For resource-poor farmers, only one simulated practice, namely live hedges with highly valuable seeds, would be profitable after 25 years. From a 'tradeoff' perspective, management options involving 4 to 10 metric tons of manure and short-term improved fallows would satisfy global needs and local economic interests, at least for better-endowed farmers. However, given the low investment capital of most smallholders and the high upfront costs associated with most 'best' practices, most will probably not be able to afford their implementation. Hence, carbon activities will not automatically reduce poverty. If, however, external and well-targeted funding or cost-sharing mechanisms are set in place to cover initial investment costs, as also proposed by Izac (1997) and Ayuk (2001), household budgets are likely to increase with time. In some cases, the increases may be substantial, thus playing a major role in rural poverty reduction.

The study suggests that increasing C stocks, improving food security, and reducing rural poverty are inevitably linked. The key to achieve all three is to strengthen and diversify local livelihoods. The premise, then, is twofold. First, carbon sequestration should be perceived as an additional element in people's livelihood portfolios and not as a goal *per se*. Whether the main objective is improved soil fertility, increased yields, additional income, enhanced food security, or the restoration of degraded lands will depend on the specific priorities of a household. Second, the fact that farmers are not a homogeneous group should be acknowledged. It is highly unlikely that one and the same 'best' practice will work for every single participant. As observed in other cases of farming system research in the Sahel, well-intentioned technologies for an 'average' farmer on an 'average' field are not only inappropriate, but also counterproductive since they risk undermining the very flexibility of the system (Defoer et al., 1998).

In order to address complex and multiple livelihood needs through carbon sequestration, a basket of management choices rather than one 'best' practice will be required. This includes a mix of farm and nonfarm activities. One possibility is to actually shift the emphasis on food crops to other income-generating activities that could be used to purchase food, thereby also broadening livelihood strategies. These income-generating activities could include cash cropping, the sale of wood products and valuable seeds, and animal fattening. Under conditions of land scarcity, smallholders may be very reluctant to convert their cropland to alternative types of land use, even if this conversion results in increased C stocks. Other sources of income and reliable food programs will be crucial.

Carbon sequestration in drylands can be a win–win strategy. It has the potential to address global needs of reduced C_2 concentration and livelihood needs of rural populations. Focusing on merely one or two aspects, for the sake of simplicity — accountability and verification — may satisfy the needs of external agencies. Complex livelihoods, however, require complex solutions.

ACKNOWLEDGMENTS

I am grateful for the support of U.S. Agency for International Development/Africa Bureau/Office of Sustainable Development (AFR/SD), the Earth Resources Observation Systems Data Center/U.S. Geological Survey, the Centre for Environmental Studies at Lund University, Sweden, and a Dean's

Fellowship from the University of Arizona. Special thanks go to Agatha Thiaw, Djibril Diouf, and AlHassan Cissé for their invaluable contribution to a year of fieldwork; Lennart Olsson for his input to the STELLA model; Bill Parton and Steve DelGrosso for their help with CENTURY; and all farmers in the Old Peanut Basin who participated in the study.

REFERENCES

Ayuk, E.T. 2001. Social, economic, and policy dimensions of soil organic matter management in Sub-Saharan Africa: challenges and opportunities. *Nutrient Cycling Agroecosyst.*, 61:183–195.

Badiane, A.N., M. Khouma, and M. Sène. 2000. Région de Diourbel: Gestion des sols. Drylands Research Working Paper, Somerset, United Kingdom.

Badiane, A.N. and B. Lesage. 1996. *Rapport Annuel. 1995: Gestion des ressources naturelles en zone sèch.* Opération fixation azote et recyclage matière organique. Bureau C.N.R.A., Bambey, Senegal.

Batjes, N.H. 2001. Options for increasing carbon sequestration in West African soils: an exploratory study with special focus on Senegal. *Land Degradation Dev.*, 12:131–142.

Bellon, M.R. 2001. *Participatory Research Methods for Technical Evaluation: A Manual for Scientists Working with Farmers.* Centro Internacional de Mejoramiento de Maízy Trigo (CIMMYT), Mexico City.

Bernard, R.H. 1994. *Research Methods in Anthropology: Qualitative and Quantitative Approaches.* Sage, Thousand Oaks, CA.

Centre de Suivi Ecologique. 2000. *Annuaire sur l'environnement et les ressources naturelles du Sénég.* Centre de Suivi Ecologique, Dakar, Senegal.

Chambers, R. 1997. *Whose Reality Counts? Putting the First Last.* Intermediate Technology Publications, London.

Christopherson, A.K. and A. Faye. 1999. Farmer-Perspective Financial and Economic Analysis, Annex C, Impact Assessment of the AG/NRM Strategic Objective of USAID Senegal (Old SO2). Vol. 2. Prepared for U.S. Agency for International Development/ Senegal, Environmental Policy Indefinite Quality Contract (EPIQ).

Copans, J. 1988. *Les marabouts de l'arachide: La confrérie mouride et les paysans du Séné*. L'Harmattan, Paris.

Davies, S. 1996. *Adaptable Livelihoods: Coping with Food Insecurity in the Malian Sahel*. Macmillan Press, Houndmills, Basingstoke, United Kingdom; St. Martin's Press, New York.

Defoer, T., H. De Groote, T. Hilhorst, S. Kante, and A. Budelman. 1998. Participatory action research and quantitative analysis for nutrient management in southern Mali: a fruitful marriage? *Agric. Ecosyst. Environ.*, 71:215–228.

Devereux, S. 1993. *Goats Before Ploughs: Dilemmas of Household Response Sequencing During Food Shortages*. Institute for Development Studies Bulletin 24. Institute for Development Studies, Brighton, United Kingdom.

Dilley, M. and T.E. Boudreau. 2001. Coming to terms with vulnerability: a critique of the food security definition. *Food Policy*, 26:229–247.

Direction de l'Analyse, de la Prévision et des Statistiques/Division de Statastiques, Documentation et Information Agricole (DAPS/DSDIA). 2001.

Food and Agriculture Organization. 2002. *The State of Food and Agriculture*. FAO, Rome.

Food and Agriculture Organization, and UN Educational Scientific and Cultural Organization. 1974. *Soil Map of the World, 1:5000000*. Vol. 1, *Legend*. UNESCO, Paris.

Freudenberger, K. and M. Schoonmaker. 1994. Livelihoods, livestock and change: the versatility and richness of historical matrices. *RRA* (Rapid Rural Appraisal) *Notes,* 20:144–148, IIED (International Institute for Environment and Development), England.

Gladwin, C.H., A.M. Thomson, J.S. Peterson, and A.S. Anderson. 2001. Addressing food security in Africa via multiple strategies of women farmers. *Food Policy*, 26:177–207.

Guijt, I. and J.N. Pretty. 1992. *Participatory Rural Appraisal for Farmer Participatory Research in Punjab, Pakistan*. International Institute for Environment and Development, London.

High Performance Systems. 2000. *Stella: An Introduction to Systems Thinking*. High Performance Systems, Hanover, NH.

Hunt, D. 1991. Farm system and household economy as frameworks for prioritising and appraising technical research: a critical appraisal of current approaches. In M. Haswell and D. Hunt, Eds. *Rural Households in Emerging Societies: Technology and Change in Sub-Saharan Africa*. Berg, Oxford, United Kingdom, pp. 49–76.

Izac, A.-M.N. 1997. Developing policies for soil carbon management in tropical regions. *Geoderma,* 79:261–276.

Kimble, J.M., L.R. Everett, R.F. Follett, and R. Lal. 2002. Carbon sequestration and the integration of science, farming, and policy. In J.M. Kimble, R. Lal, and R.F. Follett, Eds. *Agricultural Practices and Policies for Carbon Sequestration in Soils*, Lewis Publishers, Boca Raton, FL, pp. 3–11.

Lal, R. 1999a. Global carbon pools and fluxes and the impact of agricultural intensification and judicious land use. In *Prevention of Land Degradation, Enhancement of Carbon Sequestration and Conservation of Biodiversity Through Land Use Change and Sustainable Land Management with a Focus on Latin America and the Carribbean*. World Soil Resources Report 86. Food and Agriculture Organization, Rome, pp. 45–52.

Lal, R. 1999b. Soil management and restoration for C sequestration to mitigate the accelerated greenhouse effect. *Prog. Environ. Sci.,* 1:307–326.

Lal, R. 2002a. Carbon sequestration in dryland ecosystems of West Asia and North Africa. *Land Degradation Dev.,* 13:45–59.

Lal, R. 2002b. Soil carbon sequestration in China through intensification and restoration of degraded and desertified ecosystems. *Land Degradation Dev.,* 13:469–478.

Lericollais, A., P. Milleville, and G. Pontié. 1998. Terrains anciens, approches renouvelées: analyse du changement dans les systèmes de production sérères au Sénégal. In R. Cligne *Observatoires du développement, observatoires pour le développeme*. ORSTOM, Paris.

Lightfoot, C. and R. Noble. 1993. A participatory experiment in sustainable agriculture. *J. Farming Syst. Res. and Extension*, 4(1):11–34.

Manu, A., A. Bationo, and S.C. Geiger. 1991. Fertility status of selected millet producing soils of West Africa with emphasis on phosphorus. *Soil Sci.*, 152:315–320.

Maxwell, S. and T.R. Frankenberger. 1992. *Household Food Security: Concepts, Indicators, Measurements*. UNICEF/International Fund for Agricultural Development, New York/Rome.

Mortimore, M. and W.M. Adams. 1999. *Working the Sahel: Environment and Society in Northern Nigeria*. Routledge, New York.

Parton, W.J., D.S. Ojima, C.V. Cole, and D.S. Schimel. 1994. A general model for soil organic matter dynamics: sensitivity to litter chemistry, texture, and management. In Proceedings, Quantitative Modeling of Soil Forming Processes, Soil Science Society of America, Minneapolis, MN, pp. 147–167.

Pelissier, P. 1966. *Les Paysans du Sénégal, les civilisations agraires du Cayor à la Casaman*. Fabrègue, Saint-Yrieix (Haute-Vienne).

Pelissier, P. and G. Laclavère. 1984. *Atlas du Senegal*, 4th ed. Les Editions Jeune Afrique, Paris.

Rabot, C. 1984. *Vingt ans de successions de cultures dans la moitié sud du Sénégal, impacts écologiq*. DEA, Ecologie tropicale. Université des sciences et techniques du Languedoc, Montpellier, France.

Ringius, L. 2002. Soil carbon sequestration and the CDM: opportunities and challenges for Africa. *Climatic Change,* 54:471–495.

Sanchez, P.A. 2000. Linking climate change research with food security and poverty reduction in the tropics. *Agric. Ecosyst. Environ.*, 82:371–383.

Sanchez, P.A., K.D. Shepherd, M.J. Soule, F.M. Place, R.J. Buresh, A.-M.N. Izac, A.U. Mokwunye, F.R. Kwesiga, C.G. Ndiritu, and P.L. Woomer. 1997. Soil fertility replenishment in Africa: an investment in natural resource capital. In R.J. Buresh, P.A. Sanchez, and F. Calhoun, Eds. *Replenishing Soil Fertility in Africa*. Soil Science Society of America and American Society of Agronomy, Madison, WI, pp. 1–46.

Scoones, I., C. Chibudu, S. Chikura, P. Jeranyama, D. Machanja, W. Mavedzenge, B. Mombeshora, M. Mudhara, C. Mudziwo, et al. 1996. *Hazards and Opportunities. Farming Livelihoods in Dryland Africa: Lessons from Zimbabwe.* Zed Books, London.

Sen, A.K. 1981. *Poverty and Famines: An Essay on Entitlement and Deprivation.* Clarendon Press and Oxford University Press, Oxford and New York.

Stomal-Weigel, B. 1988. L'évolution récente et comparée des systèmes de production serer et wolof dans deux villages du vieux Bassin Arachidier (Sénégal). *Cah. Sci. Hum.*, 24(1):17–33.

Thomson, A.M. and M. Metz. 1997. *Implications of Economic Policy for Food Security.* Training Materials for Agricultural Planning 40. Food and Agriculture Organization, Rome.

Tschakert, P., M. Khouma, and M. Sène. 2004. Biophysical potential for soil carbon sequestration in agricultural systems of the Old Peanut Basin of Senegal. *J. Arid Environ.*, 59:511–533.

Tschakert, P. 2004a. Carbon for farmers: assessing the potential for soil carbon sequestration in the Old Peanut Basin of Senegal. *Climatic Change*, 67(2-3), December (I, II):273:290.

Tschakert, P. 2004b. The costs of soil carbon sequestration: an economic analysis for smallholder farming systems in Senegal. *Agric. Syst.*, 81:227–253.

Watson, R.T., M.C. Zinyowera, and R.H. Moss. 1996. *Impacts, Adaptations, and Mitigation for Climate Change: Scientific-Technical Analysis.* Intergovernmental Panel on Climate Change. Cambridge University Press, London; New York.

Watts, M. 1991. Entitlements or empowerment? Famine and starvation in Africa. *Rev. Afr. Political Econ.*, 51:9–27.

23

Hillside Agriculture and Food Security in Mexico: Advances in the Sustainable Hillside Management Project

JOSE I. CORTÉS, ANTONIO TURRENT, PRÓCORO DÍAZ,
LEOBARDO JIMÉNEZ, ERNESTO HERNÁNDEZ, AND
RICARDO MENDOZA

CONTENTS

23.1 INTRODUCTION

It is estimated that 791 million people in 98 developing countries were undernourished in 1996–1998. They were geographically distributed as follows: 86 million in Sub-Saharan Africa, 36 million in the Near East/North Africa, 55 million in Latin America and the Caribbean, 348 million in China (including Taiwan) and India, and 166 million in other nations of Asia (Food and Agriculture Organization [FAO], 2000).

In Latin America and the Caribbean, the number of undernourished people in total populations varies considerably. Brazil has the largest number of malnourished citizens, followed by Colombia, Mexico, Haiti, Peru, Venezuela, Guatemala, Dominican Republic, Cuba, Bolivia, Nicaragua, Honduras, Paraguay, El Salvador, Chile, Ecuador, Argentina, Panama, Jamaica, Trinidad and Tobago, Costa Rica, Guyana, Uruguay, and Suriname (FAO, 2000). The FAO has reported that there are 5 million undernourished people in Mexico of a total population of approximately 101 million people.

Mexico has a rural population of around 25 million people, and almost half of them are indigenous. The 1991 census indicates that more than two-thirds of the farms are less than 5 ha. Typically, farmers commonly have several land parcels with various soil types located at various distances from the farmstead.

Indigenous farmers live in the sierras (highlands), where most of the maize (*Zea mays*), beans (*Phaseolus* spp.), pumpkins (*Cucurbita* spp.), and fruit trees are produced on hillsides, under rainfed conditions, and using local production technology. Most of them are native people whose ancestors resided in the same areas at the time of Spanish conquest. Their traditional farming system is slash-and-burn agriculture. Great genetic diversity is evident in the local varieties of maize, bush beans, pole beans, and pumpkin. Most of these genetic resources are well adapted to local soil and climatic conditions, since most have been selected by farmers over many generations.

Burgeoning populations, however, have forced hillside farmers into an almost permanent agriculture system.

Typically, soils lie fallow for only 3 to 5 years, instead of 15 to 50 years as in the past. Furthermore, farmers do not attempt to reduce soil erosion, which is a major problem in hillside agriculture. These farmers typically obtain very low yields. In many cases, their net incomes are negative, and soil quality continues to decline.

For some officials, technicians, and farmers, these conditions are a colossal barrier to increased food production and inhibit the modernization of agriculture. The sierras play a key role in the terrestrial cycling of water. This process represents the replenishment of fresh water resources and meeting the water demand of rural and urban populations. When soil erosion leads to the destruction of surface soil cover, runoff overtakes water infiltration. It affects the water-holding capacity of soil, which results in more water stress during the growing season and lower yields on indigenous farmers' small parcels. Thus, major challenges for these populations are how to achieve food security, and how to maintain and/or improve soil quality, which is a critical issue for atmospheric concentration of greenhouse gases.

23.2 SUSTAINABLE HILLSIDE MANAGEMENT PROJECT

The Sustainable Hillside Management Project (SHMP), financed by the Global Environment Facility through the World Bank and the Mexican government, is a research/development project. It is related to climate change and involves indigenous Cuicateca, Mazateca, and Mixe populations who live in the sierras of northern Oaxaca State, Mexico.

These native rural populations are characterized by a high degree of economic and social isolation. Most of them practice hillside agriculture to produce staple crops using the slash-and-burn agricultural system (Figure 23.1). They obtain low yields of maize, beans, pumpkin, and other species, and they impact negatively on the quality of natural resources. The project target population includes 62 municipalities, 745 communities, and 325,000 people (Jiménez, 2001).

Figure 23.1 Hillside plots of maize and beans grown using the slash-and-burn agricultural system in the Mixe region, Oaxaca, Mexico. (Photos courtesy of Leobardo Jiménez S. and Aurelio León M.)

Families in these regions average five members. Adults typically speak their native language in the Mazateca and Mixe communities. Over 25% of parents in these communities are illiterate. Typically, these inhabitants, including those who can read and write, have only completed the third or fourth grade. Houses in the communities consist of two to three rooms, comprised of adobe walls, metal sheet ceilings, and dirt floors. Most farms vary in size between 2 and 3.5 ha. They are divided into two to three parcels on which maize, beans, coffee, and fruit trees are cropped. Generally, parcels are not cropped for 2 to 3 years, and then are cropped for the next 3 to 7 years. Maize, with an average yield of 0.7 Mt ha^{-1} in the three regions, does not contribute to household income, but many households earn cash from coffee bean production (León et al., 2001).

Research objectives of the SHMP are to (1) design of a sound methodology to measure carbon sequestration related to principal land use types in hillside ecosystems; (2) assess soil erosion and runoff; (3) characterize hillsides; microsheds, and regions geographically; (4) design and generate alternative technologies for a sustainable agriculture; and (5) characterize the socioeconomic conditions of target populations. The project also deals with development issues through an SHMP training and technology transfer unit. This unit focuses on technological innovations that have been derived from research. It is committed to farmer training to use and transfer the technological innovations, and improvement of target population living standards, particularly those related to nutrition, food production, and family income. The SHMP operational strategy is participatory, interdisciplinary, and fosters cooperation among participant farmers, both women and men, and their families, communities, and municipal organizations; scientific and technical staff devoted to basic tasks in research and development; and public and private institution staff who are participating at the community, municipal, state, and federal levels. Operationally, the SHMP works on a continuum from microsheds to subregions, to regions, and its development activities constitute a continuum (Jiménez, 2001).

23.3 ALTERNATIVE TECHNOLOGIES

Since the crucial issues in the three regions of SHMP are lack of sufficient food and soil conservation, the principal objective of the SHMP research unit is the design and generation of sustainable alternative technologies for the "milpa" system, assuming a permanent agricultural system as opposed to the current shifting system (Cortés et al., 2001).

Sustainable farming systems conserve and protect the essential agroecosystem resource base including soils, water, and genetic diversity; provide enough quality food and fiber to meet present and future requirements; optimize crop output; and are profitable enough to provide farmers with adequate living standards to support viable rural communities (Merwin and Pritts, 1993).

The design and generation of alternative technologies gives priority to terrestrial C sequestration, soil and water conservation, staple crop production, land and labor productivity, sufficient and distributed net income throughout the year, efficient land use, employment opportunity, and opportunity costs of land and water use. Since staple crops are in crisis in Mexico, an alternative technology for the milpa system for small farmers, such as those who live in the Cuicateca, Mazateca, and Mixe communities, based on maize and beans only, would not be able to achieve both SHMP objectives and farmer expectations.

Field research conducted in the Puebla Valley indicates that a rural family producing maize on 4 ha of land typically yields 5 and 7 Mt ha^{-1} of grain and corn stalk fodder. This is equivalent to a net income well below the minimum poverty-level wage of US $4.00 per day (Turrent and Cortés, 2002). Thus, an alternative technology for small farmers must include an economic alternative to support continued production of staple crops, and to enhance rural development.

Other studies in the same Puebla Valley show that intercropped maize with deciduous fruit trees is a farming system that will significantly increase family net income derived from cropping maize only. Fruit trees such as peaches, apricots, apples, pears, and other deciduous species represent cash crops, while annual crops guarantee family food security. This system, called "milpa intercropped in fruit trees" (MIFT), is being developed by an interdisciplinary research group of the Colegio de Postgraduados and Instituto Nacional de Investigaciones Forestales, Agrícolas y Pecuarias.

The MIFT system takes advantage of complementary relationships between fruit trees and maize and bean annual crops regarding the integral use of soil, water, sunlight, and family labor. Maize and beans can be rotated annually, and can take advantage of agronomic interactions by cropping them in alternate and microrotating strips (Turrent and Cortés, 2002). Thus, the MIFT system consists of integrated maize and bean crops in alternate strips of two rows for each species, intercropped between rows of fruit trees that are trained in Tatura trellis systems as shown in Figure 23.2.

Tree spacing is 1 to 2 m in rows, with 14.4 m between rows on level soils or on moderate slopes. For this model, each species occupies one-third of the land in the parcel, in strips of 4.8 and 1.6 m width for fruit trees and maize and beans, respectively. Fruit trees are planted in the middle of the strip, and maize and beans in six alternate strips of two rows spaced at 0.8 m on both sides of tree rows.

This model under rainfed conditions has yielded 6 Mt of fresh peaches, 2.5 Mt of corn, and 0.5 Mt of beans per hectare per year. This means that compared to monocropping systems, peach trees and maize yields are at 50%, and beans at 33%. The global land equivalent ratio (LER) for this system is greater than 1. Labor demand is distributed during the year according to species being cropped, and its accumulated demand is up to five times greater than that required for monocropped maize. Incomes derived selling several harvests over the course of a year can be 400% or more than those derived from staple crops. The peach, corn, and bean yields discussed above have a per hectare gross value equal to US $4800, US $375, and US $200, respectively. This gross income, which amounts to US $5375 per hectare, contrasts sharply with the US $1100 gross value of 5 Mt of corn plus 440 straw bales obtained when producing maize alone on the same land parcel. Thus, staple crops, such as maize and beans, when

Figure 23.2 The milpa system intercropped with peach trees on level soils in the Puebla Valley, Mexico. (Photos courtesy of José I. Cortés F. and Antonio Turrent F.)

produced using the MIFT system are less sensitive to low prices in global markets. Incomes generated by selling staple crops are relatively secondary to ensure a healthy local economy for small farming units, as compared with income derived from fruit sales.

Field research, conducted in the semihumid tropics of Mexico, has shown that terraces formed by planting *Glyricidia sepium* as living walls in contour rows effectively increase rain water infiltration, and diminish soil erosion to acceptable ranges for sustainable production of maize under hillside conditions (Turrent et al., 1995; Turrent and Moreno, 1998). This is possible because *Glyricidia sepium* are planted close together in the row, which forms a living wall that supports the runoff filter composed of whole corn stalks. These are placed horizontally along and on the upper side of tree rows immediately after the harvest of maize as illustrated in Figure 23.3. This technology, however, has a major drawback. In contrast to fruit trees, *Glyricidia sepium* uses land, but has no economic value. Small farmers would be unlikely to adopt it for soil erosion control.

Experiences described above can lead to the design and generation of alternative technologies in the SHMP. Research that is currently being conducted on alternative farming systems integrate the principles of living-wall terrace technology with the MIFT system practiced under level soil conditions to develop a MIFT system for hillside agriculture.

23.4 MILPA/FRUIT TREE INTERCROPPING SYSTEM IN SUSTAINABLE HILLSIDE MANAGEMENT PROJECT

A major objective of this project was to identify alternative technology options for the target regions. Since no previous agronomic research had been conducted on intercropped staples and fruit trees, development of the MIFT system took into account the management of peach trees in the Puebla Valley and local cropping patterns for maize and beans, and assumed that soils on slopes greater than 20% would be a common condition. These slopes are in contrast to moderate

Figure 23.3 Living wall terrace technology in the semihumid tropics of Mexico, developed by the Instituto Nacional de Investigaciones Forestales, Agrícolas y Pecuarias. (Photos courtesy of Antonio Turrent F.)

slopes in semihumid regions where the living wall terrace technology was developed.

The MIFT system consists of maize and beans cultivated either in association with one another or in a relay cropping pattern, intercropped between rows of peach trees in temperate regions, and between rows of industrial trees, such as coffee in semitropical areas. Peach trees are trained in the Tatura trellis system, and coffee trees in the central leader system. Both species are fertilized with N, P, and K every year. Fruit and coffee trees cannot be planted very close to each other in rows as is the case for *Glyricidia sepium.* Spacing in rows is between 0.75 m and 1.0 m, with 9 m between rows in contour. Peach and coffee trees are planted in the middle of the rows, which have a width of 3.0 m. The free strip of 6.0 m in width between two rows of trees is occupied by maize and beans in eight rows of 0.75-m width (see Figure 23.4).

Thus, trees occupy one-third, and maize and beans two-thirds, of plot land surface. Maize is fertilized with N, P, and poultry waste, and beans with N and P only.

Relay cropping patterns consist of beans, which are planted in February, and maize, which is planted 1 month later between two rows of beans. Land preparation and cultural practices are done by hand. Peaches bloom from January to February and are picked from late May to early June. Beans are harvested in June just after the rainy season begins. Then maize is grown by itself and harvested in October. Subsequently, corn stalks are cut and used to form the runoff filter discussed previously. This system is pictured in Figure 23.5.

In association cropping pattern, land preparation is completed using animal traction or manual labor. If land preparation is done in late fall, maize and beans can be planted in mid-April. Residual soil moisture from rainfall of the previous year allows seed germination and plant growth before the onset of the May rainy season. Harvest time is in October. However, if land is not prepared on time, planting is delayed until the rainy season begins, and the crops are harvested in December. When the rainy season is delayed, as in 2003, crops fail because of drought stress, and farmers have to replant

Figure 23.4 The milpa system intercropped with peach and coffee trees on hillside plots in the Mazateca and Mixe regions, Oaxaca, Mexico. (Photos courtesy of José I. Cortés F. and Mariano Morales G.)

Figure 23.5 The milpa system intercropped with fruit trees illustrating the runoff filter as a key control of soil erosion on hillsides. (Photos courtesy of José I. Cortés F.)

the crops, using shorter-season local maize cultivars. Under these conditions, however, the growing season for peaches is not altered. The MIFT system, described in this chapter, is being compared with a system of maize cultivated as a monocrop and managed under traditional (control), improved, and no-tillage systems.

23.5 RESULTS

Results obtained on two hillsides with a 35% slope in the Mazateca region during the first 3 years, indicate that the average yield of maize varies widely between treatments. It ranged from 0.63 to 6.62 Mt ha^{-1}, as reported in Table 23.1. Yields for the traditional slash-and-burn system were ten times lower than those for milpas intercropped between rows of peach trees with a spacing of 1.0 m in the rows, and using poultry waste manure for the maize along with mineral fertilizers. Yields for other treatments of monocropped maize were also improved greatly, especially under no-tillage conditions. The main difference in yields between monocrop maize treatments is due to mineral and organic fertilization. In slash-and-burn systems, maize is not fertilized, while in the other three treatments, N rates range from 80 to 120 kg ha^{-1} and P from 35.2 to 44 kg ha^{-1}. In the case of no-tillage treatment, poultry waste is applied at a rate of 2.0 Mt ha^{-1} every year in addition to the application of N and P. Then it can be concluded that the higher yields associated with this treatment are primarily due to supplemental organic fertilizers.

Yields of maize under MIFT system treatments confirm the response to poultry waste, applied at an equivalent rate of 2.0 Mt ha^{-1}. Yield response was 1.0 Mt ha^{-1} higher for maize intercropped between rows of peach trees with a spacing of 1.0 m than for maize intercropped between rows with a spacing of 0.75 m. Reasons for this variation are currently being analyzed.

Maize yield responses for field trials conducted for the project indicate that it is possible to produce sufficient quantities of this staple crop to sustain small farm families simply by improving the monocrop system. However, improvements

Table 23.1 Yields (Mt ha^{-1}) of Maize and Peaches, and Economic Parameters with Two Cropping Systems Under Several Treatments in Mazateca Region

Treatment	Year				Average Yield Maize	Accumulated Yield		Accumulated Cost (US $)	Accumulated Gross Income (US $)	B/C Ratio
	2000 Maize	2001 Maize	2002 Maize	2002 Peach		Maize	Peach			
MMSB[1]	1.10	0.40	0.40	N/A2	0.63	1.9	N/A	1791	420	0.23
MM[3]	2.49	4.32	1.81	N/A	2.87	8.63	N/A	4137	1898	0.46
MMI[4]	1.85	2.69	2.71	N/A	2.38	7.25	N/A	3273	1580	0.48
MMNT[5]	2.70	4.68	4.79	N/A	4.06	12.18	N/A	3344	2679	0.80
MIFT-P1.0[6]	3.43	4.38	4.90	2.6	4.29	12.72	2.6	4951	4098	0.83
MIFT-P1.0(p)[7]	4.96	7.90	7.00	2.6	6.62	19.86	2.6	5746	5668	0.99
MIFT-P0.75[8]	3.37	6.00	3.71	3.7	4.36	13.07	3.7	5763	4727	0.82
MIFT-P0.75(p)[9]	3.06	8.71	4.86	3.7	5.54	16.62	3.7	6554	5507	0.84
LSD$_{0.05}$[10]					0.58					

[1] Maize in monocrop under slash and burn system.

[2] Non applicable.

[3] Maize in monocrop under traditional management in roturated soil.

[4] Maize in monocrop under improved management in roturated soil.

[5] Maize in monocrop under no tillage system in the same site where soil is roturated.

[6] Milpa intercropped between rows of peach trees with a spacing of 1.0 m in the row in roturated soil.

[7] Same as 6 but maize receiving poultry waste in addition to mineral fertilization.

[8] Milpa intercropped rows of peach trees with a spacing of 0.75 m in the row in roturated soil.

[9] Same as 8 but maize receiving poultry waste in addition to mineral fertilization.

[10] Least significant difference for last seven treatments in the same plot only.

Source: Adapted from Cortés F., et al. 2003. Proyecto Manejo Sostenibles de Laderas. Subproyecto III: Tecnologías Alternativas Sostenibles. Informe de actividades. Colegio de Postgraduados, Mexico City.

in maize production alone for small farmers with 2- to 3.5-ha plots will not greatly improve their socioeconomic situation. In the next section, we analyze economic parameters of several treatments discussed above.

Accumulated cost data in Table 23.1 indicated that monocropping with rotated soils and in a no-tillage system was 1.8 to 2.3 times more expensive than monocropping in a slash-and-burn system, and the MIFT system was 2.8 to 3.7 times more expensive. Accumulated gross income for the alternative systems, however, showed a reverse situation. Incomes were 3.8 to 6.4 times higher for the alternate monocropping in no-tillage systems, and 9.7 to 13.5 times higher in the MIFT system. These differences are reflected in the benefit–cost (B/C) ratio, which varied widely between treatments. Monocropped maize for the slash-and-burn system had the lowest B/C ratio, equal to 0.23. This value was twice as high for monocropped maize under crop rotation, and 3.7 to 4.3 times higher for monocropped maize in a no-tillage system and for maize grown in the MIFT system.

These results are consistent with reports of socioeconomic evaluation studies carried out by staff members of the SHMP, which indicate that small farmers, cropping maize alone under traditional systems are in a critical socioeconomic situation (León et al., 2001). Maize production in no-tillage systems appears to be a viable alternative since its B/C ratio was very close to the B/C ratio for the MIFT systems. Although future maize yields in no-tillage systems could be improved, maize prices are going to be a major limiting factor in obtaining a higher B/C ratio. Economic analyses were based on US $0.22 kg^{-1} for maize, which is average for the region.

For MIFT systems, it is expected that the B/C ratio will increase as peach tree yields increase in the coming years. The assumption used for this project is that yields will be 6 to 8 kg tree^{-1} during a productive life of 15 to 17 years. Consistent with recent field observations, yields averaged about 5 kg tree^{-1} in 2003. It is assumed that in 2004 and subsequent years that potential yields will be actualized. Economic analyses used an average price of US $0.50 kg^{-1} for peaches. During late spring, however, consumers pay from US

$1.00 to US $2.50 kg^{-1}, depending on fruit quality. Target regions of the SHMP are able to produce quality peaches that can compete in the market.

Intercropped peach trees, with tree spacing of 0.75 m and 1.0 m in the row, resulted in LERs of 0.48 and 0.39, respectively. A yield hypothesis for the MIFT system was that intercropped peach trees, in one-third of the land parcel, would produce 50% of the yield of monocropped peach trees, which would mean a LER of 0.50. Thus, the LERs obtained thus far tend to support this yield hypothesis, and suggest also the capability of the MIFT system to increase land use efficiency in hillside agriculture.

Peach trees in the MIFT system are more vigorous than those in the monocrop system. This difference is observed early in the growing season, when peach trees are growing alone, since annual crops are not still planted or are growing slowly. This is also the dry season. Thus, it is reasonable to assume that peach trees in the MIFT system are growing without any competition for soil water, and under better soil moisture and nutrition conditions, because of the runoff filter along the row of trees that increases water infiltration and diminishes soil erosion.

Research on carbon sequestration and soil erosion, which is also being undertaken by researchers involved with the SHMP, indicates that the MIFT system improves also soil quality (Etchevers et al., 2003; Figueroa et al., 2003).

Results in the other two SHMP regions follow similar trends. However, local maize varieties are susceptible to diseases at the end of the growing season in areas where maize is intercropped between rows of coffee trees, thus affecting yields. Conventional breeding research is resolving this problem, as well as height and lodging problems observed in the three regions.

In addition to peaches, other deciduous fruit trees can be included in the MIFT system in temperate zones. Thus, apples are also being introduced into the MIFT system in order to advance its diversification as soon as possible. Fruit tree diversification rates will depend on the availability of fruit cultivars adapted to the study regions, and their ability to become cash crops for small farmers.

In semitropical areas, it will be important to identify fruit tree species that can be trained under the Tatura trellis system to form a living wall. Research on this topic has been initiated in the states of Veracruz and Chiapas through a joint project between the Instituto Nacional de Investigaciones Forestales, Agrícolas y Pecuarias and the Colegio de Postgraduados.

The no-tillage system adapted to the highland conditions of small farmers in the Cuicateca, Mazateca, and Mixe regions is also needed for improvement and diversification of the MIFT system for steeper hillsides and shallower soils. Some work has been done on identification of cover crops, and a field experiment on methods of land preparation, planting dates, and mulching has been initiated in the Cuicateca region, where soil erosion is more critical.

Access to inputs and/or services required to support technological innovations in these isolated areas is another topic that needs to be addressed. The SHMP is currently establishing family micronurseries in rural communities in order to propagate peach trees and other fruit species. It is also working on alternate methods to establish fruit trees in the field in order to diminish as much as possible initial investments related to the MIFT system. Planting stratified peach seeds in contour rows at recommended row spacing, in order to establish rootstocks, seems to be a viable alternative. They can later be grafted by farmers themselves.

Farmer training about the MIFT system and other technological innovations is another step in order to achieve SHMP objectives. Specialists in training and technology transfer have proposed a field school approach to train selected small farmers in their own communities. Today, there are several field schools functioning in the three SHMP target regions. Small farmers, after receiving training, teach their neighbors how to adopt the MIFT system (Jiménez et al., 2003).

23.6 CONCLUSIONS

Results from the SHMP indicate that it is possible to achieve food security and maintain and/or improve soil quality in smallholder hillside agriculture systems. However, there are many challenges to overcome in the coming years. Small farmers and researchers and development experts who are working on these problems tend to agree about what needs to be done. They are all interested in finding solutions. More people from several indigenous rural communities are applying to participate in the project. And community, municipal, state, and federal institutions are willing to support small-farmer initiatives and to support the field research activities that are required to improve and diversify the MIFT system in several ecosystems. The future of food security for small farmers undertaking hillside agriculture in Mexico depends in part on the degree of cooperation generated between small farmers and national institutions in the coming years.

REFERENCES

Cortés, J.I., A. Turrent, E. Hernández, R. Mendoza, L.A. Lerma, E. Aceves, H. Mejía, G. Narváez, P. Díaz, and A. Ramos. 2001. Proyecto Manejo Sostenibles de Laderas. Subproyecto III: Tecnologías Alternativas Sostenibles. Informe de Actividades. Colegio de Postgraduados, Montecillo, Mexico.

Cortés, J.I., A. Turrent, P. Díaz, E. Hernández, H. Mejía, R. Mendoza, A. Ramos, and E. Aceves. 2003. Proyecto Manejo Sostenibles de Laderas. Subproyecto III: Tecnologías Alternativas Sostenibles. Informe de Actividades. Colegio de Postgraduados, Montecillo, Mexico.

Etchevers, J.D., C. Hidalgo, J. Padilla, R.M. López, C. Monreal, C. Izaurralde, B. Rapidel, F. deLeón, M. Acosta, and M.A. Vergara. 2003. Proyecto Manejo Sostenible de Laderas. Subproyecto II: Metodología de la Medición de la Captura de Carbono. Informe de Actividades. Colegio de Postgraduados. Mexico.

Figueroa, B., M. Martínez, L. Aceves, et al. 2003. Proyecto Manejo Sostenible de Laderas. Subproyecto I: Caracterización Geográfica y Medición de Escurrimientos. Informe de Actividades. Colegio de Posgraduados, Montecillo, Mexico.

Food and Agriculture Organization. 2000. *Food Insecurity: When People Live with Hunger and Fear Starvation*. FAO, Rome.

Jiménez, L. 2001. Manejo Sostenible de Laderas: Presentación. Mid-Term Review. Colegio de Postgraduados, Montecillo, Mexico.

Jiménez, L., M. Morales, J. Zamora, A. Ramos, and N. Ortiz. 2003. Proyecto Manejo Sostenible de Laderas. Subproyecto V: Capacitación y Divulgación. Informe de Actividades. Colegio de Postgraduados, Montecillo, Mexico.

León, A., M. Hernández, and L. Jiménez. 2001. Proyecto Manejo Sostenible de Laderas. Suproyecto III: Evaluación Socio-económica de Comunidades Indígenas. Mid-Term Review. Colegio de Postgraduados, Montecillo, Mexico.

Merwin, I.A. and M.P. Pritts. 1993. Are modern fruit production systems sustainable? *Horticul. Technol.*, 3:128–136.

Turrent, A., and J.I. Cortés. 2002. La milpa intercalada en árboles frutales. Campo Experimental Valle de México. Memoria Técnica No. 2:20–24. Instituto Nacional de Investigaciones Forestales, Agrícolas y Pecuarias, El Horno, Mexico.

Turrent, A., and R. Moreno. 1998. Sustainable production of food from crops in the world. *Terra*, 16:93–111.

Turrent, A., S. Uribe, N. Francisco, and R. Camacho. 1995. La terraza de muro vivo para laderas del trópico subhúmedo de México. I. Análisis del desarrollo de las terrazas durante 6 años. *Terra*, 13:276–298.

24

Soil Organic Carbon, Quality Index, and Soil Fertility in Hillside Agriculture

JORGE D. ETCHEVERS, MIGUEL A. VERGARA,
MIGUEL ACOSTA, CARLOS M. MONREAL, AND
LEOBARDO JIMÉNEZ

CONTENTS

24.1 INTRODUCTION

Global climate change is a result of high atmospheric concentration of greenhouse gases (mainly CO_2, N_2O, CH_4, and fluorochlorocarbons). Experts predict an increment of the average global temperature of 3°C to 5°C by the end of the present century if the accumulation of these greenhouse gases

(GHGs) continues (Intergovernmental Panel on Climate Change [IPCC], 2001).

Agricultural activities take place on approximately 35% of the world's land. Land use and changes in land use constitute an important source of GHGs, being responsible for 25% of CO_2, 50% of CH_4, and 70% of N_2O emitted by all human activities (Agriculture and Agri-Food Canada, 1998; IPCC, 2001). The adoption of new management practices by farmers may help lower global GHG emissions associated with food production. It has been demonstrated that beneficial management practices (BMPs) help to restore air quality by converting atmospheric carbon (C) and nitrogen (N) into soil organic matter (SOM) (Agriculture and Agri-Food Canada, 1998). Carbon constitutes nearly 58% of the SOM, and improves the physical, chemical, and biological conditions of soil, and favors higher yields in crops (Herrick and Wander, 1998).

Little information exists on soil carbon stored under different climate conditions and land use systems, and on changes in soil organic carbon (SOC) content induced by management practices on sloping land or on hillside agriculture. Also, a standard methodology to monitor SOC changes in hillside farming systems does not exist. Studies conducted around the world indicate it is possible to capture atmospheric C in soils through BMPs that increase the aboveground dry matter, and reduce both soil erosion and the microbial oxidation of SOM. These practices include minimum tillage, crop rotation with legumes, cover crops, afforestation, tree barriers, pastures, and the addition of compost and animal and vegetal waste (Lal, 1998). Mexico has great potential to capture C in original forestland and hillside landscapes that are now being cultivated. Simulation models have been developed to estimate changes in C stocks; however, these models have not been validated for hillside agriculture conditions.

Concern expressed by national and international agencies about global climate change and certain agricultural activities practiced by rural communities on hillsides in Mexico, led to the establishment in 1999 of the Sustainable Management on Hillside Agriculture Project (PMSL) in three socially marginal regions in Oaxaca State (Mazateca,

Cuicateca, and Mixe). This project was sponsored by the Mexican Ministry of Agriculture, Livestock, Rural Development, Fisheries, and Foodstuffs (SAGARPA), Oaxaca State, and the Global Environment Facility with the World Bank. Several indigenous groups reside in the selected target areas under conditions of severe poverty. The objectives of this project were (1) increase understanding of C stocks associated with secondary-growth forest and agricultural hillside systems managed under traditional and modern agricultural practices, (2) determine actual soil fertility conditions, and (3) characterize the spatial variability of SOC and of soil properties associated with soil fertility.

24.2 MATERIALS AND METHODS

This study was carried out in three microwatersheds (Mazateca, Cuicateca, and Mixe regions) in the northern sierra of Oaxaca State, Mexico (Table 24.1). The hillside area represented by these microwatersheds is approximately 1 million ha in size. Names and biophysical characteristics of these experimental watersheds are San Jerónimo Tecoatl (319 ha, 1300 to 1900 m.a.s.l.) in the Mazateca region (latitude 18° 08' 57" and 18° 10' 13" north; and longitude 96° 53' 30" and 96° 54' 43" west); Concepción Pápalo (147 ha, 1700 to 2200 m.a.s.l.) in the Cuicateca region (latitude 17° 50' 20" and 17° 51' 25" north; and longitude 96° 51' 55" and 96° 52' 35" west); and Tzompantle (32 ha, 1280 and 1520 m.a.s.l.) in Cacalotepec in the Mixe region (latitudes 17° 00' 45" and 17° 01' 16" north; and longitudes 95° 53' 53" and 95° 54' 10" west). Principal land use systems in these regions are secondary-growth forest (*Alnus*, *Liquidambar*, and *Quercus* spp. mixed with *Pinus* spp.); coffee (*Coffea arabica*); permanent *milpa* (traditional practice of cultivating maize [*Zea mays* L.] interspersed with other food crops such as beans [*Phaseolus vulgaris* or *Phaseolus coccineus*] and Cucurbitacea); and milpa for 2 or 3 years followed by a variable number of years of *acahual*. Acahual is a local name for a secondary-growth vegetation characterized by a mixture of young tree species, herbaceous species, and shrubs, which are usually more than 3 years old, that

Table 24.1 Characteristics of Experimental Regions and
Microwatersheds

	Region		
Characteristics	Mazateca	Cuicateca	Mixe
Area (km^2)	2301.33	1329.67	6470
Dominant Soils [a]	Luvisol, Rendzina, Feozem, and Acrisol	Luvisol, Rendzina, Feozem, and Acrisol	Acrisol and Cambisol
Land use	Agriculture, forest (pine-oak) high sylva, pasture	Forest (pine-oak), sylva, pasture	Agriculture, low sylva, pasture, forest (pine)
Altitude (m)	200–3250	200–3000	200–3200
Slope gradient (%)	>15 (25-45)	>15	>25
Precipitation (mm)	>2000	500–700	1500–2000
Temperature (°C)	16–27	18–20	17–27

	Microwatersheds		
	San Jerónimo Tecoatl	Concepción Pápalo	Zompantle
Area (ha)	319	149	32
Altitude (m)	1380–1910	1700–2200	1280–1525
Slope gradient (%)	30 to 60	30 to 65	30 to 65
Land use (ha)			
Milpa[b]	41.04	65.31	3.00
Pasture	16.49	27.66	—
Forest	20.29	49.00	0.52
Coffee	169.14	—	8.39
Acahual (second growth)	72.79	—	20.77
Bushes	—	5.20	—

[a] The Food and Agriculture Organization–UNESCO and Mexican soil classification systems are similar, but they are not directly equivalent. The following are approximations: Luvisol = Alfisol; Rendzina = Borol; Cambisol = Inceptisol; Feozem = Mollisol; and Acrisol = Alfisol.

[b] A traditional Mexican practice of cultivating maize (*Zea mays* L.) interspersed with other food crops such as beans (*Phaseolus vulgaris* or *Phaseolus coccineus*) and Cucurbitacea.

grow after farmers cultivate the land using the traditional slash-and-burn system. Burning the biomass accumulated during the fallow cycle contributes to soil fertility. If slash-and-burn practices are eliminated, the acahual may become secondary-growth forests comprising *Liquidambar, Alnus*, or *Quercus* spp.

Observation sites representing various combinations of secondary forest vegetation, dominant perennial crops (coffee and pastures), milpa, and acahual were selected for study in the watersheds. In addition, technologically improved production systems (plots with peach or coffee plantings, depending on the altitude, in rows 9 m apart) were established at a very high density of 1400 trees ha^{-1}. Milpa cropping was located between tree rows either under traditional (LT) or minimum tillage (LC) management. This improved technology was named "MIAF" (Cortés et al., 2004).

In the experimental secondary-growth forest and perennial agricultural crop sites, five 4 × 25-m plots were established at random (Woomer and Palm, 1998; Kotto-Same et al., 1997). Three plots 9 m long and of variable width (4 to 8 m) were established for each of the milpa, acahual, and MIAF systems. In each plot, the aboveground (tree, bush plus weeds, and litter) and underground biomass was measured. Detailed procedures for sampling were described by Acosta et al. (2001a, 2001b, 2002). Soil samples were collected from 0 to 1.05 m deep in 15-cm increments at 18 to 20 georeferenced points within each plot. The latter permits monitoring changes in C pools over time (Acosta, 2001b).

Soil samples were taken with a handheld cylindrical sampler of 19-cm length and 4.17-cm internal diameter. The soil sampler allowed taking uncompressed soil cores (15 cm long) that were used to determine bulk density.

Soil samples were air dried and sieved to less than 2 mm before chemical analyses. Carbon concentration was measured in aboveground (tree, bush plus weeds, litter) and underground biomass, and in soil samples after removing roots and stones. A Shimadzu 5050A automatic carbon analyzer was used to measure total C. In this study, total C represents SOC, as soil pH was less than 7 (Table 24.5).

Table 24.2 Assumed Optimal Values for Selected
Properties

Indicator	Ideal Value
pH	7.0
Exchangeable acidity, c mole kg^{-1}	0.2
Organic matter, %	4.2
Olsen-P, ppm	12.0
Exchangeable K, c mole kg^{-1}	0.3
Exchangeable Ca, c mole kg^{-1}	10
Exchangeable Mg, c mole kg^{-1}	3
Cation exchange capacity, c mole kg^{-1}	30
Bulk density, g cm^{-3}	1.1
Available soil moisture, %	20
Hydraulic conductivity, cm hour $^{-1}$	10
Microorganism activity, μL 0_2 $hour^{-1}$ 100 g^{-1}	150

Source: From Vergara S., 2003. Identificación y Selección de Indicadores de Calidad del Suelo y Sustentabilidad en Sistemas Naturales y Agrícolas de Ladera en Oaxaca. Ph.D. thesis, Colegio de
Postgraduados, Montecillo, Mexico.

A separate independent soil-sampling scheme was implemented in selected plots to measure chemical properties associated with soil fertility, and to assess spatial variability of
SOC. Table 24.2 presents assumed optimal levels of various
properties obtained from the literature, as cited by Vergara
(2003). Assumed optimal levels for each characteristic
(assigned a 100% value) were used to build radar-type graphs
and to compare different systems. Eighteen soil samples were
taken for this study from the 0- to 20-cm and 20- to 40-cm
depths in each plot. The sampling design permitted the calculation of some semivariograms and spatial variability maps
(Vergara et al., 2003).

24.3 RESULTS AND DISCUSSION

Microwatersheds presented various land use and management systems. Tables 24.3, 24.4, and 24.5 show the total C
content aboveground (trees, shrubs plus grasses, litter, and/or

stalks) and underground (roots and soil) from 0- to 105-cm depth for the main land use systems of the Mazateca, Cuicateca, and Mixe regions (Etchevers et al., 2003).

The C stock in the aboveground plus roots plus soil pools was highest (306 Mg ha^{-1}) in the Mixe region, and lowest (54 Mg ha^{-1}) in the Cuicateca region. Carbon stocks corresponding to various land uses (secondary-growth forest, permanent agricultural crops, milpa, and MIAF, etc.) were similar in each watershed. Soil and prevailing climate conditions appear to determine the amount of accumulated carbon in each region. In general, areas with deeper soils and higher precipitation presented higher C stocks. The data suggest that under hillside conditions, some agricultural systems accumulate as much C as secondary-growth forest systems under hillside conditions. In all cases, C stored underground pools was higher than that stored in aboveground pools. More than 90% of the C in agricultural systems was stored in the soil, yet this percentage was lower in the secondary forest. Carbon content decreased with soil depth, and about 60% of the C stock was found in the first 50 cm of the soil profile (Etchevers et al., 2001). Older secondary forest stored a large proportion of C in the aboveground pool. The estimated rate of C accumulation in the secondary forest was between 1.5 to 3 Mg ha^{-1} year^{-1}. Greatest C accumulation occurred in the young acahual and the lowest accumulation in mature forests. Annual increments of C in the MIAF system trees were approximately 1 to 2 Mg ha^{-1} year^{-1}. This rate of C sequestration is considered comparable to rates reported in the literature for tropical forests (Szott et al., 1994; Houghton, 1997).

Weeds growing on residual soil water after the milpa was harvested contributed between 1 and 2.5 Mg ha^{-1} yr^{-1} of C to both the traditional and conservation tillage systems. Crop residues contributed between 2 to 4 Mg C ha^{-1} yr^{-1} (data not shown).

Selected chemical properties associated with soil fertility in three microwatersheds are presented in Table 24.6. The surface soil layers (0- to 20-cm depth and 20- to 40-cm depth) had in all cases acid pH, high exchangeable acidity, and in

Table 24.3 Carbon Content in Aboveground (Litter, Grasses and Shrubs, Trees, and/or Stalks) and Underground (Roots and Soil, 0- to 105-cm Depth) Components, in Predominant Vegetation Systems, Mazateca Region, Oaxaca, Mexico (Mg ha^{-1})

| Component | Forest Systems | | | Permanent | | Agricultural Systems[a] | | | | | |
| | BL | BA15 | BA10 | CA | PR | Mixed | | Annual | | | |
						Mv>30	Mv<30	LC>30	LC<30	LC>30	LC<30
Aboveground	99.5	46.3	31.0	34.5	5.4	2.6	1.6	0.6	1.2	1.0	1.4
Litter	5.9	8.4	12.6	9.2	NA	NA	NA	NA	NA	NA	NA
Grass and shrubs	0.5	1.00	3.0	0.7	5.4	1.3	1.6	0.6	1.2	1.0	1.4
Trees	93.1	36.9	15.4	24.7	NA	1.3	0	0	0	0	0
Straw	NA	NA	NA	NA	NA	a	a	a	a	a	a
Underground	155.3	158.9	244.3	151.7	175.2	159.4	131.1	267.0	276.0	235.9	200.3
Root	3.5	2.5	4.1	4.00	1.3	1.12	2.9	1.4	3.1	1.3	5.3
Soil	151.9	156.4	240.2	147.7	173.9	158.3	128.2	265.6	272.9	234.7	195
Total	255	205	275	186	181	162	133	268	277	237	202
SE[b]	(80)	(45)	(47	(69)	(43)	(13)	(15)	(27)	(99)	(44)	(38)

a Information was not available, as replanting occurred in 2001.

b SE is the standard error of the mean of total carbon in the system.

Notes: BL = Liquidambar forest; BA15 = Alnus forest of 15 years; BA10 = Alnus forest of 10 years; PR = prairie; Mv = living wall; NA = not applicable; LC = minimum tillage; and LT = traditional tillage.

Table 24.4 Carbon Content in Aboveground (Litter, Grasses and Shrubs, Trees, and/or Stalks) and Underground (Root and Soil, 0- to 105-cm Depth) Components in Predominant Vegetation Systems of Cuicateca Region, Oaxaca, Mexico (Mg ha⁻¹)

| | Forest Systems | | Agricultural Systems | | | | | |
| | Permanent | | Mixed | | Annual | | | |
Component	BE	PR	Mv(M)	Mv(G)	LC(M)	LC(G)	LT(M)	LT(G)
Aboveground	37.6	2.2	4.3	3.4	4.2	3.8	3.3	2.7
Litter	7.6	NA	NA	NA	NA	NA	NA	NA
Grass and shrubs	0	2.2	0.2	0.5	0.6	0.6	0.3	0.4
Trees	30.0	NA	0.7	0.1	NA	NA	NA	NA
Straw	0	0	3.4	2.8	3.5	3.1	3.0	2.2
Underground	58.9	97.2	63.5	113.9	67.5	50.1	57.8	65.7
Root	14.3	6.2	0.7	1.0	1.9	1.0	0.6	0.6
Soil	44.6	91.0	62.8	112.8	65.6	49.1	57.2	65.1
Total	97	99	68	117	72	54	61	68
SE[a]	(20)	(23)	(7)	(46)	(11)	(8)	(4)	(3)

[a] SE is the standard error of the mean of total carbon in the system.

Notes: BE = Quercus forest; LC = minimum tillage; LT = traditional tillage; Mv = living walls; NA = not applicable; PR = prairie.

Table 24.5 Carbon Content in Aboveground (Litter, Grasses and Shrubs, Trees, and/or Stalks) and Underground (Root and Soil, 0- to 105-cm Depth) Components in Predominant Vegetation Systems of Mixe Region, Oaxaca, Mexico (Mg ha^{-1})

				Agricultural Systems			
	Forest Systems			Permanent	Mixed	Annual	
Component	AC10	AC7	AC2	CA	Mvc	LC	LT
Aboveground	25.0	24.1	9.9	11.2	5.6	3.1	4.8
Litter	7.3	6.7	3.3	2.0	NA[§]	NA	NA
Grass-shrubs	4.3	1.9	6.6	0.3	0.9	1.0	1.0
Trees	13.4	15.5	0	8.9	0.2	NA	NA
Straw	NA	NA	NA	NA	4.5	2.1	3.8
Underground	128.1	174.5	123.2	163.7	267.9	281.0	300.3
Root	7.8	5.1	4.0	4.0	1.9	2.8	2.3
Soil	120.4	169.3	119.2	159.7	266.0	278.1	298.0
Total	153	199	133	175	273	284	305
SE[a]	(21)	(23)	(15)	(31)	(8)	(23)	(17)

[a] SE is the standard error of the mean of total carbon in the system.
Notes: AC2, AC7, and AC10 = Acahuales (second growth) aged 2, 7, and 10 years, respectively; CA, LC = minimum tillage; LT = traditional tillage; Mvc = living walls of coffee plants; NA = not applicable.

general low percentage base saturation and Olsen-P. Mazateca and Mixe region soils presented similar chemical characteristics. Soils in the Cuicateca region differed somewhat due to high erosion. Low soil productivity in the region is associated with low soil fertility. An appropriate management of soil fertility results in significant yield increments in the Cuicateca region (Cortés et al., 2004).

The spatial variability of most soil chemical properties was high in the 0- to 20-cm depth, as indicated by Vergara (2003). Figure 24.1 illustrates the spatial variability of soil C. The high spatial variability of C necessitates redesigning sampling strategies specifically designed for obtaining soil samples under hillside conditions and other ecosystems with variable soils. The SOC measurements were not always spatially correlated, as indicated by geostatistical parameters (Vergara et al., 2003). The extrapolation of data from a

Table 24.6 Mean Values of Selected Chemical Properties Determined in Soil Samples from Experimental Plots in Mazateca, Cuicateca, and Mixe Microwatersheds, Oaxaca, Mexico

Region and Watershed	Samples (n)	pH	Organic Matter (%)	Organic Carbon (%)	N (%)	Olsen-P (ppm)	Al^{+3} (c mole (+)/kg)	Ca^{+2} (c mole (+)/kg)	K^+ (c mole (+)/kg)	Mg^{+2} (c mole (+)/kg)	Na^+ (c mole (+)/kg)	Total (c mole (+)/kg)	Cation Exchange Capacity (c mole (+)/kg)	Base Saturation (%)
					Means by MicroWatershed (0- to 40-cm depth)[a]									
Mazateca	372	5.2 b	6.4 a	3.9 a	0.3 a	5.8 a	1.9 b	3.9 b	0.2 b	0.7 b	0.28 a	5.1 b	7.1 b	63.3 b
Cuicateca	166	5.6 a	3.1 b	1.9 b	0.2 b	6.0 a	0.4 c	9.6 a	0.7 a	6.5 a	0.01 b	16.7 a	17.1 a	94.8 a
Mixe	214	4.9 c	6.9 a	4.2 a	0.3 a	4.4 b	3.9 a	2.6 c	0.2 b	0.7 b	0.05 b	3.6 c	7.4 b	41.6 c
LSD		0.1	0.5	0.3	0.3	0.7	0.3	0.9	0.1	0.9	0.08	1.5	1.4	5.6
				Means by Microwatershed and Depth (0- to 20-cm and 20- to 40-cm depths)[a]										
Mazateca (0–20 cm)	186	5.2 b	7.7 b	4.7 b	0.4 b	6.5 a	1.7 b	4.8 b	0.3 b	0.9 b	0.29 a	6.3 b	8.0 b	73.1 b
Cuicateca (0–20 cm)	96	5.7 a	3.7 c	2.3 c	0.2 c	7.2 a	0.2 c	9.7 a	0.7 a	5.2 a	0.01 b	15.6 a	15.8 a	97.3 a
Mixe (0–20 cm)	107	5.0 c	9.4 a	5.6 a	0.5 a	5.2 b	3.8 a	3.5 c	0.3 b	1.1 b	0.07 b	5.0 b	8.8 b	51.7 c
LSD		0.14	0.74	0.44	0.03	1.08	0.45	1.22	0.10	0.95	0.12	1.82	1.64	6.57
Mazateca (20–40 cm)	186	5.2 b	5.1 a	3.1 a	0.3 a	5.1 a	2.2 b	3.1 b	0.2 b	0.4 b	0.27 a	3.9 b	6.2 a	53.6 b
Cuicateca (20–40 cm)	70	5.4 a	2.2 c	1.4 c	0.1 c	4.3 ab	0.7 c	9.4 a	0.7 a	8.2 a	0.01 b	18.3 a	19.0 b	91.3 a
Mixe (20–40 cm)	07	4.9 c	4.4 b	2.7 b	0.2 b	3.6 c	4.0 a	1.6 c	0.1 b	0.4 b	0.03 b	2.1 b	6.1 b	31.5 c
LSD		0.16	0.61	0.36	0.03	1.0	0.49	1.19	0.15	1.52	0.11	2.43	2.27	9.29

[a] Means followed by a different letter in the same column (0 to 40-cm) and (0 to 20-cm and 20 to 40-cm) are significantly different ($p = 0.05$), according to Tukey test.
Note: LSD = least significant difference ($p = 0.05$).

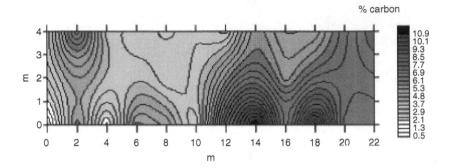

Figure 24.1 Spatial variability of soil carbon content for the 0- to 20-cm depth in a coffee plantation. (From Vergara et al. 2004. *Terra Latinoamericana*, 22:359–367.)

point/pedon level must be evaluated carefully to avoid misinterpretations.

The chemical characteristics of the soil were used to generate graphical indexes of actual quality for the various systems. This graphical index is useful to compare the different management systems. Near-optimum values for each variable were calculated based on data from the literature. Figure 24.2 is an example of this type of exercise. In general, severe deviations from ideal values were observed. Correction of these deviations may lead to improved C sequestration in the long run.

24.4 CONCLUSIONS

Carbon stored by the aboveground and underground components of land use systems on the northern highland hillsides of Oaxaca State, Mexico ranged from 54 to 306 Mg ha^{-1}. The amount of stored C was related more to soil depth and climate conditions than to vegetation systems.

Most C was stored in underground pools, with approximately 50% of the total C content stored in the top 50 cm of soils. The C content in all pools of annual agricultural systems

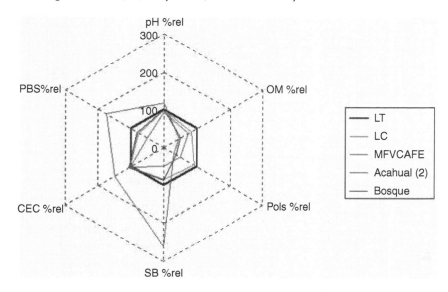

Figure 24.2 Radar-type graphical index used to establish deviation from "ideal" values (100%). CEC = cation exchange capacity; LC = milpa under conservation tillage; LT = milpa under traditional tillage; MFVCAFE = MIAF system with coffee plants as living wall; bosque = forest; OM = organic matter; Pols = extractable soil phosphorus by sodium bicarbonate solution; PBS = percentage base saturation; SB = sum of exchangeable bases.

(milpa) was similar to that found in the secondary-growth forest systems.

ACKNOWLEDGMENTS

This chapter is based on a paper submitted to the Workshop on Climate Change, Carbon Dynamics, and Food Security, Columbus, OH, June 11–12, 2003. We are grateful to the Global Environment Facility, World Bank, Mexican Ministry of Agriculture, Livestock, Rural Development, Fisheries, and Foodstuffs (SAGARPA), and the Oaxaca State government for collaborating in this research.

REFERENCES

Acosta, M., K. Quednow, J.D. Etchevers, and C. Monreal. 2001a. Un método para la medición del carbono almacenado en al parte aérea de sistemas de vegetación natural e inducida en terrenos de ladera en México. Paper presented at International Symposium on Measurement of Carbon Sequestration in Forest Ecosystems, Universidad Austral de Chile, Valdivia, November. Available in CD-ROM and at *www.iufro.org/iufro/publications/ws13 papeles1.pdf.*

Acosta, M., J.D. Etchevers, C. Monreal, K. Quednow, and C. Hidalgo. 2001b. Un método para al medición del carbono en los compartimentos subterráneos (raíces y suelo) de sistemas forestales y agrícolas de ladera. Paper presented at International Symposium on Measurement of Carbon Sequestration in Forest Ecosystems, Universidad Austral de Chile, Valdivia, November. Available in CD-ROM and at *www.iufro.org/iufro/publications/ws13 papeles1.pdf.*

Acosta, M., J. Vargas, A. Velásquez, and J.D. Etchevers. 2002. Estimación de la biomasa aérea mediante el uso de relaciones alométricas en seis especies arbóreas en Oaxaca, México. *Agrociencia*, 36:752–736.

Agriculture and Agri-Food Canada. 1998. *The Health of Our Air: Towards Sustainable Agriculture in Canada*. Compiled and edited by H. Janzen, R.L. Desjardins, J. Asselin, and B. Grace. Research Branch, Publication 1981/E. Agriculture and Agri-Food Canada, Ottawa.

Cortés, J.I., A. Turrent, P. Díaz, L. Jiménez, E. Hernández, L. Jiménez, and R. Mendoza. 2004. Chapter 23, this volume.

Etchevers, J.D., M. Acosta, C. Monreal, K. Quednow, and L. Jiménez. 2001. Los stocks de carbono en diferentes compartimentos de la parte aérea y subterránea, en sistemas forestales y agrícolas de ladera en México. Paper presented at International Symposium on Measurement of Carbon Sequestration in Forest Ecosystems, Universidad Austral de Chile, Valdivia, November. Available in CD-ROM and at *www.iufro.org/iufro/publications/ws13 papeles1.pdf.*

Etchevers, J.D., M. Acosta, C. Monreal, C. Hidalgo, J. Padilla, and L. Jiménez. 2003. Underground (roots and soil) compartments of carbon in forest and agricultural systems on hillsides in Mexico. In Scott Smith, C.A., Ed. *Soil Organic Carbon and Agriculture: Developing Indicators for Policy Analyses.* Proceedings of an OECD expert meeting, Ottawa, Canada, October 2002. Agriculture and Agri-Food Canada, Ottawa, and Organization for Economic Cooperation and Development, Paris, pp. 163–174.

Herrick, J.E. and M.M. Wander. 1998. Relationships between soil organic carbon and soil quality in cropped and rangeland soils: the importance of distribution, composition, and soil biological activity. In Lal, R., J.M. Kimble, R.F. Follett, and B.A. Stewart, Eds. *Soil Processes and the Carbon Cycle.* CRC Press, Boca Raton, FL, pp. 405–425.

Houghton, R.A. 1997. Terrestrial carbon storage: global lessons from Amazonian research. *Ciencia e Cultura,* 49:58–72.

Intergovernmental Panel on Climate Change. 2001. *Climate Change 2001: Mitigation. Technical Summary.* Report of Working Group III of the Intergovernmental Panel on Climate Change. Available at: *www.ipcc.ch/pub/wg3TARtechsum.pdf.*

Lal, R. 1998. Land use and soil management effects on soil organic matter dynamics on alfisols in Western Nigeria. In Lal, R., J.M. Kimble, R.F. Follett, and B.A. Stewart, Eds. *Soil Processes and the Carbon Cycle.* CRC Press, Boca Raton, FL, pp. 109–126.

Kotto-Same, J., P.L. Woomer, A. Maukam, and L. Zapfpack. 1997. Carbon dynamics in slash-and-burn agriculture and land use alternatives of humid forest zone in Cameroon. *Agric. Ecosyst. Environ.,* 65:245–256.

Szott, L.T., C.A. Palm, and C.B. Davey. 1994. Biomass and litter accumulation under managed and natural tropical fallows. *For. Ecol. Manage.,* 67:177–190.

Vergara, M.A. 2003. Identificación y Selección de Indicadores de Calidad del Suelo y Sustentabilidad en Sistemas Naturales y Agrícolas de Ladera en Oaxaca. Ph.D. thesis, Colegio de Postgraduados, Montecillo, Mexico.

Vergara, M.A., J.D. Etchevers, and M. Vargas. 2004. Variabilidad del carbono orgánico en suelos de ladera del sureste de México. *Terra Latinoamericana*. 22:359–367.

Woomer, P.L. and C.A. Palm, 1998. An approach to estimating system carbon stocks in tropical forests and associated land uses. *Commonwealth For. Rev.*, 77:181–190.

25

Terrestrial Carbon Sequestration in Zambia

ROBERT B. DADSON, JAGMOHAN JOSHI, FAWZY M. HASHEM,
ARTHUR L. ALLEN, CATHERINE S. BOLEK,
STEVEN W. MULIOKELA, AND ALBERT CHALABESA

CONTENTS

Carbon (C) is one of the most important and abundant elements on Earth, occurring in five general pools, that is, soil organic C (SOC) and soil inorganic C (SIC), oceanic, geologic, atmospheric, and biotic components. Carbon pools are of immense significance to the growing human population, which depends on soil quality for agricultural sustainability, poverty alleviation, and improved nutritional and health status. Carbon is a constituent part of humus, which improves soil quality by binding soil particles into aggregates. It enhances the chemical and physical properties of the soil and crop productivity. Carbon also combines with oxygen to form carbon dioxide (CO_2), a raw material for photosynthesis and, hence, an important component of dry matter for the production of food and fiber. Carbon dioxide absorbs heat from sunlight, thus helping to keep the Earth warm. Increased concentrations of CO_2 and other greenhouse gases will raise mean annual temperatures and cause excessive global heating, and melting of ice, glaciers, and permafrost. In turn, higher temperatures are expected to result in increased flooding in coastal areas from rising sea levels and droughts in low-rainfall areas, adversely affecting climate and agricultural production, especially in tropical and subtropical zones (Lal, 2001).

The SOC pool occurs as a complex mixture of nonhumic substances, such as carbohydrates, proteins, and amino acids, and humic products of secondary synthesis. The SIC pool consists of carbonates and bicarbonates. Together, the SOC and SIC pools form terrestrial C, and have a great impact on soil quality, and thus, on plant and animal life.

SOC is important for various reasons:

- It is a major sink/source of essential plant nutrients because of its strong impact on the effective cation exchange capacity in soils with low-activity clays.
- The release of plant nutrients through mineralization of soil organic matter (SOM) is essential for agronomic productivity.
- Mineralization of organic matter leads to emission of CO_2 under aerobic conditions and methane (CH_4) under anaerobic environments.
- Depletion of SOC causes soil degradation that results in crusting, compaction, reduced water retention and transmission properties, accelerated soil erosion, and decline of soil fertility (Baldock and Nelson, 2000; Rochette et al., 2000).

When accentuated by plow tillage, clean cultivation, and residual biomass removal and burning, SOC also significantly reduces agronomic productivity through soil crusting, compaction, accelerated erosion, water and nutrient imbalance, leaching, and acidification (Sanchez et al., 1982; Cassel and Lal, 1992).

SOM has numerous functional roles within ecosystems that range from the molecular to global, such as complexation of toxic cations in the soil solution (Hue et al., 1986), where the soil serves as an important source, sink, and buffer of greenhouse gasses (Hall, 1989; Post et al., 1990). Generally, SOM is closely associated with soil fertility as a source and sink of mineralizable nutrients (Duxbury et al., 1989), with the retention of nutrients and water (Russell, 1973; Lal, 1986), as an agent of soil structural stability (Oades, 1984), and with detoxification of naturally occurring and human-manufactured substances. Invariably, SOM declines once land is converted to agriculture in tropical and subtropical regions of the world.

SOC has both on-site and off-site beneficial effects. On-site beneficial effects can be derived through:

1. Increased aggregation and structural properties
2. Increased available water holding capacity
3. Improved macro porosity
4. Increased infiltration capacity
5. Decreased crusting, compaction, and soil erosion
6. Improved cation exchange capacity (CEC) and nutrient retention capacity
7. Decreased nutrient losses by leaching
8. Increased soil's trafficability and tilth
9. Increased number and diversity of soil biota
10. Increased soil capacity to biodegrade chemicals

Off-site benefits of SOC include:

1. Lower rates of sediment transport in natural waters
2. Lower siltation of waterways and reservoirs
3. Lower rates of transport of pollutants of natural waters
4. Lower emissions of CO_2, CH_4, and N_2O, from C and N displaced by soil erosion

Thus, increased SOC content and retention help enhance soil properties, and ultimately, contribute to increased agronomic productivity and sustainability.

Recognizing the significant role that C plays in agriculture, this chapter examines some of the factors that affect terrestrial C sequestration in Zambia. The agroecological zones, anthropogenic factors that affect biomass production, causes of C emissions, and strategies to sequester C in soils in Zambia are also discussed.

25.1 CHARACTERISTICS OF AGRO-ECOLOGICAL REGIONS IN ZAMBIA

Zambia is a landlocked country located in the tropics between latitude 8°-18°S and longitude 22°-33°E (Figure 25.1). Its tropical continental climate has clearly defined dry and rainy seasons. Three agroecological regions (see Figure 25.2 for locations), based on climatic characteristics with annual rainfall ranging between 700 and 1400 mm as the dominant factor, have been defined (Veldkamp et al., 1987). These zones are described below.

Figure 25.1 Location of Zambia in Southern Africa.

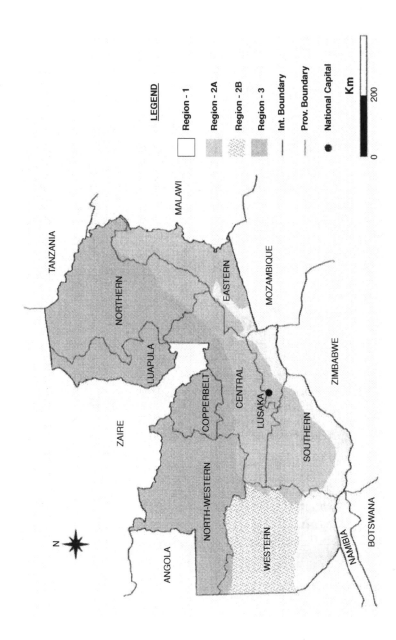

Figure 25.2 Geopolitical locations of Zambia and agroecological zones. (Courtesy of Soils Research Team, P.B. 7, Mt. Makulu, Chilanga.)

25.1.1 Region 1

This region covers major valleys of Zambia such as Gwembe, Lunsemfwa, and Luangwa, and the southern parts of Western and Southern provinces. Mean annual rainfall in Region 1 is low, up to 800 mm, and is generally well distributed. It is the driest and most drought-prone region. Temperatures in the growing season are generally high, and range from 20°C to 38°C.

The soil groups are either clay soils with fine loam topsoil and slightly acid, or alkaline with minor fertility limitations and good agricultural potentials, or shallow and not suitable for cultivation. Other soil groups are poorly drained, coarse, sandy, and are medium to very strongly acid. They are unsuitable for crop production. The predominant vegetation types in Region 1 are mopane, acacia woodlands, and deciduous thickets.

25.1.2 Region 2

This region covers the central part of Zambia extending from east to west, and comprises the sandveld plateau of Central, Eastern, Lusaka, and Southern provinces (Subregion 2a), and the Kalahari sand plateau and the Zambezi flood plains in Western province (Subregion 2b). Region 2 receives a well-distributed annual rainfall of between 800 and 1000 mm. The growing season ranges from 100 to 140 days with one to three 10-day dry periods of less than 30 mm of rainfall, which reduces crop yields, especially on sandy soils. However, in normal years, favorable climatic conditions and good soils make the region the most productive in Zambia. Mean daily temperatures during the growing season range from 23°C to 25°C. Severe frost may be experienced in some parts of Region 2 in June to August. The main soils in agroecological Region 2 have slight to severe chemical and physical limitations for crop production. They include low water-holding capacities, shallow rooting depth, low organic matter, low nutrient reserve and acidity, crusting, and coarse-textured topsoils that increase erosion hazard. The soils are clayey, loamy, or sandy, and strong to medium acidity. The vegetation type is varied and dominated by *miombo* and *munga* woodland, and to a lesser degree Kalahari woodland (dry deciduous forests), and dry evergreen forest.

25.1.3 Region 3

Region 3 covers Northern, Luapula, Copperbelt, and North-Western provinces, and parts of Serenje and Mkushi. It is part of the degraded Central African plateau. Mean annual rainfall exceeds 1000 mm with a growing season that varies from 120 to 150 days. Crop production is limited by fewer sunshine hours. The mean daily temperature in the growing season ranges from 16°C to 25°C.

The soils are highly weathered and leached, with pH levels of less than 4.5 and very low reserves of primary minerals. They are usually deficient in P, N, and other major plant nutrients and some micronutrients. The low soil pH and the associated high levels of aluminum and manganese are often toxic to plant growth. Soil physical properties, however, are favorable. Soils are microstructurally stable and deep, and possess well-drained profiles as well as high levels of biological activity.

The vegetation of Region 3 is broadly divided into miombo and a mixture of chipya and dry evergreen forest. The miombo — a two-storied woodland with an open or lightly closed canopy of semideciduous trees 15 to 21 m high — is characterized by species of *Brachystegia, Julbernardia*, and *Isoberlina* (Fanshawe, 1969).

25.2 ANTHROPOGENIC PRACTICES AND SOC SEQUESTRATION IN ZAMBIA

25.2.1 Farming Systems

Zambia's agriculture is predominantly rainfed. The major crops grown in the country are maize, sunflower, soybeans, groundnuts, sorghum, cotton, beans, bambara groundnuts, cowpea, sugarcane, finger millet, bulrush millet, rice, sweet potato, cassava, tobacco, and wheat. The main tools for land clearing in Zambia are the machete, hoe, and fire. Cultivation methods are mainly based on use of the manual hoe. Livestock raising is also an important traditional practice in the Southern, Eastern, and Western, and parts of the Central provinces of Zambia, where animal draft power is common. Tractors are mainly used by commercial farmers.

A form of slash-and-burn agriculture and adaptations to it are practiced. In the slash-and-burn *chitemene* system, tree offshoots from an area 12 times the size of the area to be cropped and the tree limbs and branches are heaped in a circular area in the middle of the field and burned. The ash from the burned wood acts as an ameliorator of soil acidity and supplier of nutrients, mostly P and K. Nitrogen, C, and S are lost in the combustion. The area is used for 3 to 5 years to produce millet and sorghum mixed with legumes and cassava. It is then left fallow for 15 to 20 years to regenerate before it is cropped again (Chidumayo and Chidumayo, 1984).

In Zambia, inversion tillage is the conventional practice, especially for resource-poor farmers. Although inversion tillage has its benefits, it is a major contributor to SOM/SOC loss. This is because the soil is turned over, exposing the SOM to high temperatures and erosive agents that bring the (mostly infertile) subsoil to the surface. Because of high temperatures, SOM is oxidized quickly and lost. This leads to a host of soil quality problems, such as structure destruction, crusting, hard setting, poor transmission properties, and loss of plant nutrients.

Conservation tillage in Zambia started as early as the 1930s (Environmental Conservation Association of Zambia, 1998). A traditional form, known as *galauza,* is practiced in the Eastern province. At the end of the dry season, the farmers scrape all dry plant residues off the soil and arrange them in a line along old ridges by hand hoeing. Crop residues are placed in the furrows between the previous season's ridges. The old ridge soil is turned over to cover the residues and to form new ridges in the position where the old furrows were located in the past season. Some farmers burn residues before the soil is turned over on them, in which case the beneficial effects of galuza are probably negated, and SOM and SOC are lost.

An adaptation to the chitemene system, due to diminishing woodlands, is the grass-mound *fundikila* system, where grass is buried in mounds and left to decompose at the end of the rainy season. The mounds are spread open at the beginning of the next rains, and beans, maize, groundnuts, cucurbits, cowpeas, and millet are planted. The land under this system is cropped for about 8 years before it is abandoned for another 8 to 10 years to regenerate.

The chitemene system was able to support 2.8 people per km^{-2} 30 years ago. It now supports 12 to 15 people km^{-2} (Chidumayo, 1987). The fundikila, which supported 10 people km^{-2} 30 years ago, now supports 20 to 30 people km^{-2}. These systems are breaking down under population pressure, and intervention is needed to rebuild soil fertility.

In the central plateau of the Southern and Eastern provinces, continuous cultivation, improper use of inorganic fertilizers, poor crop residue management, and generally poor land husbandry practices have caused acidity, as well as declines in soil fertility and SOM. Since the best land is already taken, farmers who are under survival pressures must continue farming on the same land.

25.2.2 Soil Carbon Dynamics in Smallholder Agriculture

In smallholder agriculture, a three-step process drives C changes. Initially, in shifting cultivation, C and nutrient stocks accumulated in natural vegetation are mobilized through land conversion (Nye and Greenland, 1960; Sanchez and van Houten, 1994). Clearing land for agriculture involves removing large volumes of vegetation, often followed by burning to enhance access to the land for cultivation. This causes considerable loss of above-ground biomass C to the atmosphere. The nonvolatile plant nutrients are concentrated in the ash, and incompletely combusted materials cover or enter the soil (Ewel et al., 1981; Araki, 1993; Kauffman, et al., 1995). The soil resource base is utilized for crop production for a number of years while nutrient-rich mineralized organic matter and root residues decompose. As the SOM declines, yields also decline. The land is then abandoned to fallow for reaccumulation of C and nutrients in vegetation and soils. With increasing population pressure, the fallow period is curtailed almost to continuous cropping. As a final step in soil C dynamics, the labile SOM becomes mineralized as the lower-level equilibrium of the SOM. External inputs become increasingly crucial in crop productivity (Swift et al., 1994), or else the farmers abandon the land.

25.2.3 Impact of Farming Systems and Agroforestry on Soil Organic Carbon

The three major land use categories in Zambia are cropland, forest reserves, and national parks. Forest reserves and national parks cover about 10% and 8% of the country, respectively, and are managed by the Forest Department and the National Parks and Wildlife Services (German Technical Cooperation [GTZ], 1995).

Forests in Zambia are used for agriculture, wood fuel, and timber harvesting. Out of 753,000 km^2 total land area, about 420,174 km^2 or 56% of the land area is potentially available for agriculture. However, the land area suitable for crop production is between 110,000 to 150,000 km^2, that is, between 15% to 20% of the total land area (GTZ, 1995). Currently, average woodland clearing for chitemene is estimated at 4.15 ha per household, culminating in an estimated annual deforestation of 600,000 ha (Centre for Energy Environment & Engineering Zambia [CEEEZ], 1999).

Wood fuel and charcoal are the main sources of energy in rural and urban areas, respectively. Charcoal was introduced in the Copperbelt in 1947, and has since become a major urban household fuel in Zambia. In 1990, at least 40,000 ha were cleared for urban wood fuel, causing deforestation in forest reserves and unprotected areas (CEEEZ, 1999; Chidumayo and Chidumayo, 1984). This type of deforestation does not affect forests in the national parks because these are preserved for wildlife conservation.

Timber harvesting is another major use of forests in Zambia. Although only a few valuable indigenous timber species exist, indigenous hardwoods are used for various subsistence and commercial purposes (CEEEZ, 1999).

Data on various land uses in Zambia from 1990 to 1994 indicate that there has been no increase in any category or total land area usage (CEEEZ, 1999). Thus, C loss or sequestration should not vary significantly. However, two- to threefold increases in deforestation activities for various enterprises are projected by 2010 and 2030 (CEEEZ, 1999). Therefore, SOC losses in the future are anticipated.

Decline in soil fertility in Zambia is making traditional farming unsustainable. Poor land husbandry is practiced by small-scale farmers. Data show that:

1. A total of 35 metric tons ha^{-1} of topsoil or 160 million metric tons of soil are lost annually.
2. An estimated 339,090 metric tons of nitrogen and 10,915 metric tons of phosphate, valued at \$300 million are lost annually.
3. Average maize yields are declining by 4% to 11% annually.
4. Sixty percent of the population is under the poverty level defined by the UN Development Programme (Zambia Department of Agriculture, 1999).

When the forest is cleared or fallow land brought into continuous production, SOM declines and reaches new steady-state levels depending on the quantity and quality of annual organic inputs and tillage methods. In the conversion of forest to cropland, SOC is reduced from 43 to 25 metric tons ha^{-1} in 15 years. It is replenished at 2 metric tons ha^{-1} instead of 11 metric tons ha^{-1} through organic dry matter per year (Greenland et al., 1992). For drier savannah conditions, a decrease of SOC from 18 to 14 metric tons ha^{-1} was predicted as a consequence of converting natural vegetation to cropland.

When natural vegetation is converted into cropland, soils rapidly become more acidic and lose nutrients through crop harvests, residue removal, and/or burning and leaching. Physically, the soil seems to collapse on itself. It becomes more dense, and erosive forces often cause the finer particles to disappear, due to poor tillage practices that leave a sandy or gravely material. The soil loses its capacity to form stable aggregates because the binding material, the SOM, has been lost. These degradation processes result in reduced biomass production and reduced amounts of organic matter returned to the soil, depletion of the SOC pool, decline in soil quality, and greater emissions of CO_2, CO_4, and N_2O to the atmosphere, where soil biota play an important role in all of these processes (Metting, 1993; Mendes et al., 1999). Emissions of these greenhouse gases caused by traditional farming practices lead to mining of soil C and N reserves. Since the SOC pool is very labile and highly dynamic, and its amounts depend on the input–output balance of the system, the result of the traditional farming system is a rapid decline of soil productivity, food production, food shortages, and malnutrition (Stewart et al., 1991). Intervention with improved farming methods, such as integrated nutrient management to rebuild SOM becomes very important. This is especially true since SOM affects the dynamics of various nutrients, such as N and P; influences soil physical and chemical properties and microbial biomass; and directly impacts the atmospheric concentration of CO_2.

25.2.4 Integrated Soil Nutrient Management

The aim of integrated soil nutrient management (INM) is to utilize available organic and inorganic nutrient sources judiciously and efficiently. This involves all aspects of nutrient cycling to achieve tight nutrient balances that synchronize with crop demand for nutrients and nutrient release in soil by the decomposing organic resource while minimizing losses through leaching, runoff, volatilization, and immobilization. Loss of nutrients by immobilization is, however, a temporary loss, which can serve as reservoir for holding nutrients in the soil that would otherwise be lost via leaching or in runoff (Brady and Weil, 1996). Nutrients that are lost to the soil through harvested biomass and other processes must be replaced with nutrients from external sources, or from the currently unavailable reserves in the soil since, for example, there are 10 to 15 times more P in the soil than is usually measured as available P. Declines in soil productivity can be attributed, in part, not only to the quantities of nutrients removed by plants compared to the quantities of nutrients being put back into the system, but also to the unavailability of nutrients in soils due to soil environments such as soil pH, leaching, mineralization, and other processes. Extensive agricultural production with tillage encourages continuous mining of SOM and release of N. Soil organic matter has a C:N ratio of 11:1; thus, sequestration of SOM requires high opportunity cost of N. Therefore, symbiotic N_2 fixation through the use of legumes seems the most likely cost-effective means of achieving this N input (Olness et al., 2002). Further, addition of plant nutrients through chemical fertilizers and organic amendments may be necessary to set in motion the restoration of these highly depleted soils. The SOC content cannot be improved without availability of additional quantities of N, P, and K in soils of low inherent fertility (Lal, 2000a; Mokwunye et al., 1996; Renard et al., 1997). An important aspect of INM is the maintenance of SOM, which improves soil structure, increases the soil's water-holding capacity, and improves soil biota populations. Integrated nutrient management practices improve soil biological and physiochemical properties, the buffering capacity and the cation retention capacity of the soil; it is a sink as well as a source of major and minor nutrients. The adoption of INM strategy on low acid clay soils of the tropics is ecologically necessary, economically desirable, and a realizable goal (Franzluebbers et al., 1998).

Traditional farming systems have generally resulted in substantial losses of soil fertility through the loss of SOC, due mainly to organic matter degradation by soil biota (Metting, 1993), and also due to the loss of micro and macro nutrients. With improved land use management and INM, soil biota in Zambian soils can be a sink for atmospheric C (Lal, 2000b). Traditional practices, akin to mining of soil C and N reserves, are major causes for the loss of SOC and SOM from agricultural ecosystems. Thus, balanced nutrient management through sustainable land husbandry practices can play a regulatory role of sequestering C (Franzluebbers et al., 1998).

Soil biota plays an extensive role in the decomposition of organic matter and the production of humus, cycling of nutrient and energy and elemental fixation, soil metabolism, and the production of compounds that cause soil aggregates to form (Metting, 1993; Mendes et al., 1999). Many of the soil biota such as rhizobia and bradyrhizobia are in symbiotic relationships with plants (legumes) (Kuykendall et al., 2000), and also have a beneficial role in nutrient cycling and atmospheric CO_2 enrichment (Baldock and Nelson, 2000; Rilling et al., 1999). This affects nitrogen dynamics, and thus, stimulates plant growth (Biswas et al., 2000; Gorissen and Controfo, 1999; Keeling et al., 1995).

25.2.5 Soil Organic Carbon Sequestration Strategy

Several practices and strategies that have been recommended to improve soil fertility for farming systems in Zambia also enhance SOC. These strategies and practices include:

1. Returning organic materials to the soil to replenish SOM lost through decomposition
2. Ensuring minimum disturbance of the soil surface
3. Reducing soil temperature and water evaporation through mulching with plant residues
4. Integrating planting of multipurpose trees and perennials
5. Improving fallows to increase production of organic materials
6. Using biofertilizers in legume production in order to improve biological nitrogen fixation (Lal, 1975; Peat and Brown, 1962; Sanchez et al., 1989)

These strategies will also enhance N supply, and SOC sequestration. The rate of SOC sequestration by adoption of recommended agricultural practices in tropical Africa is reported to be 0.1% to 0.2%. By expanding cropland area and adopting agroforestry techniques and recommended agricultural practices, it is possible to increase the potential SOC sequestration in croplands of the arable land in tropical Africa from 4 to

700 Tg (10^{12} g) over 50 years (Lal, 2001). In soils of the tropics a larger potential of SOC sequestration can be realized through restoration of degraded soils. This is especially true when the soils are used to grow trees and other perennials that have the potential to produce larger aboveground biomass usable as fossil fuel offset (Lal, 2001).

In ecosystems where soils such as acid savanna Oxisols and Ultisols are characterized by toxicity of Al and Mn and deficiency of P, Ca, Mg, and other bases, corrective strategies such as adopting recommended agricultural practices can enhance the SOC pool (Lal, 2001; Sanchez and Buol, 1975). In arid and semi-arid regions of Zambia, the inadequate supply of water can be mitigated through irrigation to enhance net primary productivity, and hence, raise the equilibrium level of the SOC pool higher than under natural ecosystem conditions (Lal, 2001).

25.2.6   Carbon Stocks in Natural Ecosystems and C Loss with Land Conversion in Africa

Total C stocks in Southern and Eastern Africa have been estimated to be 55.4 to 275 metric tons per hectare, respectively, in the humid forest and miombo woodland (a savannah-like deciduous woodland), and varies with precipitation (White, 1983; Woomer et al., 1998). Generally, the dead to live biomass and belowground to aboveground C ratios increase with aridity (White, 1983; Woomer et al., 1998). Of the total system C, 23%, 50%, and 65%, respectively, occurs below ground in the humid forest, Guinea savanna, and broadleaf (woody) savanna (Woomer, 1993). Although the drier areas tend to be more susceptible to burning, the amount of system C lost from burning is less because a greater proportion of biomass is present as coarse roots and in soil. Although there is biomass destruction prior to cultivation, the total soil C pools remain relatively stable throughout the slash-and-burn cycle, with only a slight decrease in the initial 4 years, and then increase to initial levels within 13 additional years of natural re-growth (Woomer et al., 1998). In Northern Zambia, where the practice of slash-and-burn agriculture is on the increase due to expensive mechanical clearing costs, the loss of C, biota, and crop biomass has accelerated in the past decade (Goma, 1999). SOM loss may be small, but the system C loss may be massive under slash-and-burn and continuous cultivation. Hence, as sometimes practiced in East and Southern Africa, continuous cultivation has caused large and more permanent changes in soil C, ranging from 0.9 metric tons ha^{-1} year^{-1} to 10 metric tons ha^{-1} year^{-1}. Lands that have been recently converted to agriculture show the greatest decline (Wissen, 1974; Woomer, 1993).

25.3 POTENTIAL FOR SEQUESTERING SOIL ORGANIC CARBON IN ZAMBIA SOILS

The SOC contents in most agricultural soils are below their potential levels, especially soils in developing agricultural systems (Lal, 1999). Carbon content of Zambian soils is low, ranging between 0.5% to 1% organic C (Singh et al., 1990), with an average of 0.8% (Silanpaa, 1982). In the three agroecological regions of Zambia, benchmark data of organic C content (Table 25.1) range from a low of 0.15% to 1.11% for bottom soils (38- to 70-cm depth) to a high of 2.45% to 3.0% for topsoils (0- to 8-cm depth) of clay soil (Lungu and Chinene, 1999). Other soil texture types show similar variations. When cultivated, these soils rapidly lose considerable amounts of organic C. The historic SOC loss can be regained by converting to an appropriate land use and adopting recommended agricultural practices (Lal, 1999). These include conservation tillage, mulch farming, cover crops, INM consisting of widespread use of biofertilizers for soil fertility replenishment, crop rotation, use of multipurpose agroforestry and green manures, soil water conservation, water table management, efficient use of irrigation water, erosion management using vegetative barriers, and other cultural practices in areas with high erosion hazards (Figure 25.3).

The rate of SOC sequestration through adoption of these measures has been estimated to be between 50 to 200 kg C ha^{-1} year^{-1} for warm and semiarid regions and negative or 50 kg C ha^{-1} year^{-1} for hot and dry regions. There is also a

Table 25.1 Organic C Contents (by Walkey–Black Method) of Selected Benchmark Soils from Three Agroecological Regions of Zambia (Topsoil 0–40 cm)

Series	Region	Order/Suborder	Depth (cm) and Horizon	Organic C (%)	Texture*
Mufulira	III	Typic Kandiustult	0–10A	0.85	SL
			10–20AB	0.57	SC
			20–43 Bt$_1$	0.37	C
Misamfu	III	Typic Kandiustult	0–10A	1.50	SL
			10–21AB	0.81	SCL
			21–48Bw	0.39	SCL
Misamfu Red III	III	Typic Kandiustult	0–16A	1.22	SL
			16–37BA	0.48	C
			37–63 Bw$_1$	0.30	C
Katito	III	Typic Kandiustult	0–10A	1.20	SC
			10–20 A$_2$	0.95	SC
			20–46 BA	0.66	C
Malashi	III	Typic Kanhaplustult	0–8 Ap	1.84	C
			8–23 Bt$_1$	1.04	C
			23–60 Bt$_2$	0.46	C
Mpongwe	III	Rhodic Kandiustult	0–7A	3.00	C
			7–24 AB$_1$	1.48	C
			24–42 Bw$_1$	0.74	C
Mushemi	II	Typic Kandiustalf	0–11A	1.60	SL
			11–34 AE	0.26	SL
			34–61 EB	0.23	SCL
Mukumbi	II	Typic Kandiustalf	0–8A	1.01	SL
			8–27 AB	0.31	SCL
			27–45 BA	0.25	C
Liteta	II	Udic paleustoll	0–21 A	2.03	C
			21–29 AB	1.41	C
			39–70 Bt$_1$	1.11	C
Kafue	II	Udic Chromustert	0–8 A	2.45	C
			8–24 Bw	0.81	C
			24–74 Bk$_1$	0.57	C
Makeni	II	Udic Paleustoll	0–20 Ap	1.67	SC
			20–38 AB	1.10	C
			38–70 Bt$_1$	0.70	C
Chelston	II	Typic Kandiustalf	0–24 Ap$_1$	0.63	SC
			24–38 Ap$_2$	0.55	C
			38–66 Bt$_1$	0.49	C

(continued)

TABLE 25.1 Organic C Contents (by Walkey–Black Method) of Selected Benchmark Soils from Three Agroecological Regions of Zambia (Topsoil 0–40 cm) (Continued)

Series	Region	Order/Suborder	Depth (cm) and Horizon	Organic C (%)	Texture
Kabuyu	I	Ustoxic	0–17 A	0.84	LS
		Quartzipsamment	17–32E	0.27	LS
			32–54 Bw_1	0.09	LS
Choma	I	Typic Kandiustalf	0–19 Ap_1	0.46	LS
			19–31 Ap_2	0.24	LS
			31–72 Bt_1	0.15	SCL

* C = clay; L = loam; S = silt
Source: From Lungu, O.I.M., and V.R.N. Chinene. 1999. Carbon sequestration in agricultural soils of Zambia. In Zambia Status Paper, presented at Workshop on Carbon Sequestration in Soils and Carbon Credits: Review and Development of Options for Semi-Arid and Sub-Humid Africa, Earth Resources Observation Systems Data Center, Sioux Falls, SD. Available at: *www.kafuku.org/wwwboard/messages/5.html*. With permission.

potential for SIC sequestration in soils of arid and semi-arid regions from formation of secondary carbonates (5 to 50 kg ha^{-1} $year^{-1}$) and leaching of carbonates into groundwater (50 to 500 kg ha^{-1} $year^{-1}$) in irrigated soils (Lal, 1999, 2001).

25.3.1 Soil Organic Matter, Soil Carbon, and Carbon Sequestration in Zambia

Soils under almost all farming systems in Zambia are low in N, P, and SOM (Stromgaard, 1984). It has been easy to detect and demonstrate evidence of N deficiency in cropped soils, but declines in SOM have been rather difficult to identify and quantify (Lungu, 1987). On the other hand, when crop residues are returned to the soil and legume cover crops are grown, they lead to increased organic C content and improved water retention.

The contributions of decomposed legume litter to SOC were assessed in a study of soil chemical and physical properties in a 7-year-old alley cropping trial containing *Leucaena leucocephala* and *Flemingia congesta* in northern Zambia. Trees had a

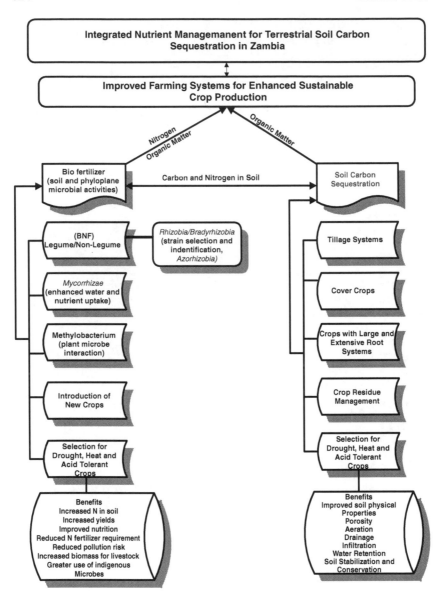

Figure 25.3 Relationship between integrated nutrient management, soil carbon sequestration, and improved farming systems.

Table 25.2   Litter Production and Maize Yields from 3-Year Improved Agroforestry Fallows in Eastern Zambia (metric tons/ha)

Treatment	Fallow Surface litter (12% moisture)	Maize Grain	Stover
Sesbania sesban	9.30	8.20	8.33
Fully fertilized	—	7.67	7.00
Gliricidia sepium	8.20	6.75	7.17
Leucaena leucocephala	5.60	4.72	6.52
Grass fallow	3.85	3.53	4.76
Senna siamea	11.70	2.31	3.85
Maize without fertilizer	—	2.61	3.23

Source: From 1996 report by International Centre for Research in Agroforestry for Zambia, cited in Lungu, O.I.M., and V.R.N. Chinene. 1999. Carbon sequestration in agricultural soils of Zambia. In Zambia Status Paper, presented at Workshop on Carbon Sequestration in Soils and Carbon Credits: Review and Development of Options for Semi-Arid and Sub-Humid Africa, Earth Resources Observation Systems Data Center, Sioux Falls, SD. Available at: *www.kafuku.org/wwwboard/messages/ 5.html*. With permission.

beneficial effect on soil chemical properties under hedge rows, particularly under *Leucaena*, and high levels of organic C, Mg, K, ECCE, and pH values (Dalland et al., 1993). Higher levels of organic C in the alley crop treatments were responsible for improvements observed in soil physical properties.

In Eastern province, maize grain yields following 3-year fallows were consistently greater than under conventional grass fallow of similar age (Table 25.2) in improved agroforestry fallows experiments (Lungu and Chinene, 1999). Improved fallows generated substantial surface leaf litter and rhizosphere biomass that contributed to C and N, and resulted in improved crop yields. The rhizosphere biomass contributed by *Sesbania sesban* root mass was 1.7 metric tons ha^{-1} for a 2-year fallow with 90% concentrated in the top 50 cm of soil.

A study of the effect of burning and ash fertilization in chitemene demonstrated that surface soils of protected woodlands had the highest values of C and lowest values of bulk density, compared to burned or cleared land (Araki, 1993). The relationship between bulk density and C content of both infield (burned) and outfield (cleared) soils of the miombo woodland were negatively correlated. Infield soils had less and higher values of bulk density than outfield soils. The range of sequential changes in the chitemene practice fluctuated between 0.5% and 2.5% for C, and between 0.9% and 1.4% for bulk density. The C content of infield soils decreased to 1% after cultivation and gradually increased with time. The C content of the outfield decreased to 1.1% a few years after clearing and gradually increased to 1.7% in 50 years. This decrease in SOM might be due to a reduction in the litter fall, and accelerated decomposition rate from increased exposure to solar radiation. Levels of 1.1% to 1.3% may be a minimum requirement for sustainable agriculture capable of utilizing weak plateau soils of the *miombo* woodland (Araki, 1993).

The impact of topsoil depth (TSD) and management on soil properties has been assessed using farmyard manure (FYM), N and P fertilizer, tie ridging, and ordinary farmer practices in Tanzania. SOC and P generally declined with a decrease in TSD, while SOC, N, P, K, and Mg were significantly increased by FYM application. The application of N and P fertilizers also had significant effects on SOC and P. Applying FYM increased SOC by about 0.55% in comparison with normal farmer practices, improved available water capacity (AWC), and root growth in soils with unstable structure and low SOC content (Gajri et al., 1994). The SOC pool was increased by adopting conservation tillage, using crop residue (mulch), incorporating cover crops in the rotation cycle, and using improved agricultural practices. The total potential for SOC sequestration in arable land of the tropics was estimated to be about 4 to 8 Pg (10^{15} g, or 1 gigaton) over the next 50 years (Lal, 1995a, 1995b).

25.3.2 Conservation Tillage, Cover Crops, and Residue Management

25.3.2.1 Conservation Tillage

Tillage plays a major role in agricultural sustainability through its effects on soil processes, soil properties, and crop growth (Food and Agriculture Organization/World Bank, 1988). It may enhance or curtail these processes depending on initial conditions and type of tillage tools used. Hand hoeing, a common form of minimum tillage practiced by smallholder farmers in Zambia, causes minimum disturbance of the soil, creates a rough soil surface, and provides for surface retention, increased infiltration time, and reduced runoff. Its drawback is in the incomplete inversion of the soil, thus exposing the SOM to other detrimental biotic and abiotic factors.

25.3.2.2 Management of Residual Organic Matter and Soil Organic Matter Sequestration

The culture of proper residue management is not found among smallholder farmers. The common practice is to completely harvest (remove residue from fields), burn residue, or graze off the residue by livestock. This leaves the soil with no external input in the form of organic materials. Burning, as much as it destroys pests and diseases, also destroys residues and SOM, and kills active beneficial microflora and fauna, making the soil inert.

In a maize/bean rotation subjected to various management practices in Kenya, on a Kikuyu Red Clay Loam (Humic Nitisol, Food and Agriculture Organization; Alfisol, U.S. Department of Agriculture), the C content of the soil ranged from 1830 kg C ha^{-1} for stover application to 7940 kg C ha^{-1} for stover/fertilizer/manure management practice (Woomer et al., 1998; Swift et al., 1994). Addition of fertilizer, in this case, reduced soil C. The efficiency of C sequestration ranged from 1.4% for stover-fertilizer management to 6.9% for fertilizer management (Woomer et al., 1998).

25.3.3   Increasing Carbon Sequestration through Nutrient Recapitalization and Agroforestry

Throughout Sub-Saharan Africa there is a constant threat of famine because of declining crop yields, depletion of soil nutrients, and reduced C stover of soils. The direct subsidy of inorganic fertilizers and nutrient or nutrient recapitalization has been adopted to combat nutrient depletion in smallholder farming systems (Mokwunye, 1995). Nutrient capitalization enhances crop productivity, which then contributes to soil C sequestration from stover, and stubble and root incorporation, as well as applied manures resulting from stover use as livestock feed.

Agroforestry provides fruit and wood fuel, sequesters C in tree biomass, and reduces C emissions through accumulation of C in wood biomass and soil. It has been estimated that agroforestry in Sub-Saharan Africa would accumulate an average of 15 metric tons soil C ha^{-1} (Woomer et al., 1998) or 59 metric tons C ha^{-1} (Houghton et al., 1993) in 20 years.

The SOM in tropical soils is of greater agricultural importance than any other soil property except water availability. Nutrient release from leaf litter and SOM accumulated in the underbrush on forest fallow is the main source of nutrients in subsistence cropping systems in the tropics. Soils in Zambia are among the lowest in organic C content; however, the soil C status can be considerably increased through adoption of improved fallowing with the use of agroforestry multipurpose tree planting to sequester C. The rapid release of SOM after clearance and cultivation promotes rapid SOM decomposition, energy availability to microorganisms, and nutrients released for uptake by microorganisms and plants, as well as SOC accumulation.

25.3.4   Mitigation Options in Zambian Forest Sector

Mitigation options to reduce soil C loss consist of maintaining existing stocks and expanding C sinks as described below.

25.3.4.1 Maintaining Existing Stocks

Forest protection and conservation, and increased efficiency in forest management, and in harvesting and utilization of forest products are the main ways to maintain existing stocks. Measures in this category aim to increase biomass stocks so as to reduce the current deficit. These measures are expected to increase forest land area (or biomass stocks) by 42,000 ha (2 million metric tons) by 2010 and 10 million ha (516 million metric tons) by 2030. Reduced land clearing for agricultural purposes and timber and wood fuel production can also contribute to maintenance of existing stocks (Chidumayo and Chidumayo, 1984).

25.3.4.2 Expanding Carbon Sinks

Enhanced natural regeneration and reforestation in Zambia can increase C sequestration. The high rate of forest land clearing that has occurred over time has left substantial pieces of land bare. This land can be replanted with trees. Using carefully selected exotic species to establish plantations offers a mechanism for rehabilitation of damaged lands and for increasing SOC accumulation (Lugo and Brown, 1993). Areas cleared for wood fuel, for example, can be successfully regenerated if proper management is undertaken to aid coppicing stumps. In degraded or damaged lands, rehabilitation can begin with the establishment of grasses and pastures. The plants that modify site conditions do not preclude tree establishment, and immediately favor SOC accumulation (Lugo and Brown, 1993). Proper control of forest fires is also considered important.

Under natural regeneration, it is estimated that 24,000 ha can effectively be regenerated by 2005 (CEEEZ, 1999). This figure is about half the current estimated annual forest land area cleared for charcoal production. It is estimated, however, that effective natural regeneration will cover 7 million ha by 2030.

25.3.5   Characteristics of Mitigation Options

Forest protection as a strategy can ensure that the area under national parks remains at 5,942,000 ha up to the year 2030. Increased efficiency in forest management, harvesting, and utilization of forest products will also be necessary. The goal is to reduce deforestation to the barest minimum by reducing the production of firewood, charcoal, timber, and clearing for agriculture.

Factors that contribute to the magnitude of C losses from soil are initial C stocks, texture, climate, slope, and management (Woomer et al., 1994). Thus, 52% of 91.3 metric tons ha^{-1} total organic C loss at 120-cm soil depth occurred in the surface horizon — that is, 0 to 15 cm (Moshi et al., 1974) — and manure application reduced C loss (Woomer et al., 1998).

25.5   SUMMARY AND CONCLUSION

Carbon is an important component of the soil and is essential for the maintenance of soil quality. As a constituent part of humus, C serves as a binding agent that holds soil particles into aggregates, and thereby enhances the soil physical properties, particularly porosity, aeration, infiltration, and drainage. Further, polysaccharides secreted by soil microflora can have the same effect, but not to the extent as C. Carbon also serves as a major sink/source of essential plant nutrients. Chemically, it is very important for enhancing soil fertility through improvement in CEC and nutrient retention, and as a source of energy for microorganisms.

Soil C dynamics are influenced by environmental factors such as temperature and rainfall along with anthropogenic factors that include the need for clearing forests and grasslands for farming, farming systems in place, and harvesting forests for timber and wood fuel. In Zambia, the chitemene and fundikila systems of land clearing and preparation, and the felling of forest trees for charcoal and wood fuel, result in a serious loss of soil organic and inorganic C.

Soil C sequestration practices have been recommended to ameliorate these conditions. They include the use of cover crops, green manure, crop rotation, integrated nutrient management systems, reforestation, and conservation tillage to reduce soil disturbances and hence soil erodibility.

Soil organic C sequestration is a highly recommended approach to sequester organic C. Increasing SOM content under current farming systems in Zambia maybe an easier approach to sequester SOC. In the northern part of the country, the chitemene and fundikila systems could be made more productive by increasing biomass accumulation in the miombo woodlands and reducing the burning of vegetation. The introduction of suitable multipurpose trees, herbaceous woody legumes that contain high lignin, polyphenols, and condensed tannins like *Calliandra calothyrsus* and *Gliricidia sepium*, would be one way to sequester C, as these materials are C sinks and release C slowly over long periods of time (Malama, 1998). Encouraging water harvesting may help to sequester soil C in the drier areas of Region 1, where the soils are prone to erosion and water is a limiting factor to crop production. In all three agroecological regions, enhancing biomass production and accumulation, and reducing deforestation when clearing land for agricultural production, and lumber and wood fuel harvesting will enhance SOC sequestration for the betterment of soil quality, agricultural sustainability, food security, and a reduction in global warming.

REFERENCES

Araki, S. 1993. Effect on soil organic matter and soil fertility of the chitemene slash-and-burn practice used in northern Zambia. In K. Mulongoy and R. Merckx, Eds. *Soil Organic Matter Dynamics and Sustainability of Tropical Agriculture*. John Wiley, Chichester, United Kingdom, pp. 367–375.

Baldock, J.A. and P.N. Nelson. 2000. Soil organic matter. In M.E. Sumner, Ed. *Handbook of Soil Science*. CRC Press, Boca Raton, FL, pp. 25–84.

Biswas, J.C., J.K. Ladhs, and F.B. Dazzo. 2000. Rhizobia inoculation improves nutrient uptake and growth of lowland rice. *Soil Sci. Soc. Am. J.*, 64:1644–1650.

Brady, N.C. and R.R. Weil. 1996. *The Nature and Properties of Soils*, 11th ed. Prentice Hall, Upper Saddle River, NJ, pp. 404–405.

Cassel, D.K. and R. Lal. 1992. Soil physical properties of the tropics: common beliefs and management restraints. In R. Lal and P.A. Sanchez, Eds. *Myths and Science, of Soils of the Tropics*. Special Publication 29, Soil Science Society of America, Madison, WI, pp. 61–89.

Centre for Energy Environment & Engineering Zambia. 1999. *Climate Change Mitigation in Southern Africa: Zambia Country Study*. Ministry of Environment and Natural Resources, Zambia, UNEP Collaborating Centre on Energy and Environment, Risø National Laboratory, Denmark.

Chidumayo, E.N. 1987. A shifting cultivation land use system under population pressure in Zambia. *Agrofor. Syst.*, 5:15–25.

Chidumayo, E.N. and S.B.M. Chidumayo. 1984. *The Status and Impact of Woodfuel in Urban Zambia*. Department of Natural Resources, Lusaka, Zambia.

Dalland, A., I.P. Vaje, R.B. Matthews, and R.B. Singh. 1993. The potential of alley cropping in the improvement of farming systems in the high rainfall areas of Zambia. III. Effects on soil chemical and physical properties. *Agrofor. Syst.*, 21:117–132.

Duxbury, J.M., M.S. Smith, and J.W. Doran. 1989. Soil organic matter as a source and sink of plant nutrients. In D.C. Coleman, J.M. Oades, and G. Vehara, Eds. *Dynamics of Soil Organic Matter in Tropical Ecosystem*. University of Hawaii Press, Honolulu, pp. 33–67.

Environmental Conservation Association of Zambia. 1998. *Conservation Tillage in Zambia*. Environmental Conservation Association of Zambia, Lusaka.

Ewel, J., C. Berish, B. Brown, N. Price, and J. Raich. 1981. Slash and burn impacts on a Costa Rican wet forest site. *Ecology*, 62:816–829.

Fanshawe, D.B. 1969. *The Vegetation of Zambia*. Research Bulletin No. 7. Ministry of Rural Development, Lusaka, Zambia.

Food and Agriculture Organization/World Bank. 1988. *Soil Fertility Initiative*. Exploratory Mission Report No. 98/063 cp-ZAM, July. Food and Agriculture Organization/World Bank Cooperative Program, Lusaka, Zambia.

Franzluebbers, A.J., F.M. Hans, and D.A. Zuberer. 1998. Long-term changes in soil C and nitrogen pools in wheat management systems. *Soil Sci. Soc. Am. J.,* 85:59–80.

Gajri, P.R., V.K. Arora, and K. Kumar. 1994. Maize growth responses to deep tillage straw mulching and farmyard manure in coarse textured soils of NW India. *Soil Use Manage.,* 10:15–20.

Goma, H.C. 1999. Soil fertility management and sustainable agriculture in Northern Province. Proceedings of Stakeholders Workshop, Lusaka, Zambia, 28–29 February.

Gorissen, A., and M.F. Cotrufo. 1999. Elevated C dioxide effects on nitrogen dynamics in grasses, with emphasis on rhizosphere process. *Soil. Sci. Soc. Am. J.,* 63:1695–1702.

Greenland, D.L., A. Wild, and D. Adams. 1992. Organic matter dynamics in soils of the tropics: from myths to complex reality. In R. Lal and P.A. Sanchez, Eds. *Myths and Science of Soils of the Tropics*. Special Publication 29. Soil Science Society of America, Madison, WI, pp. 17–33.

German Technical Cooperation. 1995. *Zambia Country Study on Climatic Change, Inventories, and Mitigation Analysis*. German Technical Assistance and Ministry of Energy and Water Development, Lusaka, Zambia.

Hall, D.O. 1989. Carbon flows in biosphere: present and future. *J. Geog. Soc.,* 146:176–181.

Houghton, R.A., J.D. Unruh, and P.A. Lefebvre. 1993. Current land use in the tropics and its potential for sequestering carbon. *Global Biogeochem. Cycles*, 7, 305–320.

Hue, N.V., G. Craddock, and F. Adams. 1986. Effects of organic acids on aluminum toxicities in subsoils. *Soil Sci. Soc. Am. J.,* 50:28–34.

Kauffman, J.B., D.L. Cummings, D.E. Ward, and R. Babbitt. 1995. Fire in the Brazilian Amazon: I. Biomass, nutrient pools, and losses in slashed primary forests. *Oecologia*, 104:397–408.

Keeling, C.D., T.P. Whort, M. Wahlen, and J. van der Plicht. 1995. Interannual extremes in the rate of rise of atmospheric carbon dioxide since 1980. *Nature (London)*, 375:666–670.

Kuykendall, L.D., F.M. Hashem, R.B. Dadson, and G.H. Elkan. 2000. Nitrogen fixation. In J. Lederberg, Ed. *Encyclopedia of Microbiology*, 2nd ed. Vol. 3. Academic Press, San Diego, CA, pp. 392–406.

Lal, R. 1975. *Role of Mulching Techniques in Tropical Soils and Water Management*. Technical Bulletin 1. International Institute of Tropical Agriculture, Ibadan, Nigeria.

Lal, R. 1986. Soil surface management in the tropics for intensive land use and high and sustained productivity. B.A. *Adv. Soil Sci.*, 5:1–97.

Lal, R. 1995a. Global soil erosion by water and carbon dynamics. In R. Lal, J. Kimble, E. Levine, and B.A. Stewart, Eds. *Soils and Global Change*. CRC Press/Lewis Publishers, Boca Raton, FL, pp. 132–142.

Lal, R. 1995b. The role of residue management in sustainable agricultural systems. *J. Sustainable Agric.*, 5:51–78.

Lal, R. 1999. Soil management and restoration for carbon sequestration to mitigate the accelerated greenhouse effect. *Prog. Environ. Sci.*, 1:307–326.

Lal, R. 2000a. World cropland soils as a source or sink for atmospheric carbon. *Adv. Agron.*, 71:145–191.

Lal, R. 2000b. Restorative effects of *Mucuna utilities* on soil organic carbon pool of a severely degraded Alfisol in Western Nigeria. In R. Lal, J.M. Kimble, and B.A. Stewart, Eds. *Global Climate Change and Tropical Ecosystems*. CRC Press, Boca Raton, FL, pp. 147–156.

Lal, R. 2001. Conservation tillage in the tropics for soil carbon sequestration. In *GART 2001 Year Book*. Golden Valley Agricultural Trust, Golden Valley, Zambia, pp. 50–54.

Lugo, A.E., and S. Brown. 1993. Management of tropical soil as sinks or sources of atmospheric carbon. *Plant Soil,* 149:27–41.

Lungu, O.I. 1987. *A Review of Soil Productivity Research in High Rainfall Areas of Zambia.* Occasional Paper No. 8, Series A. Zambia Soil Productivity Research Programme, Department of Soil Science, School of Agricultural Sciences, University of Zambia, Lusaka, Zambia.

Lungu, O.I.M. and V.R.N. Chinene. 1999. Carbon sequestration in agricultural soils of Zambia. In Zambia Status Paper presented at Workshop on Carbon Sequestration in Soils and Carbon Credits: Review and Development of Options for Semi-Arid and Sub-Humid Africa, Earth Resources Observation Systems Data Center, Sioux Falls, SD. Available at: *www.kafuku.org/www-board/messages/5.html.*

Malama, C. 1998. *Manipulating Nitrogen Release from Tropical Trees Prunnings Through Interactions Mediated by Their Resource Qualities.* Wye College, University of London, Ashford, Kent, United Kingdom.

Mendes, I.C., A.K. Bandick, R.P. Dick, and P.J. Bottomley. 1999. Microbial biomass and activities in soil aggregates affected by winter cover crops. *Soil Sci. Soc. Am. J.,* 63:873–881.

Metting, F.B. 1993. *Soil Microbial Ecology.* John Wiley & Sons, New York.

Mokwunye, A.U. 1995. Phosphate rock as capital investment. In H. Gerner, and A.U. Mokwunye, Eds. *Use of Phosphate Rock for Sustainable Agriculture in West Africa.* Miscellaneous Fertilizers Studies No. 11. International Fertilizer Development Center, Muscle Shoals, AL, pp. 100–106.

Mokwunye, A.U., A. de Jager, and E.M.A. Smaling. 1996. *Restoring and Maintaining the Productivity of West Africa Soils: Key to Sustainable Development.* International Fertilizer Development Center, Muscle Shoals, AL.

Moshi, A.O., A. Wild, and D.G. Greenland. 1974. Effect of organic matter on the charge and phosphate absorption characteristics of Kikuyu red clay from Kenya. *Georderma,* 11:275–285.

Nye, P.H. and D.J. Greenland. 1960. *The Soil Under Shifting Cultivation.* CAB International, Wallingford, United Kingdom.

Oades, J.M. 1984. Soil organic matter and structural stability: mechanisms and implications for management. *Plant Soil,* 76:319–337.

Olness, A., D. Lopez, J. Cordes, C. Sweeney, N. Mattson, and W.B. Voorhees. 2002. Application of a management decision aid for sequestration of carbon and nitrogen in soil. In Kimble, J.M., R. Lal, and R.E. Follet, Eds. *Agricultural Practices and Policies for Carbon Sequestration in Soil*. CRC Press, Boca Raton, FL, pp. 245–263.

Peat, J.E. and K.J. Brown. 1962. The yield responses of rain grown cotton and Ukiriguim the lake province Tangayika. 1. The use of organic manure, inorganic fertilizers, and cotton seed ash. *Empire J. Exp. Agric.*, 30:215–231.

Post, W.M., T. Peng, W.R. Emmanuel, A.W. King, V.H. Dale, and D.L. De Angelis. 1990. The global carbon cycle. *Am. Scientist*, 78:310–326.

Renard, G., A. Neef, K. Becker, and M. von Oppen, Eds. 1997. *Soil Fertility Management in West African Land Use Systems*. Margraf Verlag, Werkersheim, Germany.

Rilling, M.C., C.B. Field, and M.F. Allen. 1999. Soil biota responses to long term atmospheric CO_2 enrichment in two California annual grasslands. *Oecologia*, 119:572–577.

Rochette, P., D.A. Angers, and D. Cote. 2000. Soil carbon and nitrogen dynamics following application of pig slurry for the 19th consecutive year: I. Carbon dioxide fluxes and microbial biomass carbon. *Soil Sci. Soc. Am. J.*, 64:1389–1395.

Russell, E.W. 1973. *Soil Conditions and Plant Growth*, 10th ed. Longman, London/New York.

Sanchez, P.A., and S.W. Boul. 1975. Soils of the tropics and world food crisis. *Science*, 188:598–603.

Sanchez, P.A., M.P. Gichuru, and L.B. Katz. 1982. Organic matter in major soils of the tropical and temperate regions. In Proceedings of 12th International Soil Science Congress, New Delhi, India, vol. 1, pp. 99–114.

Sanchez, P.A., C.A. Palm, L.T. Scott, E. Cuevas, and R. Lal. 1989. Organic input management in tropical agroecosystems. In D.C. Coleman, J.M. Oades, and G. Uehara, Eds. *Dynamics of Soil Organic Matter in Tropical Ecosystems*, University of Hawaii Press, Honolulu, pp. 125–152.

Sanchez, P.A. and H. van Houten. 1994. Alternatives to slash-and-burn agriculture. Symposium at 15th International Soil Science Congress, Acapulco, Mexico.

Silanpaa, M. 1982. *Micronutrients and the Nutrient Status of Soils: A Global Study.* FAO Soils Bulletin 48. Food and Agriculture Organization, Rome.

Singh, R.P., J.F. Barr, and B.A. Stewart, Eds. 1990. *Dryland Agriculture: Strategies and Sustainability.* Vol. 13. Springer-Verlag, New York.

Stewart, B.A., R. Lal, and S.A. El-Swaify. 1991. Sustaining the resource base of our expanding world agriculture. In R. Lal and F.J. Pierce, Eds. *Soil Management for Sustainability.* Soil and Water Conservation Society, Ankeny, IA, pp. 125–144.

Stromgaard, P. 1984. The immediate effect of burning and ash-fertilization: cut-and-burn shifting systems in woodland ecosystems of Zambia, *Plant Soil,* 80:307–320.

Swift, J.J., L. Bohren, S.E. Carter, A.M. Izac, and P.L. Woomer. 1994. Biological management of tropical soils: integrating process research and farm practice. In P. Woomer and M.J. Swift, Eds. *The Biological Management of Tropical Soils.* John Wiley, New York, pp. 209–217.

Veldkamp, W.J., M. Muchinda, and H.A.P. Delmol. 1987. *Agroecological Zones of Zambia.* Research Branch, Ministry of Agriculture, Lusaka, Zambia.

White, F. 1983. *The Vegetation of Africa.* UN Educational, Scientific, and Cultural Organization, Paris.

Wissen, H.L.M. van. 1974. The influence of cultivation practices on the organic carbon content of some deep soils in Kissii District. Preliminary Report, Training Project in Pedology Kissii No. 4. Winard Staring Centre, Agricultural University Wageningen University, Netherlands.

Woomer, P.L. 1993. The impact of cultivation on carbon fluxes in woody savannas of Southern Africa. *Water Air Soil Pollut.,* 70:403–412.

Woomer, P.L., A. Martin, A. Albrecht, D.V.S. Resck, and H.W. Scharpenseel. 1994. The importance and management of soil organic matter in the tropics. In P. Woomer and M.J. Swift, Eds. *The Biological Management of Tropical Soils*. John Wiley, New York, pp. 47–80.

Woomer, P.L., C.A. Palm, and J.N. Qureshi. 1998. Carbon sequestration and organic resource management in African smallholder agriculture. In R. Lal, J.M. Kimble, R.F. Follett, and B.A. Stewart, Eds. *Management of Carbon Sequestration in Soils*. CRC Press, Boca Raton, FL, pp. 153–173.

Zambia Department of Agriculture. 1999. *Agricultural Statistics Bulletin*. Policy and Planning Division. Department of Agriculture, Lusaka, Zambia.

Section V

Policy and Economic Issues

26

Policy and Economic Issues Dealing with Global Warming

G. EDWARD SCHUH

CONTENTS

Debates about global warming have tended to be rather intense. The contributions to this volume have, however, moved beyond the rather frustrating debates about causes of global warming, and have sought instead to improve our understanding of such issues as the consequences of any warming for agricultural and terrestrial productivity, the potential contributions of carbon sequestration to reducing global warming, and the effects of mitigation efforts expected at the farm level. We are offered a cornucopia of scientific analyses and evidence on these issues, together with some constructive debate.

It is important to address the policy and economic dimensions of this subject. Both aspects have received too little attention in the discussions of global warming. It is true that some of the economic costs and implications have received attention in recent years (see, e.g., Nordhaus, 1994), but economic *policy* issues have still received only limited attention. Technological solutions to the problem have also received some attention, as evidenced in the contributions to this volume, but even then, the range of alternatives considered has been rather limited.

This chapter has three main focuses: (1) global warming as an international policy issue, with important implications; (2) some optimal degree of warming or temperature change as a matter for policy analysis; and (3) international cooperation in a world of uneven scientific capability, uneven institutional development, and uneven impacts from projected global changes. This discussion will lead to certain conclusions for policy consideration.

26.1 GLOBAL WARMING AS AN INTERNATIONAL POLICY ISSUE

We should begin by noting that global warming is, by definition, a global problem. Because it extends beyond national boundaries, it must be addressed by global policy measures. The fashioning and implementation of global policy measures is a novel challenge in that it implies the need for creating international public goods. Public goods are in general provided by governments, yet at the international level there is no government.

This is a long-standing problem in foreign affairs and international relations. When addressing this issue some years ago, Kindleberger (1986) proposed that in general, the problem can be solved by two means. The first is for the reigning hegemonic power to provide the appropriate public good. The United Kingdom did this for the international economy during much of the 19th century. Similarly, in the second half of the 20th century the United States provided monetary stability for the global economy by serving as central banker for the world, and sustaining an effective dollar standard for the international trade and financial system.

An alternative approach is what Kindleberger referred to as the "realist" solution, where groups of countries agree to come together to form institutional arrangements that provide the needed public good or goods. There are currently a number of such arrangements on the international scene, perhaps the most significant at this point being the World Trade Organization (WTO).

The origins of the WTO are to be found in the General Agreement on Tariffs and Trade (GATT), a rather modest framework of rules governing international economic relations established in 1947 after the U.S. Senate failed to ratify the treaty creating the more ambitious International Trade Organization. There are many lessons derivable from the GATT and its evolution into the WTO. In the first place, membership is strictly voluntary. Neither the GATT nor the WTO had (have) an army, navy, or air force. The successive organizations have negotiated the rules for international trade, and member states have agreed to follow them. Moreover, from a modest group of industrialized countries that agreed to negotiate reductions in tariffs on manufactured products, the scope of trade issues subject to negotiation by the WTO has increased significantly, and membership has grown to over 180 countries.

The Kyoto Treaty was a "realist" attempt to find a workable solution to the global warming problem. (The Montreal Treaty for dealing with the problem of atmospheric ozone change was a predecessor.) The withdrawal of the United States from the Kyoto Treaty may have delivered it a fatal blow, especially if Russia follows the U.S. lead in withdrawing its support.

It may be that the drafters of the Kyoto Treaty were too ambitious. A more modest beginning might have made success more likely. The modest initial beginnings of the GATT may provide some important lessons. I will return to this issue later.

26.2 AN OPTIMAL DEGREE OF GLOBAL WARMING?

Global warming issues are hard to resolve, in part because their discussion is so open-ended. Contemporary discussions have concentrated on ways to *mitigate* or *reduce* global warming, rather than looking beyond those objectives. In the prevailing estimates of the costs and benefits of global warming, there is an implicit assumption that there is some optimal degree of global warming, or some optimal temperature for the Earth, and that there is only one direction to move in attaining it — to mitigate or stop the current trend. However, do we really know what this temperature is?

R. Lal has implied that identifying an optimal degree of warming is possible when he noted that average global temperature has ranged between –18° C to +15°C. That is quite a wide range, and if there are costs and benefits in going from one average temperature to another, then there must be an optimal temperature for the Earth — something that C. Rosenzweig referred to, when commenting on my proposition, as a stabilization level.

Certainly, if one is willing to commit significant public and private resources to affect the global temperature, then why should those resources not be committed to moving the world toward what would be an optimal temperature? In fact, when establishing quotas and other targets for policies designed to affect global warming, some target temperature would seem to be an imperative. Is it appropriate to expend resources to mitigate or reduce global warming when we have no notion of the ultimate goal?

I would argue that it makes no sense to insist on mitigating policies without having some notion of the ultimate target. The failure to establish such an optimal temperature follows from the tendency in much of the discussions of global warming to ignore people — similar to the approach by many environmentalists. Yet surely, if Nordhaus can estimate the costs and benefits of global warming, we can move beyond those calculations to ascertain some optimal temperature for the Earth.

It may be helpful to introduce the concept of an optimal global temperature as we address these problems, even if we currently lack the data to estimate the optimum as a practical matter. First, it would focus our attention on assembling the necessary data and undertaking the appropriate analyses to arrive at such a conclusion, which should give us a better understanding of all the important issues involved. Second, it would help us avoid committing massive resources in what could turn out to be misguided directions.

26.3 TOWARD INTERNATIONAL COOPERATION

Consideration of the broader discussion of policy issues related to global climate change and of measures to modify this can usefully begin with a look at the Kyoto Treaty. We also need, however, to adopt a broader policy perspective, to address food security issues, the fact of variation in climatic conditions, and the institutional capability to address global warming problems. A discussion of these issues follows.

26.3.1 The Kyoto Treaty as a Starting Point

The United States has a great deal to offer to global society by rejoining the efforts to implement the Kyoto Treaty, and by recommitting to the objectives of that treaty. The United States has substantial scientific, technological, and analytical capability needed to address these issues. We could use these resources to help identify the optimal acceptable degree of global warming. Such a concept should guide the Kyoto Treaty initiative. However, we also might scale down the scope of this initiative, and focus it on a more modest set of objectives that would help mobilize the political support for what is proposed. The beginnings of the GATT/WTO serve as an attractive precedent.

Americans are rightly concerned about their national security. That national security is ultimately rooted in our economic security. We can add to our economic security by deploying our considerable economic power to strengthen international institutions. That power should be deployed in support of international cooperation with national and international organizations and institutions.

26.3.2 The Broader Policy Perspective

Implementation of the Kyoto Treaty should take place within the context of a broader perspective on the evolution of international policy institutions. Rapid and significant initiatives are taking place in this broader context. Globalization, for example, is causing economic policymaking and implementation to be increasingly beyond the reach of national economic policymakers. The loss of national sovereignty over economic policy is the source of much contemporary concern in this country, as it is elsewhere.

When policymaking and implementation move beyond the reach of national economic policy, a process of bifurcation follows. Some part of economic policymaking and implementation shifts up to the international level and becomes embedded in international organizations and institutions. Another part shifts down to the state and local level and becomes embedded in organizations and institutions at that level.

The policy milieu is currently in a rapid state of flux as shifts occur as to where policymaking and implementation take place. Reform is needed at all levels of the process if we are to make efficient use of our resources and provide for an equitable distribution of income (Schuh, 2003). This provides an opportunity to do the institutional design needed at each level to address global warming. In addition to the design work that is a policy imperative, there is much analytical work to be done in sorting out the problems unique to the international, national, and state and local levels. That analysis has to be the starting point for the institutional work.

26.3.3 Food Security

World food security is a concern in many contributions to this volume, some focused at the local level, and some at the national level. This concern incorporates the "people" dimension into discussion of problems usually framed in biophysical terms. However, I would note that many if not most of these presentations regard the essence of the food security problem as a matter of production. We learned long ago that famine and malnutrition are mainly results of poverty, rather than production shortfalls *per se*. In fact, Sen's famous analysis (1981) of the classic famines of China and India showed that during the most dire part of the famines, local food prices actually declined. Demand dropped under desperate circumstances, so that those with purchasing power could get food more cheaply — but most lacked purchasing power.

This observation in no way diminishes the importance of agricultural modernization and its contribution to alleviating poverty. However, it does provide an important perspective on how such modernization contributes to poverty alleviation. The contribution occurs through broad-based increases in consumers' real incomes rather than through an effect on producers as is commonly assumed. That becomes clear once one takes into consideration the general equilibrium effects of agricultural modernization (Schuh 1999).

Understanding the contribution of agricultural modernization to poverty alleviation in society directs increased attention to rural development. The modernization of agriculture makes it almost inevitable that much rural labor has to leave the agriculture sector if the per capita income of that labor is to keep pace with that in the nonfarm sector. That makes the exit of labor from agriculture a measure of policy success, assuming that such movement results from the "pull" of urban opportunity rather than the "push" of rural immiseration.

Because gainful employment outside of agriculture typically entails migration to alternative employment at long distances, the labor market becomes very imperfect, with wide disparities in wages for quality-equivalent labor. One way of promoting mobility is to promote expansion of nonfarm employment in rural areas, which is understood by the term "rural development." Ironically, the natural process of selection of migration and most of the policy measures implemented to promote economic development operate on the opposite development. The process of migration is highly selective of human capital, and thus drains this most important resource from the part of the economy where it is much needed. Similarly, policy measures tend to concentrate on subsidies for the expansion of the nonfarm sector in urban areas. This has the effect of imposing negative externalities in both the supplying and receiving region.

Our consideration of how different measures can deal most effectively with the challenge of global warming to food security would have benefited from more serious consideration of rural development policies. No matter what is done about climate change itself, we need to design more effective institutional arrangements that will not just mobilize the agricultural surplus generated by agricultural modernization, but will use it to generate an expansion of nonfarm activities in rural areas.

Much of the discussion of agricultural modernization has appeared misguided, as it has focused on raising the productivity of land — for instance, this perspective underlaid the Green Revolution in Asia. However, raising the productivity of land may contribute little to raising labor productivity, and this is what is critical to increasing the per capita incomes of rural people.

Hayami and Ruttan (1985) made the important point long ago that the adoption of new technology follows the relative resource scarcity prevailing in the agricultural sector. In land-scarce Japan, the beginning of technological progress was built on technological innovations that enhanced land productivity, the most limiting factor. Conversely, in labor-scarce United States, the process was the opposite, with labor productivity-enhancing technological innovations emphasized in the beginning. Moreover, later when the conditions of resource scarcity reversed themselves, the processes of technological innovation reversed themselves in both countries. That analysis, supported by extensive empirical evidence, implies that the subproduction functions underlying the land and labor sides of the basic production functions are separable. What happens on one side of the production function is largely independent of the other.

The work of Hayami and Ruttan has provided important guidance to policymakers all around the world. It is sad that both policymakers and policy analysts have ignored this central part of their analysis. In Sub-Saharan Africa in particular, labor is the relatively scarce resource. Policy needs to be directed to raising labor productivity at this stage of economic development, not primarily to raising land productivity.

To conclude this section, I should underscore the need for greater emphasis on rural development and a more rational science and technology policy if the productivity and the per capita incomes of rural people are to be raised and their food security problem addressed. Increases in per capita income are critical to addressing the food security problem, as most of the poverty in the developing countries is found in rural areas.

26.3.4 Geographic Variability

The impacts of global warming will vary a great deal from one region to another. That means that the policies appropriate for one region may not be appropriate for another. Similarly, the institutional arrangements available for addressing the problems of global warming will also vary a great deal. This includes the capability of markets to perform efficiently, and thus the efficiency and effectiveness with which markets can bring resources to bear on the problem. In the developed Organization for Economic Cooperation and Development (OECD) countries, for the most part the institutional arrangements available to address global warming problems are more than adequate, although the political processes for consolidating constructive action may have less capability. In both domains, the capability in many developing countries is less than what is expected to be necessary for effective action in the face of climatic difficulties.

26.4 A NEW POLICY PERSPECTIVE

The world continues to drift along with growing evidence that global warming is a fact, even though we do not fully understand what is causing that warming. Moreover, it has been difficult to mobilize globally the political support to do anything about it. For the most part, we rely on the volunteer efforts of national governments to take whatever steps they can find political support for. The Koyoto Protocol was designed to move us one step further toward collective action, but the reticence of the United States and other countries to collaborate effectively has resulted in stalled efforts.

Revitalizing efforts toward more effectively managing the process of climate change would benefit from three actions or initiatives at the international level: (1) a stronger knowledge base, (2) a stronger global institutional capability, and (3) an incentive system that rewards nations and their citizens for participating in something along the lines of the Koyoto Protocol.

26.4.1 Strengthening the Knowledge Base

The lack of knowledge on important issues becomes obvious as one participates in debates surrounding global warming. A number of priorities quickly emerge. The first, and perhaps most fundamental and glaring, is the lack of a consensus among scientists on the underlying causes behind the observed changes in global temperatures. We know that historically these temperatures have experienced both short-term fluctuations and longer-term trends. Despite the substantial additions to our knowledge, there continues to be a great deal of uncertainty about the underlying causes that are driving those changes.

Having a better understanding of the causal factors behind global warming is fundamental to designing policy responses to this phenomenon. If the cause is anthropomorphic, and this is where a significant part of the lack of agreement centers, some of the remedies are obvious. If instead it is not anthropomorphic, but due to physical phenomena in the larger environment, some of the same policy recommendations may apply, while others may not.

Understanding the underlying causes of global changes in temperature is also important if serious attempts are to be made to calculate an optimal temperature or stabilization value for the global temperature. Such knowledge will be especially important once one recognizes the need to reflect regional differences in such calculations.

Knowledge is also important if "quotas" are ultimately assigned to individual countries for changes in their contributions to global temperature changes. That knowledge will be critical to estimating the marginal benefits and the marginal costs of remedial efforts for individual countries. Ultimately, such allocations are needed. Moreover, knowledge of the marginal benefits and costs of individual policy actions is needed if rational policies and their associated institutional arrangements are to be designed.

26.4.2   Strengthening Global Institutional Arrangements

As noted above, the institutional capacity to deal with global temperatures varies a great deal from country to country. For the most part, OECD countries have an adequate capacity not only to identify the problem, but also to design and implement policy responses to it. At the same time, most of the developing countries are sorely deficient in their institutional capacity. Similarly, institutional capacity is also sorely limited at the international level, especially if we should be able to design incentive arrangements that will make national participation viable on a broad scale.

Perhaps the most effective way to deal with the deficiencies in the developing countries is by way of scientific and technological cooperation. If those efforts are directed to strengthening scientific and technological capacity in the developing countries, both the developed and developing countries that collaborate will benefit. Both sides will benefit from the new knowledge they generate in the process of cooperating. They will also benefit from the investments they make in their own human capital by collaborating on the research. The exchange of faculty and students will be a critical feature of such cooperation.

The variability of conditions and of capabilities to address the consensus building and operational problems poses a major problem because of the existing complexity. Having a wide variety of approaches to solving the problem should be accepted and embraced. Helping to develop local capabilities to address the problem and to design institutional arrangements for solving the problem presents a major opportunity for international cooperation. To the extent that such cooperation makes it possible to find more rapid and efficient solutions to the problem, this may help to mobilize political support for an international solution to the problem.

26.4.3   Devising an Effective Incentive System

The goals of public policy are frequently articulated as (1) efficiency, (2) equity, and (3) security. Briefly, efficiency refers to making rational use of available resources so that both individual countries and the global economy grow at the optimal rate. Equity refers to obtaining a distribution of income consistent with what the body politic desires. And security refers to an absence of, or minimal exposure to, risk and uncertainty.

A major problem with the Kyoto Protocol is that there are no incentives for national governments or policymakers to participate. Whether governments participate or not is strictly a voluntary issue. That, in turn, depends on the degree of political support and consensus that they can generate to mobilize support for what can in some cases be painful processes.

If there is to be progress in addressing the issue of global temperature change, it seems likely that there will have to be incentives for policymakers and national governments to participate and comply with what is mutually agreed upon. This is an area in which more creativity is needed. To offer one speculative suggestion on the possibility for incentives, it would seem that one could obtain a fruitful cooperation and linkage between the issues of global warming and international trade. In principle, the idea would be to link progress in dealing with the global temperature problem with potential benefits from trade.

This is not the occasion to sketch out in full detail what such a system might look like. However, it would seem that the critical component would be a means of linking performance on environmental modification with potential benefits from trade liberalization. The tarrification of all barriers to trade that was associated with the establishment of the WTO would seem to be an observable parameter with which one could work. The problem would be to have an institutional mechanism for granting trade liberalization for individual countries as the means to reward sound performance in reducing greenhouse gases, together with a means of enforcing it.

The obvious choice would be the WTO, but whether that organization can be credible in such a situation is questionable. An alternative would be to create a new international institution specifically to broker such negotiations. Finally, there is the possibility that the OECD countries could muster sufficient political will and commitment to provide the leadership for brokering such arrangements.

26.5 THE POTENTIAL FOR DISCIPLINARY SYNERGISM

This volume provides an example of the potential synergism from interdisciplinary collaboration in research on global warming. Given the degree of disciplinary specialization that prevails in academic circles, it is difficult to realize the synergism that comes from cooperation. The most difficult cooperation tends to be that between the biological and natural sciences on one side, and the social sciences on the other. It is across these boundaries that the communication problems are greatest, but it is at this point that collaboration is most needed.

In today's world, it is not feasible for disciplines working alone to develop adequate solutions to contemporary policy problems. No matter whether one is thinking about health policy, research and technology policy, environmental policy, or agricultural policy, disciplinary specialists by themselves can contribute only modestly to solving the problems. This is in part because better policy requires and involves the reform of old institutional arrangements and the design of new ones. Such institutional arrangements cannot be designed without inputs from multiple disciplines. When the problem to be solved extends across national borders, the need for institutional arrangements designed to facilitate such cooperation becomes clearer.

26.6 CONCLUDING COMMENTS

Two points should be made in conclusion. First, at the start of this chapter, I argued for trying to identify an optimal temperature for the Earth, or the optimal degree of global warming, but the following discussion dealt with how we might more effectively address the problem of global warming. While I believe that our efforts at dealing with global warming would be more persuasive and effective if we knew more about our objective function — the most desirable position to occupy — that is no reason to wait for more evidence on that function. Many of the changes and much of the knowledge that we need to bring about to address the problem will come about only with the passage of time. The direction in which we need to move is, however, not in significant doubt. We need to begin moving in the proper direction in the near future.

Second, economic policy does not necessarily have to be first-best (i.e., most efficient) in order for us to make progress in attaining our goals. There are many policies that if reformed would move us closer to some global optimum. An important example is the commodity programs in the United States and European Union. A reduction in the subsidies provided through these programs can have a substantial effect in reducing greenhouse gases. If political support can be mobilized for moving toward reform, that is sufficient reason for moving in that direction even if we do not know yet what the optimum would be. Delays aimed at having more complete information impose their own costs and make future reversal of undesirable dynamics more difficult. Given what we already know, the benefits in terms of slowing and attenuating global warming should be significant.

ACKNOWLEDGMENTS

I wish to thank the following people for helpful comments on an earlier version of this paper, without holding them responsible for remaining errors of judgment or fact: Julio Mario Barragan, Vernon W. Ruttan, and John Schaus.

REFERENCES

Hayami, Y. and V.W. Ruttan. 1985. *Agricultural Development: An International Perspective.* Johns Hopkins University Press, Baltimore and London.

Kindleberger, C.P. 1986. International public goods without international government. *Am. Econ. Rev.,* 76:1–13.

Nordhaus, W.D. 1994. *Managing the Global Commons: The Economics of Climate Change.* MIT Press, Cambridge, MA.

Schuh, G.E. 1999. Agriculture and economic development. *Chicago Policy Rev.,* 3:57–66.

Schuh, G.E. 2003. *Globalization, Governance, and Policy Reform.*

Sen, A.K. 1981. *Poverty and Famines: An Essay on Entitlement and Deprivation.* Clarendon Press and Oxford University Press, Oxford and New York.

27

Confronting the Twin Problems of Global Warming and Food Insecurity

LUTHER TWEETEN

CONTENTS

27.1 INTRODUCTION

The three components of the conference title, Climate Change, Carbon Dynamics, and World Food Security (Ohio State University, Columbus, Ohio, June 10–11, 2003) are interrelated. This chapter makes a case that careful public policy is required to simultaneously address global warming (including carbon dynamics) and global food insecurity, which are two major problems of our time.

Judging by where people choose to vacation and permanently reside (if they can afford the "luxury"), warmer climates are preferable to cooler climates. Thus, global warming that brings balmy temperatures to more people would seem to be welcome.

One problem with that assertion is that global warming's geographic impacts are uneven, as is apparent for agriculture. To be sure, atmospheric carbon dioxide acts like a fertilizer to plants. Despite this, poor countries at or near the tropics fare badly, although temperatures are expected to rise more in regions distant from the tropics. With moderate adaptation of varieties and cultural practices by farmers, global warming of a 2.5°C to 5.2°C magnitude expected in the next century is projected to reduce cereal output in developing countries 6% to 7%, and to raise cereal output in developed countries 4% to 14%, which suggests virtually no net change in cereal production for the world as a whole (Rosenzweig and Parry, 1994; Mendelsohn and Neumann, 1999). Thus, global warming does not threaten *global* food availability in the 21st century.

Uneven regional impacts are troubling, however. With global warming, food production could expand in Canada and Russia, while it falls in Nigeria and Indonesia. Lowlands such as in Bangladesh would be inundated, and many people would need to be relocated to higher ground. One conclusion is that global warming is especially threatening to food security in tropical and semitropical regions, where chronic undernourishment is and will continue to be concentrated.

In the following sections, I first outline the food security problem and selected public policies to address it and global warming. Of special concern is whether measures to promote global food security promise to be compatible with measures to restrain global warming.

27.2 FOOD SECURITY AND THE FOOD SECURITY POLICY SYNTHESIS

Chronically undernourished individuals rarely receive dietary energy considered by nutritionists to be necessary for light activity and good health. The Food and Agriculture Organization (2002) defines *food security* as a goal achieved "when all people at all times have physical, social, and economic *access* to sufficient, safe, and nutritious food that meets their dietary needs and food preferences for an active and healthy life." The World Bank and the U.S. Agency for International Development employ similar definitions of food security.

Defining food security as *access* to food omits another important dimension of food needs, namely availability. Food *availability* refers to the supply of food from production, imports, or stocks. Global food availability has not been a problem for decades. International food supply per capita has been rising and has been more than adequate to serve nutritional needs of all — not just of those who have the means to acquire food. Food availability will not be a problem for the foreseeable future at the global level, although projections indicate that the global food supply–demand balance will be tighter and the reduction in real food prices will be less in the future than during the past several decades (Tweeten, 1998).

Food *access* refers to the ability of individuals to acquire food through purchasing power or through transfers from family or others. With perennial availability of food in world markets for anyone, perhaps it is not surprising that food *accessibility* is the major constraint to food security.

Most food-insecure people live in rural areas of developing countries. Many are subsistence farmers. For them, food production (availability) is also buying power (accessibility). Food consumers everywhere outnumber producers, and poverty is widespread among landless peasants, hired agricultural workers, and smallholders in poor countries. Thus it is not surprising that most hungry individuals are net food buyers (Barraclough, 1991). It follows that it is important to lower food prices and raise food production along with productivity and *real incomes* to promote food security.

27.2.1 Setting

At the beginning of the 21st century, some 30% of the world's population suffered from one or more forms of undernutrition (FAO, 2002). Approximately 800 million persons remain undernourished, facing daily shortfalls of 100 to 400 kilocalories in meeting basic energy requirements. Micronutrient deficiencies are even more widespread, with an estimated 2 billion people afflicted by anemia mainly from iron deficiency, thus causing mental retardation and other disorders. Some 740 million people suffer disorders such as mental retardation and delayed motor development from iodine deficiency. Vitamin deficiencies also are widespread.

Considerable progress in addressing world hunger is apparent from data in Table 27.1 compiled by FAO. The number of individuals who on average do not consume enough dietary energy for normal activity and health fell steadily from 917 million persons annually in 1969–1971 to 777 million persons in 1997–1999, or by 5 million persons per year. Overall progress in reducing the incidence of undernutrition is spectacular. The proportion of population undernourished in the developing world was halved, falling from 35% in

Table 27.1 Actual and Projected Undernutrition in Developing Countries, Selected Years, 1969 to 2010

Region	Million Persons Actual (% region total)				Million Persons (% region total)
	1969–1971	1979–1981	1990–1992	1997–1999	Projected 2010
Latin America and Caribbean	53	48	59	54	40
	(19)	(14)	(13)	(10)	(7)
South Asia	238	303	289	303	200
	(33)	(34)	(26)	(24)	(12)
East and Southeast Asia	475	378	275	193	123
	(41)	(27)	(16)	(10)	(6)
Near East and North Africa	48	27	25	32	53
	(27)	(12)	(8)	(9)	(10)
Sub-Saharan Africa	103	148	168	194	264
	(38)	(41)	(35)	(34)	(30)
Total developing countries	917	904	816	777	680
	(35)	(28)	(20)	(17)	(12)

Source: From Food and Agriculture Organization (FAO). 1996. *Food and Agriculture 1996.* FAO, Rome; and Food and Agriculture Organization. 2002. *The Way Ahead: The State of Food Security in the World, 2001.* FAO, Rome. Projections for 2010 from FAO (1996).

1969–1971 to 17% in 1997–1999. Continuing progress is suggested by projections for year 2010 in Table 27.1.

Undernourishment differs widely among regions, as shown in Table 27.1. Ninety percent of the undernourished people in developing countries resided in South Asia, East Asia, Southeast Asia, and Sub-Sarahan Africa in 1997–1999. Thus, progress in reducing hunger will depend in great part on what happens in these regions. South Asia (Indian subcontinent) alone accounted for two-fifths of the undernourished. In that region, numbers of undernourished people have not changed much in recent decades, but the proportion of undernourished people fell from 34% in 1979–1981 to 24% in 1997–1999.

Economic progress in China accounts for most of the 282 million fewer undernourished people in East and Southeast Asia in 1997–1999 compared to 1969–1971. The reduced incidence of undernutrition in that region is even more striking, going from 41% to 10% during the same period.

Progress toward food security in Sub-Sarahan Africa is mixed, at best. In that region, the number of undernourished people has increased, and the proportion of people undernourished remains high and well above proportions in other regions.

Average dietary (food) energy supplies data in Table 27.2, like data in Table 27.1, show considerable progress toward food security. Food supplies are more available than ever. Kilocalories per capita increased in all major food-short regions between 1990–1992 and 1997–1999. The East and Southeast Asia together comprised the only region to exceed the overall average increase of 5.5% for the developing world during that period. Among countries with more than 25 million people, China led the developing world, having experienced a 12.2% rise in food energy supply per capita between 1990–1992 and 1997–1999.

The dietary energy supply in Sub-Sarahan Africa increased by 3.3%, a rate below that for developing countries as a whole, but higher than the rates for South Asia and for the Near East and North Africa. Optimism must be tempered, however, because calories per capita remain the lowest in Sub-Sarahan Africa.

Table 27.2 Average Per Capita Dietary Energy Supply, 1990–1992 and 1997–1999

Region	1990–1992 (kilocalories/ day)	1997–1999 (kilocalories/ day)	Change (%)
Latin America and Caribbean	2710	2830	4.4
South Asia	2330	2400	3.0
East and Southeast Asia	2647	2899	9.4
Near East and North Africa	3010	3010	0.0
Sub-Saharan Africa	2120	2190	3.3
Total, developing countries	2540	2680	5.5

Source: From Food and Agriculture Organization. 2002. *The Way Ahead: The State of Food Security in the World, 2001.* Food and Agriculture Organization, Rome.

Fortunately, food supplies do not depend only on local food production. Food can be imported to meet dietary needs. Thus, the evidence that numerous Sub-Sahara African countries have diminishing capacity to produce enough to feed themselves is not decisive — they can import food. But that conclusion raises another troubling issue: Can nations raise overall economic productivity enough to finance food imports if they cannot produce enough food at home?

World leaders meeting at the World Food Summit in Rome in 1996 pledged to halve the number of chronically undernourished people in the world, approximately 766 million at that time, by 2015. At the 1990–1992 to 1997–1999 rate of 5.6 million persons lifted out of undernutrition per year, the World Food Summit target will not be met until 2064. One reason for slow progress is inadequate understanding of the causes and cures for food insecurity.

27.2.2 Food Security Synthesis

The following *food security synthesis* helps to structure our thinking regarding causes and policy cures for transitory and chronic undernourishment (Tweeten et al., 1992; Tweeten, 1999). The seven-step logical framework is outlined below.

1. Transitory and chronic food insecurity is caused mainly by *poverty*. The nearly 800 million undernourished people noted in Table 27.1 are mostly part of the 1.2 billion people in abject poverty, defined as people living on less than $1 per day. People with adequate buying power overcome the frictions of time (e.g., unpredictable, unstable harvests from year to year) and space (e.g., local food shortages) to be food secure.

2. Poverty is best alleviated through broad-based, sustainable *economic development*. Altruism is commendable and plays a critical role in feeding members of the family. But for families with no "pie" to divide, issues of redistribution are moot. According to the FAO (1997): "The need is for policy measures that address all aspects of food insecurity with a view

to providing safety nets for the vulnerable and to creating the conditions that can lead to an eradication of endemic hunger. *This has to mean economic growth* [emphasis added].... Improving the equitableness of the income distribution can only achieve so much [in countries with low and falling income], and, as seen time and again, will be strongly resisted by the potential losers. So growth is necessary, and against a background of economic growth, experience shows that it is easier (although never easy) to implement measures that increase equity, particularly if the growth is broadly based to include the agricultural sector." Interregional and international charity to redistribute food has a proven record of responding to acute hunger (famine). It is never likely to be a dependable source of food for the chronically undernourished, however, in part because of donor fatigue, and in part because food aid diminishes market incentives facing farmers in poor countries. Few families or countries would wish to become dependent on the caprice of fickle donors for their long-term daily sustenance.

3. The most effective and efficient means to broad based economic sustainable development is to follow the *standard model* that ensures an economic "pie" to divide among people and among functions such as human resource development, infrastructure, family planning, a food safety net, and environmental protection. The standard model, outlined later, is applicable to any culture and provides a workable prescription for economic progress that ensures buying power for self-reliance and food security. (Food self-reliance emphasizes building agricultural and/or industrial productivity to produce food at home or the ability to purchase it abroad. It contrasts sharply with food self-sufficiency, a reckless policy if it compromises buying power so that a nation cannot afford to purchase food abroad when local production fails as it is prone to do.) Eventually, in conjunction with

family planning, the standard model brings zero population growth. It is not prized for its ideology, but because it works. It is not a one-size-fits-all model. All components of the model need not be followed, but some key features are essential for a sound economy. It is a checklist of economic measures that will offer a different policy reform prescription for each failed economy, depending on which items on the checklist a country fails to satisfy. Although no country has adopted every component, many countries have adopted enough components of the standard model to bring economic success.

4. *Political failure* explains why some countries do not adopt enough components of the proven standard model to end poverty and food insecurity. Individuals and groups with authority often oppose reform because they lose power with political change — even if current policies egregiously compromise the public interest.

5. Political failure is inseparable from *institutional failure*. Food insecurity and economic stagnation are not the result of limited natural resources, lazy and fecund people, greedy corporations, environmental degradation, or rapacious rich nations. Rather they are the result of misguided domestic public policies, which in turn are the product of weak, mismanaged, and corrupt institutions, especially national government. Thus, the standard model is inseparable from institutional change.

6. Poorly structured, inadequate institutions often trace to *cultural factors* such as tolerance of the public for unrepresentative, corrupt, incompetent government, and indifference to the broad-based involvement of citizens in government. Government leaders of poor countries are themselves products of the culture and too often view their position as an opportunity for personal aggrandizement rather than to serve the public interest. Socio-institutional change is blocked by cultural characteristics such as tribal animosities

that provide a fertile climate for governments not representing the public interest to play one group against another.

7. The core challenge to attain food security is socio-institutional change. How to bring about such change deserves attention by the best minds in economics, sociology, political science, and other disciplines.

27.3 THE STANDARD MODEL

A seminal development in economics since World War II is validation of the *standard model* for economic progress. The model has been presented elsewhere (Tweeten et al., 1992; Tweeten and McClelland, 1997; Tweeten, 1999), and is outlined only briefly in this section. Any country following the standard model can be assured of sufficient national product to be food secure. Of course, economic progress differs among countries adopting the standard model because countries differ in natural endowments (including location), institutions, and attitudes. It is important to emphasize that following the growth-promoting standard model is most important for food security in countries held back by unfavorable attitudes and sparse natural resources.

The standard model is prized for its success rather than for its neoclassical economic origins. Economic planners of Hong Kong and Chile have followed it and have succeeded spectacularly, while those of nations rejecting it such as the Soviet Union, North Korea, and Cuba have failed spectacularly. Anecdotal evidence for the standard model is telling, but numerous analytical studies cited by Tweeten (1999) make the case even more compelling.

The standard policy model calls for a lean public sector doing a few things well. Thus, it may seem incongruent that the following outline of the standard model is mostly about the role of government in the institutional framework. The standard model calls for markets to allocate market goods and services, defined as those that are rival, exclusionary, and transparent. (See Tweeten, 1989, chapter 2 for definitions of public and market goods.) Although the market makes most decisions regarding when, what, how, and where to produce in any successful economy, not much need be said about the market because it works on "autopilot" if the proper institutional framework is in place. The challenge is to establish the proper institutional framework.

Some countries appropriately choose the tradeoff of more safety net and less economic growth. Several Asian economies have experienced rapid economic growth and poverty reduction, with the public sector accounting for as little as 10% of their economy. At the same time, some countries of Western Europe have chosen slower growth and a larger safety net, with the public sector accounting for half of more of their capitalist economies. Thus, there is no one optimal size for the public sector.

27.3.1 Public Functions Required for Economic Progress and Food Security

27.3.1.1 Public Administration

- Honesty and competence. Corruption in government undermines economic progress (Sachs, 1997). Like sin, corruption cannot be eliminated but it can be diminished.
- Security, stability, order. The rule of law and order needs a judicial system to administer justice and interpret laws. Rules of the "game" are established by government so as to create an environment where business contracts can be made and efficiently carried out.
- Property rights. To encourage investment and improvements in property, investors must be able to "reap what is sown." Property rights allow property to be used as collateral for loans. A favorable investment climate avoids capital flight and attracts foreign direct investment.
- Competition. Governments need to avoid giving protection to firms exercising monopoly power. Open trade is often the most effective option to countervail the economic power of domestic firms. Parastatals (state-owned enterprises) need to be avoided where possible.

27.3.1.2   Sound Macroeconomic Policies

- Fiscal responsibility. Countries need to avoid operating account deficits. A deficit is justified in the capital account only for investments with the strong chance of a return that will more than cover the principal and interest.
- Monetary restraint. A rough rule of thumb is to increase money supply slightly in excess of the real gross domestic product (GDP) growth rate, with appropriate adjustments for foreign exchange and direct investment flows. A central bank at "arm's length" from political pressure, and with the sole objective of price stability is useful. However, because wages are inflexible, inflation of up to 3% annually can be tolerated and may even be helpful.
- Appropriate taxation. One of the most difficult challenges in any developing country is to collect taxes to pay for even minimal public services. Successful governments tax "bads" (consumption, tobacco, alcohol, emissions), not "goods" (investment, exports). They charge user fees to cover costs for electricity, irrigation water, and the like provided by the public sector. Sales, value-added, and property taxes distort the economy less than taxes on industry profits and exports.

27.3.1.3   Liberal Trade Policy

- Properly valued, preferably a market determined, foreign exchange rate.
- Open economy to trade and investment.

27.3.1.4   Infrastructure Investment

- All-weather roads for food security and for commercial activity consistent with comparative advantage.
- Bridges, seaports, airports, electricity, and the like.

27.3.1.5   Public Services

- Agricultural research offers unusually high returns on public and private investment. More will be said of this later.
- Human resource investments are essential for broad-based development for women and men, including minorities. Because rates of return on elementary schooling investment are especially high, universal elementary schooling is a priority for food security and development.
- Sanitation for food security requires attention to water and waste. Parasites and bacteria interfere with digestion of food and sap vitality.
- Health clinics staffed by volunteers and paraprofessionals can provide cost-effective services such as immunization, vitamin supplements, and can educate for HIV/AIDS prevention, family planning, and pre- and post-natal care education.

27.3.1.6   Environment

- For development to be sustainable, attention must be given to the environment. This is especially urgent where poverty attends high population growth and density. Agriculture can benefit from integrated crop management (forage legume rotations, alley cropping, etc.), conservation tillage (no tillage, ridge till, mulch till), integrated pest management (economic threshold intervention, biological pest controls, pest-resistant crops and livestock, best management practices), integrated crop–livestock systems including forage legumes, and integrated forest management (plantation forests, ecotourism, etc.).

27.3.1.7   Food and Income Safety Net

- The public safety net is for those unable to depend on themselves, the market, family, or other private sources. Landless peasants, smallholders, and the urban poor are especially vulnerable. Options include targeted food assistance and food for work. A country following the standard model will have the economic means to leave no person hungry. Transfers to serve humanitarian needs are as critical as investment in elementary education serving both economic equity and efficiency. Beyond that, the height and breadth of the safety net is a political decision that is not dictated by the standard model. Countries choosing high social safety nets ("welfare states") tend to have slow economic growth, and lack "creative economic destruction" that prunes economic deadweight.

- A free press, independent judiciary, and democracy can be very helpful. The free press and independent judiciary along with other checks and balances in government help to curb corruption. Democracy contributes to stability by diminishing problems of succession in government.

27.3.2 Sequencing Development

The issue of which sector should lead economic growth of poor countries has been controversial since R. Nurske (1953) made a case for balanced growth among sectors and A. Hirshman (1958) rebutted the argument by making the countercase for unbalanced growth. The lesson of the standard model is that public policy must not be fragmented, but, realistically, not all elements of the standard model can be pursued simultaneously. The general rule is to allocate resources and achieve growth in services and sectors offering the highest social returns on resources. Priorities for the public sector in poor countries are policy reform, agricultural technology development/transfer and adoption, elementary schooling, and infrastructure (Tweeten and McClelland, 1997). An important and challenging role for government is to gather and apply information to properly sequence public investments as the nation progresses through its various phases of growth.

In regard to sectors, international competition suggests that returns will be highest in sectors with a comparative advantage. Rather than arguing over which sector comes first, a wise approach is to view growth in the farm and nonfarm sectors as synergistic — each sector benefits from the other.

To be sure, the process of development in poor, food-insecure countries having agriculture as their economic base cannot begin in earnest without improvements in agricultural productivity. That productivity provides a surplus of output and earnings over needs that can be invested in high-return human capital, infrastructure, and agricultural research. The labor-saving and output-increasing improved capital inputs supplied to farms by the nonfarm sector free labor from farming. Investments by farmers in human capital are transferred to the nonfarm sector, thus raising the quantity and quality of industrial output and standards of living. Improvements in agriculture and industry are simultaneous.

This iterative process continues until human and material resources accumulate to support labor-intensive manufacturing and, eventually, high-paying service industries. That development process takes many years and much patience. The journey is aborted before it starts when governments neglect the first step — investment in agriculture — as T.W. Schultz (1964) perceptively noted four decades ago.

Pinstrup-Andersen (2002) highlights how little has changed, observing recently that developing countries devote only 7.5% of government expenditures to agriculture, although on average they derive three-fifths of their GDP from agriculture. Sub-Saharan Africa on average devotes only 0.5% of its agricultural GDP to agricultural research, compared to the United States, which devotes 3% of its agricultural GDP to agricultural research.

27.3.3   Economic Growth and the Environment

Would pursuit of economic development as called for by the food security synthesis compromise the environment and contribute to global warming as massive carbon-based energy is used to fuel expanding economies? Numerous empirical estimates of environmental Kuznets curves indicate that environmental degradation increases as per capita income rises up to a turning point, after which degradation per capita falls (Meier and Rauch, 2000; Hervani and Tweeten, 2002). Economic growth also reduces population growth, eventually to a point where absolute population count falls. The turning point at which environmental damage begins to drop differs by environmental variable, but for several variables including petroleum energy, the turning point is at $15,000 to $20,000 of income per capita (Hervani and Tweeten, 2002). The *environmental dilemma* is how to get poor countries through the turning point to lower greenhouse gas emissions and other environmental problems without irreversibly damaging the environment in the process.

27.4 AGRICULTURAL PRACTICES FOR FOOD AND ENVIRONMENTAL SECURITY

Policies to address global warming must be multifaceted, encompassing all industries and rich and poor nations and consumers. Taxes that bring private marginal costs up to social marginal costs induce private firms to act in the public interest. A carbon emissions tax, for example, would apply to agriculture and nonagricultural enterprises. Parry (2002) has estimated that a gasoline tax of $1.00 per gallon, 2.5 times the current U.S. rate, is required to bring private cost up to the social cost. That externality tax correction applied to all carbon fuels, not just gasoline, could significantly reduce carbon emissions and encourage a shift to cleaner energy.

Other policies would encourage sequestering carbon in agricultural soils. Farming practices that sequester carbon in soils also reduce atmospheric carbon. When carbon is sequestered as organic matter, it is kept out of the atmosphere where it causes global warming. Organic matter holds water and nutrients such as phosphate, potash, and nitrogen for slow release to growing plants. Organic matter improves soil structure to diminish soil compaction and erosion. High-yield agriculture in general helps to confine cropping to environmentally safe areas, thereby enhancing soil conservation, wildlife, and biodiversity.

Tanner (2000) measured the cost of sequestering additional carbon in soils with crops, livestock, and trees under specific resource situations for northern Ohio. The cost of carbon sequestration was measured by the net income foregone by following enterprises and practices that build carbon reserves in the soil. Not surprisingly, she found considerable complementarity — some measures that increased organic matter in soils in perpetuity also increased farm income in perpetuity. Her study found that the cost in foregone net receipts to farmers per metric ton for modest levels of carbon sequestration achieved with no-tillage practices is less than Nordhaus's (1993) estimates of benefits of carbon sequestered per year, ranging from $6 per metric ton of carbon (MTC) in 2001 to $21 per MTC by 2100.

Continuous no-tillage corn, the most profitable rotation for farmers, also ranked at or near the top in sequestering carbon. On the other hand, Tanner's (2000) results indicated that the cost per metric ton of carbon sequestered is several times higher than estimated benefits, and hence is economically prohibitive for other tillage practices under the resource situations considered in her study.

Agricultural crops and practices other than no-tillage could not compete with forestry and fossil fuel conservation measures in cost-effective carbon sequestration. Tanner (2000) calculated the one-time cost, which is equivalent to total net present cost in perpetuity at a 5% discount rate, to be $5 per metric ton of carbon sequestrated in Ohio forests. Nordhaus (1991), Adams et al. (1999), and Plantinga et al. (1999) report that the cost for sequestering carbon in U.S. forests ranges from a low of $5 to a high of $120 per MTC depending on the region and forestry sector. Annual cost is also below the discounted benefit of $6 (short run) to $21 (long run) for storing a metric ton of carbon in perpetuity as reported earlier by Nordhaus (1993). Furthermore, compared to crops in northern Ohio, forests have the potential to sequester approximately 100 times as much carbon per hectare. Thus, it pays to store high levels of carbon in forests rather than in croplands. Forest costs would rise, and could become uneconomic if a significant amount of cropland were to be lost to forests, driving up the cost of food, cropland, and forestland.

No-tillage continuous corn was found to be the most profitable crop per acre and per ton of carbon sequestered in the northern Ohio. However, it is less profitable in areas of the Corn Belt having cold, wet, and tight soils in the spring, and in areas troubled by perennial weeds that must be controlled by cultivation. Banded tillage and other technologies are being developed to address such problems. Each resource situation is unique; study is needed of agriculture's potential to profitably sequester carbon under the diverse conditions found around the world.

27.5 CONCLUSIONS

Several conclusions follow from the foregoing analysis:

- Poverty and the high population growth that attends it are inimical to food security and environmental protection. The favorable news is that the rather simple, straightforward *standard model* offers a proven economic policy prescription for any country to have the means to be food secure while addressing environmental problems such as global warming. The unfavorable news is that seemingly intractable political, institutional, and cultural barriers have precluded implementation of that model in poor countries.
- Exuberance that standard model policies raise income is dampened by the fact that environmental Kuznets curves show that carbon emissions and other environmental "bads" increase up to $15,000 to $20,000 of per capita income. Higher per capita income levels are then associated with an improved environment. Technology is reducing that turn-around income threshold, and policies to develop and adopt improved technology such as cleaner, affordable energy will continue to lower that threshold.

- Poor countries cannot afford to subsidize farmers to build organic matter. Hence, it will be important to find or to develop complementarities between crop and livestock profitability on the one hand, and environmental and food security on the other hand. Opportunities for such complementarities need to be exploited, as in the case of no tillage in Ohio.
- The Ohio study suggests that some practices used to sequester carbon forego no farm income and hence are "free." However, the scope to profitably store carbon in cropland is limited. Beyond no tillage, farming practices to sequester carbon are expensive in that they sacrifice considerable farm profits. Sequestration in forests, energy conservation through carbon taxes, and other measures offer more promise.
- High-yield agriculture contributes to the environment by concentrating crop production on "safe" land, and leaving environmentally sensitive land to trees and grasses that sequester more carbon while providing recreation, wildlife protection, and biodiversity.
- Although research results from this study revealed that little to no policy intervention is needed to induce farmers to sequester modest amounts of carbon in northern Ohio, the sequestration process can be quickened at low public and private cost through technology development and diffusion programs. Greater public outlays for such efforts promise high payoff. Massive carbon sequestration in cropland appears to be prohibitively expensive. However, modest sequestration is free with no tillage and many farmers operate close to the margin. Thus, small payments could tip the scale of profitability from conventional to no-tillage or other carbon-sequestering practices. Conservation compliance policies currently are in disarray, but the public has much latent potential to obtain more conservation tillage and other environmental protection in return for the billions of

dollars distributed to support farm income each year in the United States and European Union.

- Research is critical to enhance now-limiting complementarities among crop yields, profits, and carbon sequestration. Research to increase crop yields so that more land can be freed for carbon-sequestering forests and grazing rather than crops will be financed largely by countries that have prospered by following the standard model.

REFERENCES

Adams, D.M., R.J. Alic, B.A. McCarl, J.M. Callaway, and S.M. Winnettt. 1999. Minimum cost strategies for sequestering carbon in forests. *Land Econ.*, 75:360–374.

Barraclough, S. 1991. *An End to Hunger?* Zed Books, London.

Food and Agriculture Organization. 1996. *Food and Agriculture 1996*. FAO, Rome.

Food and Agriculture Organization. 1997. *Food Security: Some Macroeconomic Dimensions*. FAO, Rome.

Food and Agriculture Organization. 2002. *The Way Ahead: The State of Food Security in the World, 2001*. FAO, Rome.

Frisch, R. 1958. *The Strategy of Economic Development*. Yale University Press, New Haven, CT.

Hervani, A. and L. Tweeten. 2002. Kuznets curves for environmental degradation and resource depletion. In L. Tweeten and S. Thompson, Eds. *Agricultural Policy for the 21st Century*. Iowa State Press, Ames, pp. 204–230.

Hirschman, A. 1958. *The Strategy of International Development*. Yale University Press, New Haven, CT.

Meier, G. and J. Rauch. 2000. *Leading Issues in Economic Development*, 7th ed. Oxford University Press, New York.

Mendelsohn, R. and J.E. Neumann, Eds. 1999. *The Impact of Climate Change on the United States Economy*. Cambridge University Press, London; New York.

Nordhaus, W.D. 1991. The cost of slowing climate change: a survey. *Energy J.,* 12:37–65.

Nordhaus, W.D. 1993. Rolling the 'DICE': an optimal transition path for controlling greenhouse gases. *Resour. Energy Econ.,* 15:27–50.

Nurske, R. 1953. *Problems of Capital Formation in Underdeveloped Countries.* Oxford University Press, Oxford.

Parry, I.W. 2002. *Is Gasoline Undertaxed in the United States?* Resources for the Future, Washington, DC.

Pinstrup-Andersen, P. 2002. More research and better policies are essential for achieving the World Food Summit goal. Speech to World Food Summit: Five Years Later, Rome, June. International Food Policy Research Institute, Washington, DC.

Plantinga, A.J., T. Mauldin, and D.J. Miller. 1999. An econometric analysis of the costs of sequestering carbon in forests. *Am. J. Agric. Econ.,* 81:812–824.

Rosenzweig, C. and M. Parry. 1994. Potential impact of climate change on world food supply. *Nature,* 367:133–138.

Sachs, J. 1997. The limits of convergence. *The Economist,* July 14–20, pp. 19–22.

Schultz, T.W. 1964. *Transforming Traditional Agriculture.* Yale University Press, New Haven, CT.

Tanner, M. 2000. Potential for Carbon Sequestration on Ohio Cropland. Master's thesis, Ohio State University, Columbus.

Tweeten, L.G. 1989. *Farm Policy Analysis.* Westview Press, Boulder, CO.

Tweeten, L.G. 1998. Dodging a Malthusian bullet in the 21st century. *Agribusiness,* 14:15–32.

Tweeten, L.G. 1999. The economics of global food security. *Rev. Agric. Econ.,* 21:473–488.

Tweeten, L., J. Mellor, S. Reutlinger, and J. Pines. 1992. Food Security Discussion Paper. PN-ABK–883. U.S. Agency for International Development, Washington, DC.

Tweeten, L.G. and D.G. McClelland, Eds. 1997. *Promoting Third World Development and Food Security.* Praeger, Westport, CT.

28

Policies and Incentive Mechanisms for the Permanent Adoption of Agricultural Carbon Sequestration Practices in Industrialized and Developing Countries

JOHN M. ANTLE AND LINDA M. YOUNG

CONTENTS

Scientists agree that soil degradation continues to be a key limiting factor for agricultural sustainability, despite decades of research to improve soil conservation and other sustainable practices (Barrett et al., 2002; Scherr, 1999; Hudson, 1991; Sanchez, 2002). The prevailing economic explanation for the continuing loss of natural resources in many parts of the world is that economic incentives often encourage degradation and discourage conservation. In industrialized countries, with well-developed market institutions, these incentive problems are generally associated with what are known as "externalities" and "market failures." In developing countries, with less-developed market institutions, these incentive problems are often attributed to high discount rates that poor farmers apply when assessing the future, lack of capital markets, high transport costs and other market imperfections, adverse government policies, insecure property rights, and limited availability of fodder for livestock and domestic uses (Lutz et al., 1994; Scherr, 1999; Antle and Diagana, 2003). Thus, numerous important factors must be considered in assessing the potential for soil carbon sequestration in order to address the problem of agricultural sustainability and associated problems of soil degradation and poverty.

This chapter addresses two sets of issues related to soil carbon sequestration. First, greenhouse gases (GHGs) cause a global externality in the form of an enhanced greenhouse effect. Policies are needed to create appropriate incentives in order for carbon sequestration to be undertaken in industry or agriculture. We provide an overview of the current policy environment and deal with the following questions: What kinds of national and international policies are needed to create adequate incentives for carbon sequestration? How will those policies affect opportunities for agricultural carbon sequestration? Second, mechanisms must be devised to incorporate farmers into national or international efforts to reduce greenhouse gas emissions once incentives for farmers are in place. Appropriate and efficient incentive mechanisms for soil carbon sequestration need to be designed for both industrialized and developing countries. These incentive mechanisms are likely to take different forms, depending on the local institutions operating in each country. We discuss the forms that these incentive mechanisms are likely to take, recognizing that appropriate mechanisms must provide for the permanent sequestration of carbon in aboveground biomass or in soils.

28.1 POLICIES AND INCENTIVES FOR AGRICULTURAL CARBON SEQUESTRATION

International concern about increased concentrations of carbon dioxide and other GHGs in the Earth's atmosphere led to the establishment of the Intergovernmental Panel on Climate Change in 1988. The panel concluded that human activities were responsible for increased emissions of GHGs and the accompanying climate change. While acknowledging a degree of scientific uncertainty, the UN Framework Convention on Climate Change (UNFCCC) concluded that "where there are threats of serious or irreversible damage, lack of full scientific certainty should not be used as a reason for postponing such measures [to reduce GHG emissions]" (Climate Change Secretariat, 2002). Having been ratified by 175 nations, including the United States, the UN Framework took effect in 1994. The primary objective of the convention is to reduce GHG concentrations in the atmosphere to levels that would prevent dangerous anthropogenic interference with the climate system. All parties to the UNFCCC agreed to prepare and update national climate change mitigation and adaptation programs. These would include measures to reduce emissions and enhance sinks for carbon, promote the development and application of climate-friendly technology, undertake research on climate change and its mitigation, and compile and submit a national inventory of GHG emissions.

In addition, parties to Annex I (41 industrialized countries and economies in transition [EITs]) agreed to the non-legally binding aim of reducing emissions to 1990 levels by the year 2000. Annex II parties (industrialized countries minus the EITs) further agreed to provide financial assistance to other parties for the acquisition of appropriate GHG-reducing technology. Annex I parties are encouraged to undertake projects in other countries to reduce emissions or increase removal of GHGs.

Since the U.N. Framework took effect in 1994, stronger scientific evidence on the existence of climate change, as well as concern that emissions are continuing to increase, prompted negotiation of the Kyoto Protocol, which concluded in December 1997. The key provision of the protocol requires Annex I parties to collectively reduce emissions of GHGs to a level 5% below the 1990 level by the first commitment period of 2008 to 2012. The protocol also calls for establishment of policies to reduce emissions, including the phasing out of subsidies for energy-intensive technologies, the creation of regulatory standards encouraging the adoption of alternative energy sources and the taxing of emissions, the reduction of emissions from transport systems, and the control of methane emissions through innovative waste management.

The issue of establishing credits for forestry and agriculture as carbon sinks was hotly debated during the negotiations. Technical difficulties in verifying the amount of carbon sequestered by agriculture was one contentious issue. Nevertheless, guidelines for carbon sequestration were agreed to in the 2001 Marrakesh accords, which provide rules for land use, land use change, and forestry (UNFCCC, 2002a). The accords recognize revegetation and improved management of cropland and grazing land as carbon sinks, and allow parties to receive credits for carbon sequestered in excess of 1990 levels. Scientific bodies supporting the Kyoto Protocol continue to work on protocols to verify emissions removals achieved by these activities.

The Kyoto Protocol allows Annex I countries to meet their commitments at least cost within some constraints. They can receive credit for implementing projects that reduce emissions or remove carbon from the air in other countries, and they can trade credits with other Annex I parties, subject to conditions. The Clean Development Mechanism allows Annex I parties to implement projects that reduce emissions in non–Annex I countries and to receive credits toward meeting their emissions-reduction targets. However, agricultural sequestration activities cannot be used in the Clean Development Mechanism for the first implementation period of the Kyoto Protocol.

A recent decision limited eligibility for this mechanism to afforestation and reforestation activities. However, the World Bank launched the BioCarbon Fund to sponsor a broad range of carbon sequestration projects in developing countries, including agricultural sequestration projects. The motivation for sponsoring non–Kyoto compliant projects is that agricultural sequestration projects can achieve both emissions reductions and sustainable development goals (World Bank, 2002).

The Kyoto Protocol will enter into force when it has been ratified by 55 Annex I governments representing 55% of the total Annex I 1990 carbon dioxide emissions (UNFCCC, 2002b). Currently, 128 parties accounting for 61.6% of emissions have ratified the protocol, which became effective on February 16, 2005. Prolonged uncertainty over the ratification of the Protocol existed due to indecision by Russia on whether or not to ratify.

Only parties that have ratified the protocol and have met methodological and reporting requirements may engage in emissions trading that gets counted toward meeting Kyoto Protocol requirements. Annex I parties can purchase credits only from other ratifying Annex I parties. Because they have not ratified the Kyoto Protocol, the United States and Australia cannot sell emissions or removal credits to other Annex I parties. This means that the carbon trading market will be fractured into a market for sellers who have ratified the protocol and for sellers who have not. International demand is likely to be severely limited for carbon credits from sellers who have not ratified the Kyoto Protocol, as the only buyers would be countries that have not themselves ratified, and thus are not under any binding commitments to reduce emissions.

28.1.1   U.S. Response to Climate Change and Kyoto Protocol

Although the Clinton administration negotiated the Kyoto Protocol, the Bush administration has thus far refused to ratify it. In his public remarks, President Bush has disagreed with both the form of the Kyoto Protocol and the scientific evidence that prompted other governments to ratify the treaty. In 2001, he stated that the emissions targets established by the Kyoto Protocol "were arbitrary and not based on science," and further claimed that "no one can say with any certainty what constitutes a dangerous level of warming, and therefore what level must be avoided" (O'Neill and Oppenheimer, 2002). In addition, President Bush has been dissatisfied with the division of responsibility for carbon emissions reductions between developed and developing countries (White House, 2001).

The Bush administration has proposed a voluntary program of reducing GHG "intensity" by 18% in the next 10 years (U.S. Department of State, 2002). GHG intensity is defined as the ratio of GHG emissions to economic output. The administration proposes to lower the current GHG intensity of 183 metric tons of carbon equivalent (MTCE) per million dollars of GDP to 151 MTCE per million dollars by 2012 through voluntary and incentive-based measures. A key component of the administration's proposal is the creation of tax incentives for the development of renewable energy, hybrid and fuel cell–powered vehicles, co-generation and landfill gas, and other new technologies. In response to this proposal, some businesses have developed their own voluntary initiatives to reduce GHG emissions (White House, 2003).

A major criticism of the administration plan is that it allows U.S. total emissions to continue to increase along its current trend. Total U.S. GHG emissions increased from 1.671

billion MTCE in 1990 to 1.907 billion MTCE in 2000, a 14% increase. Under the administration's plan, 2012 emissions (2.155 billion MTCE) would be 30% above 1990 levels (Pew Center on Global Change, 2002). If the United States had ratified the Kyoto Protocol, it would have been required to reduce its emissions to 93% of 1990 levels by 2008–2012. The Pew Center on Global Climate Change notes that GHG intensity fell by 21% in the 1980s and by 16% in the 1990s, so the Bush plan will result at best in only very slight improvements over existing trends.

The Bush administration proposal mandates improvements to the current federal registry of GHG emissions. The goal of registry improvements is to ensure that voluntary actions taken by industry to reduce GHG emissions will be rewarded in the future with transferable credits for emissions reductions. Currently, few businesses participate in the registry because there is no third-party verification of reductions for buyers of carbon credits (Chartier, 2002). Registry improvements are currently expected to be slow due to concerns over the legality of binding future Congresses to the current administration's plan to grant transferable credits.

The U.S. administration's failure to adopt binding national emissions limits and to ratify the Kyoto Protocol is a signal to U.S. industries that Bush does not consider climate change to be a serious national problem that requires international cooperation. While some firms may voluntarily reduce emissions to take advantage of new technology and/or to enhance their reputations, other firms will not find incentives strong enough to bear the cost. An example in point is the U.S. failure to meet the voluntary goals agreed to in the UNFCCC. Current U.S. policy is a departure from previous policy solutions to national pollution problems.

The U.S. national program to reduce acid rain includes emissions limits and trading. It has been widely considered to be successful in meeting environmental goals in a cost-effective manner (U.S. Environmental Protection Agency [USEPA], 2002a). In 2003, discontented with Bush administration climate policy, Senators McCain and Lieberman introduced a bill in Congress to mandate emissions reductions and create an emissions trading market. This bill did not pass, but it was supported by 44 senators.

Because GHG emissions emanate from all over the globe, it is widely recognized as an international problem that requires broad-based international solutions (Antle, 2004). We should note that the United States has ratified other multilateral environmental agreements to reduce international emissions of pollutants. An example is the 1987 Montreal Protocol on Substances that Deplete the Ozone Layer. This treaty has been more successful than anticipated in promoting technological solutions, achieving reduction goals, and productively involving developing countries in the multilateral effort (UN Environment Programme, 2000; USEPA, 2002b).

28.1.2 Strength of Incentives for Agricultural Sequestration

Other countries and trading blocs are implementing emissions limits and trading schemes. Denmark has implemented a cap and trade scheme targeted at its electricity sector, with potential fines on firms that exceed their emission targets. The United Kingdom has an industry-wide pilot program that includes incentives for firms to meet energy-efficiency or emissions-reductions targets. In addition, the European Union has announced that it will start a carbon trading scheme in 2005, with member governments setting emissions caps for their respective countries. While trading will begin among EU members, it is expected to expand to include other European countries and neighbors. Australia is considering a national emissions trading scheme. New Zealand is currently negotiating binding emission limits with major emitters, but has not yet decided on a trading scheme.

The future of carbon markets is unclear. Implementation of the Protocol will generate demand for carbon credits, although the outlook for carbon markets would be much stronger with U.S. participation. However, even with implementation of the Kyoto Protocol, demand for agricultural sequestration in developing countries is likely to remain weak due to its exclusion from the Clean Development Mechanism. Programs like the World Bank's BioCarbon fund will provide incentives for agricultural sequestration in developing countries; however, as this program has just been launched, it is not possible to forecast how strong this incentive will be.

28.1.3 Implications for Carbon Trading

We know that the atmosphere is a public good, and that greenhouse gas emissions intensify the greenhouse effect and thus cause a negative global externality. Although this externality may be costly, individuals, businesses, and even countries lack incentives to take action on their own to curtail greenhouse gas emissions. From the preceding discussion, it is clear that the world currently lacks effective global policies that limit greenhouse gas emissions. Until such limits exist, a well-functioning market for carbon and other greenhouse gas emissions reduction credits will not exist, and carbon dioxide and other greenhouse gas emissions will not be priced at their marginal social cost.

Interestingly, attempts have been made to create "markets" for carbon — such as the Chicago Climate Exchange — before governments created legally binding caps on emissions. Some recent "trades" of carbon contracts have occurred at prices less than $10 per metric ton, leading some analysts to conclude that carbon sequestration would not be profitable for farmers. Yet, it should be clear that these prices for carbon emissions credits are heavily discounted below the marginal social value. Indeed, studies of the opportunity cost of reducing carbon emissions find values that range from $10 to $100 per metric ton and higher, depending on the quantity of emissions reduced (Antle et al., 2002).

28.2 INCENTIVE MECHANISMS FOR AGRICULTURAL CARBON SEQUESTRATION

Some policies that limit total GHG emissions have been put in place. Assuming that all scientifically credible forms of

GHG emissions reductions, including carbon sequestration, are allowed to be counted towards national GHG emissions reductions, we should consider how incentive mechanisms can be designed to implement agricultural carbon sequestration.

For industrialized countries, which have existing and efficiently operating financial markets, we assume that tradable emissions reductions credits will be used to implement caps on GHG emissions. We also assume that sources of emissions reductions will have to satisfy criteria which ensure that the reductions are verifiable over the duration of the contract. Under these conditions, contracts for agricultural carbon sequestration can be formulated along the lines presented in Antle et al. (2003). We will now review carbon contract design, and discuss whether this type of carbon contract would be likely to work in developing countries.

Soil science has established that the amount of soil C at a given point in time and space is a function of the biophysical conditions at the site, including soils, topography, microclimate, and the land use history at the site. We let C_j^i denote the soil C stock (metric tons of C per hectare) on land unit j that has been managed with production system i. Thus, if a farmer uses a production system i associated with a relatively low equilibrium soil C stock (e.g., conventionally tilled corn, or wheat in a crop–fallow rotation), then the stock of soil C can be increased over time to a new level of $C_j^s > C_j^i$ by adopting an alternative system s that is associated with a higher equilibrium level of soil C (e.g., corn produced with reduced tillage, or wheat in a continuous rotation).

The time path between C_j^i and C_j^s is generally nonlinear and may follow a hyperbolic or logistic-shaped trajectory, converging on a maximum level attainable with practice s in T years. Simulations with the CENTURY model show that, in a number of cases, this attainable maximum is reached in 20 to 30 years after the change in practices occurs. We assume that, for purposes of implementing soil C contracts, the annual average rate of soil C accumulation per hectare, Δc_j^{is} = $(C_j^s - C_j^i)/T$, will be used to estimate the amount of C that a particular change in practices will provide.

Another important issue is the size of the spatial unit over which Δc_j^{is} is either estimated before a contract is agreed to, or measured to verify compliance with a contract. Thus, the index j may refer to the spatial unit at which a farmer makes management decisions (i.e., a single field), or to a larger spatial unit at which measurements may be made with a sampling scheme, as discussed in Mooney et al. (2002).

We consider two types of contracts for soil C sequestration, namely, per-hectare contracts and per-ton contracts. The per-hectare contract provides incentive payments to producers for each hectare of land that is switched from a production system associated with a relatively low equilibrium level of soil C to a system associated with a higher equilibrium level of soil C. Thus, the key feature of the per-hectare contract is that the payment per hectare is the same for all land under contract that uses a specified technology — often referred to as a best management practice — regardless of the amount of C that is actually sequestered as a result.

Typically, we would expect that per-hectare contracts (1) will require that farmers establish what practices they have used in the past, (2) will specify which practices the farmer must adopt over the duration of the contract, and (3) will specify the payments made for compliance with the terms of the contract. In order to enforce compliance with the contract, land use and management practices specified in the contract will be monitored on a periodic basis.

The per-ton contract pays farmers a specified price P for each metric ton of C that is accumulated and maintained in the soil for the duration of the contract, regardless of what management practices are used. Allowing farmers to choose the most efficient production technology at each site, rather than specifying a best management practice, means that the per-ton contract is more efficient than the per-hectare contract, as demonstrated by Antle et al. (2003). To implement per-ton contracts, it is necessary to quantify the amount of C added to the soil over the duration of the contract. Hence, it is necessary to establish the baseline amount of C in the soil at the beginning of the contract, and the time path of soil C accumulation over the duration of the contract. Because soil

C cannot be observed directly, procedures for measuring the baseline levels of soil C and the amount accumulated must be established. Moreover, because of the typically low annual rates of soil C accumulation, it is only possible to measure soil C changes periodically, such as every 5 years, with a reasonable degree of accuracy (Watson et al., 2000). Therefore, farmers entering into per-ton soil C contracts face the problem of estimating how much they will earn from the contracts, and buyers of C credits from farmers face a similar challenge of estimating how much soil C that they can expect to take credit for.

To resolve this *ex ante* uncertainty about the amount of soil C that will be produced under a per-ton contract, we assume that the contracts operate as follows:

1. Buyers of soil C credits specify a price per metric ton of carbon, P, that they are willing to pay. For this discussion, we assume this is a constant for the duration of the contract, but it could also be specified to change over time.

2. Based on available data from independent entities such as a government agency, farmers, and buyers agree upon a schedule of expected C accumulation rates for all production systems i actually in use, and for all feasible production systems s that farmers could adopt. Based on this schedule, farmers choose what practices they will use and receive P dollars for each expected metric ton C they produce per time period according to this schedule. Subsequently, measurements are made to estimate actual C rates Δc_j^{is}, and farmers receive additional compensation if Δc_j^{is} is greater than the expected accumulation, or refund some of the payments that they have received if Δc_j^{is} is less than the expected accumulation.

28.2.1 Farmer Participation in Carbon Contracts

We now consider the farmer's decision to participate in carbon contracts in a heterogeneous region, following Antle and Diagana (2003). They show that, if expected net returns and the price of carbon are constant over time and the fixed cost of changing practices is equal to zero, the farmer formerly using practice i will adopt the carbon-sequestering practice s if

$$NR(p, w, z, s) + g(i,s) - m(i,s) - fc(i,s) > NR(p, w, z, i)$$
$$(28.1)$$

where $NR(p, w, z, j)$ is the net return to practice $j = i,s$; p and w are crop and input prices, and z represents fixed factors of production; $g(i,s)$ is a carbon payment for switching from practice i to practice s; $m(i,s)$ is maintenance costs of changing practices per time period; and $fc(i,s)$ is the cost of financing the change in practices.

This expression has several implications for analysis of adoption of soil carbon sequestration practices. First, suppose that there are no payments for carbon sequestration, so that $g = 0$. In this case, a farmer adopts the conservation practice only if it provides higher net returns than the conventional practice. If the productivity benefits of the conservation practice are realized with a time lag, Equation 28.1 shows that if a farmer is uncertain about future productivity benefits or highly discounts future benefits, she or he would bear the costs of adopting the practice, but would not be aware of or attach value to the benefits. Therefore, a lag between adoption and the realization of productivity benefits may create an adoption threshold. Furthermore, if farmers do not have access to well-functioning capital markets, they cannot finance the fixed component of the investment cost, and so the annualized investment cost term $fc(i,s)$ would be replaced with the full investment cost in Equation 28.1, thus exacerbating the threshold effect.

Second, if there is a payment for adoption of practices that sequester carbon, Equation 28.1 can be rearranged to:

$$g(i,s) > NR(p, w, z, i) - NR(p, w, z, s) + m(i,s) + fc(i,s)$$
$$(28.2)$$

The expression on the righthand side is the farm opportunity cost for switching to system s from system i. Farmers will likely switch practices when the farm opportunity cost is less than the payment per period. If a per-ton contract, $g(i,s) =$

PΔc(i,s), the condition for participation in the contract can be expressed as

$$P > (NR(p, w, z, i) - NR(p, w, z, s) + M(i,s) + fc(i,s))/\Delta c(i,s)$$

$$(28.3)$$

The term at the right is now the farm opportunity cost per metric ton of soil C, and thus the farmer will participate when the price per metric ton of soil C is greater than the farm opportunity cost per metric ton. This last expression shows that when farmers are being paid per metric ton of carbon sequestered, as would be the case when they participate in a market for carbon emissions reduction credits, the market price per metric ton of carbon plays a key role in determining which farmers would participate.

Equation 28.2 shows that when incentive payments are made, it is no longer necessary for the conservation practice to be more profitable than the conventional practice for adoption to take place. When farmers are informed about the benefits and costs of conservation investments, and they choose not to adopt them, we can infer that the conservation practices are less profitable. A positive financial incentive will be required to induce and maintain adoption.

28.2.2   Contract Duration and Carbon Permanence

In public debates about soil carbon sequestration, much has been made of the idea that soil carbon is not necessarily permanently stored in soils (Marland et al., 2001). However, these discussions have considered permanence as a biophysical phenomenon. From an economic point of view, the permanence of carbon in the soil is also an issue of how farmers choose to manage land, and it is related to adoption thresholds, as now explained.

In public debates about soil carbon sequestration, much has been made of the fact that biological forms of carbon sequestration, including soil carbon, are not necessarily stored permanently in soils. Forests can be harvested and used for fuel; abandonment of soil carbon–sequestering practices could likewise cause carbon to be released back into the atmosphere through the same processes that cause soil carbon to be released when agriculture was first introduced.

Soil science has established that there is a steady-state level for soil C that can be stored in the soil for a given soil under specified practices, referred to by Ingram and Fernandes (2001) as the "attainable maximum." Thus, under a specified set of conditions, carbon can be said to accumulate at an average annual rate only for a certain number of years, at which time the soil becomes "saturated." In addition, soil research has shown that sequestered carbon is volatile, and that if soil C sequestering practices are discontinued, the C stored in the soil can be released back in to the atmosphere in a short period of time. For example, if a farmer were to revert from reduced tillage to conventional plowing, the accumulated soil C may be released back to the atmosphere and the soil C level may return to the level it was at before reduced tillage was adopted.

A simple way to address the permanence issue is to view farmers who enter into soil C contracts as providing a *service* in the form of accumulating and storing soil C. During the time period in which C is being accumulated, the farmer is providing both accumulation and storage services. Once the soil C level reaches the saturation point, the farmer is providing only storage services. However, *both* accumulation and storage services would depend on the farmer continuing to maintain the land use or management practices that make the accumulation possible. Therefore, *if carbon storage practices are less profitable than the conventional practices that release carbon for the duration of the contract*, farmers will have to be provided an incentive for the time that the carbon sequestering and storing practices are to be maintained, not just for the sequestration period.

The discussion in the previous section noted that in many cases, practices that store carbon may also yield higher returns to farmers over time, although these higher returns may come with a lag. When this is the case, the model presented in the previous section (Equation 28.1) can be used to show that during the initial phase of adoption, when the conventional practice yields higher returns than the conservation practice, farmers may require a positive financial incentive to bear the fixed and variable costs of adopting it. However, this model also can be used to show that once the practice becomes more profitable, and once the fixed costs of adoption have been borne, farmers should be willing to maintain the practice indefinitely without additional financial incentives for carbon sequestration. Therefore, we can conclude that if the carbon sequestration practices become profitable at some point before the contract expires, the carbon sequestered through adoption and maintenance of these practices is likely to be permanent as long as the practices remain profitable. Thus, with conservation practices that enhance the productivity of the production system, thus making the practices profitable, carbon permanence can become an emergent property of the system once the practice is adopted and maintained for a sufficient amount of time. This may occur despite farmers' unwillingness to adopt or maintain the practices initially without incentives.

28.2.3 Designing Soil Carbon Contracts for Farmers in Developing Countries

There has been considerable discussion in the literature about how contracts for soil sequestration might be designed in the context of the United States (Antle et al., 2003; Antle and McCarl, 2002). Some discussion of how carbon sequestration incentives could be created for farmers in developing countries (Antle and Diagana, 2003) is also now in the literature. In a country with well-defined property rights and corresponding financial institutions, farmers can participate in domestic or international markets for tradeable emissions reductions credits. Farmers might enter contracts, with either private or public entities, to adopt specified practices for a specified period of time. They could earn per hectare payments or could be paid per metric ton of carbon sequestered. To verify compliance with contracts, it would be necessary to monitor farmers' practices and to measure the quantities of carbon being sequestered over time. It is likely that third-party intermediaries would consolidate contracts with farmers, thus aggregating enough land units to make a commercially tradeable contract. Such arrangements could possibly deal with various problems affecting the feasibility of this kind of scheme.

In developing countries, the participation of small-scale farmers in this kind of carbon credit market would be inhibited by several factors. First, the transactions costs associated with aggregating land units to create a marketable contract would be large because of the smaller scale of production. In addition, verifying compliance with contracts by, for example, monitoring land use and management practices, could be more costly for a number of small farms. Second, significant issues would arise if land property rights are not formalized. It is not clear how contracts would work if farmers did not hold legal title to the land they manage. For example, in many parts of the developing world, farmers have use rights given by village authorities that can change over time. Third, many parts of the developing world lack well-functioning legal and financial institutions. If contracts are not enforceable, buyers of carbon contracts will have little recourse if farmers are found to be violating the terms of their contract.

In countries that lack financial markets, farmers may not be able to borrow to make the investments needed to adopt practices that sequester carbon. The carbon market could function as a form of financing for these investments, by paying in advance all or part of the capitalized value of the carbon expected to be sequestered. But this would only be attractive to carbon buyers if they are confident that farmers possess property rights and/or other sufficient assets for collateral. If governments incorporate the promotion of soil carbon sequestration into their agricultural policies, they could provide "carbon loans" that could be repaid either by adopting practices that sequester certain amounts of carbon, or by repaying in cash. This type of program could serve as a financing mechanism for adoption of carbon sequestration practices.

28.2.3   Co-Benefits and Costs

Many soil conservation and other management practices that would increase soil organic carbon (SOC) were not developed and promoted to enhance the accumulation of soil carbon *per se*, but rather to increase agricultural productivity, reduce soil erosion, and reduce off-farm impacts of soil erosion on water quality. In regions with predominantly subsistence or semisubsistence agriculture, the various soil management practices that contribute to carbon sequestration also are likely to have important impacts on the level and stability of farm production and food consumption. These impacts translate into improvements in health and nutrition of rural households and ultimately into increased rural economic development. Measuring these co-benefits of carbon sequestration requires analysis that goes beyond the models of agricultural production considered in this discussion. For example, additional data would be needed to characterize farm and nonfarm rural households and to analyze the market and nonmarket effects of improvements in agricultural production. Partial or general equilibrium economic models would be needed to assess fully the rural development impacts in economic terms. Antle (2002) presents an integrated assessment framework that could be used to address the on-farm and immediate off-farm environmental consequences of adoption of management practices that sequester soil C. Likewise, partial or general equilibrium models would be needed to assess the broader economic and rural development impacts of this (e.g., McCarl and Schneider, 2001).

There is a major difference between the benefits of sequestering carbon and most other co-benefits that should be taken into consideration when designing incentives for farmers to adopt practices that sequester soil carbon. Reducing atmospheric concentrations of greenhouse gases produces a global benefit by reducing the risks and rate of climate change, whereas most other environmental and social impacts are local. Therefore, a market for carbon emissions reductions will not take into account the local co-benefits produced by farmers. This implies that incentives provided to farmers through a greenhouse gas emissions trading system will not be as large as they could be if they incorporated the social value of other environmental and social co-benefits. An important topic for further research is to assess how appropriate incentives can be created that consider and incorporate the value of local co-benefits.

28.3 CONCLUSIONS

The present status of policies for carbon sequestration and how they relate to agricultural sequestration was reviewed in the first part of this chapter. The current mix of local, national, and international policies does not create any well-defined global cap on acceptable emissions, and therefore it does not provide the basis for a well-functioning market for GHG emissions reductions credits. A market price for carbon that reflects the social opportunity cost of reducing emissions will exist only when such policies have been implemented, and even then, the price will depend on how much emissions are reduced. The prices at which carbon is currently being "traded" are heavily discounted, and do not reflect the social opportunity cost of reducing greenhouse gas emissions.

Incentive mechanisms for agricultural soil carbon sequestration were explored in the second part of this chapter. Economically rational farmers can be expected to adopt practices that enhance soil carbon when they believe that it is in their economic interest to do so. If farmers have not adopted those practices before carbon contracts are offered to them, then they will face the opportunity costs of adopting them. If these opportunity costs are positive for the duration of the time that carbon is accumulated and stored, then incentives will have to be provided to farmers during the total time period. However, if yields increase over time, the opportunity cost will decline and may even become negative, meaning that a positive incentive will not be required for them to maintain the practice. Thus, permanence may be an emergent property of agricultural soil carbon when it is true that carbon sequestering practices enhance productivity.

While farmers in the industrialized countries may be able to enter fairly readily into carbon sequestration contracts, the situation is likely to be quite different in developing countries. In effect, the transactions costs are likely to be higher in developing countries, both because farms are smaller, and because local financial and market institutions are less well developed. However, the co-benefits of carbon sequestration in developing countries in the form of enhancing food security, sustainability, and combating poverty, are potentially high. Therefore, special incentive mechanisms need to be devised for carbon sequestration contracts involving farmers in developing countries. These arrangements should take into account the broader social benefits that may be associated with adoption of more sustainable agricultural production systems. Incentive mechanisms for agricultural carbon sequestration could be designed in ways that overcome certain adoption thresholds associated with a lack of financial markets, and uncertainty about future productivity benefits of conservation practices.

REFERENCES

Antle, J.M. 2002. Integrated assessment of carbon sequestration in agricultural soils. In *A Soil Carbon Accounting and Management System for Emissions Trading*. Special Publication SM CRSP 2002-04. Soil Management Collaborative Research Support Program, University of Hawaii, Honolulu, pp. 69–98.

Antle, J.M., S.M. Capalbo, S. Mooney, E.T. Elliott, and K.H. Paustian. 2002. A comparative examination of the efficiency of sequestering carbon in U.S. agricultural soils. *Am. J. Alternative Agric.*, 17:109–115.

Antle, J.M., S.M. Capalbo, S. Mooney, E.T. Elliott, and K.H. Paustian. 2003. Spatial heterogeneity, contract design, and the efficiency of carbon sequestration policies for agriculture. *J. Environ. Econ. Manage.*, 46(2):231–250. Available at: *www.climate.montana.edu*.

Antle, J.M. and B. Diagana. 2003. Creating incentives for the adoption of sustainable agricultural practices in developing countries: the role of soil carbon sequestration. *Am. J. Agric. Econ.*, 85:1178–1184. Available at: *www.tradeoffs. montana.edu.*

Antle, J.M. and B.A. McCarl. 2002. The economics of carbon sequestration in agricultural soils. In T. Tietenberg and H. Folmer, Eds. *The International Yearbook of Environmental and Resource Economics 2002/2003.* Edward Elgar Publishing, Cheltenham, United Kingdom and Northampton, MA, pp. 278–310.

Barrett, C.B., F. Place, and A.A. Aboud, Eds. 2002. *Natural Resources Management in African Agriculture: Understanding and Improving Current Practices.* CAB International, Wallington, United Kingdom.

Chartier, D. 2002. From the EMA president. *The Emissions Trader*, June. Available at: *www.emissions.org/publications/ emissions_trader/0206/index.html.*

Climate Change Secretariat. 2002. A Guide to the Climate Change Convention and Its Kyoto Protocol. Preliminary version. UN Framework Convention on Climate Change, Bonn, Germany, preliminary version. Available at: *http://unfccc.int/resource/guide-convkp-p.pdf.*

Hudson, N.W. 1991. *A Study of the Reasons for Success or Failure of Soil Conservation Projects.* FAO Soils Bulletin 64. Food and Agriculture Organization, Rome.

Ingram, J.S.I. and E.C.M. Fernandes. 2001. Managing carbon sequestration in soils: concepts and terminology. *Agric. Ecosyst. Environ.*, 87:111–117.

Lutz, E., S. Pagiola, and C. Reiche. 1994. Lessons from economic and institutional analyses of soil conservation projects in Central America and the Caribbean. In E. Lutz, S. Pagiola, and C. Reiche, Eds. *Economic and Institutional Analyses of Soil Conservation Projects in Central America and the Caribbean*, World Bank Environment Paper Number 8, Washington, DC, March.

Marland, G., K. Fruit, and R. Sedjo. 2001. Accounting for sequestered carbon: the question of permanence. *Environ. Sci. Policy*, 4:259–268.

McCarl B.A. and U.A. Schneider. 2001. The cost of greenhouse gas mitigation in U.S. agriculture and forestry. *Science,* 294:2481–2482.

Mooney, S., J.M. Antle, S.M. Capalbo, and K.H. Paustian. 2002. Contracting for Soil Carbon Credits: Design and Costs of Measurement and Monitoring. Staff Paper 2002-01. Department of Agricultural Economics and Economics, Montana State University–Bozeman. Available at: *www.climate.montana.edu/pdf/ mooney.pdf.*

O'Neill, B.C. and M. Oppenheimer. 2002. Dangerous climate impacts and the Kyoto Protocol. *Science,* 296:1971–1972.

Pew Center on Global Climate Change. 2002. Pew Center analysis of President Bush's February 14th climate change plan. Arlington, VA. Available at: *www.pewclimate.org/policy/response_ bushpolicy.cfm.*

Sanchez, P. 2002. Soil fertility and hunger in Africa. *Science,* 295:2019–2020.

Scherr, S.J. 1999. *Soil Degradation: A Threat to Developing-Country Food Security by 2020?* Food, Agriculture, and the Environment Discussion Paper 27. International Food Policy Research Institute, Washington, DC.

UN Environment Programme. 2000. *Montreal Protocol on Substances that Deplete the Ozone Layer.* Nairobi, Kenya.

UN Framework Convention on Climate Change. 2002a. Issues in the Negotiating Process: Land Use, Land-Use Change, and Forestry Under the Kyoto Protocol. Bonn, Germany. Available at: *unfccc.int/issues/lulucf.html.*

UN Framework Convention on Climate Change. 2002b. Kyoto Protocol: Status of Ratification. Bonn, Germany, as of September 17. Available at: *http://unfccc.int/resource/kpstats.pdf.*

U.S. Department of State. 2002. U.S. Climate Action Report 2002: Third National Communication of the United States of America Under the United Nations Framework Convention on Climate Change. Washington, DC, May.

U.S. Environmental Protection Agency. 2002a. Acid Rain Program: Overview. EPA Clean Air Markets – Programs and Regulations. Washington, DC, October 25. Available at: /*www.epa.gov/air-markets/arp/overview.html*.

U.S. Environmental Protection Agency. 2002b. EPA Marks 15th Anniversary of Landmark Environmental Treaty. EPA Newsroom, Washington, DC, September 16. Available at: *www.epa.gov/epahome/headline_091602.htm*.

Watson, R.T., I.R. Noble, B. Bolin, N.H. Ravindranath, D.J. Verardo, and D.J. Dokken. 2000. *Land Use, Land-Use Change, and Forestry*. Special report of Intergovernmental Panel on Climate Change, Cambridge University Press, London; New York.

White House. 2001. Bush Will Not Require Power Plants to Reduce Carbon Emissions: Letter to Senators Reiterates Opposition to the Kyoto Protocol. Office of the Press Secretary, March 13. Available at: *www.usis.usemb.se/Environment/letter.html*.

White House. 2003. Statement by the President. February 12. Available at: *www.whitehouse.gov/news/releases/2003/02/20030212.html*.

World Bank. 2002. World Bank Launches Biocarbon Fund. November 12. Available at: *http://web.worldbank.org/WBSITE/EXTERNAL/NEWS/0,contentMDK:20074429~menuPK: 34457~pagePK:34370~piPK:34424~theSitePK: 4607,00.html*.

29

The Impact of Climate Change in a Developing Country: A Case Study from Mali

TANVEER BUTT AND BRUCE McCARL

CONTENTS

Greenhouse gas–induced climate change may worsen climatic conditions in many developing countries. Rosenzweig and Iglesias (1994) argue that in low-latitude regions, where most of the developing world is located, crop yields are likely to be hard hit by climate change. However, information on the economic consequences of such shifts remains sparse (Intergovernmental Panel on Climate Change [IPCC], 2001). This chapter provides an economically based climate change impact assessment for the Malian agricultural sector, highlighting food security implications.

In addition to possible climate change, many other factors are relevant to Mali's future food security and the implications of climate change. Food production intensification has caused resource degradation with slow technological change and continuing high population growth have caused cultivation of marginal lands and shortened fallow periods (Coulibaly, 1995; Kuyvenhoven et al., 1998; Benjaminsen, 2001; Vitale, 2001). In turn, declining soil fertility and yields are being observed. In addition, Malian farmers face high climatic risk that influences their decision making. Thus, climate change needs to be considered in conjunction with technical progress, resource degradation, cropland expansion, and climatic risk.

29.1 GENERAL FINDINGS ON CLIMATE CHANGE IMPACT

Across the large number of climate change–related studies several key findings have emerged regarding physiological effects on crops and, to a lesser extent, livestock. Following McCarl et al. (2003), the main findings are:

Table 29.1   Ranges of Estimated Effects on Country Crop Yields

Location	Impact (Crop: % Change in Yield)
Indonesia	Rice: −2.5% and +5.4%; soybeans: −2.3%; maize: −40%
Malaysia	Rice: −22% to −12%; maize: no change; rubber: −30% to −3%
Pakistan	Wheat −60% to −10%
Sri Lanka	Rice: −2.1% to +3%
Bangladesh	Rice: −6% to +8%
Mongolia	Spring wheat: −74.3% to +32.0%
Kazakhstan	Spring wheat: −56% to −44%; winter wheat: −35% to +15%
Czech Republic	Winter wheat: −3% to +16%
United Kingdom	Crop productivity: +5% to +15%
The Gambia	Maize: −26% to −15%; early millet: −44% to −29%; late millet: −21% to −14%; groundnuts: +40% to +52%
Zimbabwe	Maize: −13.6% to −11.5%
Brazil	Maize: −27% to −7%; wheat: −46% to −17%; soybeans: −6% to +38%
Argentina	Maize: −17% to +4%; wheat: −12% to +6%
Uruguay	Barley: −40%; wheat: −31% to −11%
United States	Wheat: −14% to −2%; maize: −29% to −15%; rice: −23% to +1%

Source: From Rosenzweig, C., and Iglesias, A. 1994. *Implications of Climate Change for International Agriculture: Crop Modeling Study.* U.S. Environmental Protection Agency, Climate Change Division, Washington, DC; and Intergovernmental Panel on Climate Change. 1995. *Climate Change: The IPCC Second Assessment Report, Volume 2: Scientific-Technical Analyses of Impacts, Adaptations,* and *Mitigation of Climate Change.* Watson, R.T., M.C. Zinyowera, and R.H. Moss, Eds. Cambridge University Press, Cambridge, MA.

- Dryland and irrigated crop yields are altered as is irrigation water use. Table 29.1 presents crop yield implications for selected countries, while Table 29.2 shows yield impacts for irrigated and rainfed crops in the United States projected for 2030.
- Production effects vary by crop, location, temperature, and precipitation change (as reviewed in Adams et al., 1998a, and Lewandrowski and Schimmelfennig, 1999).

Table 29.2 Projected National Average Data on Irrigated and Dryland Crop Yield and Irrigation Water Use Sensitivity to Global Climate Change for 2030, United States

Crop	Without Adaptation			With Adaptation		
	Dryland Yield	Irrigated Yield	Irrigated Water Use	Dryland Yield	Irrigated Yield	Irrigated Water Use
Cotton	+18% to +32%	+36% to +56%	−11% to +36%	NA	NA	NA
Corn	+11 to +19%	−1% to +21%	−30% to +57%	+11 to +20%	+1% to +21%	−32% to +57%
Soybeans	+7% to +34%	+16% to +17%	−12% to 0%	+7% to +49%	+23%	0% to +18%
Hard red spring wheat	+15% to +20%	−10% to +4%	−28% to −17%	+17% to +23%	−6% to +10%	−12%
Hard red winter wheat	−16% to +24%	−4% to +5%	−8% to +5%	−9% to +24%	−1% to +8%	−6% to +9%
Soft wheat	−5% to +58%	−6% to +3%	−12% to +5%	−3% to +58%	−5% to +5%	−26% to +3%
Durham wheat	+10% to +21%	−10% to +5%	−25% to +15%	+10% to +22%	+2% to +9%	−10% to −5%
Sorghum	+15% to +17%	−1% to +1%	−9% to −7%	+32% to +43%	+22% to +22%	+2% to +3%
Rice	NA[a]	−2% to +9%	−10% to −2%	NA	+7% to +9%	+2% to +5%
Potatoes	+6% to +7%	−6% to −3%	−5% to −1%	+7% to +8%	−4% to −1%	−3% to 0%
Tomatoes	NA	−9% to +1%	−9% to −5%	NA	+1% to +10%	−8% to +2%
Citrus	NA	+32% to +40%	−21% to −6%	NA	NA	NA
Hay	−10% to +43%	+3% to +37%	−29% to +44%	NA	NA	NA
Pasture	+3% to +22%	NA	NA	NA	NA	NA

Note: The data provide percentage changes in the item from a no climate change case.

[a] NA indicates that results were not developed for this case due to the small acreage involved or that the simulation was not done (in the adaptation cases as only selected crops were studied).

Source: From Reilly, J., et al. 2000b. Report of the Agricultural Sector Assessment Team. In U.S. Global Change Research Program, National Assessment Report on Changing Climate and Changing Agriculture. Available at: *www.nacc.usgcrp.gov/sectors/agriculture/*.

- Different crops exhibit different degrees of sensitivity. Treatment of only selected crops can bias the results. For example, early U.S. studies only examined corn, soybeans, and wheat, in contrast to later studies that included many more heat-tolerant crops. Economic implications were moderated as a result. (For an example of such an effect, see the experiment on cotton in McCarl [1999], which showed that inclusion of the differential response by this more heat-tolerant crop caused a reversal in sign of the total welfare impact, thus showing a beneficial effect rather than a detrimental effect of climate change.)
- CO_2 fertilization is an important factor. Inclusion of this effect in yield studies significantly raises the estimates of climate-affected yields. It is, however, somewhat controversial (see discussion in Reilly et al., 2000b, or Council for Agricultural Science and Technology, 1992, 2000).
- Yield effects vary latitudinally. These are generally positive in the higher latitudes, but are frequently negative in low-latitude and semi-arid areas. (See reviews in Adams et al., 1998a, and Lewandrowski and Schimmelfennig, 1999.) Mohamed et al. (2002a, 2002b) estimate that by 2025, climate change might lower Niger millet yields by 13%, groundnuts by 11% to 25%, and cowpeas by 30%.
- Human adaptations help mitigate climate change effects. Adaptations can be made in planting and harvest dates, crop choice, crop rotations, crop varieties, irrigation, fertilization, and tillage practices. Livestock producers can adapt by the provision of shading, sprinklers, improved airflow, lessened crowding, altered diets, and more care in handling animals. In the longer term, new crop varieties and livestock breeds may be developed that perform better under the anticipated future climate regime. (See reviews in Adams et al., 1998b, 1999; Rosenzweig and Hillel, 1998; Kaiser et al., 1993; Reilly et al., 2002; and Yates and Strzepek, 1998.)

- Livestock effects can be significant. Adams et al. (1999) and a recent U.S. national assessment (Reilly et al., 2000a, 2000b) estimated livestock productivity reductions ranging from 1.5% to 5%.
- Changes in agricultural supply result from the combined effect of changes in yields and changes in crop acreage, with livestock herd size and location, livestock diets, human consumption, international trade adjustments, and many other factors also adjusting to the new conditions (Adams et al., 1999; Reilly et al., 2000a, 2000b).
- Commodities that decline in supply will rise in price. Higher prices reduce consumption and consumer welfare. The negative consumer effects are offset by producer gains, but in general, total welfare tends to decline (Adams et al., 1999; Reilly et al., 2000a, 2000b). Yates and Strzepek (1998) found minor gains in economic welfare for Egypt attributable to lower projected international prices of food imports, and mildly harsh to mildly beneficial climate change projections for the country.
- Downing (1992) studied the implications of climate change for the water balance in Zimbabwe and found that over the entire surface of the country, water evaporation would be increased by 15% for a 1°C increase in temperature, while with a 2°C temperature increase, the country's core agricultural zone would be reduced by 67% due to high evaporation rates.

29.2 MALI CLIMATE CHANGE PROJECTIONS

Mali lies just south of the Sahara. The country is broadly categorized into three climatic zones (Wang'ati, 1996). The Saharo-Sahelian zone is extremely dry with an annual rainfall of 100 to 200 mm having 50% to 100% interannual variability. The Sahelo-Sudanian zone and Sudano-Guinean zones have annual rainfall of 200 to 400 mm and 400 to 800 mm, respectively, with 25% to 50% interannual rainfall variability.

Gommes and Petrassi (1994) and Jenkins et al. (2002) argued that since the early 1960s, regional rainfall patterns in West Africa have shifted mainly toward lower annual average rainfall.

The global circulation model (GCM)–based projections for climate change in Mali predict an extension, or perhaps continuation, of recent observed changes in the climate of West Africa. Figure 29.1 presents projected changes in temperature and precipitation for 2030 as different from the base temperature and precipitation of the 1960–1991 period. (GCM projections are provided for three time periods: 2010 to 2040, 2040 to 2070, and 2070 to 2100. In this study, we focus on the 2010–2040 time period.) Climate change projections made with two global circulation models — the Hadley Center coupled model (HadCM), and the Canadian coupled general circulation model (CGCM) — were obtained from the IPCC's Data Distribution Center. (The HadCM and CGCM were used

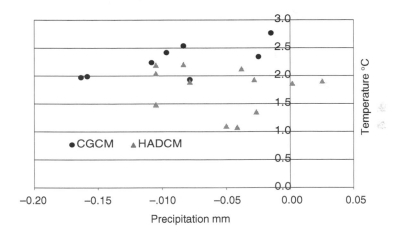

CGCM = Canadian Coupled Model; HADCM = Hadley Coupled Model.

Figure 29.1 Projected changes in precipitation and temperature for Mali. CGCM = Canadian coupled general circulation model; HadCM = Hadley Center coupled model.

since they were employed in the U.S. national assessment, and we planned to use international market data arising in that assessment. We used the data from the IPCC's greenhouse gas integration scenario.) Projections indicate that Mali is likely to face a hotter, drier future. Twelve HadCM and eight CGCM grids cover Mali. Temperature increases are predicted in every grid cell. Five of the eight CGCM grids show increases in temperature of more than 2°C, as do four of the 12 HadCM cells. CGCM temperatures are generally higher than those in HadCM projections. On the precipitation side, 10 of the 12 HadCM cells show a decrease in precipitation, while all eight CGCM cells show a decrease.

29.3 ANALYTICAL FRAMEWORK

To assess the impact of climate change and adaptations on food security in Mali, we integrated a number of modeling frameworks. First, to assess yield impacts, we used the following biophysical models: (1) the erosion–productivity impact calculator (EPIC) crop growth simulator (Williams et al., 1989) for crops; (2) the PHYGROW plant growth model (Rowan, 1995) for forage yields; and (3) the nutritional balance (NUTBAL) model (Stuth et al., 1999) for livestock feed demand and yield. Second, to assess changes in production, prices, and trade, plus adaptation strategies, we used the Mali agriculture sector model (MASM). MASM, developed by Chen et al. (1999), was adapted for use as described in Butt (2002). The model shows economic conditions in various agricultural regions in Mali. It is a price-endogenous mathematical programming model based on the concepts reviewed by McCarl and Spreen (1980) and Norton and Schiefer (1980). MASM incorporates climate variability following Lambert et al. (1995), and crop subsistence behavior as discussed in Calkins (1981).

MASM was augmented by a methodology to compute a risk of hunger measure (ROH) (Food and Agriculture Organization [FAO], 1996) indicating the incidence of malnourishment, and hence, food insecurity in a country. The ROH measure estimates the percentage of the population whose daily caloric intake falls below requirements for a healthy life.

For the climate change analysis, the biophysical models were run under base HadCM and CGCM climate projections for 2030. (Model predictions are available under two scenarios: greenhouse gas integrations [GG], and greenhouse gas plus sulphate aerosol integrations [GS]. In this study, we used GG, as this scenario has captured the observed signal of global mean temperature changes better than the GS scenario for the past 100-year record.) In turn, the biophysical responses were incorporated into MASM, along with changes in world trade conditions derived from the U.S. national assessment (Reilly et al., 2000a, 2000b) to obtain simulated changes in production, exports/imports, and food for consumption. MASM results provided changes in prices, production, consumption, and risk of hunger in Mali in response to the projected climate change.

29.4 ASSESSMENT METHODOLOGY

Four principal steps were used to complete the analysis. First, the climate change projections from HadCM and CGCM were superimposed on the data from 72 weather stations across Mali.

Second, the biophysical models were run with the CO_2 level set at 1.5 times the current level of 330 ppmv (parts per million by volume). The mean and standard deviation of yields across 85 agroecological zones in Mali were computed and weighted into sensitivity data for nine MASM production zones. (Information needed to set up EPIC simulations was not available for Tombouctou, the northernmost region of Mali. EPIC responses from the nearby Segou region were used as representative of Tombouctou.)

Besides estimating climate change impact, the EPIC was also used to estimate the consequence of degradation on yields. In particular, 40 years of simulated yield were used to estimate the regression

$$Y_{ijst} = a_{tjst} + b_{tjst}\, T + \varepsilon_{ijst}$$

where Y_{ijst} is the yield of item i, in simulation zone j, on soil type s, for technology t; and T is a time trend variable. The intercept a in this regression shows the base year yield; the coefficient b shows how yields change over time; and ε_{ijst} shows the error term in the regression. The coefficient b was generally found to be negative, showing degradation over time. Yield for 2030 was estimated by setting T equal to 34 in the estimated regression. Degradation rates estimated through regression were applied to the MASM base yields. (The information available restricted our calibration of EPIC to using observed crop yields only in Sikasso region. Hence, the trend regressions were estimated for 43 agroecological zones in Sikasso. The degradation estimates averaged across agroecological zones in Sikasso-North, a relatively dry area, were applied to other areas in Mali.)

Third, the biophysical climate change and degradation scenarios were imposed on MASM. In particular, yields across the 12 states of nature were adjusted to reflect changes in both the mean and the dispersion of yields, using the procedure employed by Lambert et al. (1995). Therein, the new yields became

$$Yield_i^{cc} = MeanYld^b\left(\frac{\mu^{cc}}{\mu_b}\right) + (Yield_i^b - MeanYld^b)\left(\frac{\sigma^{cc}}{\sigma^b}\right)$$

where superscripts cc and b, respectively, represent with climate change/degradation and without (base) parameters; MeanYld is the mean of the yields in the MASM base model; $Yield_i^b$ and $Yield_i^{cc}$ are yields for the i-th uncertain weather year under base conditions and climate change/degradation; μ and σ are, respectively, the mean and standard deviation of the simulated biophysical yields.

Fourth, the possible influences of climate change on international grain prices were imposed on MASM using the results from the international market component of the 2002 U.S. national assessment study of climate change impacts in the United States (Reilly et al., 2002).

29.4.1   Incorporating Adaptation

While climate change may adversely affect agriculture, human adaptations are almost certain to mitigate impacts, and ignoring them may over- or under-estimate climate change damages (Adams et al., 1999; Rosenzweig and Hillel, 1998; Kaiser et al., 1993). Adaptations may be categorized as biophysical (Adams et al., 1999; Rosenzweig and Hillel, 1998; Kaiser et al., 1993), resource and market adjustments (Adams et al., 1999; Reilly et al., 2002), and changes in policy regimes. We, therefore, consider a set of adaptations and examine climate change impacts with and without adaptations.

29.4.1.1   Biophysical Adaptations

Three types of biophysical adaptations are considered. The first involved alteration of crop choice (Adams et al., 1999; Reilly et al., 2002). The base solution required MASM to choose regional crop mixes from an array of historically observed regional crop mixes. We relaxed this requirement, and allowed the model to choose crop mixes from warmer, northern regions in addition to its own historically observed crop mixes (Reilly et al., 2002; Adams et al., 1999). The second involved adjustments in planting and harvesting dates to adapt the crop schedule to changes in temperature and precipitation (Rosenzweig and Hillel, 1998; Kaiser et al., 1993). The third involved use of heat-resistant varieties that may mitigate climate-change impact (Kaiser et al., 1993). The alternative planting and harvesting dates along with the heat-resistant varieties were simulated in EPIC, and the resultant data were included in MASM.

29.4.1.2   Resource Adaptations

Adaptations to climate change may also come about through expansion of cropland and public investment in the development of improved cultivars. The latter two adaptations are possible given government actions that aim to facilitate adaptation to climate change.

With a total area of 1.22 million square kilometers (slightly less than twice the size of Texas), and with a population of about 11 million, Mali is a low-population-density country (about nine people per square kilometer). For 25% of the land area, grazing is the main agricultural use. Less than 3% of total land area is under cultivation by farmers, many or most of whom cultivate government-owned land. A possible adaptation is that the government allows planting of a larger area, that is, extensifying production. However, much of the country's unused land, such as desert or sand dunes, is not suitable for cultivation. Therefore, we assume that expansion of cropland comes at the expense of rangeland; we consulted local experts to estimate the amount of existing rangeland that might shift to cropland. The suggested scale of the rangeland shift to cropland varied from none in Tombouctou to 25% in the northern Sikasso region.

29.4.1.3 Improved Varieties

For the wider adoption of improved varieties, we considered improved cultivars for sorghum, millet, cotton, maize, cowpeas, and rice. Data used for these cultivars were based on experience with cultivars that have already been developed and are in the experimental phase. Data on yield, cost, and returns for these varieties, and on the extent of their applicability to Malian conditions, were provided by a local expert (Kergna, 2002).

29.4.1.4 Market Adaptations

MASM by its very nature also depicts a number of other adaptations that include farmers' and consumers' responses to changes in market conditions caused by yield alterations, and also changes in foreign and domestic trade patterns. For example, changes in yields alter the relative profitability of various crops, and this, in turn, affects farmers planting decisions. As a result, crop production patterns may change, leading to changes in market prices, which then affect consumer spending decisions and also the overall flow of trade due to changes in relative prices across borders.

29.4.2 Scenarios Used for Impact Assessment

To assess climate change effects under adaptation we set up five scenarios: (1) base cropping patterns, (2) regional shifts in cropping patterns, (3) allow both crop pattern shift and trade adaptation, (4) introduce heat-resistant crop varieties, and (5) consider the effects of 1 though 4 simultaneously. We did not consider changes in planting and harvest dates, as the EPIC results did not show favorable adaptations from that source. In Mali, planting schedules are heavily dependent on when the rainy season begins. EPIC simulations under early planting to mitigate the impact of warming, before the beginning of rain, resulted in further yield reduction.

Three additional scenarios were also superimposed on scenario 5: (1) expansion of cropland at the expense of rangeland (extensification), (2) a wider adoption of existing improved cultivars (intensification), and (3) combining extensification and intensification strategies.

29.5 RESULTS

The results are presented in two sections: (1) biophysical results involving climate change implications for crop yield, forage yields, and changes in animal weight; and (2) economic and food security implications of the biophysical impacts described in the first section.

29.5.1 Biophysical Results

EPIC simulations show that sorghum was the most susceptible crop to the projected climate change, as its yields decreased significantly, while groundnuts and cowpeas were relatively less susceptible. Cotton yields increased in most cases, while millet was largely unaffected. Results showed that relatively humid areas of Mali would be less affected by climate change compared to drier areas. Table 29.3 shows nationally averaged changes in crop yields under the projected climate change.

Under the HadCM scenario, 33 of the 48 crop-region cases show yield losses. In drier and less productive regions

Table 29.3 Nationally Area
Weighted Averages for Percentage
Changes in Crop Yields Under
Projected Climate Conditions

	Global Circulation Models	
Crop	HadCM	CGCM
Cotton	+7.63	+5.34
Cowpeas	−2.57	−6.45
Groundnuts	−2.96	−6.61
Maize	−5.52	−8.47
Millet	+2.61	−4.38
Sorghum	−9.39	−16.94

Notes: Estimated with erosion–productivity
impact calculator (EPIC) model. CGCM =
Canadian coupled general circulation model;
HadCM = Hadley Center coupled model.

of Mali, sorghum yield decreased up to 18%, while in subhumid areas of Sikasso, projected sorghum yield decreases were modest (5% to 7%). Cowpeas were only somewhat affected, showing decreases in the range of 1% to 6%.

Losses in yields are higher under CGCM projected climate change than under HadCM. CGCM exhibited lower yields in 42 of the 48 cases. Yield increases were found in only 11 cases, compared to 15 under the HadCM simulations. High losses were found for sorghum, with yield decreases up to 30% in Segou and Koulikoro.

Livestock impacts resulted from changes in forage yield (Table 29.4) and livestock performance (Table 29.5). Forage yields are projected to decrease between 5% to 36%. The primary reason is the combined loss of grass basal area due to overgrazing and reduced woody plant cover, coupled with increased temperatures during the growing season. The results show small ruminants to be more resilient to climate change compared to cattle.

Rate of degradation in cropland productivity was estimated using EPIC. The rate of degradation, in terms of yield losses varied between 19% to 25%, compared to the base

Table 29.4 Percentage Changes in Forage Yields Under Projected Climate in Sikasso Region

Region	Global Circulation Models	
	HadCM	CGCM
North	−18	−26
Central	−5	−12
West	−26	−36

Notes: Estimated with Phygrow plant growth (PHYGROW) model. CGCM = coupled general circulation model; HadCM = Hadley coupled model.

Table 29.5 Changes in Animal Intake and Rate of Weight Gain (Estimated From NUTBAL Model)

Animal	Intake (%)		Weight (%)	
	HadCM	CGCM	HadCM	CGCM
Cattle	−12.8	−13.3	−13.6	−15.7
Sheep	−3.4	−5.9	0.0	0.0
Goats	−4.1	−4.7	0.0	0.0

Notes: Estimated with nutritional balance (NUTBAL) model. CGCM = Canadian coupled general circulation model; HadCM = Hadley Center coupled model.

conditions for the three subregions in Sikasso by 2030 (Table 29.6).

29.5.2 Economic and Food Security Implications

The collective effect of the crop yields, forage yields, and animal weight changes is presented in Table 29.7, and shows an overall decrease in production, resulting in an overall increase in price levels. The lost economic welfare (consumers' and producers' surplus from domestic and international markets) under climate change and degradation is $103 million

Table 29.6 Decrease in Crop
Yield by 2030 as Result of
Projected Cropland
Degradation in Main Sikasso
Regions (% Decrease from Base)

	North	Central	West
Maize	23	22	25
Groundnuts	23	21	22
Cotton	19	19	19
Sorghum	21	21	24
Millet	21	20	18
Cowpeas	24	22	23

Table 29.7 Economic and Food Security Indicators Under
Projected Climate Change

Scenarios	Price Index	CS ($ millions)	PS ($ millions)	TS ($ millions)	ROH (%)
Base (1996)[a]	100	521	515	1036	32
Base (2030)[b]	121	476	520	975	42
HadCM (2030)	229	409	647	933	64
CGCM (2030)	274	384	689	910	70

[a] Existing conditions.
[b] Productivity loss due to land degradation.
Notes: CGCM = Canadian coupled general circulation model; CS = consumer
surplus; HadCM = Hadley Center coupled model; PS = producer surplus;
ROH = risk of hunger measure; TS = total surplus.

to $126 million, respectively. The price index, set at a value
of 100 for the base year (1996), was projected to increase to
229 and 274. The ROH increased from 32% of the population
to 64% and 70%.

29.5.2.1 Effects of Adaptation

A significant amount of the climate change and degradation
loss can be reduced by adaptation (Table 29.8). More than
one-third of the welfare loss can be mitigated by crop mix,
trade, and heat-resistant cultivar adaptations. ROH falls to
a range of 37% to 46% as compared to 65% to 70% when

Table 29.8 Potential of Economic and Technological Adaptations

Adaptation Type	HadCM (2030) TS (% Rec)	HadCM (2030) ROH (%)	CGCM (2030) TS (% Rec)	CGCM (2030) ROH (%)
Crop mix	14.0	−11.2	15.6	−9.7
Trade	1.5	−1.7	5.3	−2.8
Technology	5.8	−2.3	7.0	−1.6
Full adaptation	35.8	−28.0	39.5	−26.9

Notes: CGCM = Canadian coupled general circulation model; HadCM = Hadley Center coupled model; ROH (%) = percentage reduction in risk of hunger; TS (% Rec) = percentage of total welfare loss recovered.

adaptations are not considered. However, adaptations did not fully mitigate the climate change impact. The gains due to adaption do not include the cost of adaptation.

Overall economic welfare and ROH show considerable improvement under land expansion and adoption of improved cultivars (Table 29.9). Under HadCM, adaptations reduced ROH from 65% to 36%, while under CGCM, adaptations

Table 29.9 Economic and Food Security Indicators Under High-Yield and Area-Expansion Scenarios with Adaptations

Indicators	1996 Base	HadCM 2030	HadCM Exp	HadCM HYld	HadCM Exp+ HYld	CGCM 2030	CGCM Exp	CGCM HYld	CGCM Exp+ HYld
Price index	100	107	87	92	66	120	91	95	68
CS ($ millions)	521	409	536	521	562	384	526	511	533
PS ($ millions)	515	647	497	481	457	689	504	509	468
TS ($ millions)	1036	933	1114	1039	1138	910	1104	1020	1112
ROH (%)	32	36	32	26	18	43	33	30	18

Notes: 2030 = only climate is changed; CS = consumer surplus; CGCM = Canadian coupled general circulation model; Exp = climate is changed and land area is expanded; ExpHYld = climate is changed, land is expanded, and high-yielding cultivars are made available; HadCM = Hadley Center coupled model; HYld = climate is changed, and high-yielding cultivars are made available; PS = producer surplus; ROH =risk of hunger; TS = total surplus.

reduced ROH from 70% to 43%. When considered simultaneously, area expansion and adoption of high-yielding varieties reduced ROH by 18%.

29.5.2.2 Climate Change Impacts on Variance
of Welfare Measures

MASM simulates market conditions across 12 weather years
in Mali, so we were able to compute the variance of welfare
measures. Table 29.10 summarizes the coefficient of variation
for some of the indicators presented in Table 29.9. Generally,
the variation in consumer surplus and producer surplus
increased with climate change, while it decreased as adaptations were considered. The results indicate a further increase
in the riskiness of Malian agriculture as a result of climate
change.

29.5.2.3 Effects of Population Expansion

A critical factor to the future food security conditions in Mali
will be population dynamics. In the results reported above,
we did not consider any change in population. Here we use
population projections by the FAO for 2030 (FAO, 2002),
where population is projected to double in urban areas and
increase by over 50% in rural areas.

Table 29.10 Changes in Variance of Welfare Estimates
Under Climate Change With and Without Adaptation Cases
($ Millions)

		Without Adaptation		With Adaptation	
	Base	HadCM	CGCM	HadCM	CGCM
Consumer surplus	0.15	1.32	1.59	0.09	0.1
Producer surplus	6.28	38.7	43.69	5.12	5.74
Foregin surplus	0.56	0.08	0.12	0.02	0
Total surplus	0.38	22.15	25.13	0.34	0.57

Notes: CGCM = Canadian coupled general circulation model; HadCM =
Hadley Center coupled model.

Results show quite a precarious food security situation in Mali attributable to population growth alone. Under the relatively optimistic scenario considered in this study — full adaptation plus land expansion along with the adoption of high-yielding cultivars — ROH increases to 84% and 85% under the climates projected with HadCM and CGCM. Also, the results show high dependence on the import of cereals, as the cereal import index rises to 242 and 319, respectively, under the HadCM and CGCM projections, compared to the base level of 100.

29.6 CONCLUSIONS

We anticipate that Mali may experience definite but moderate economic losses under the magnitude of climate change projected with the HadCM and CGCM models if population does not greatly expand. Losses will be much more sever if population expands.

Without population changes, annual losses range from $103 million to $126 million, which is roughly 2% of agricultural gross domestic production. Producers gain at the expense of consumers, due to rises in food prices. ROH increases from 32% to 64% to 70% of the population. In terms of the FAO's ranking, Mali moves from category 4 to category 5, that is, into the highest risk category.

However, adaptations can greatly mitigate overall economic losses and ROH. The results highlight the importance of land expansion and the adoption of high-yielding varieties in providing higher economic benefits and improving food security conditions in Mali. Other measures to mitigate climate change impacts might include reversal of land degradation by improving soil organic matter and the promotion of research and development for heat-resistant cultivars.

REFERENCES

Adams, R.M., Hurd, B., Lenhart, S., and Leary, N. 1998a. The effects of global warming on agriculture: an interpretative review. *J. Climate Res.*, 11:19–30.

Adams, R.M., McCarl, B.A., Segerson, K., et al. 1998b. The economic effects of climate change on U.S. agriculture. In R. Mendelsohn and J. Neumann, Eds. *The Economics of Climate Change*, Cambridge University Press, London; New York, pp. 18–54.

Adams, R.M., Hurd, B.H., and Reilly, J. 1999. *Agriculture and Global Climate Change: A Review of Impacts to U.S. Agricultural Resources*. Pew Center on Global Climate Change, Washington, DC.

Benjaminsen, T.A. 2001. The population-agriculture-environment nexus in the Malian cotton zone. *Global Environ. Change*, 11:283–95.

Butt, T.A. 2002. The Economic and Food Security Implications of Population, Climate Change, and Technology: A Case Study for Mali. Ph.D. thesis, Texas A&M University, College Station.

Butt, T.A., McCarl, B.A., Dyke, P., Angerer, J.P., and Stuth, J.W. 2003. Economic and food security implications of climate change — a case study for Mali. *Climatic Change*.

Calkins, P. 1981. Nutritional adaptations on linear programming for planning rural development. *Am. J. Agric. Econ.*, 63:247–54.

Chen, C.C., McCarl, B.A., and Eddleman, B. 1999. *Mali Agriculture Sector Model*. Impact Assessment Group, Agriculture Program Office, Texas A&M University, College Station, TX.

Coulibaly, O.N. 1995. Devaluation, New Technologies, and Agricultural Polices in Sudanian and Sudano-Guinean Zones of Mali. Ph.D. thesis, Purdue University, Lafayette, IN.

Council for Agricultural Science and Technology. 1992. *Preparing US Agriculture for Global Climate Change*. Task Force Report 119. Council for Agricultural Science and Technology, Ames, IA.

Council for Agricultural Science and Technology. 2000. *Storing Carbon in Agricultural Soils to Help Mitigate Global Warming*. Issue Paper 14, April. Council for Agricultural Science and Technology, Ames, IA.

Data Distribution Center, Intergovernmental Panel on Climate Change. 2001. Available at Climate Scenario Gateway: *http://ipcc-ddc.cru.uea.ac.uk/*.

Downing, T. 1992. *Climate Change and Vulnerable Places: Global Food Security and Country Studies in Zimbabwe, Kenya, Senegal and Chile.* Research Report No. 1. Environmental Change Unit, University of Oxford, Oxford.

Food and Agriculture Organization. 1996. *The Sixth World Food Survey.* FAO, Rome.

Food and Agriculture Organization. 2002. FAOSTAT. FAO Statistical Database. Available at: *www.fao.org.*

Gommes, R. and Petrassi, F. 1994. *Rainfall Variability and Drought in Sub-Saharan Africa Since 1960.* Working Paper 9. Environment and Natural Resource Division, Food and Agriculture Organization, Rome.

Hahn, G.L. 1995. Global Warming and Potential Impacts on Cattle and Swine in Tropical and Temperate Areas. In Proceedings of First Brazilian Conference on Biometeorology, pp. 136–173.

Hahn, G.L. 2000. Potential Consequences of Climate Change on Ruminant Livestock Production. Draft USDA/ARS report. U.S. Meat Animal Research Center, Clay Center, NE.

Intergovernmental Panel on Climate Change. 1995. *Climate Change: The IPCC Second Assessment Report, Volume 2: Scientific-Technical Analyses of Impacts, Adaptations,* and *Mitigation of Climate Change.* Watson, R.T., Zinyowera, M.C., and Moss, R.H., Eds. Cambridge University Press, London; New York.

Intergovernmental Panel on Climate Change. 2001. *Climate Change: The IPCC Third Assessment Report: Impacts, Adaptations, and Vulnerability.* McCarthy, J.J., Canziani, O.F., Leary, N.A., Dokken, D.G., and White, K.S., Eds. Cambridge University Press, London; New York.

Jacobs, K., Adams, D.B., and Gleick, P. 2000. Potential consequences of climate variability and change for the water resources of the United States. In US Global Climate Change Program National Assessment Report on The Potential Consequences of Climate Variability and Change. Available at: *www.gcrio.org/nationalassessment/14WA.pdf.*

Jenkins, G.S., Adamou, G., and Fongang, S. 2002. The challenges of modeling climate variability and change in West Africa. *Climatic Change,* 52:263–86.

Kaiser, H.M., Riha, S.J., Wilks, D.S., Rossier, D.G., and Sampath, R. 1993, A farm-level analysis of economic and agronomic impacts of global warming. *Am. J. Agric. Econ.,* 75:378–98.

Kergna, A. 2002. Personal communication, Malian Agricultural Ministry, Bamako.

Kuyvenhoven, A., Ruben, R., and Kruseman, G. 1998, Technology, market policies and institutional reform for sustainable land use in southern Mali, *Agric. Econ.,* 19:53–62.

Lambert, D.K., McCarl, B.A., He, Q., Kaylen, M.S., Rosenthal, W., Chang, C.C., and Nayda, W.I. 1995. Uncertain yields in sectoral welfare analysis: an application to global warming. *J. Agric. Appl. Econ.,* 27:423–436.

Lewandrowski, J., and Schimmelpfennig, D. 1999. Economic implications of climate change for U.S. agriculture: assessing recent evidence. *Land Econ.,* 75:39–57.

McCarl, B.A. 1999. Results from the National and NCAR Agricultural Climate Change Effects Assessments. Report on U.S. Global Change Research Program National Assessment. Department of Agricultural Economics, Texas A&M University, College Station, TX. Available at: *http://ageco.tamu.edu/faculty/mccarl/papers/778.pdf.*

McCarl, B.A., Adams, R.M., and Hurd, B. 2003. Global climate change and its impact on agriculture. In C. Chang and C. Huang, Eds. *Encyclopedia of Life Support Systems.* Institute of Economics Academia Sinica, and UN Educational Scientific and Cultural Organization,Taipai, Taiwan.

McCarl, B.A. and Reilly, J.M. 1999. Water and the Agricultural Climate Change Assessment: Issues from the Standpoint of Agricultural Economists. Proceedings of American Water Resources Association Special Conference on Potential Consequences of Climate Variability and Change to Water Resources, Atlanta, GA.

McCarl, B.A. and Spreen, T.H. 1980. Price endogenous mathematical programming as a tool for sector analysis. *Am. J. Agric. Econ.,* 62:87–102.

Mendelsohn, R., Nordhaus, W.D., and Shaw, D. 1994. The impact of climatic change on agriculture: a Ricardian analysis. *Am. Econ. Rev.,* 84:753–771.

Mohamed, A.B, Duivenbooden, N.V., and Abdoussallam, S. 2002a. Impact of climate change on agricultural production in the Sahel, Part 1: Methodological approach and case study for groundnut and cowpea in Niger. *Climatic Change*, 54:327–348.

Mohamed, A.B., Duivenbooden, N.V., and Abdoussallam, S. 2002b. Impact of climate change on agricultural production in the Sahel. Part 2: Methodological approach and case study for millet in Niger. *Climatic Change*, 54:349–368.

Norton, R. and Schiefer, G.W. 1980. Agricultural sector programming models: a review. *Eur. Rev. of Agr. Econ.*, 7:229–264.

Reilly, J., Tubiello, F., McCarl, B.A., and Melillo, J. 2000a. Climate Change and Agriculture in the United States: Sectoral Vulnerability. In U.S. Global Change Research Program National Assessment Report on Climate Change Impacts on our Nation. Available at: *www.gcrio.org/NationalAssessment/*.

Reilly, J., Graham, J., Abler, D.G., et al. 2000b. Report of the Agricultural Sector Assessment Team. In U.S. Global Change Research Program, National Assessment Report on Changing Climate and Changing Agriculture. Available at: *http://www.nacc.usgcrp.gov/sectors/agriculture/*.

Reilly, J.M., Hrubovcak, J., Graham, J., et al. 2002. *Changing Climate and Changing Agriculture: Report of the Agricultural Sector Assessment Team, U.S. National Assessment*. Prepared as part of USGCRP National Assessment of Climate Variability. Cambridge University Press, London; New York.

Rosenzweig, C. and Hillel, D. 1998. *Climate Change and the Global Harvest: Potential Impacts of the Greenhouse Effect on Agriculture*. Oxford University Press, New York.

Rosenzweig, C. and Iglesias, A. 1994. *Implications of Climate Change for International Agriculture: Crop Modeling Study*. U.S. Environmental Protection Agency, Climate Change Division, Washington, DC.

Rosenzweig, C., Parry, M.L., and Fischer, G. 1995. World food supply. In K.M. Strzepek and J.B. Smith, Eds. *As Climate Changes: International Impacts and Implications*. Cambridge University Press, London; New York, pp. 27–56.

Rowan, R.C. 1995. *PHYGROW Model Documentation Version 2.0.* Ranching Systems Group, Department of Rangeland Ecology, and Management, Texas A&M University, College Station, TX.

Stuth, J.W., Freer, M., Dove, H., and Lyons, R.K. 1999. Nutritional management for free-ranging livestock. In H. Jung, Ed. *Nutrition of Herbivores.* American Society of Animal Science, Savoy, IL, pp. 696–751.

U.S. Global Change Research Program. 2001. Climate Change Impacts on the United States: The Potential Consequences of Climate Variability and Change. Available at: *www.usgcrp.gov.*

Vitale, J.D. 2001. The Economic Impacts of New Sorghum and Millet Technologies in Mali. Ph.D. thesis, Purdue University, Lafayette, IN.

Wang'ati, F.J. 1996. The impact of climate variation and sustainable development in the Sudano-Sahelian region. In J.C. Ribot, A.R. Magalhaes, and S.S. Panagrides, Eds. *Climate Variability, Climate Change and Social Vulnerability in the Semi-Arid Tropics.* Cambridge University Press, London; New York, pp. 71–91.

Williams, J.R., Jones, C.A., Kiniry, J.R, and Spaniel, D.A. 1989. The EPIC crop growth model. *Trans. Am. Soc. Agric. Eng.,* 32:497–511.

Yates, D.N. and Strzepek, K.M. 1998. An assessment of integrated climate change: impacts on the agricultural economy of Egypt. *Climatic Change,* 38:261–287.

Section VI

Toward Research and Development Priorities

30

Researchable Issues and Development Priorities for Countering Climate Change

RATTAN LAL, B.A. STEWART,
DAVID O. HANSEN, AND NORMAN UPHOFF

CONTENTS

The globe has experienced a 31% increase in the atmospheric concentration of carbon dioxide (CO_2) and substantial increases in other greenhouse gases (GHGs) since the industrial revolution (Intergovernmental Panel on Climate Change [IPCC], 2001). The current rate of increase of CO_2 is about 0.5% or 1.5 ppm per annum. At this rate, the concentration of atmospheric CO_2 will double by the end of the 21st century. Environmental and related agricultural impacts of this increased concentration of CO_2 and other GHGs are subject to debate. But most would concur that several impacts will result, namely:

- A rise in the mean global temperature, which will cause alterations in the amount and distribution of precipitation, and local, regional, and global changes in water and energy balances
- A fertilization effect of increased atmospheric CO_2 on plant growth, with probable increases in biological productivity and water-use efficiency
- A decrease in soil organic carbon (SOC) pools, accompanied by a decline in soil quality and an increase in soil erosion and other degradation processes
- An increase in the incidence of pests and pathogens with attendant adverse effects on crop yields and food production
- Adverse effects on global food security, especially in tropical and subtropical regions that are characterized by soils prone to degradation, large numbers of resource-poor farmers, and high demographic pressures

Climate shifts have occurred almost constantly during the Earth's history. However, the rate of projected change during the 21st century may be unprecedented. Interacting factors involved in this process are complex. However, it is important to assess whether global agricultural production will increase or decrease, whether the quality of soil and water resources will improve or decline, whether the beneficial effects of CO_2 fertilization will be enhanced or nullified by other adverse impacts of global warming such as decline in soil quality, and whether food security will be jeopardized in regions with fragile soils and high population density.

Anthropogenic activities, especially land use change and conversion of natural to agricultural ecosystems, have contributed to enrichment of GHGs in the atmosphere since the dawn of civilization (Ruddiman, 2003). Land use conversion and agricultural activities also adversely impacted soil quality. Further, the atmospheric concentration of GHGs is also closely related to soil quality. A decline in soil quality, which results from accelerated erosion and the reduced soil fertility associated with subsistence farming, contributes to the release of CO_2 and other GHGs into the atmosphere. Since 1850, global terrestrial ecosystems have released 136 ±55 Pg (billion metric tons) of C, while fossil fuel emissions have contributed 270 ±30 Pg (IPCC, 2000). Regarding emissions from terrestrial ecosystems, reductions in the SOC pool represent 78 ±12 Pg (Lal, 1999). The conversion of natural ecosystems to agricultural ecosystems may have depleted as much as 30% to 50% of the SOC pools in the soils of temperate regions and 50% to 75% of those in the tropics (Paul et al., 1997; Lal, 1999, 2000). Depletion of the SOC pool is exacerbated by erosion and soil degradation. Degraded soils in Sub-Saharan Africa and elsewhere in developing countries have low SOC pools because of nutrient mining and accelerated erosion. Enhancement of SOC pools through soil restoration would reverse degradation trends, improve soil quality, increase agronomic/biomass productivity, and mitigate climate change.

30.1 ISSUES

Several major issues related to projected climate changes need to be given greater attention in national and international fora. These issues are summarized below.

30.1.1   Climate Change and Net Primary Production

Three globally significant issues are as follows:

30.1.1.1   Climate Change: CO_2 Fertilization Impacts on Terrestrial Ecosystem Production

The relative contributions of CO_2 fertilization and climate change effects on terrestrial CO_2 sources and sinks have been estimated for different ecoregions. They suggest that the CO_2 effect on total net primary production (NPP) content will be positive, and that NPP will decrease without it. On the other hand, the effect of projected climate changes on total NPP appear to be negative. Aggregate global impacts of CO_2 fertilization and climate change on CO_2 appear to be positive.

Modeling results also suggest that the geographic distribution of NPP will change along with changes in CO_2 and climate. Ecosystems with the highest NPP show the greatest change. The effects of other factors, such as land use history and nitrogen cycling, have generally not been considered in these simulation models. They need to be carefully assessed.

30.1.1.2   Climate Change Impacts on Forest Ecosystems

Free-air CO_2 enrichment (FACE) experiments have shown that trees grow faster under elevated levels of CO_2. These findings suggest that temperate forests may be stimulated by higher atmospheric CO_2 levels. However, data also suggest that tree growth may be negatively affected by increasing temperatures. In warmer climates, tropical forests could become a source rather than a sink for CO_2. If all forests in the world were to experience stimulated growth from increased atmospheric CO_2 levels, the effect would be minor relative to CO_2 emissions from continued fossil fuel burning.

30.1.1.3 Climate Change Impacts on Water Supply

The potential impacts of climate change on global water supplies are of great importance. Results of a general circulation model (GCM) based on watershed data suggest that water flows will increase during the fall, winter, and spring seasons, but decrease during the summer. However, the increased demand for water associated with projected population increases will offset any additional water supplies over time. Future climate changes may reduce the carrying capacity of major reservoirs and water flows in rivers around the world because of large water demands for irrigated agriculture.

30.1.2 Global Warming and Food Insecurity

Projected climate changes may adversely affect global and regional food security.

30.1.2.1 Greenhouse Gases and Food Security in Low-Income Countries

Climate change may impact food security in different ways. Some parts of the world — notably West Africa — have already been adversely impacted by climate change. Significant differences in regional food gaps may be explained at least in part by higher levels of agricultural productivity in Latin America and Asia as compared to Sub-Saharan Africa. A major challenge facing food-deficit regions is to address long-term issues, such as global climate change, while also dealing with shorter-term issues such as availability of fertilizers, farm machinery, land tenure, and so on.

30.1.2.2 Effect of Global Climate Change on Agricultural Pests

The manner in which climate change will impact the incidence of pests and diseases is a potentially important but understudied problem. Pest incidences may shift in response to climate change. The "standard model," based on temperature and precipitation, may be useful for studying impacts of climate change, but much more research needs to be done. Most crop models that are used to predict impacts of climate change on production fail to incorporate pests. The problem is compounded by the complexity of the climate processes. Nitrogen appears to have important effects on insect herbivores and on important species interactions that are not well understood. It is also important to disconnect the scale at which tests are impacted by climate change from the scale used for GCM models. Pest–climate interactions with climate change need to be assessed.

30.1.2.3 Impact of Climate Change on Agricultural Production in Different Regions

Projected climate changes could affect crop yields and soil carbon in different regions of the world. In fact, severe adverse impacts in the tropics could occur. Modeling studies based on data from Brazil show that changes in temperature and rainfall may result in soybean production increases and maize and wheat production decreases by 2050. In a case study of the Amazon Basin in which forested areas were converted to pastureland, the soil carbon pool was projected to decrease by 30% over a 100-year period. A decline in the SOC pool would have adverse impacts on soil quality and result in a decline in agronomic productivity and the capacity of the environment to moderate changes.

30.1.2.4 Modeling Future Climate Changes and Crop Production Scenario Challenges

The prediction of impacts of climate change on agricultural research and the use of crop models to assess the potential impacts of climate change suffer from several limitations. Crop models can result in accurate predictions when they are based on observed data, but are much less reliable when based on data that are downscaled from GCMs. A key limitation of GCM data appears to be an inability to represent extreme weather events, particularly those associated with precipitation. The hydrological cycle in the U.S. Midwest is not well represented by downscaled GCM data, because of nonlinear and feedback effects in more detailed regional climate models.

30.1.2.5 Policy Considerations Related to Twin Problems of Global Warming and Food Insecurity

There are major policy issues related to aspects of climate change. The world faces important food insecurity and global warming challenges in the 21st century. Previous research failed to suggest that global warming will have a large impact on aggregate food availability. However, it does suggest that global warming could have some major regional effects on food security, especially in the tropics, in which 800 million people are already at risk. The solution to the food security problem is to address the more fundamental problem of poverty. The "standard model," based on the "Washington consensus," may be the best starting point for economic development. This model emphasizes markets and institutions in infrastructure, education, and agricultural research. This model is not always used for political and cultural reasons. Nevertheless, past efforts indicate that an important step in promoting national food security is to develop economies to the point at which nations can afford to address environmental sustainability as well as economic growth.

30.1.3 Terrestrial Carbon Sequestration and Food Security

Carbon sequestration in soil and vegetation, as well as temporal and spatial variations in relation to land use and management and their related policy issues have major implications for global food security, climate change, and environmental quality.

30.1.3.1 Environmental and Socioeconomic Context for Soil Carbon Sequestration

Land degradation is a constant major threat to food security, especially in Africa and Asia. Soil carbon sequestration (SCS) can be a major way to counter increases in atmospheric CO_2 and to reduce land degradation. The conservation of tropical forests and reforestation activities can offset fossil fuel use. In fact, drastic reductions in rates of deforestation are needed to protect the positive functions of tropical ecosystems. A serious problem of land degradation exists in the humid tropics, and there is an urgent need to find alternatives to slash-and-burn agriculture in this region. Similarly, SCS in countries in the Sahel region may be an important way to increase carbon sinks, control desertification, and promote sustainable agriculture and improved livelihoods for its small farmers.

Degraded lands have low soil C content. SOC concentration in degraded topsoils of Africa varies from 5 to 20 MT ha^{-1}. An urgent need exists to implement SCS practices in order to improve soil quality and farmer livelihoods while removing CO_2 from the atmosphere.

30.1.3.2 Land Use, Soil Management, and Soil Carbon Sequestration

There are several examples of SOC sequestration from experiments that combine SCS practices with alternative food production systems. The rate of aboveground C sequestration ranges from 1 to 5 MT C ha^{-1} year^{-1} in the tropics, but C stocks associated with the traditional peanut-based cropping system in Sub-Saharan Africa ranged from 5 to 25 MT C ha^{-1}. The clay content in soils has a strong effect on C stocks.

Overall rates of SCS in the tropics are lower than those found in higher latitudes (Lal, 2002). Most C accrual is accounted for by a limited number of plant species in agroforestry systems that use fruit and palm trees, timber–pasture combinations, and secondary-growth forest. Soil C accrual rates are difficult to estimate due to the presence of charcoal in fire-prone or fire-dependent ecosystems, such as those associated with traditional land managers in Asia, Africa, and Latin America.

Examples from the humid tropics suggest that ecosystem C stocks may vary in size, being highest in natural systems and then declining in relative terms for agroforestry, fallow, tree crops, and annual cropping systems. Improved fallow practices (with *Tephrosia* spp.) can lead to substantial SCS, but sequestration rates appear to depend on soil type. High SCS rates can be achieved by adopting recommended practices on clay soils, but rates may be only half as high for coarse-textured soils.

30.1.3.3 Modeling and Extrapolating Soil Carbon Sequestration

Simulation models for crop production and soil processes are imperfect but valuable tools. Useful models can be continuously improved through testing. Models can be particularly useful for testing hypotheses related to management strategies designed to reduce atmospheric CO_2 and to soil improvement practices.

The methodology related to SCS modeling in developing countries and to extrapolation from these studies includes: (1) collection of experimental data, (2) use of these data to improve model parameters and thus simulation results, (3) extrapolation of results using remote sensing data, and (4) use of data assimilation techniques that are designed to improve soil carbon estimates and to evaluate prediction uncertainties. Long-term studies are needed to validate the models thus developed, and to relate SCS rates to land use, soil management, and agronomic productivity.

30.1.3.4 Environmental and Socioeconomic Analysis of Soil Carbon Sequestration

SCS is related to the environment in numerous ways. It is a resource conservation practice that needs to be competitively and practically justified. A need exists to include full C accounting procedures in order to determine the suitability of specific SCS practices. Full C accounting should consider

C emissions resulting from the use of farm machinery and other agricultural inputs, as well as relative differences in N_2O and CH_4 emissions. No-tillage agriculture may have more C-based input than conventional tillage. However, any comparative C accounting procedure must use baseline data for the traditional and improved farming systems that are being compared.

Data also need to be collected on attempts to assess the biophysical and socioeconomic dimensions of SCS in developing countries. Some experimental data and models are available to identify best management practices leading to SCS. However, the best solutions predicted by models are difficult to implement because of related economic and social constraints. Implementation costs are generally the highest for poor and small landholders. Thus, their adoption by these farmers will require subsidies or the identification of farming systems that raise incomes while implementing the SCS practices.

30.1.4 Policy and Economic Issues

Policy and economic issues are important attempts to simultaneously address food insecurity, climate change, and reduced SCS. Because these topics are interrelated, they must be addressed holistically using a multidisciplinary approach.

30.1.4.1 Policies and Incentives for Permanent Adoption of Agricultural Carbon Sequestration Practices in Industrialized and Developing Countries

Policies and incentives to trade C can be assessed along two dimensions, namely, policy recommendations and implementation practices. A global cap on C emissions is a major prerequisite for C trading systems to develop. The Kyoto Accord still lacks the signatures of several large nation states, including the United States and Russia. Several major limitations related to the current U.S. policy of voluntary compliance lead to the conclusion that it is probably a nonstarter. The current price of fossil fuel energy does not represent its full opportunity cost if C emissions and sequestration are not considered. Furthermore, C markets may not fully capture secondary costs. For example, the destruction of forests reduces C sequestration, but it may also increase downstream flooding, reducing wildlife habitat and biodiversity, and accelerating the depletion of underground aquifers.

Several major implications of C trading exist. For example, because of spatial variations in distribution impacts of emission reductions, some farmers will benefit from an existing cap on CO_2 and trading mechanisms. In fact, distribution impacts are likely to be more important than economic efficiency impacts. Carbon trading has important implications for other GHG emissions as well. However, there is a need to support efforts to reduce GHG emissions. Beginning with reductions in CO_2 may help identify financial incentives that permit the evolution of more holistic and sustainable systems.

30.1.4.2 Climate Change, Poverty, and Resource-Friendly Agriculture

A consensus exists regarding the need for changes in traditional agricultural practices that lead to soil degradation, including reductions in soil fertility and carbon content. However, a major challenge faced by practitioners is the need to develop alternative farming systems for resource-poor farmers that enable them to achieve food security and more satisfactory standards of living. This challenge becomes more urgent under conditions of climate change which imply increased negative consequences related to continuation of current nonresource-friendly agricultural practices.

All farmers experience a hierarchy of needs. At the most basic level, poor farmers' primary objective is achieving minimal levels of food security. At the following level, they are concerned about improving the standard of living for themselves and their families. Perhaps beyond this level, they are concerned about protecting and improving the natural resource base on which their sustenance depends and on which enhancing its condition in turn depends. This implies the need to identify agricultural production practices that simultaneously improve income levels while contributing to carbon sequestration and other forms of soil and water quality improvement.

These practices may need to be location-specific. Simulation analyses suggest that payments for soil carbon sequestration may be insufficient to balance losses in agricultural production for small- and medium-scale farmers under tropical conditions. On the other hand, alternative production practices hold great promise for increasing small-farm incomes while simultaneously improving soil quality. More experimentation along these lines will be necessary to enable poor small farmers to participate in the process of soil carbon sequestration.

30.1.4.3 Climate Change and Public Policy Challenges

Several current treaties and conventions on climate change and biodiversity will be difficult to implement, and thus, they are unlikely to have much regional or global impact. Changes are needed to improve related existing public policy climate. These include: (1) greater cooperation among the various scientific disciplines; (2) expanded cooperation among new stakeholders; (3) increased funding support for long-term research agendas, such as those of the Intergovernmental Panel on Climate Change, U.S. National Academy of Science, and U.S. Environment Protection Agency; (4) more policy-relevant applied research; and (5) more analysis of public policy anomalies. Examples of the latter are latent policies that often conflict with official policies and special interest group objectives.

30.1.4.4 Climate Change Impacts on Developing Countries

The implications of climate change for food security in developing countries of Sub-Saharan Africa, such as Mali, may be drastic. Increased price volatility and reduced yields are leading to increased risks of hunger. However, adaptive measures can moderate them. Improved cost data associated with ways that food-insecure nations attempt to alleviate the problem are needed.

30.1.4.5   Climate Change and Tropical Agriculture: Implications for Social Vulnerability and Food Security

Globalization may impact food security in developing countries. The context for globalization and its impacts on people are frequently neglected. Food security is also related to access to food. Experience has shown that poor people in the tropics benefit less from globalization since these regions are at a comparative disadvantage. Regarding agricultural production, price supports and subsidies are typically lowered as a consequence of global trade policies. Future research on small farms in these regions needs to focus more on integrated small-farm systems, and to give more attention to indicators of instability, food sovereignty and security, local autonomy, and increased investments in appropriate technology.

30.2   IDENTIFICATION OF RESEARCHABLE PRIORITIES

Researchable priorities are region-specific, and they are defined in great measure by physiographic, soil, climate, sociocultural, economic, and political factors. Several major criteria that can be used to identify key researchable priorities are discussed below.

- Probable impacts of climate change on food production. What is the current status of food supply and demand for individual nations and for the world, and how might it change because of climate changes? Current information suggests that climate change impacts on food supply and demand vary across ecoregions. Thus, it will be important to identify regions that may become food insecure and to formulate research and development strategies to address the potential problem. In addition to direct impacts on productivity, climate change may also affect food security indirectly by increasing postharvest losses, and by reducing the availability of key production inputs such as water. Appropriate solutions may be to increase the efficiency of water use and increase soil and water conservation. Crop yields may be affected by higher grain sterility resulting from thermal damage due to increased temperatures. This may require appropriate breeding work to create more heat-resistant seed varieties.

• Soil C dynamics and soil quality. The relationship between these two factors is very strong. The soil C pool will probably decline in regions where the mean temperature increases, and/or the mean precipitation decreases. A decline in the soil C pool will increase erosion probabilities, degrade soil structure, reduce inherent soil fertility, and decrease overall productivity. For regions at risk, it is important to identify land use and management strategies that will nullify the adverse impacts of climate change, maintain or enhance soil C pools, and sustain or improve soil quality. This is an ambitious agenda, especially for regions with resource-poor farmers who cannot afford the inputs required to adopt the recommended management practices.

Understanding the magnitude and direction of the three-way interaction among climate change, soil C sequestration, and food security is crucial to addressing the global issue. These interactive effects are specific to ecoregions/biomes, and require a holistic and multidisciplinary approach in order to develop specific studies with results that can be scaled up to watershed, regional, and global levels. A great need exists for more local level studies based on analytical methods used for larger-scale analyses.

ACKNOWLEDGMENTS

We gratefully acknowledge rapporteur summaries for the sessions that were prepared by Drs. Fred Hitzhusen, John Antle, Bobby Stewart, and Carrie Stokes.

REFERENCES

Intergovernmental Panel on Climate Change. 2000. *Special Report on Land Use, Land Use Change, and Forestry.* Cambridge University Press, London; New York.

Intergovernmental Panel on Climate Change. 2001. *Climate Change: The Scientific Basis.* Cambridge University Press, London; New York.

Lal, R. 1999. Soil management and restoration for carbon sequestration to mitigate the accelerated greenhouse effect. *Prog. Environ. Sci.,* 1:307–326.

Lal, R. 2000. World cropland soils as a source or sink for atmospheric carbon. *Adv. Agron.,* 71:145–191.

Lal, R. 2002. The potential of soils of the tropics to sequester carbon and mitigate the greenhouse effect. *Adv. Agron.,* 76:1–30.

Paul, E.A., E.T. Elliott, K. Paustian, and C.V. Cole, Eds. 1997. *Soil Organic Matter in Temperate Agroecosystems: Long-Term Experiments in North America.* CRC Press, Boca Raton, FL.

Ruddiman, W.F. 2003. The anthropogenic greenhouse era began thousands of years ago. *Climatic Change,* 61:261–293.

Index

A

Abiotic extremes, tolerance of, 58
Acacia
    nilotica (*nebneb*), 544
    raddiana (*seing*), 544
    seyal (*surur*), 544
Acyrthrosiphum pisum, 340
Adams equation, 479
Adansonia digitata (*gui*), 544
Adaptation, 259, 260, 271, 286
Adult equivalent (AE), 547
AE, *see* Adult equivalent
Afghanistan, food insecurity of, 83,
    118
Africa
    carbon loss with land conversion
        in, 620
    Doubly Green Revolution in,
        6
    fertilizer prices, 5
    impacts of climate change on,
        282
    maize calorie deficit, 11
    road density, 5
    soil nutrient depletion in, 5
Africa, climate change and
        reducing hunger in, 3–19
    adapting to thermal damage,
        10–12
    coping with climate change, 7–8

estimating biomass of young
    tropical vegetation, 8–9
how to measure soil carbon, 9
mitigation, 12–16
    carbon sequestration by
        smallholder farming
        communities, 15–16
    high carbon sequestration
        potential of tropical
        agroecosystems, 12–15
reducing hunger in Africa, 6
sharpening predictive tools for
    key agroecosystems, 10
African agriculture, technological
        options needed in, 51
Agricultural carbon sequestration
    practices, policies and
    incentive mechanisms for
    adoption of, 679–701
incentive mechanisms, 687–697
    co-benefits and costs,
        696–697
    contract duration and carbon
        permanence, 692–694
    designing soil carbon
        contracts for farmers in
        developing countries,
        694–695
    farmer participation in
        carbon contracts,
        690–692